PREFACE

The present agriculture growth rate is very low and the pace is required to be maintained at least double of the existing one enabling us to achieve the targets to feed the burgeoning population world over which may surge to 10 billion by 2030. The world leaders have alread· been struggling to cope up with the food demands, whereas o.. other hand malnutrition is afflicting several million people in developing countries, more specifically African and Asian countries. There remains nonetheless a big challenge for plant scientists to fight with the problems such as food starvation, malnutrition, post harvest conservation evolving technologies to achieve food security. Simultaneously, redressal of a watchdog for monitoring the contingent plans to cover over the emergencies in wake of prolonged droughts havocs, tsunamis, epiphytotics etc. These are not the times to sit in compartmentalized disciplines, instead an integrated dynamic approach to major problems will inevitably be adequately required for solving complexities.

The subject of Plant Pathology has not remained an isolated piece of cake, but however, in recent past it has shown its Oceania characteristics by dissolving the elusive partition walls. It is relevant to quote Toffler's dictum "future has already arrived", which argues well in context to the subject of Plant Pathology where symptomatologically this vast and dynamic discipline has already coalesced with molecular biology on one hand and biotechnology/nanotechnology on the other hand. In essence marvels of research in plant sciences shall offer a bountiful of fascinating conceptual theories, which would put us on lever of challenges to accommodate these comfortably in teaching and research agenda. Evidently as the legend goes, dynamism has traditionally been so inherent to the subject of Plant Pathology that the paraphyses of it ever remain receptive to advanced sciences.There can not be any doubt, whatsoever, that the edifice of developments in modern biotechnology in totality as one of its stupendous achievements of the collective scientific wisdom of mankind perhaps is best exemplified by producing GM crops.

This review volume we are presenting to the readers is encompassing wide spectrum interests inclusive of mycology (fungi of aquatic nature and human pathogens, bioagents etc.), biochemical and molecular plant pathology, integrated disease management, root/soil microbiology etc. We have considered several aspects to sift, analyse and arrange the articles to provide best possible to readers and hope that it will serve the purpose of Reviews in Plant Pathology. We thank all the learned authors for contributing the novel and up-to-date articles for inclusion in the review book. Thanks to ISMPP for entrusting us with the task of completing the volume 5 of the review series.

Editors

CONTENTS

POLYGALACTURONASE INHIBITING PROTEINS: TITANIC PLAYERS IN PLANT IMMUNITY

Praveen Mamidala, Guguloth Mahesh Kumar and Appa Rao Podile*

Department of Plant Sciences, School of Life Sciences,
University of Hyderabad, Hyderabad, Andhra Pradesh-500 046, India.
*Corresponding author: arpsl@uohyd.ernet.in

KEY WORDS: Plant-Pathogen interactions, Polygalacturonase-inhibiting Protein (PGIP), leucine-rich repeat (LRR), Plant innate immunity, Protein-Protein interactions, Molecular interactions, Transgenic Plants.

ABSTRACT

Following pathogen attack, plants respond with a wide spectrum of defence reactions, such as the induction of antifungal proteins and enzymes involved in plant cell wall reinforcement or breakdown of pathogen infection structures. Polygalacturonase-inhibiting proteins (PGIPs) are extracellular plant protein inhibitors of endopolygalacturonases (PGs) that belong to the leucine-rich repeat (LRR) protein family which retards the invasion of plant tissues by phytopathogenic fungi. PGs are variable in terms of shape and electrostatic surfaces as well as desolvation energy surfaces, although maintaining the same overall fold and single PGIPs can inhibit different PGs with different mechanisms, either competitive or noncompetitive. The PG-PGIP interaction studies would contribute to design effective inhibitors for crop improvement and also help in understanding the molecular basis of the interaction in other plant LRR proteins. The current review brings an update of properties of PGIP, analysis of PGIP *cis* elements, structural elucidation and molecular docking of PGIP and PG-PGIP interactions which contribute to develop transgenic plants resistant to fungal diseases.

INTRODUCTION

Plants routinely combat pathogenic microbes and are able to discriminate between self and non-self, which is the basis for the activation of innate defense mechanisms against pathogens. The innate immune system comprises the cells and mechanisms that defend the host from infection by other organisms, in a non-specific manner i.e., the cells of the innate system recognize and respond to pathogens in a generic way, but unlike the adaptive immune system, it does not confer long-lasting or protective immunity to the host. Innate immune systems provide immediate defense against infection. Generally, pathogen recognition and the subsequent activation of disease resistance responses in plants occur either at the non cultivar-specific level (non-host resistance) or at the cultivar level (host resistance).

Early identification of pathogen upon contact is an important defense mechanism in plants. This is achieved by microbial patterns by germline encoded receptors termed as pattern recognition receptors (PRRs). After the detection event, PRRs activate several, complex signalling cascades, which ultimately regulate the transcription of target genes that encode effectors and regulators of the immune response (49). The perception and activation of innate immune responses in plants is often mediated by receptors of pathogen-associated molecular patterns (PAMPs) and these motifs are not restricted to pathogens, but are rather characteristic of a whole class of microorganisms in general. They are also referred as microbe associated molecular patterns (MAMPs). The detection of PAMPs and danger signals constitutes the basis of innate immunity. With the exception of vertebrates, all other organisms (invertebrates, plants and fungi) rely exclusively on innate immunity (32).

Fungal disease has been one of the main causes for crop losses since the time of plant cultivation. For decades plant scientists have recognized that, components of fungal and bacterial cell walls can activate defences in plants (23,9). In addition, pathogens can release polysaccharide oligomers from the plant surface, which can induce defenses (14,50,51,). Perception of elicitors takes place in receptors located either at the cell surface or inside the cell (16). Upon perception of pathogen-derived molecules, collectively called elicitors, plants which include (glyco) proteins, peptides, carbohydrates, and lipids, all of which can activate inducible defense responses, including the production of reactive oxygen species (ROS) and the accumulation of antimicrobial compounds (phytoalexins) and pathogenesis-related proteins (24,40,51).

Many of the recognition events occur in the plant cell wall, which is the first barrier to come into contact with the invading organisms. To counteract fungal invasion, plants have a preexisting or induced defense mechanisms and one of the barriers against pathogenic fungi is plant polysaccharide rich cell wall made of pectin in dicotyledonous and non-graminaceous monocotyledonous cells. Changes in the structure and composition of the cell wall accompany different developmental stages of the plant cell and local modifications of the wall are produced upon a pathogen attack. The cell wall is strengthened in such a way that pathogen cannot colonize plant tissue and cannot draw the nutrients from it.

The cell wall is often the place where the recognition of pathogens takes place and molecular signals are generated that set in motion subsequent transcriptional activation of defense-related genes. Plant-pathogen recognition in the cell wall may determine the so-called innate immunity response which in plants is often achieved through the recognition of PAMPs by specific receptors of the plant surveillance system.

In order to break the cell wall majority of fungi secrete a number of hydrolytic enzymes which degrade the cell wall to colonize the plant tissue and obtain nutritional requirements. The important enzymes which break the first line of defense in plant cells are cell wall degrading enzymes (CWDE) which includes exo - endo-polygalacturonase, pectate lyase, and pectin methyl esterase involved in pectin modifications (54). Pectin-degrading enzymes, including endopolygalacturonases (EPG), are among the first enzymes to be secreted during fungal infection. PG is the first cell-wall-degrading enzyme synthesized by phytopathogenic fungi cultured on isolated cell walls. PGs are known to be secreted by phytopathogenic fungi, bacteria and nematodes. PGs that break down the polygalacturonate chain in a random manner into small chains of oligogalacturonate are called endo polygalcatouranses (EPG) and other PGs that cleave the polygalacturonate chain in terminal manner and release monomeric products, i.e. galcturonic acids are called exo polygalacturonase (XPG). EPG (poly-α-1,4-galacturonide glycanohydrolase) degrades unesterified regions of homogalacturonan (HGA), the 1,4 linked α-D-galactosyluronicacid polymer found in pectin of plant cell wall (29). The CWDEs macerate plant cell walls and release nutrients for bacterial growth and PGs degrade plant cell walls which allow other hydrolyzing enzymes to also degrade cell walls more easily (39).

The elicitor active principle was not the enzymes themselves, but oligosaccharides released from the plant cell wall by these enzymes (57). Oligogalacturonides (OGs) are released due to the action of PGs which act as elicitors (53) that elicit SA-independent defenses (64,65). The most intensively studied oligosaccharides are the oligo-α-galacturonides (OGAs), which are produced from degraded homogalacturonans by fungal endo-polygalacturonases (65,66). Because they are plant-derived, OGAs are endogenous elicitors. Their perception by plants nicely illustrates the concept of "danger signal" perception by the innate immune system (47). These elicitors transmit the signals into the cell through the signal transduction pathways, leading to production of polygalacturonases-inhibitory proteins (PGIPs).

PGIPs which are induced early during an attempted infection are members of leucine rich repeat (LRR) protein family, a class of proteins specialized in protein-protein interactions (15, 17). Plants have selected the LRR fold for their immune functions and recognition of non-self molecules (37). PGIPs are the only plant LRR proteins with known ligands, are pectinase inhibitors, bound by ionic interactions to the extracellular matrix (ECM) of plant cells and have high affinity for fungal EPGs. PGIPs decrease the rate of enzymatic depolymerization of polygalacturonate and result in an increase in the concentration of elicitor-active OGs in apoplast (17,18).

PGIP-PG Interactions

PGIPs are cell wall associated glycoproteins, which have been isolated from a variety of plants with different range of molecular weight. Until now, the reported PGIP fall in the range of 34 to 54kDa with smallest PGIP (15kDa) from peach and largest was reported in pear with 91kDa (2, 27). The PGIPs are relatively heat-stable up to 50^{0}C and slowly lose their activity when heated above 55^{0}C (1,5,11). PGIPs have been identified in many dicot and monocot plants. Some of these plants, such as raspberry, tomato, pear, apple, grape, soybean, bean and *Arabidopsis*, possess small gene families coding for closely related but different PGIP isoforms. PGIPs are specific inhibitors and do not inhibit other CWDEs of plant origin (29). Cervone et al., (12) showed the inability of *Phaseolus vulgaris* PGIP (*Pv*PGIP) in inhibiting EPG and XPG of tomato and water oak. Intriguingly, *Pv*PGIPs have inhibited XPG of corn pollen and sorghum (12). Similarly, PGIP of pear inhibited *Penicillium expansum*, *Botrytis cinerea* and *Dothiorella gregaria*, while it did not inhibit PG of pear fruit (1). Many plants possess more than one PGIP with differential abilities to inhibit different PGs of pathogens (29).

The wide degree of functional redundancy and recognition specificity of the PGIP gene families provides higher protection and a selective advantage for the activation of defense responses (25), while the expression of PGIPs is regulated by different stress-related molecules through separate signal transduction pathways. PGIP is a constitutive protein; the amount of PGIP gene transcripts varies at different stages of maturity and at different distances from the diseased region. Expression of PvPGIP3 is induced by oligogalacturonides but not by fungal glucan or salicylic acid or wounding; expression of *Pv*PGIP4 is not altered by any of these treatments. Expression of *Pv*PGIP1 is induced by wounding only (20), whereas *Pv*PGIP2 is up-regulated by oligogalacturonides, salicylic acid and wounding (22). Expression of PGIPs is also regulated by various stress stimuli in *Arabidopsis* (26). The two genes of *Arabidopsis* are activated by wounding or *B. cinerea* infection and are responsive to different signals: *At*PGIP2 responds to jasmonate, whereas *At*PGIP1 is up-regulated by oligogalacturonides but is unaffected by salicylic acid, jasmonate or ethylene (26). In apple fruits, high level of PGIP gene was expressed in the decayed area and in adjacent areas of the fruit inoculated with *P. expansum* and *B. cinerea*. However, there was no apparent increase in PGIP in tissues away from the decayed region, suggesting that the PGIP gene is readily activated by fungal infection (29).

PGIPs specifically interact with PGs by forming a bimolecular complex. The PG–PGIP interaction varies in terms of inhibition kinetics and strength, and reflects the counter adaptation occurring in both enzymes and inhibitors. Pathogens have evolved different PGs to maximize their offensive potential and, conversely, plants have evolved various PGIPs with different specificities to counteract the many forms of PG existing in nature. The 3-D structure of *Pv*PGIP2 revealed a positively charged patch located between the two *β*-sheets B1 and B2, directly below the negative pocket putatively involved in PG binding. The positive patch consists of a cluster of regularly spaced Arg and Lys residues (R183, R206, K230, R252) that protrude into the solvent creating a regular distribution of charges. LRRs of PGIPs which are concavely folded in the crystal structure have a wide negative pocket involved in binding with PGs (46).

Docking studies has revealed that active sites of *Fm*PGI (PG of *Fusarium moniliforme*) was blocked by bean PGIP (*Pv*PGIP2) (Figure 1a) (34,58). It is interesting to note that *Fm*PG1 is inhibited by *Pv*PGIP2 but not by *Pv*PGIP1, even though these proteins only differ in eight residues (41,48). With a single lysine-to-glutamine mutation at position 224, *Pv*PGIP1 acquired the ability to interact with *Fm*PG1 (41). Importantly, this position is among the seven sites under diversifying selection that were found around the negatively charged pocket implicated in PG binding (blue and red in Figure 1b). Sequence analysis of PGs from different *Fusarium* spp. uncovered diversifying selection at two positions (59) both located in the lid (red and orange in Figure 1b). Site directed mutagenesis at position 121 containing lysine residue in the lid has revealed that *Fm*PG have become sensitive to *Pv*PGIP2 (8). Interestingly, in the docking model this residue 121 of *Fm*PG1 would be in direct contact with *Pv*PGIP2 residue 224, illustrating the importance of single residues in enzyme–inhibitor interactions, and explaining the diversifying selection at both positions (Figure 1c).

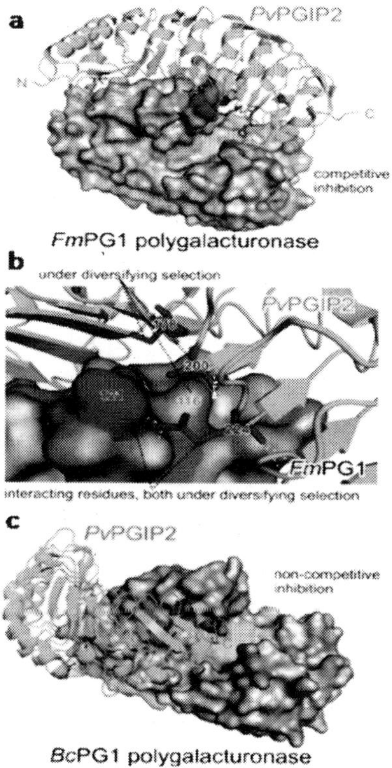

Figure 1. PG-PGIP enzyme inhibitor interactions a. Docking model of the crystal structures of *Pv*PGIP2 and *Fm*PG1. competitively by binding the lid (red/pink) of *Fm*PG1 and blocking the active site (yellow) from the top. Blue, red and orange residues in *Fm*PG1 and blue and red residues in *Pv*PGIP2 are variant and often under positive selection for variance. **b.** Close-up of the *Pv*PGIP2–*Fm*PG1 docking model from the side showing the interaction between *Pv*PGIP2 (green) and the lid of *Fm*PG1 (red/pink). Residue 121 in *Fm*PG1 (red) and residue 224 in *Pv*PGIP2 (red) physically interact and contribute to the interaction and are under strong diversifying selection. Please also note the close proximity between the second positively selected residue in *Fm*PG1 (orange) and other sites under diversifying selection in *Pv*PGIP2 (blue). Other variant sites in *Fm*PG1s are indicated in blue. **c.** Docking model of the crystal structure of *Pv*PGIP2 and a homology model of *Bc*PG1. *Pv*PGIP2 binds to the side of *Bc*PG1 keeping the substrate-binding site open for binding substrates, explaining the noncompetitive, mixed mode of inhibition (Adapted from Johanna and Renier 2008).

*Pv*PGIP-1 and *Pv*PGIP-2 of *P. vulgaris* are classified on the basis of their distinct specificity, where *Pv*PGIP-2 completely inhibits the PG activity whereas *Pv*PGIP-1 was not able to interact with the PG of *F. moniliforme.* EPGs and *Pv*PGIP interactions make these two proteins an excellent system to study the basic recognition specificity of a LRR protein (17). Single mutation of amino acids residues of *Pv*PGIP-2 into the correspon-ding residues present in *Pv*PGIP-1 cause a loss of affinity for *F. moniliforme* EPGs. Variations in amino acid induced by site-directed mutagenesis is excellent model system to study the basis of recognition specificity of PGIPs (34). This work helps us to understand the important residues present in the PGIPs and to develop a protein which inhibits a broad range of fungal pathogens by making the susceptible plants resistant towards fungal pathogens by transgenic approach.

PGIPs appear to bind PGs in different orientations, due to which PG-PGIP interaction studies are difficult (25). *Pv*PGIP2 is a competitive inhibitor of *Fm*PG1 but it inhibits *Bc*PG1 in a noncompetitively mixed mode, allowing the substrate to bind with reduced affinity and reduced kinetics of hydrolysis (58). Docking studies of *Pv*PGIP2 with a model of BcPG1 indicates that *Pv*PGIP2 binds *Bc*PG1 in a different orientation, for substrate interaction (Figure 1c) (58). The docking studies further suggest that a single PGIP can be versatile in binding and inhibiting their targets in different orientations; because of this PGIPS adapt quickly and diversify their spectrum of inhibition (34).

Transgenic approach using PGIPs

The first gene encoding a PGIP was cloned from *P.vulgaris* (63) and since then PGIPs were reported in many plant species such as *Brassica napus*, *Brassica rapa*, *Medicago sativa*, *Vitis vinifera*, *Arabidopsis thaliana*, *Oryza sativa*, *Lycopersicon esculentum* etc., and these genes were transformed into different plants to develop resistance to different fungal pathogens (Table 1). In all the transgenic plants reported so far, plants expressing various PGIPS grew much larger and more vigorously than the wild type during fungal infections and limited fungal invasions.

The PGIP1 gene of *P. vulgaris,* encoding a polygalacturonase inhibiting protein expressed under control of the cauliflower mosaic virus 35S promoter in tomato plants via *Agrobacterium tumefaciens*–mediated transformation. No evident enhanced resistance, compared with the resistance of untransformed plants, was observed. The PGIP1 gene was also transiently expressed in *Nicotiana benthamiana* with potato virus X (PVX) as a vector. PGIP-1 purified from transgenic tomatoes and PGIP-1 in crude protein extracts of PVX-infected *N. benthamiana* plants were tested with several fungal polygalacturonases (PGs). PGIP-1 from both plant sources exhibited specificity different from that of PGIP purified from *P. vulgaris* (bulk bean PGIP). The experiments carried out by Desiderio *et al.* (19) showed that PGIPs with different specificities are expressed in *P. vulgaris* and that the high-level expression of one member (PGIP1) of the PGIP gene family in transgenic plants is not sufficient to confer general, enhanced resistance to fungi.

Transgenic tomato plants expressing the pear fruit polygalacturonase inhibitor protein (pPGIP) were used to demonstrate that this inhibitor of fungal pathogen endopolygalacturonases (endo-PGs) influences disease development. Transgenic expression of pPGIP resulted in abundant accumulation of the heterologous protein in all tissues and did not alter the expression of an endogenous tomato fruit PGIP (tPGIP). The pPGIP protein was detected, as expected, in the cell wall protein fraction in all transgenic tissues. Despite differential glycosylation in vegetative and fruit tissues, the expressed pPGIP was active in both tissues as an inhibitor of endo-PGs from *B. cinerea*. The growth of *B. cinerea* on ripe tomato fruit expressing pPGIP was reduced, and tissue breakdown was diminished by as much as 15%, compared with non-transgenic fruit. In transgenic leaves, the expression of pPGIP reduced lesions of macerated tissue approximately 25%, a reduction of symptoms of fungal growth. Heterologous expression of pPGIP has demonstrated that PGIP inhibition of fungal PGs slows the expansion of disease lesions and the associated tissue maceration (55).

Bean PGIP1 gene was cloned and expressed in tomato and the PGIP was purified from the extract by affinity chromatography and this inhibited the *Stenocaprella maydis* a fungal pathogen of maize. PGIP extracts from the transgenic tomato inhibited both *Aspergillus niger* and *S. maydis* PG, whereas extracts from a control untransformed tomato did not (7).

The inhibition of fungal PGs by PGIPs suggests that PGIPs have a role in plant tolerance to fungal infections and this has been observed in transgenic grape plants expressing pPGIPs under the control of the CaMV 35S promoter. *Xylella fastidiosa*, the causal agent of Pierce's disease (PD) in grapevines, has genes that encode cell-wall-degrading enzymes, including a putative PG. Substantial pear PGIP (pPGIP) activity was found in crude extracts from leaves and in xylem exudate of transgenic lines obtained from independent transformation events, but not in untransformed controls. pPGIP activity was detected in xylem exudate of untransformed scions grafted on to transgenic rootstocks expressing *pPGIP*. Leaves of transgenic plants infected with *B. cinerea* had reduced rates of lesion expansion. The development of plant disease (PD) was delayed in some transgenic lines with increased pPGIP activity. PD-tolerant transgenic lines had reduced leaf scorching, lower *Xylella* titres and better re-growth after pruning than the untransformed controls (4).

With the target of increasing disease resistance to fungal pathogens, Szankowski et al (60) reported transgenic apple plants containing the *Vst1* gene from *V. vinifera* and the PGIP gene from *Actinidia deliciosa.* However the resistance of transgenic apple plants harboring AdPGIP was not reported. Fungal PG activity was inhibited by crude extracts from shoots transformed with the *PGIP* gene of *Pyrus communis* into Persimmon (62).

The importance of PGIPs in plant defence has been corroborated by *in vivo* studies. The overexpression of the genes *At*PGIP1 and *At*PGIP2 in *Arabidopsis* limits the colonization by *B. cinerea* and reduced disease symptoms (26). A significant increase of PG-inhibitory activity and a decrease in susceptibility to *B. cinerea* has been found in transgenic tomato and grapevine plants overexpressing a pear PGIP(44) in Tobacco and *Arabidopsis* plants overexpressing a bean *PGIP* (*Pvpgip2*) (44).

AtPGIP1 and *AtPGIP2* encode functional inhibitors of similar activity, both of which are able to enhance resistance against *Botrytis* when they are overexpressed in transgenic plants. Importantly, although *AtPGIP1* and *AtPGIP2* are induced coordinately during fungal infection, they are regulated by independent signal transduction pathways. Leaves of wild-type and transgenic plants were inoculated with *Botrytis*, and the extent of disease symptoms was determined after 3 days. No significant differences were observed between the average diameters of the lesions formed by *Botrytis* on the leaves of untransformed Plants. By contrast, lesion size was reduced by 30% in leaves overexpressing *At*PGIP1 and in leaves overexpressing *At*PGIP2 (26,31) has analyzed sixteen PGIP promoter sequences and found that the expression of *Bn*PGIP genes highly responded to infection by the necrotrophic pathogen, *Sclerotinia sclerotiorum* and also it was found that PGIP genes were induced either by Jasmonic acid, wounding and salicylic acid or in combinations thereof. Enhanced resistance to *B. cinerea* and *Verticillium dahliae* was observed among transgenic tobacco, potato and grapevine expressing PGIPs from bean, apple and pear (3,28,29) and found reduced rates of lesion expansion compared to wild. Similar observations were recorded by Manfredini et al. (44) and Joubert et al. (36) when *Pv*PGIP2 and *Vv*PGIP2 were overexpressed in tobacco.

A reduction in disease susceptibility of 47–69% was observed for a homogeneous group of transgenic lines that was statistically clearly separated from untransformed control plants following infection with *Botrytis* over a 15-day-period (36). Overexpression of *Ri*PGIP1 and *Md*PGIP in Pea and Tobacco resulted in resistance to *C lupine* and *S maydis* (52,56). Jani et al. (33) reported transgenic wheat lines expressing a bean PGIP (*Pv*PGIP2) having a wide spectrum of specificities against fungal PGs. Three independent transgenic lines were characterized in detail, including determination of the levels of *Pv*PGIP2 accumulation and its subcellular localization and inhibitory activity. Results show that the transgene-encoded protein is correctly secreted into the apoplast, maintains its characteristic recognition specificities, and endows the transgenic wheat with new PG recognition capabilities. As a consequence, transgenic wheat tissue showed increased resistance to digestion by the PG of *F. moniliforme*. These new properties also were confirmed at the plant level during interactions with the fungal pathogen *Bipolaris sorokiniana*. All the three lines showed significant reductions in symptom progression.

Not always does overexpression of PGIPs in transgenic plants result in protection against *B. cinerea*. Indeed the first study carried out in this regard showed a lack of protection when the inhibitor *Pv*PGIP1 was expressed in transgenic plants (19). Because it is now clear that *Pv*PGIP1 is a rather weak inhibitor of fungal PGs and it can be argued that the capability of a PGIP of controlling infection *in vivo* depends on its biochemical capability of inhibiting PGs acting as virulence factors (19).

These results suggest that the expression of proteins with similar defensive functions is not simply a reflection of functional redundancy. Rather, the differential regulation of functionally redundant proteins may ensure the expression of important defense activities regardless of the defense-related transduction pathway activated by a particular pathogen. Thus, these studies shed new light on the evolution of plant defense strategies against pathogens. These findings suggest that PGIPs are important players in plant innate immunity and genes encoding PGIPs can be used to develop transgenic plant resistant to fungal diseases.

Bioinformatics approach into PGIP

Most plant defence proteins are encoded by families of related genes that usually have specific structural and regulatory features (21). PGIP genes are organized in families encoding proteins with similar physical characteristics but different specificities. The evolutionary advantage of having multiple genes would derive from their different modes of regulation, which can ensure the expression of at least one gene member in different infection scenarios. Alternatively, the advantage resides in the functional redundancy, which can confer a higher level of protection and a selective advantage in the acquisition of new recognition specificities. In the case of PGIPs, both of these aspects are important. The multiple sequence alignment of twenty seven PGIPs has shown that LRRs are conserved throughout (Figure 2). The phylogenetic tree of 27 PGIPs shows that many of the PGIPs are closely related (Figure 3). After aligning twenty seven sequences of various PGIPs using Mega 4.0 with boostrap value of 1, it was found that all the sequences are closely related within the species, however *Ms*PGIP was found in separate cluster whereas *Vv*PGIP and *Os*PGIP1, OsPGIP4 are also found in separate cluster, but were closely related with *Bn*PGIPs. Though *Br*PGIPs found in separated clustersare found to be close with *Bn*PGIPs (Figure 3). The LRR elements in all the 27 PGIP matches the consensus sequence LxxLxx LxxLxLxxNxLT/sGxIPxx where residue of β-strand/β-turn structure, responsible for binding PGs are highlighted in yellow color (Figure 3). The LRR structure is shared by many plant proteins involved in recognition, such as the majority of the resistance gene products (45), and several receptors involved in development, perception of hormones, elicitors and defense response against insects (6,30,61). Plants use the LRR fold for their immune functions and recognition of non-self molecules. Several plant resistance gene products or defense related receptors display LRR motifs of the extracytoplasmic type homologous to that of PGIP (30,35).

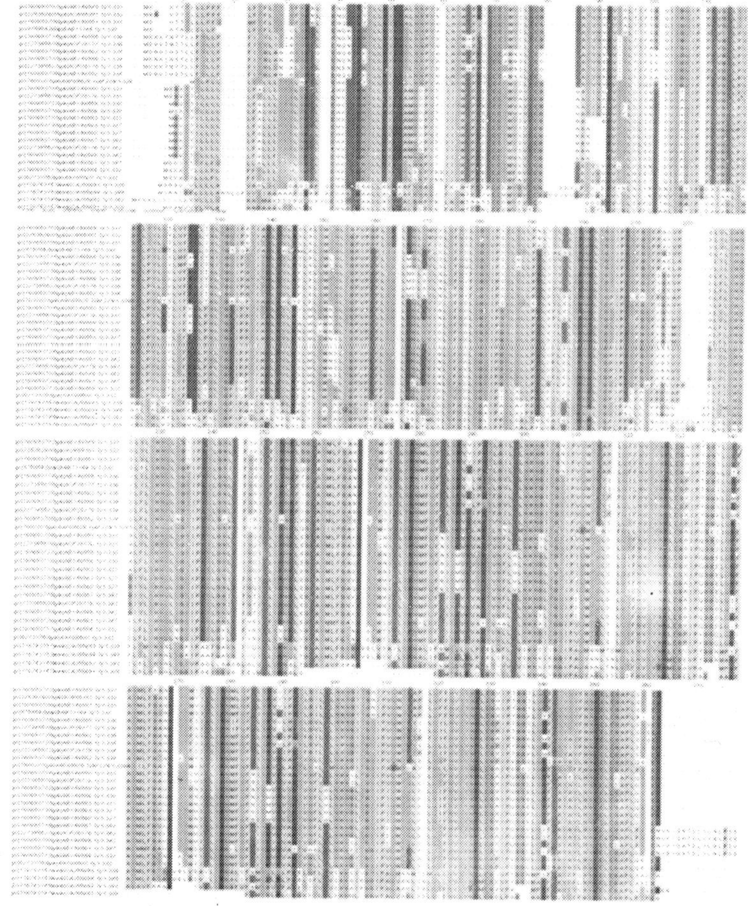

Figure 2. Multiple sequence Alignment of polygalacturonase inhibiting proteins of *Arabidopsis thaliana, Oryza sativa, Brassica napus, Brassica rapa, Lycopersicon esculentum, Vitis vinifera* and *Medicago sativa* using Clustal W.

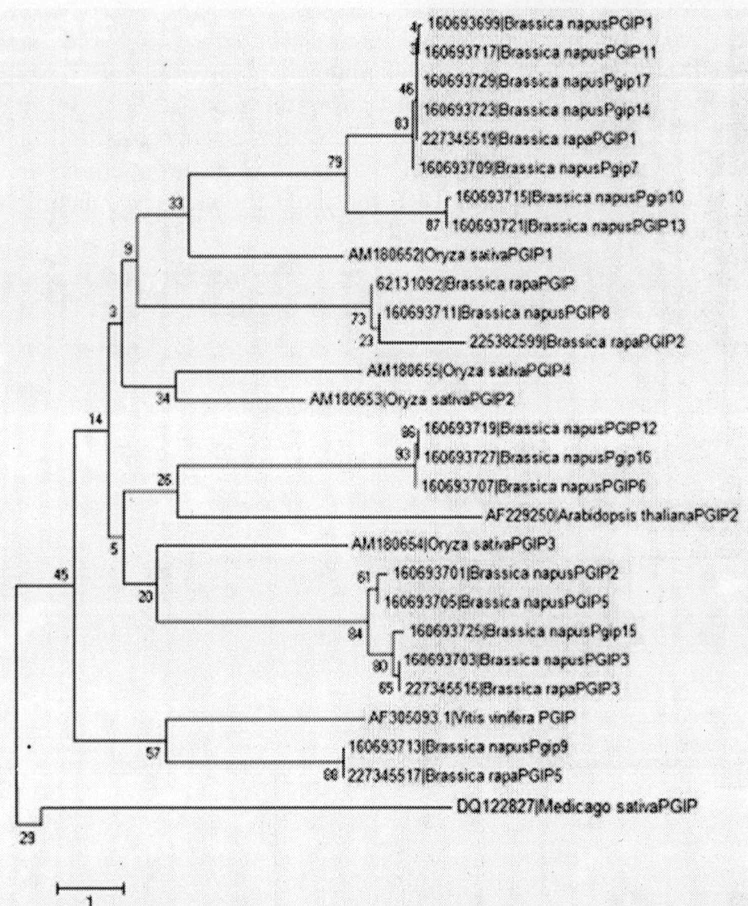

Fig 3. Phylogenetic tree of various polygalacturonase inhibiting proteins of plant species discussed in the current study (*At*PGIP, *Os*PGIP, *Bn*PGIP, *Br*PGIP, *Ms*PGIP, *Le*PGIP, *Vv*PGIP) using MEGA 4.0

Even the PGIP upstream sequences also shared common *cis* elements which involve in the regulation of PGIP gene expression upon biotic and abiotic stress besides plant development and growth (Figure 4). Light responsive elements and *cis* elements of anaerobic responsiveness, wounding, salicylic acid, elicitor responsive elements were found with high frequency (data not shown). Based on the upstream elements of various PGIPS, putative expression of PGIPs on various responses is shown in Figure 5. It is interesting to note that all the PGIP upstream sequences contain more than two LRE *cis* elements. Light was shown to be required in the development of plant resistance during infection (38). Evidence of this was shown when *Arabidopsis* plants were infected with Turnip Crinkle Virus (13). Salicylic acid glucoside levels were lower in plants placed in dark when compared to controls suggesting that darkness effects the SA biosynthetic pathway. Consistent with these results is the finding that exogenous application of SA prior to TCV inoculation partially overcame the requirement for light. Light was also required for N gene-mediated HR and resistance to Tobacco Mosaic Virus, suggesting that it is an important factor which may be generally required during defense signaling.

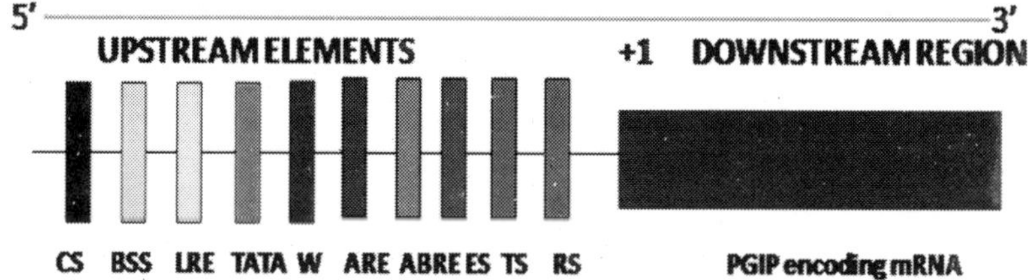

Fig 4. *Cis*-elements in upstream elements of various PGIP encoding genes; **CS**: Component specific elements; **BSS**: Binding site specific; **LRE:** Light responsive elements; **TATA:** TATA box; **W box:** Wound inducible *cis* elements; **ARE:** Anaerobic Responsive Elements; **ES:** Elicitor specific *cis* elements; **RS**: Regulation Specific Elements; **ABRE:** Absicisic acid responsive elements; **TS:** Tissue Specific *cis* elements.

Promoters often contain elements which are recognised by inducible factors which regulate the temporal and tissue-specific expression of genes. These elements are typically short sequences and are found in promoters of many (if not all) genes which respond to the same signal. Thus, in plants, analysis of promoters of PGIP genes which are expressed during biotic and abiotic stress shows that most of them are highly regulated due to light responsive and stress responsive *cis* elements (43). The current analysis helps in preparation of novel inducible promoters derived from plants and their application in the controlled expression of pathogen related protein genes upon pathogen attack. The aim of crop plant genetic engineering is to insert a gene (or genes) which alter a plant's characteristics without altering otherwise desirable elements of the genotype of the original plant. Thus, the genetic make-up of crop plants can be extended to include genes outside of the original genetic pool which are not accessible by traditional crop breeding techniques. One important aspect of such modification is the choice of promoter. Not only biotic, but abiotic factors can also induce the PGIP gene expression. Biotic elicitors are mainly pathogen fungal and insect, and abiotic elicitors are mainly wounding, salicylic acid, low temperature, jasmonate, etc (42, 43). From the Table 1, it is observed that all the transgenic plants with PGIPs are expressed under control of constitutive promoters which have deleterious effect on plant growth as well as yield. Application of inducible promoters or synthetic promoters which would express the pathogen related genes during defense mechanism would be an ideal choice to overcome the above problems.

Table 1: An overview of transgenic plants conferring resistance to biotic stress due to over expression of polygalacturonase inhibiting proteins.

S.No	Gene	Source	Promoter	Transgenic plant resistant to pathogen	Reference
1	*Pv*PGIP1	*Phaseolus vulagaris*	CaMV35S	Tomato resistant to *Botrytis cinerea*	Desiderio *et al.*, 1997 (19)
2	*Pc*PGIP	*Pyrus communis*	CaMV35S	Transgenic tomato *B.cinerea*	Powell *et al.*, 1994 (55)
3	*Pv*PGIP1	*P. vulgaris*	CaMV35S	Tomato resistant to *Stenocaprella maydis* and *Aspergillus niger*	Berger *et al.*, 2000 ()7
4	*Vv*PGIP	*Vitis vinifera*	CaMV35S	Tomato and Tobacco resistant to *B. cinerea* and *Xylella fastidiosa*	Aguero *et al.*, 2004 (3)
5	*Ad*PGIP	*Actinidia deliciosa*	CaMV35S	Resistance not reported	Szankowski *et al.*, 2003 (60)
6	*Pc*GIP1	*P. communis*	CaMV35S	Tobacco resistant to *B.cinerea*	Tamura *et al.*, 2005 (62)
7	*Pv*PGIP1	*P. vulgaris*	CaMV35S	*Arabidopsis thaliana* resistant to *B. cinearea*	Ferrari *et al.*, 2003 (26)

8	*Bn*PGIP	*Brassica napus*	BnPGIP	Transgenic *A.thaliana* for promoter analysis of BnPGIPs	Hegedus *et al.*, 2008 (31)
9	*Md*PGIP1	*Malus domestica*	E35s	Potato resistant to *Verticillium dahliae*	Gazendam *et al.*, 2004 (28)
10	*Pc*PGIP	*P. communis*	CaMV35S	Grape resistant to *X. fastidiosa*	Aguero *et al.*, 2005 (4)
11	*Pv*PGIP2	*P. vulgaris*	CaMV35S	Tobacco and *A. thaliana* resistant to *B.cinerea*	Manfredini *et al.*, 2005 (44)
12	*Vv*PGIP1	*V. vinifera*	CaMV35S	Tobacco resistant to *A.niger* and *B.cinerea*	Joubert *et al.*, 2006 (36)
13	*Ri*PGIP	*Rubus idaeus*	D35S	Pea resistant to *C. lupine* and *S. maydis*	Richter *et al.*, 2006 (56)
14	*Md*PGIP1	*M. domestica*	CaMV35s & TEV LS	Tobacco resistant to *C. lupini*	Oelofse *et al.*, 2006 (52)
15	*Pv*PGIP2	*P. vulgaris*	Ubiquitin	Wheat resistant to *Bipolaris sorokiniana*	Janni *et al.*, 2008 (33)

The crystal structure of ribonuclease inhibitor (RI) yielded the first insight into the 3D structural arrangement of LRRs and, soon thereafter, crystal structures of complexes of RI with its ligands provided the first structural views revealing how the LRR structure is used as a protein recognition motif. This structural information also formed the basis of numerous attempts to model LRR proteins and their ligand complexes. *P. vulgaris* PGIP2 is the first LRR protein belonging to the plant specific subfamily whose structure was resolved by Matteo et al. (46) and represents a model for studying the structural organization and the mode of interaction of other plants LRR proteins. It contains ten repeats of a short (24 amino acid) LRR motif. PGIP of *P. vulgaris* has three domains, the central LRR region and two cysteine- rich flanking domains.

The structure also provides a molecular basis for understanding how PGIP inhibits polygalacturonases. The overall structure shows a typical curved and elongated shape. The modelled AtPGIP1 has shown similar structure with that of PvPGIP2 crystal structure with an RMSD value of 0.246, whereas AtPGIP2 has also got RMSD value of 0.249 (Figure 6a). The central LRR domain, folded in a right handed superhelix consists of a set of ten tandemly repeating units, each made by 24 residues matching the extracytoplasmatic LRR plant specific consensus sequence LxxLxx**LxxLxLxx**NxLT/sGxIPxx. The LRR conserved residues are involved in stacking interactions that are crucial for the formation of hydrophobic core interactions in LRR domain. Both N- and C- terminal regions contribute to cap the hydrophobic core (46). An interesting region in the *Pv*PGIP2 structure is the presence of a cluster of negatively charged residues in the concave surface of the protein likely involved in the interaction with PGs. The comparative structural analysis of *At*PGIP1 with *Pv*PGIP2 has also shown a wide negative pocket involved in binding PG which is located in the middle of the inner concave surface of the protein (Figure 6b & 6c).

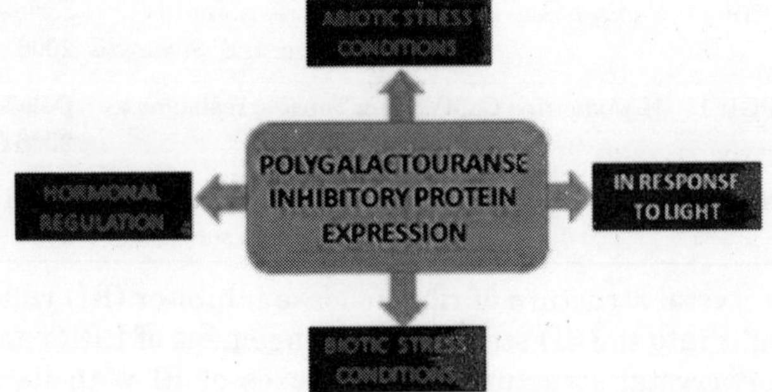

Fig. 5. Expression of PGIP genes upon induction to various responses.

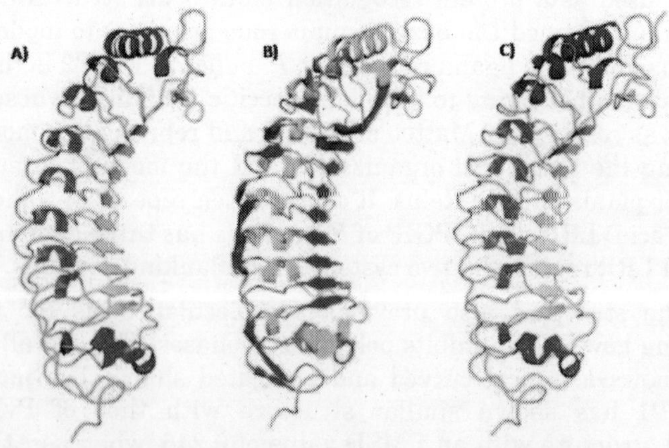

Fig 6. Structural view of *At*PGIP1 (a) and *At*PGIP2 (c) developed using Modeller with template *Pv*PGIP2 (b)

In silico studies are needed to know more about plant immunity protein

We know that the function of any protein depends on its tertiary structure and its folding. It is very important to have a thorough knowledge about the tertiary structure of any protein in order to understand how it's functionally important regions behave. The collective motions of these regions can divulge the exact interaction of a particular protein with its ligand or an enzyme with its substrate or how a protein inhibits a pathogen. All these atomistic details are not possible to obtain through experimental procedures. However, these atomistic interactions of a protein can be obtained through Molecular Dynamics (MD) simulations. MD simulation study can reveal the correlated motions of the functionally important regions of a protein like the active sites. These molecular motions can help to understand the function of the protein better. It can also help to understand the structural differences between wild type and mutant type protein. These in-silico findings can act as a supplement to understand and explain the data that is obtained through wet lab experiments.

Conventional MD simulation studies are now-a-days performed regularly on various protein structures. Various open sources, as well as commercially available MD simulation packages are available today to do these simulation studies. Among the most notable ones include GROMACS that implements the GROMOS96 force field, CHARMM and AMBER. Serial as well as parallel installation of these softwares can be done. Long MD simulation of a few hundred nano-seconds can be performed after proper energy minimization of the PDB structure which is retrieved from modelling studies. Mutant structures of PGIPs can be obtained by introducing point mutations using the MODELLER package. These wild type and mutant PGIPs can then be subjected to long simulations. Simulations can also reveal the interactions of various domains of PGIPs giving an idea about their probable functions. The differences between the wild type and the mutants can then be compared. Various structural parameters can be considered for MD analysis. The overall deviation of the mutant from the wild type structure can be obtained through RMSD or through structural superimposition using software like SPDB-Viewer. The RMS Fluctuations can reveal which regions or rather which atoms of the structures vibrate more with respect to other regions. Radius of gyration can give an idea about the compactness of the structures. Radius of gyration can also give an idea about the compactness of the hydrophobic cores of the proteins. The solvent accessible surface area can be calculated to find out information about the hydrophilic residues. The phi-psi angles of the residues can also be compared by plotting the Ramachandran plot. Differences are bound to be noticed between the wild type and mutant proteins with respect to these properties. The docking surface of PGIPs seems to be extremely versatile and capable of establishing both polar and apolar interactions with the surfaces of structurally similar but different enzymes (25).

Casasoli et al. (10) have reported a useful model to study ligand-receptor coevolution dynamics at the protein level represented by PG-PGIP interaction. They have applied codon substitution models to forty four PGIP sequences of different plant species to identify putative positively selected sites and then compared these sites with the propensity of protein surface residues to interact with protein partners, based on desolvation energy calculations and found several residues in the concave face of the leucine rich repeat domain. These residues were further mutated into alanine and their effect on the recognition of several PGs tested, leading to identification of unique hotspots for the PGIP-PG interaction.

CONCLUSION

Protein - Protein interactions play a definitive role in plant disease resistance. PGIPs are important inhibitory proteins which play important role in plant defense mechanisms against fungal pathogens due to their capacity to interact with fungal PGs. PG-PGIP interaction results in repository of oligosaccharides which act as signal molecules and have regulatory properties. Structural and functional analysis of PGIPs suggests that they are components of the plant innate immune system. A single PGIP is often capable of binding several PGs from different fungi and is not a specialized inhibitor of a single PG. More research has to be done on site directed mutagenesis of various PGIPs which eventually improve the activity against various PGs due to their new recognition specificities. Domain shuffling of two different PGIPs is a good choice of developing hyperactive PGIPs. Transformation of crop plants with these modified PGIPs or with more than two PGIPs will result in wide fungal resistance. More studies in the area of protein-protein interactions using newly developed bioinformatics tools will shed light on PG-PGIP molecular interactions. Experiments are underway with regard to simulation and docking studies of PGIPs-PGs in our laboratory.

ACKNOWLEDGEMENTS

This work was supported by University Grants Commission (UGC), New Delhi, in the form of Kothari Post Doctoral Fellowship to PM and Council for Scientific and Industrial Research (CSIR), New Delhi in the form of Senior Research Fellowship to GMK. We thank UGC-CAS, DBT-CREBB programmes of School of Life sciences and DST-FIST program of the Department of Plant Sciences, University of Hyderabad for the infrastructure support.

Literature Cited

1. Abu-Goukh AA and Labavitch JM. 1983. The *in vivo* role of 'Bartlett' pear fruit polygalacturonase inhibitors. *Physiol Plant Pathol* 23: 123–135.
2. Abu-Goukh AA, Greve LC and Labavitch JM. 1983. Purification and partial characterization of 'Bartlett' pear fruit, polygalacturonase inhibitors. *Physiol. Plant Pathol* 23: 111–122.
3. Aguero CB, Uratsu SL, Greve C, Powell ALT, Labavitch JM, Meredith CP and Dandekar AM. 2004. Evaluation of tolerance to Pierce's disease and *Botrytis* in transgenic plants of *Vitis vinifera* L. expressing the pear PGIP gene *Molecular. Plant Pathol* 6(1): 43-51
4. Aguero CB, Uratsu SL, Greve C, Powell ALT, Labavitch JM, Meredith CP and Dandekar AM 2005. Evaluation of tolerance to Pierce's disease and *Botrytis* in transgenic plants of *Vitis vinifera* L. expressing the pear PGIP gene. *Mol Plant Pathol* 6: 43–51
5. Albersheim P and Anderson AJ. 1971. Protein from plant cell wall inhibits polygalacturonases secreted by plant pathogens. *Proc Natl Acad Sci USA* 68: 1815–1819.
6. Becraft PW. 2002. Receptor kinase signaling in plant development. *Annu Rev Cell Dev Biol* 18: 163–192.
7. Berger DK, Oelofse D, Arendse MS, Plessis ED and Dubery IA 2000. Bean polygalacturonase inhibitor protein-1 (PGIP-1) inhibits Polygalacturonases from *Steno-*

carpella maydis. Physiological Mol Pl Pathol 57: 5–14.

8. Bishop JG. 2005. Directed mutagenesis confirms the functional importance of positively selected sites in polygalacturonase inhibitor protein. *Mol Biol Evol* 22: 1531-1534.

9. Boller T. 1995. Chemoperception of microbial signals in plant cells. *Ann Rev Plant Physiol Mol Biol* 46: 189–214.

10. Casasolia M, Federicib L, Spinellia F, Di Matteoc A, Vellaa N, Scalonia F, Fernandez-Reciod J, Cervonea F and De Lorenzo G 2009. Integration of evolutionary and desolvation energy analysis identifies functional sites in a plant immunity protein. *Proc Natl Acad Sci USA* 106: 7666-7671.

11. Cervone F, De Lorenzo G, Degra L, Salvi G and Bergami M. 1987. Purification and characterization of polygalacturonase-inhibiting protein from *Phaseolus vulgaris* L. *Plant Physiol* 85: 631–637.

12. Cervone F, De Lorenzo G, Presey R, Darvil AG, Albersheim P. 1990. Can *Phaseolus* PGIP inhibit pectic enzymes from microbes and plants? *Phytochemistry* 29: 447-449.

13. Chandrashekara AC, Manisha G, Duroy N, Surabhi R, Ramesh R, Daniel K and Pradeep K. 2006. Light-dependent hypersensitive response and resistance signaling against Turnip Crinkle Virus in *Arabidopsis Plant J.* 45(3): 320-334.

14. Chisholm S, Coaker G, Day B and Staskawicz B. 2006. Host-Microbe Interactions: Shaping the Evolution of the Plant Immune Response. *Cell* 124: 803-814.

15. Dangl JL and Jones JDG. 2001. Plant pathogens and integrated defence responses to infection. *Nature* 411: 826–833

16. Dardick C and Ronald P. 2006. Plant and animal pathogen recognition receptors signal through non- RD kinases. *PLoS Pathog* 2: e2.

17. De Lorenzo G and Ferrari S. 2002. Polygalacturonase-inhibiting proteins in defense against phytopathogenic fungi. *Curr Opinion in Pl Biol* 5: 295–299.

18. De Lorenzo G, D'Ovidio R and Cervone F. 2001. The role of polygalacturonase- inhibiting proteins (PGIP) in defense against pathogenic fungi. *Annu Rev Phytopathol* 39: 313–335.

19. Desiderio AB, Aracri F, Leckie B, Mattei G, Salvi H, Tigelaar J, Van Roekel, JSC. Baulcombe DC, Melchers LS, De

Lorenzo G and Cervone F. 1997. Polygalacturonase-Inhibiting Proteins (PGIPs) with Different Specificities Are Expressed in *Phaseolus vulgaris Mol Pl-Microbe Interact* 10(7): 852–860.

20. Devoto A, Leckie F, Lupotto E, Cervone F and de Lorenzo G 1998. The promoter of a gene encoding a polygalacturonase-inhibiting protein of *Phaseolus vulgaris* L. is activated by wounding but not by elicitors or pathogen infection. *Planta* 205: 165-174.

21. Dievart A and Clark SE. 2004. LRR-containing receptors regulating plant development and defense. *Development* 131: 251–261.

22. D'Ovidio R, Raiola A, Capodicasa C, Devoto A, Pontiggia D and Roberti S. 2004. Characterization of the complex locus of bean encoding polygalacturonase-inhibiting proteins reveals subfunctionalization for defense against fungi and insects. *Plant Physiol* 135: 2424–2435.

23. Ebel J and Cosio EG. 1994. Elicitors of plant defense responses. *Int Rev Cytol* 148: 1–36.

24. Espinosa A and Alfano JR. 2004. Disabling surveillance: bacterial type III secretion system effectors that suppress innate immunity. *Cell Microbiol* 6: 1027-1040.

25. Federici L, di Matteo A, Fernandez-Recio J, Tsernoglou D and Cervone F. 2006. Polygalacturonase inhibiting proteins: players in plant innate immunity? *Trends in Pl Sci* 11:65-70.

26. Ferrari S, Vairo D, Ausubel FM, Cervone F and de Lorenzo G 2003. Tandemly duplicated *Arabidopsis* genes that encode polygalacturonase-inhibiting proteins are regulated coordinately by different signal transduction pathways in response to fungal infection. The *Plant Cell* 15: 93–106.

27. Fielding AH. 1981. Natural inhibitors of fungal polygalacturonases in infected fruit tissues. *J Gen Microbiol* 123: 377–381.

28. Gazendam I, Oelofse D and Berger DK 2004. High-level expression of apple PGIP1 is not sufficient to protect transgenic potato against *Verticillium dahliae Physiol Mol Pl Pathol* 65: 145–155

29. Gomathi V and Gnanamanickam SS. 2004. Polygalacturonase-inhibiting proteins in plant defence *Curr. Sci* 87 (9): 1211-1217.

30. Gomez-Gomez L, Bauer Z and Boller T. 2001. Both the extracellular leucine rich repeat domain and the kinase activity of FSL2 are required for

flagellin binding and signaling in *Arabidopsis. The Plant Cell.* 13, 1155–1163.

31. Hegedus DD, Li R, Buchwaldt L, Parkin I, Whitwill S, Coutu C, Bekkaoui D and Rimmer SR. 2008. *Brassica napus* possesses an expanded set of polygalacturonase inhibitor protein genes that are differentially regulated in response to *Sclerotinia sclerotiorum* infection, wounding and defense hormone treatment. *Planta* 228: 241–253

32. Janeway CA and Medzhitov R. 2002. Innate immune recognition. *Annu Rev Immunol* 20: 197-216.

33. Janni M , Sella L, Favaron F, Blechl AE, de Lorenzo G, and D'Ovidio R 2008. The Expression of a bean PGIP in transgenic wheat confers increased resistance to the fungal pathogen *Bipolaris sorokiniana. Mol Pl-Microbe Interact* 21: 171–177.

34. Johana CM and Renier ALH. 2008. Enzyme–inhibitor interactions at the plant–pathogen interface *Current Opinion in Plant Biology*. 11: 380–388.

35. Jones JDG and Dangl JL. 2006 The Plant Immune System. *Nature* 444: 323-329.

36. Joubert DA, Slaughter AR, Kemp G, Becker JVW, Krooshof GH, Bergmann C, Benen J, Pretorius IS and Vivier MA. 2006. The grapevine polygalacturonase-inhibiting protein (VvPGIP1) reduces *Botrytis cinerea* susceptibility in transgenic tobacco and differentially inhibits fungal polygalacturonases. *Transgenic Res* 15: 687– 702.

37. Kajava AV. 1998. Structural diversity of leucine-rich repeat proteins. *J Mol Biol* 277: 519–527

38. Karpinski S, Gabrysy H, Mateo A, Karpinska B and Mullineauxz PM. 2003. Light perception in plant disease defence signaling *Current Opinion in Plant Biology* 6: 390–396.

39. Karr ALL and Albersheim P. 1970. Polysaccharide-degrading enzymes are unable to attack plant cell walls without prior action by a “wall-modifying enzymes”. *Plant Physiol.* 46: 69–80.

40. Kim L, da Cunha AJ, McFall Y, Belkhadir S, DebRoy JL, Dangl J and Mackey D. 2005. Two *Pseudomonas syringae* type III effectors inhibit RIN4-regulated basal defense in *Arabidopsis. Cell* 121: 749–759.

41. Leckie F, Mattei B, Capodicasa C, Hemmings A, Nuss L, Aracri B, de Lorenzo G and Cervone F. 1999. The specificity of polygalacturon- ase

inhibiting protein (PGIP): a single amino acid substitution in the solvent-exposed β-strand/β-turn region of the leucine rich repeats (LRRs) confers a new recognition capability. *EMBO J.* 18: 2352-2363.

42. Li R, Roger R, Yu MA, Sharpe G, Ginette SS, Lydiate D, and Hegedus DD. 2003. Two *Brassica napus* polygalacturonase inhibitory protein genes are expressed at different levels in response to biotic and abiotic stresses. *Planta* 217: 299–308.

43. Mahesh GK, Mamidala P and Podile AR (2009) Regulation of Polygalacturonase-inhibitory proteins in plants is highly dependent on stress and light responsive elements *Plant Omics Journal 2(6): 238-249*

44. Manfredini C, Sicilia F, Ferrari S, Pontiggia D, Salvi G, Caprari C, Lorito M and De Lorenzo D 2005. Polygalacturonase-inhibiting protein 2 of *Phaseolus vulgaris* inhibits BcPG1, a polygalacturonase of *Botrytis cinerea* important for pathogenicity, and protects transgenic plants from infection. *Physiol Mol Pl Pathol* 67: 108–115.

45. Martin GB, Bogdanove and AJ Sessa G. 2003. Understanding the functions of plant disease resistance proteins. *Annu Rev Plant Biol.* 54: 23–61.

46. Matteo AD, Federici L, Mattei B, Salvi, Johnson KA, Savino C, De Lorenzo G, Tsernoglou D, and Cervone F. 2003. The crystal structure of polygalacturonase-inhibiting protein (PGIP), a leucine-rich repeat involved in plant defense *Proc Natl Acad Sci USA* 100: 10124-10128.

47. Matzinger P. 2002. The danger model: a renewed sense of self. *Science* 296: 301-305.

48. Maulik A, Ghosh H and Basu S 2009 Comparative study of protein-protein interaction observed in Polygalacturonase-Inhibiting Proteins from Phaseolus vulgaris and Glycine max and Polygalacturonase from Fusarium moniliforme *BMC Genomics*10 (3): 1-10

49. Medzhitov R and Janeway CA. 1997. Innate immunity: the virtues of a nonclonal system of recognition. *Cell* 91: 295-298.

50. Montesano M, Brader G and Palva ET 2003. Pathogen derived elicitors: searching for receptors in plants. *Mol Plant Pathol* 4: 73-79.

51. Nurnberger T, Brunner F, Kemmerling B and Piater L. 2004. Innate immunity in plants and animals: striking similarities and obvious differ-

ences. *Immunol Rev* 198: 249–266.

52. Oelofse D, Dubery IA, Meyer R, Arendse MS, Gazendam I and Berger DK 2006. Apple polygalacturonase inhibiting protein1 expressed in transgenic tobacco inhibits olygalacturonases from fungal pathogens of apple and the anthracnose pathogen of lupins. *Phytochemistry* 67: 255–263.

53. Palva TK, Holmstrom KO, Heino P and Palva ET. 1993. Induction of plant defense response by exoenzymes of *Erwinia carotovora* subsp. *carotovora*. *Mol Plant Microbe Interactions* 6: 190-196.

54. Perombelon MCM and Kelman A. 1980. Ecology of the soft rot Erwinias. *Annu. Rev. Phytopathol* 18: 361-387.

55. Powell ALT, Stotz HU, Labavitch JM and Bennett AB. 1995. Glycoprotein inhibitors of fungal polygalacturonases. In: Daniels MJ, Downie JM, Osbourn AE, eds. *Advances in molecular genetics of plant-microbe interactions,* Vol. 3. The Netherlands: Kluwer Academic Publishers, 399:402

56. Richter A, Jacobsen HJ, de Kathen A, de Lorenzo G, Briviba K, Hain R, Ramsay G and Kiesecker H 2006. Transgenic peas (*Pisum sativum*) expressing polygala-cturonase inhibiting protein from raspberry (*Rubus idaeus*) and stilbene synthase from grape (*Vitis vinifera*) *Plant Cell Rep* 25: 1166–1173.

57. Shibuya N, Minami E. 2001. Oligosaccharide signalling for defence responses in plant. *Physiol & Mol Pl Pathol* 59: 223-233.

58. Sicilia F, Fernandez-Recio J, Caprari C, de Lorenzo G, Tsernoglou D, Cervone F and Federici L. 2005. The polygalacturonase inhibiting protein PGIP2 of *Phaseolus vulgaris* has evolved a mixed mode of inhibition of endopolygalacturonase PG1 of *Botrytis cinerea. Plant Physiol*. 139:1380-1388.

59. Stotz HU, Bishop JG, Bergmann CW, Koch M, Albersheim P, Darvill AG, Labavitch JM. 2000. Identification of target amino acids that affect interactions of fungal polygalacturonases and their plant inhibitors. *Physiol Mol Plant Pathol*. 56: 117-130.

60. Szankowski I, Briviba K, Fleschhut J, Schönherr J, Jacobsen H-J and Kiesecker H. 2003. Transformation of apple (*Malus domestica* Borkh.) with the stilbene synthase gene from grapevine (*Vitis vinifera* L.) and a PGIP gene from kiwi (*Actinidia*

deliciosa), *Plant Cell Rep* 22: 141 – 149.

61. Szekeres M. 2003. Brassinosteroid and systemin: Two hormones perceived by the same receptor. *Trends Pl Sci* 8: 102–104.
62. Tamura M, Gao R, Tao R, Labavitch JM and Dandekar AM. 2004. Transformation of persimmon with a pear fruit polygalacturonase inhibiting protein (PGIP) gene. *Scientia Horticulturae* 103 (1): 19-30.
63. Toubart P, Desiderio A, Salvi G, Cervone F and Daroda L 1992. Cloning and characterization of the gene encoding the endopolygalacturonase-inhibiting protein (PGIP) of *Phaseolus vulgaris* L. *Plant J.* 2: 367–373.
64. Vidal S, Eriksson A, Montesano M, Denecke J and Palva ET. 1998. Cell wall degrading enzymes from *Erwinia carotovora* cooperate in the salicylic acid-independent induction of plant defense response. *Mol Pl-Microbe Interact* 11: 23-33.
65. Vidal S, Ponce de Leon I, Denecke J and Palva ET. 1997. Salicylic acid and the plant pathogen *Erwinia carotovora* induce plant defense genes by antagonistic pathways. *Plant J.* 11: 115-123.
66. Vorwerk S, Somerville S and Somerville C. 2004. The role of plant cell wall polysaccharide composition in disease resistance. *Trends Pl Sci* 9: 203-209

HOST-PATHOGEN INTERACTIONS - BIOCHEMICAL AND MOLECULAR INSIGHTS TO DEFENSE RESPONSES IN PLANTS WITH SPECAIL REFERENCE TO PEARL MILLET DOWNY MILDEW

S. Niranjan-Raj[1], *S. N. Lavanya*[2], *K. N. Amruthesh*[2], *S. R. Niranjana*[1], *and H. S. Shetty*[3]

[1]Department of Studies in Biotechnology, University of Mysore, Manasagangotri, Mysore-570 006, Karnataka, India.
[2]Department of Studies in Botany, University of Mysore, Manasagangotri, Mysore-570 006, Karnataka, India.
[3]Professor and Principal Investigator, Downy Mildew Research Laboratory, ICAR-AICPMIP, Department of Studies in Biotechnology, University of Mysore, Manasagangotri, Mysore-570 006, Karnataka, India

KEY WORDS: Host pathogen interaction, plant defense responses, PR proteins, pearl millet downy mildew

ABSTRACT

Plants are often attacked by a vast array of potential pathogens like fungi, bacteria, viruses, oomycetes, nematodes, protozoa and mycoplasma. To counter these attacks, plants are well equipped with several inbuilt mechanisms which manifest as physical, biochemical and molecular defenses. The innate immune system in plants is triggered by certain signal molecules termed as elicitors of pathogen origin and are referred to as microbe associated molecular patterns (MAMPs) or pathogen associated molecular patterns (PAMPs). These molecular signatures of the pathogen are perceived by host plant receptors and activate the defense responses. Susceptibility or resistance to a particular pathogen at the molecular level is determined by the interaction of the host *R* (resistance) gene and the corresponding *Avr* (avirulence) gene of the pathogen. This gene-for-gene mediated systemic acquired/ induced resistance is central to all the subsequent defense responses. Such interaction leads to a cascade of primary and secondary defense reactions including rapid release of reactive oxygen species (ROS) /intermediates with antimicrobial effect, cell wall cross-linking, changes in cell wall architecture, induction of antimicrobial secondary metabolites like phytoalexins, synthesis and activation of defense related peptides and proteins, induction of pathogenesis related (PR) proteins, hypersensitive response (HR) and finally cell death. This orchestrated chain of defense related reactions are dependent on various signal molecules like G-proteins, salicylic acid, jasmonic acid, ethylene, nitric oxide, phosphorylation, protein recycling and also depends on cross-talk among these signal molecules. Ultimately such signaling leads to the up regulation of resistance genes and thereby the host plant is rendered resistant. Modern biotechnological tools like molecular biology and genetic engineering have shed more light on underlying mechanisms of pathogen perception, signal transduction and resistance expression in plants. In this review we have discussed biochemical and molecular aspects of the host pathogen interactions and expression of defense responses in plants in general, and the research results on defense related aspects of pearl millet – *Sclerospora graminicola* interaction in particular.

INTRODUCTION

Plants initiate an effectual battery of defense responses immediately on recognizing a potential pathogen. There are three important steps involved in plant defense responses, 1) pathogen recognition, 2) signal transduction, and 3) execution of defence programmes (61). Pathogen recognition process follows the gene-for-gene hypothesis (36) and are genetically determined by complimentary pairs of dominant pathogen-encoded avirulence (*Avr*) genes (ligand) and dominant plant resistance (*R*) genes (receptor) (49). The downstream signal transduction pathways following *R* gene-*Avr* gene recognition are not fully understood (61).

The recognition process is essentially turned on by the action of certain pathogen derived signal molecules termed as elicitors.

Pathogen derived elicitors constitute a bewildering array of compounds including different oligosaccharides, lipids, peptides and proteins (49). Elicitors, often referred to as PAMPs or MAMPs are unique microbial signatures (56, 70) that are conserved motifs essential for microbial metabolism and as such are vital for microbial survival (33, 56). Perception of PAMPs by the host receptors generates signals that trigger early defense responses (4, 56).

Following elicitor perception by the host plant, a number of biochemical changes occur in the host cells including changes in the ion permeability of the plasma membrane, activation of plasma membrane-bound enzymes, activation of kinases, phosphatases, phospholipases, and the production of signal molecules, including active oxygen species which ultimately leads to the transcriptional activation of defense-related genes. These early events are followed by other defence responses including induction of hypersensitive response (HR), a localized form of programmed cell death (PCD) limiting pathogen spread, further reinforcement of the cell walls, and production of antimicrobial compounds such as defence proteins and phytoalexins most of which are mediated by endogenous signal molecules like jasmonic acid (JA), ethylene (ET) and salicylic acid (SA) (49). These signal molecules are part of two major pathways in defence signaling, one SA-dependent, the other SA-independent but involving JA and ET (39). Other molecules such as ROS, G-proteins and nitric oxide (NO) appear also to be involved in plant defence signaling.

Pearl millet (*Pennisetum glaucum* L. Br.) production is seriously affected by many pathogens, of which the worst affecting is the oomycete *Sclerospora graminicola* (Sacc.) Schroet., an obligate biotroph, which causes downy mildew/ green ear disease. Various aspects of the pearl millet downy mildew interactions at the biochemical and molecular level and also the management strategies have been studied extensively in the past years. In this review, the host-pathogen interactions at various levels have been discussed with suitable examples of pearl millet – *S. graminicola* interactions.

PRIMARY RESPONSES OF THE PLANT TO PATHOGEN RECOGNITION

There are a host of immediate responses triggered by the recognition of a potential pathogen. These early responses include rapid plasma membrane depolarization often accompanied by altered extracellular ion concentrations resulting in ion fluxes across the plasma membrane; transcriptional activation of G proteins, protein kinases, phosphatases, and phospholipases; biosynthesis and accumulation of SA; and rapid generation and accumulation of NO and other ROS (6, 49). Active oxygen species like H_2O_2, most often catalyze cell wall strengthening and directly contain the invading pathogens. ROS also act as diffusible intracellular signals and second messengers and transcriptionally activate several genes which encode antimicrobial compounds. ROS can also trigger HR-related cell death (41, 20, 9, 32, 56).

SECONDARY RESPONSES OF THE PLANT TO PATHOGEN RECOGNITION

Primary defense responses are soon followed by activation of a battery of strong defense reactions like structural reinforcement of plant cell walls activated by allosteric enzymes; changed cytoplasmic Ca^{2+} levels; local and systemic induction of anti-microbial compounds including defensins and pathogenesis-related (PR) proteins such as proteinases, proteinase inhibitors, and hydrolytic enzymes like chitinases and ß-1,3-glucanases; and synthesis of antimicrobial secondary metabolites (phytoalexins) (10, 23). These secondary defense responses often result in the development of HR and SA-dependent systemic acquired resistance (SAR) (12, 77).

ELICITORS OF PLANT DEFENSE RESPONSES

Early recognition of the potential pathogen is vital for a plant to defend itself against pathogen attack. Arrival and attack of a pathogen is immediately sensed by an alert host by virtue of detection of various pathogen derived signal molecules often referred to as elicitors which leads defense response trigger. Pathogen derived elicitors include several compounds like oligosaccharides, lipids, peptides and proteins (4, 49, 56).

Elicitors, otherwise known as pathogen-associated molecular patterns (PAMPs), are not produced by host cells but are unique to microbial organisms (70, 56). They are conserved motifs that are essential for microbial metabolism and as such are vital for microbial survival (33, 56). Upon receptor mediated perception, PAMPs inevitably expose the invader to the plant's surveillance system, rapidly generating an internal signal that triggers early defense responses (4, 56). PAMPs or MAMPs, are classified into two groups: general or non-race specific elicitors and race-specific elicitors (6, 49). General elicitors, including cell wall polysaccharides (e.g., glucans and chitin fragments), membrane constituents (e.g., sterols like ergosterol and fatty acids such as arachidonic acid), secreted proteins, and glycopeptides signal the presence of potential pathogens to both host and non-host plants (6). Race-specific elicitors, trigger defence responses leading to disease resistance only in specific cultivars of the host plant, following the gene-for-gene hypothesis. Therefore, a race-specific elicitor of an *Avr* gene present in a particular race of a pathogen will elicit resistance only in a host plant type carrying the corresponding *R* gene (49).

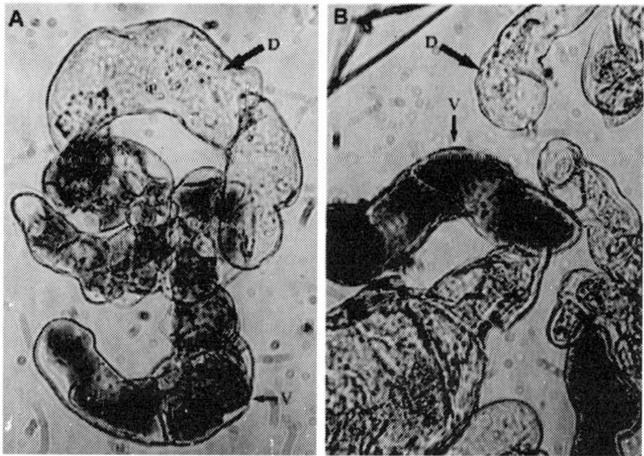

Fig. 1: Differentiation of viable and dead pearl millet cells by neutral-red staining 2 hours post elicitation with *Sclerospora graminicola* – derived crude protein elicitor. The stained cells viable after treatment showed plasmolysis while dead cells appeared light yellow without plasmolysis. (A) Susceptible cells, (B) resistant cells; D, dead cells; V, viable cells.

A peptide elicitor was isolated, purified and characterized from the zoospores of *S. graminicola.* The crude fractionated protein was able to elicit an array of defense responses in resistant and susceptible cells of pearl millet. Treatment of cultured cells of pearl millet with partially purified elicitor protein resulted in a rapid loss of cell viability in the resistant cells and the percentage of cell death was higher in the resistant than in the susceptible cells (Fig. 1). Increased oxidative burst was also recorded in the cells treated with the crude elicitor. The purified elicitor was acidic with a pI of 5.6 as revealed by isoelectric focusing (IEF) (Fig. 2) and matrix-assisted laser desorption ionization (MALDI) analysis showed that the elicitor had a molecular mass of 7040 daltons. The primary structure determined by N-terminal Edman degradation and searches with BLAST did not reveal similarities to any known plant pathogenic or oomycete elicitor. Higher activities of the important defense-related enzymes PAL and POX in the resistant cell cultures than in the susceptible cell cultures treated with the purified elicitor were clearly evident. Studies of gene expression by northern blotting with heterologous peroxidase, PAL and oxalate oxidase probes showed that the mRNA transcripts were strongly up-regulated in resistant cell cultures within 30 min of elicitor treatment. The purified elicitor also demonstrated a very strong concentration-dependent sterol binding. The purified elicitor protein belongs to a class of low molecular weight oomycete elicitors with sterol carrier properties (65).

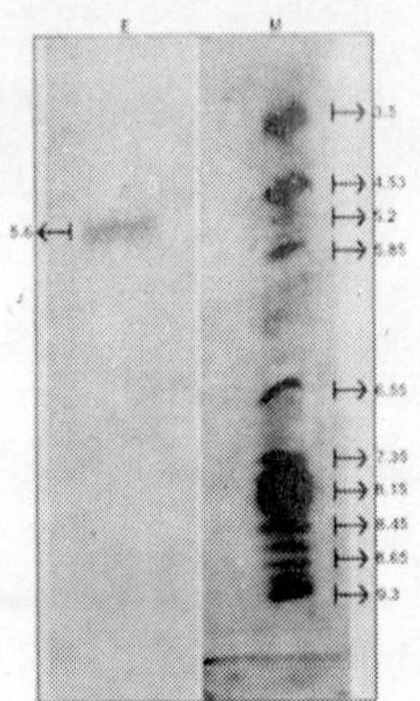

Fig. 2: Isoelectric focusing of the purified *Sclerospora graminicola* derived protein elicitor followed by Coomassie blue staining to determine the isoelectric point. Five micrograms of the purified elicitor were loaded onto the gel; M, marker; E, elicitor.

REACTIVE OXYGEN SPECIES

Singlet oxygen ($^1O_2$), superoxide anions ($^{n}O_2^-$), hydroxyl radicals (^{n}O), H_2O_2, NO and peroxynitrites ($ONOO^-$) constitute the major active oxygen species (47). ROS by-products of certain metabolic reactions, like mitochondrial electron transport (71). ROS perform multifaceted actions like intracellular signals and second messengers, regulate gene expression and mediate cell activation, and stimulate cell proliferation (20, 43). Changes in concentrations of ROS is driven by various stress conditions (80). Increase in ROS leads to oxidative stress, and will disrupt cellular components and eventual cell death (1). Decrease in ROS also act as messengers of cellular damage, severity of the oxidative stress determines activation of either programmed cell death under less intense stress or necrosis under extreme stress (66).

Treatment of cultured cells of pearl millet cultivars with salicylic acid, calcium chloride zoospore, sporangial suspension and sporangial cell wall components of *S. graminicola* elicited oxidative burst. The characteristics of the enzymatic production were similar in both compatible and incompatible interaction but the kinetics and intensity of production of ROS differed among the interactions. Key enzyme responsible for oxidative burst in pearl millet was NADPH oxidase and contribution of POX towards oxidative burst was less. Oxidative burst studied in different cultivars of pearl millet with varied degree of resistance to downy mildew disease showed the correlation of oxidative burst with host resistance (25).

CHANGES IN CELL WALL ARCHITECTURE

ROS and intermediates trigger the oxidative cross-linking of the cell wall by altering the cell wall constitution. The altered cell wall constitutes cross liking of polymers of lateral phenolic chains, like the tyrosine residues of extensins and the residues of ferulic and p-coumaric acids linked to polysaccharides. Increased H_2O_2 triggers cross-linkages among the tyrosines of extensins called hydroxyprolinerich glycoproteins (HRGPs), present in the cell wall. Cross-linking renders the HRGPs highly insoluble thus nullifying the action hydrolytic enzymes of the pathogen and thus contain the pathogen at the infection site (14)

Accumulation of HRGPs in pearl millet seedlings in response to *S. graminicola*, was monitored by estimation of hydroxyproline (Hyp) in the hydrolyzed cell wall samples. In 2-day-old seedlings of a highly resistant cultivar of pearl millet inoculated with *S. graminicola*, a marked increase in Hyp content was observed in the coleoptiles as early as 2 h compared to that in a highly susceptible cultivar. A correlation was observed between the

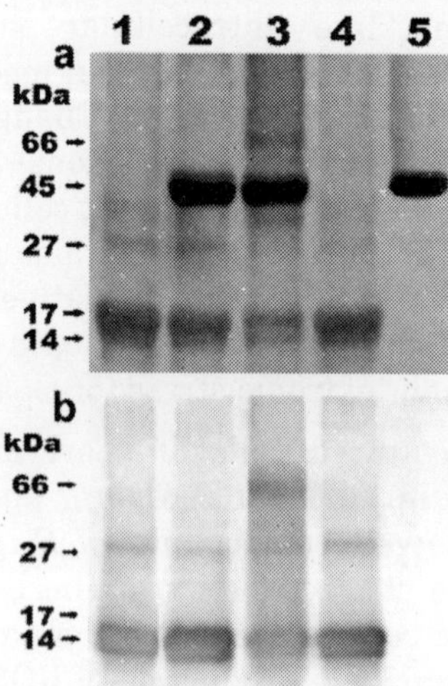

Fig. 3. *In vitro* studies on oxidative cross-linking of purified P/HGRP by H2O2 and horse radish peroxidase on SDS–PAGE followed by (A) PAS staining and (B) immunoblot using MAC 265 monoclonal antibody. Lane 1: 10 lg purified HRGP; lane 2: 10 lg purified HRGP + 1 U HRP; lane 3: 10 lg HRGP + 1 U HRP + 1 ll H2O2 (8 mM); lane 4: 10 lg HRGP + 1 ll H2O2 (8 mM); lane 5: 1 U HRP.

amount of Hyp induced in coleoptiles of different cultivars of pearl millet to *S. graminicola* infection and the host resistance. Electrophoretic separation of the acid–ethanol extracted proteins from these coleoptiles followed by periodic acid Schiff staining revealed three glycoproteins of molecular masses 27, 17 and 14 kDa (Fig. 3). These three glycoproteins also reacted positively on western blots with MAC 265, a monoclonal antibody against pea HRGP, indicating that they are HRGPs. The 17 kDa HRGP was markedly increased in highly resistant cultivar to *S. graminicola* infection compared to highly susceptible cultivar. The 14 kDa HRGP was exclusively present in the highly resistant cultivar and not in highly susceptible cultivar. Histological studies showed cross-linking of proteins in the cell walls of only highly resistant cultivar against *S. graminicola* infection. In addition, immunocytochemical studies using MAC 265 antibodies recognized these cross-linked proteins to be HRGPs. These findings indicated that accumulation of HRGPs in cell walls of highly resistant pearl millet seedlings is a component of defense reaction against downy mildew pathogen *S. graminicola* (63, 16). A resistant pearl millet cultivar, IP18292, was compared with a susceptible cultivar,

7042S, to investigate the contribution of HRGPs in the successful defense against the phytopathogenic oomycete *S. graminicola*. Northern hybridization using *MeHRGP* cDNA, a heterologous probe from cassava, indicated steady accumulation of *HRGP* transcripts, from 2 h.p.i. onwards with a maximum at 6 h.p.i., in the resistant cultivar (Fig 4). This is followed by HRGPs accumulation at about 8 h.p.i. as revealed by Western-blot analysis. Immunocytochemical localization by tissue printing and confocal immunofluorescence microscopy indicated cell walls of parenchymatic cells and the vascular tissue of coleoptile as sites of HRGP deposition (17) (Fig. 5).

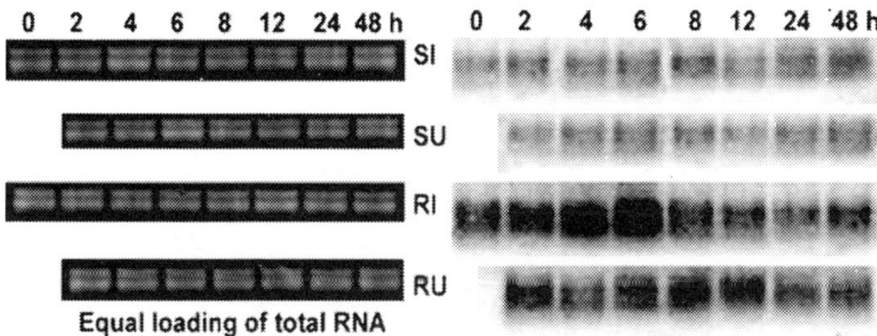

Fig.4: Northern-blot analysis. The temporal pattern of accumulation of HRGP transcripts in resistant uninoculated (RU), resistant inoculated *(RI),* susceptible uninoculated (SU) and susceptible inoculated (SI) pearl millet seedlings. Ethidium bromide staining for equal loading of total RNA. Total RNA was isolated from control and inoculated susceptible and resistant pearl millet seedlings. About 10 _g of total RNA was loaded. The RNA was hybridized with 32P-labeled HRGP cDNA probe from cassava *cMeHRGP1*

HYPERSENSITIVE RESPONSE

Hypersensitive response results from recognition of a specific pathogen-derived elicitor by a corresponding receptor in the host. Hypersensitive response is a rapid, highly localized host cell death at and around the site of pathogen entry (34). Antimicrobial effector molecules are activated by concerned defense genes due to HR (29). HR often results in the formation of dry zones of dead cells localized around or near the initial infection site that are clearly delimited from the surrounding healthy tissue (23). These so-called HR/necrotic lesions are thought to confine the pathogen to the infected tissue and thereby prevent systemic proliferation of the infection (48, 40). HR coincides with numerous metabolic changes in the affected plant cells, such as the strengthening of structural barriers by accumulation of various defense compounds like lignin, callose, guanadine and H_2O_2.

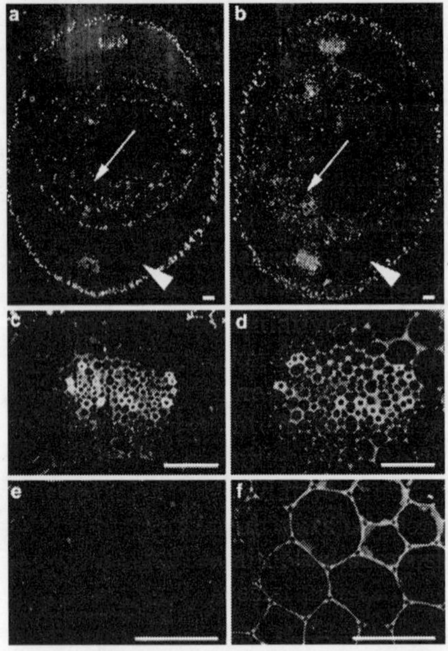

Fig. 5: a–f Immunoflorescence microscopy showing localization of HRGP. Cross sections of coleoptile region from 2-day-old pearl millet seedlings were immunostained using MAC 265 monoclonal antibody followed by an goat anti rat IgG-Alexa Fluor488 (green fluorescence). Panels show HRGP accumulation in sections of non-infected (**a**, **c**, **e**) and infected (**b**, **d**, **f**) seedlings. **a**, **b** A faint green fluorescence indicative for the occurrence of HRGPs is visible in the coleoptile (arrowheads), whereas the primary leaf exhibits only autofluorescence (arrows). **c**, **d** HRGPs were immunodecorated mainly in the cell walls of phloem cells of the coleoptile of infected seedlings; only few cells are labeled in non-infected seedlings. **e**, **f** Enlargement of parenchymatic cells of the coleoptiles showing location of HRGPs in the cell wall of infected seedlings. *Bars* 50 _m

HR has been studied during pearl millet – *S. graminicola* interaction and it has been shown that HR appears in the region of the coleoptile and also on the roots as necrotic lesions and can be used to detect resistance at an early stage of the pearl millet seedlings (Fig. 6) Higher the percentage of HR higher was the resistance and *vice versa*. HR was also correlated with accumulation of defense compounds like lignin, callose, guanidine and H_2O_2 in pearl millet seedlings infected with the downy mildew pathogen and the resistance or susceptibility was dependent on the rapidity and rate of expression of HR. HR as a defensive response and also as a marker of resistance has been demonstrated previously in pearl millet downy mildew interaction with various inducers like *Datura metel* and BTH (38, 24).

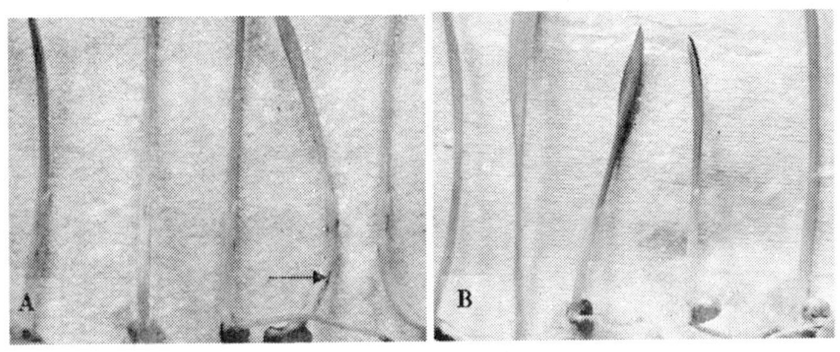

Fig. 6: Hypersensitive reaction in resistant inoculated (A) compared with resistant uninoculated (B) pearl millet seedlings.

ROLE OF HAUSTORIA IN PLANT DEFENSE

Haustoria are morphological features of an extremely succe-ssful class of plant parasites, the obligate biotrophs comprising the downy mildews (Oomycota), powdery mildew (Ascomycota) and the rust fungi (Basidiomycota) (28). These structures have arisen many times in the course of evolution and represent specific adaptations of these organisms to the close interaction with their respective host plants. Haustoria serve in nutrient uptake and also perform enormous biosynthetic duties. Haustoria play a role in suppression of host defense responses and alter host metabolism. The haustorium contains a normal complement of organelles and its own set of nuclear genes, like many *in planta*-induced genes (*PIGs*) that code for a variety of proteins involved in amino acid transport, thiamine biosynthesis, metallothioneins, cytochrome P-450 mono-oxygenases and short-chain dehydrogen-ases. Signals emanating from haustoria suppress host defense responses and facilitate disease in sensitive host plants or trigger a hypersensitive response (HR) leading to disease resistance in resistant hosts i.e., interaction between *Avr* factors and host *R* gene products (73).

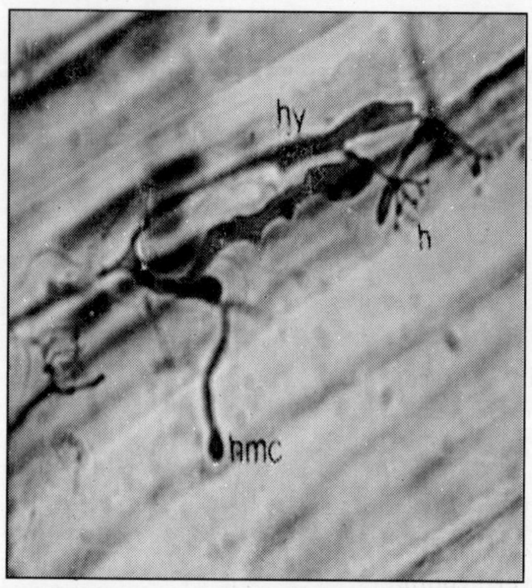

Fig. 7: A portion of the *Sclerospora graminicola* mycelium enlarged showing the formation of haustorial mother cells and haustoria in a highly susceptible cultivar.

During pearl millet – S. graminicola interaction, histological demonstration of either the number of papillae formed per 100 cells or the number of haustoria formed per 100 infected cells can be used as histological markers of downy mildew resistance/ susceptibility (Fig. 7). Fewer haustoria and more papillae indicate prolonged and durable resistance and vice versa. In most of the highly resistant and resistant pearl millet cultivars, post-haustorial resistance was seen whereby the pathogen was arrested after the formation of at least one haustorium. However, in two highly resistant cultivars pre-haustorial resistance was observed where the development of hyphae was arrested before the formation of haustoria (64). Similarly haustoria formation played an important role during nitric oxide donors mediated induced systemic resistance against pearl millet downy mildew disease. Haustoria development was inhibited in the treatments where NO level recorded 4-6.5nM during first 4 h.p.i. Inhibition of haustorial formation was probably due to enhanced resistance of the host due to NO donor treatment. Exogenous application of NO donor sodium nitro prusside (SNP) to susceptible seedlings accelerated defense responses and inhibited haustorial development (44).

SYSTEMIC ACQUIRED RESISTANCE

Of the many natural defense mechanisms plants have evolved to survive in nature, only a few can be triggered by biological or chemical agents without deleterious side effects and in a sufficiently controlled way for practical use (30, 21). The timing of these defense responses is critical and can be the difference between being able to cope or succumb to the challenge of a pathogen. During systemic acquired resistance (SAR) plant defenses are preconditioned by prior infection or treatment that results in resistance (or tolerance) against subsequent challenge by a pathogen or parasite. The compounds/agents that bring about the resistance induction are referred to as inducers/elicitors, which can be biotic or abiotic in nature. Though numerous chemical compounds were identified as effective resistance inducers, only a few like (Acibenzolar-S-methyl) (79), Probenazole (Oryzemate) (50), BTH (Benzothiodiazole) (7), Actigard (57) and Messenger (Harpin protein) (57) were commercialized due to various reasons. SAR is characterized by the induction of a number of defence-related proteins including many PR proteins (10). Thus, the expression of SAR is dependent on the accumulation of SA and is associated with the induction of PR proteins with antimicrobial activities. Induction of resistance against pearl millet downy mildew is well documented and an array of biotic and abiotic inducers have been evaluated and many promising inducers identified. The inducers of resistance against pearl millet downy mildew disease and associated defense responses are presented in Table 1.

Table 1: Inducers of resistance against pearl millet downy mildew disease and their associated defense responses in the host.

Inducer	**Defense responses**	**Reference**
ß-Aminobutyric acid (BABA)	Induced downy mildew disease resistance, accumulation of lignin, callose, HRGPs.	Shailashree *et al.* 2001
Benzothiadiazole (BTH)	Induced resistance, reduced fungal biomass, increased HR.	Geetha & Shetty 2002a
Cerebrosides (glycosphingolipids)	Elicited resistance effectively against pearl millet downy mildew.	Deepak *et al.* 2003.

Plant growth promoting rhizobacteria (PGPR) *Pseudomonas* and *Bacillus* spp.	Effectively induced systemic and durable resistance.	Niranjan Raj *et al.* 2003, 2004a
Proline	Induced systemic and durable resistance	Niranjan Raj *et al.* 2004b
L-methionine	Induced resistance, increased mRNA transcripts of Pr-1a, ß-1,3-glucanases, chitinase, POX, LOX and chalcone synthase.	Sarosh *et al.* 2005
Aspergillus niger (Carbohydrate components from spore cell wall)	Induced resistance, enhanced lignin and callose deposition, and increased POX activity.	Hindumathy *et al.* 2005
Menadione sodium bisulphite (MSB)	Induced downy mildew disease resistance.	Pushpalatha *et al.* 2007
Ind-Ile-Me synthetic (1-oxo-indanoyl-l-isoleucine methyl ester)	Induced resistance, enhanced activities of PAL, POX, and HRGPs	Deepak *et al.* 2007c
NO donors (Nitroso-R-Salt, 2-Nitroso-1-Naphthol and Sodium Nitro Prusside)	Induced resistance, enhanced HR, lignin deposition and PAL activity	Manjunatha *et al.* 2008a
Silicon (silicon dioxide)	Induced resistance, increased silicon deposition, accumulation of HRGPs	Deepak *et al.* 2008
Chitosan	Induced resistance, increased chitosanase and POX activities.	Manjunatha *et al.* 2008b
Datura metel extract	Induced resistance, increased SA, POX, b-1,3-glucanase and chitinase activities	Shivakumar *et al.* 2009

ANTIMICROBIAL PROTEINS FROM PLANTS

Several proteins and peptides with direct antimicrobial activities are induced as soon as the pathogen makes contact with the host. Among these, the main groups of antimicrobial proteins are thionins, defensins and lipid transfer proteins.

Thionins

Thionins are small basic proteins found in seeds, leaves and stems of various plants consisting of a polypeptide chain of 45–50 amino acids with 3–4 internal disulfide bonds. Thionins confer host plant resistance by exhibiting differential toxic activity to fungi and bacterial pathogens (72, 26). The toxicity of the thionins to pathogens and its synthesis triggered by pathogen stress has suggested that thionins are naturally occurring inducible proteins involved in the defense mechanism of plants against microbial infections (5). Thionin (PR 13) was also found to have a role during resistance in pearl millet against *S. graminicola.* A 5 kDa thionins with a pI of 8.2 was purified from pearl millet and it had anti-mildew activity on *S. graminicola.* Transcript expression of thionins was high in resistant compared to susceptible cultivars. Immunofluorescence studies showed intense fluorescence in epidermal regions, confirming its involvement in arresting pathogen ingress and in vascular regions, confirming its systemic translocation (Fig. 8)(8).

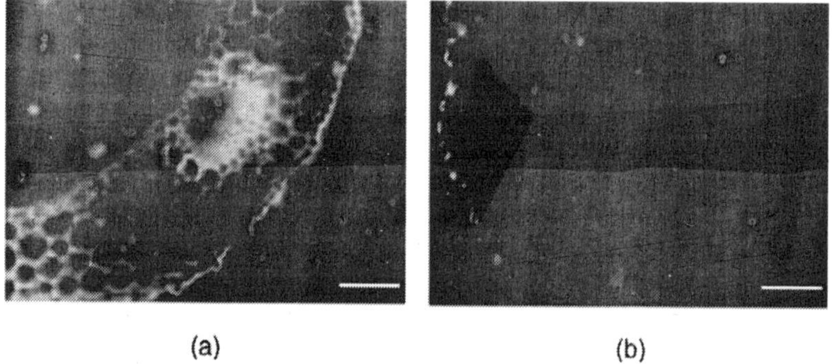

(a) (b)

Fig. 8: Immunolocalization of thionin in sections of resistant (a) and susceptible (b) cultivars of pearl millet seedlings after inoculation with *Sclerospora graminicola* as observed under UV fluorescence.

Plant defensins

Plant defensins are small (c.a. 5 kDa), basic, cysteine-rich proteins with antimicrobial activities found ubiquitous in plants and form part of the innate immunity system. They were isolated from wheat, barley, sorghum, radishes and other species of *Brassicaceae* (75) and their amino acid sequence analysis indicated them to be a new family of thionins (11). They are typically produced as preproteins, however, a small subset are produced as larger precursors with C-terminal prodomains (42). They have limited amino acid sequence identities, and relate insect defensins and toxins. Defensins though exhibit a diverse array of biological activities, are mainly involved in plant defense.

Lipid Transfer Proteins

Lipid-transfer proteins (LTP) are basic, 9-kDa proteins present in high amounts (as much as 4% of the total soluble proteinss) in higher plants found to be secreted and located in the cell wall. LTPs have a role in cutin formation, embryogenesis, defense reactions against phytopathogens, symbiosis, and the adaptation of plants to various environmental conditions (35). One of the LTPs was purified and characterized from pearl millet seeds which showed strong antimicrobial properties (78). Biomining for LTPs activity in plants and microbes and utilization of LTPs for induction of resistance against pearl millet downy mildew disease is in progress.

PATHOGENESIS RELATED PROTEINS

PR proteins are defined as proteins produced by the host plant under stress conditions particularly pathogen attack. Their induction in local as well as in distal parts of the injured or infected plant tissues and their association with the development of SAR suggest a probable role in this type of resistance. On the basis of amino acid sequence and properties, PR proteins have been classified into 17 families, denoted PR-1 to PR-17, in varying amounts detected in tobacco, tomato, cucumber, parsley, radish, *Arabidopsis*, and barley (22). Each PR protein is assigned with specific functions, PR-1 is antifungal, PR-2 is b-1,3-glucanase, PR-3 is chitinase type I,II, IV,V,VI,VII, PR-4 is chitinase type I,II, PR-5 is thaumatin-like, PR-6 is proteinase-inhibitor, PR-7 is endoproteinase, PR-8 is chitinase type III, PR-9 is peroxidase, PR-10 is 'ribonuclease-like', PR-11 is chitinase, type I, PR-12 is defensin, PR-13 is thionin, PR-14 is lipid-transfer protein, PR-15 is oxalate oxidase, PR-16 is 'oxalate oxidase-like' and PR-17 is unknown (76, 22). General characteristics of PR proteins are their accumulation in extracellular or vacuolar spaces, stability to low pH values, and resistance to proteases. Evidence that genes encoding PRs are expressed under a specific pathological condition or by a suitable elicitor is generally based on blot analysis of their related mRNAs and/or direct recognition of the induced proteins antibody methodology. Accumulation of PR-1 proteins is currently regarded as the most reliable biochemical marker for SAR.

The role of different PR proteins during pearl millet – *S. graminicola* interaction particularly during imparting resistance to the host has been well established and different PRs like β-1,3-glucanase and chitinase, (37), peroxidase (67), RNAse (69), oxalate oxidase (55), thionins (8) have been isolated, purified and characterized and it was reported that all the above mentioned PRs showed enhanced activities in resistant seedlings compared to susceptible seedlings following pathogen inoculation.

PLANT RESISTANCE EFFECTORS AND R-PROTEINS

Classical genetics has demonstrated the clustering of disease resistance proteins/genes in plant genome. These belong to a multigene family with varying specificities. Majority of the disease resistance genes isolated from several species so far share homology at the nucleotide and amino acid sequence level irrespective of the fact that they confer resistance to a variety of fungal, bacterial, or viral diseases. The most common class of resistance genes are RPS2, RPM1 from *Arabidopsis*, N gene of tobacco, L6 from flax (27). NBS class of R genes accounts for a significant proportion of the genomes of several plants (81). These R genes contain leucine-rich repeats (LRR), putative cytoplasmic signaling domains and nucleotide binding sites (NBS) (46). These structural features are identified as components of a signal transduction pathway, which is common to the different pathogenic defense responses (51, 2). The NBS domains are characters of various R proteins with ATP/GTP binding activity and comprise the P-loop, Kinase 2a, Kinase 3 and GLPL motifs. During the interactions between plants and harmful pathogens, NBS-LRR resistance proteins serve to recognize avirulence (*Avr*) gene products from extra cellular pathogens, thereby initiating a cascade of defense responses (13, 3).

R protein function may require, at least in part, the activity associated with ATP binding and/or hydrolysis (74), and by analogy, may serve as a signal transducer of cell death-related responses. The carboxy-terminal domain of this class of resistance proteins is represented by the LRR, which is typically 20–30 amino acids in length, and appears to be involved in the formation of protein–protein interactions. The NB-LRR class of R genes can be further divided into coiled-coil (CC)-NB-LRR and tollinter-leukin- 1 receptor (TIR)-NB-LRR according to their N-terminal domain. Evidence suggests that the N terminus influences the requirement for downstream defense response components.

Using specific primers designed from the conserved NBS regions, a total of twenty-two RGAs were cloned and sequenced from pearl millet (60). Phylogenetic analysis of the predicted amino acid sequences grouped the RGAs into nine distinct classes. Gene Bank database searches with the consensus protein sequences of each of the nine classes revealed their conserved NBS domains and similarity to other known R genes of various crop species (Table 2). One RGA 213 was mapped onto LG1 and LG7 in the pearl millet linkage map. Clone Uom213 (AF513969) has shown ORF of 153 amino acid and Uom301 (AY117410) has shown ORF of 144 amino acids. Partial sequences of cDNA clones AF513969 and AY117410 were used to obtain full-length genes of these clones by following Rapid Amplification of cDNA Ends (RACE) PCR. Two Protein (60kDa and 55 Kda) were expressed in E-coli using prSET expression vector. Antibodies raised against the purified recombinant R-protein were used to identify the original R-proteins from pearl millet extracts by immunoaffinity purification. The two proteins of 60 and 55 kda were subjected to MALDI-TOF analysis. The 60 kda protein did not show homology to any known protein However, the 55 kda protein showed homology to a protein kinase from brassica (Serine threonine inase) (Unpublished results).

Table 2: BlastX results indicating similarity of pearl millet partial gene sequences with known resistance genes from other crop species

Clone	Species	Accession No.	e-value
UOM 208	*Oryza sativa*	AF032690	0e-57
UOM 209	*Phaseolus vulgaris*	AF306500	0e-27
UOM 210	*Brassica napus*	AF209486	0e-12
UOM 206, 211	*Oryza sativa*	AF032697	0e-46

UOM 212, 217	*Cicer arietinum*	AJ307996	0e-15
UOM 215, 229	*Phaseolus vulgaris*	AF13553	0e-36
UOM 213	*Oryza sativa*	AJ246006	4e-48
UOM 218	*Phaseolus vulgaris*	AF363802	0e-20
UOM 219	*Phaseolus vulgaris*	AF363803	0e-18
UOM 216	*Mentha longifolia*	AF424174	0e-33
UOM 222	*Brassica napus*	AF209486	0e-11
UOM 226	*Oryza sativa*	AF220728	0e-34
UOM 234	*Glycine max*	AF325686	1e-19
UOM 235, 212	*Oryza sativa*	Y09810	0e-45
UOM 236	*Hordeum vulgare*	AF032687	1e-43
UOM 296	*Cucumis melo*	AF354505	0e—27
UOM 298	*Phaseolus vulgaris*	AF306500	0e-27
UOM 301	*Oryza sativa*	AP002540	1e-56

FUTURE PROSPECTS

Enormous number of plant derived proteins and peptides with effective antimicrobial properties are known. This knowledge is of immense potential in genetic engineering for production of disease resistant plants and also serves for exploring further research in this area of plant resistance induction. A comprehensive evaluation and utilization of the available knowledge like pathogen perception, signal transduction, gene expression etc., is the key for devising strategies for effective combat of plant pathogens. Manipulation of the pathways that result in encoding of resistance genes seems to be exciting. Transgenic plants with altered metabolic pathways leading to improved defense/tolerance capacities are to be designed by engineering the key components of plant defense. Identification of effective elicitors/inducers of both biotic and abiotic origin and their effective formulation and delivery system is also vital for management of plant pathogens. Utilization of biotechnology tools like proteomics, genomics, metabolomics and transcriptomics will lead to a better understanding of various host-pathogen inter-actions, which may ultimately contribute to the development of novel disease resistant varieties of agriculturally-significant crops. Cloning of genes controlling quantitative traits is now a major research frontier in terms of understanding traits of agronomic importance in crops, and it has been proposed that the structural domains of *R* gene families are essential for specific recognition between *R*-gene product and the corresponding *Avr* gene product and also ensuing plant defense responses. Therefore isolation and characterization of *R* genes and determination of their organization and evolution in plant genomes are of significance for understanding of plant-pathogen interactions and development of novel approaches to effective control of plant pathogens in agricu-lture. However, there is an ever present plant–pathogen arms races at the molecular level due to the coevolution between plants and their enemies involving many interactions at the molecular level. Therefore, there is a continuous need to update the knowledge on the host-pathogen interactions and developing novel strategies for combating the pathogens. The function of *R* proteins to confer *Avr* recognition and their interaction at various levels is still to be completely understood. With the rapid advancement of modern biotechnology and bioinformatics plants will be tailor-made for adaptation to both biotic and abiotic stresses and also other desirable traits will soon be a reality which cater to the needs of the farmers in particular and the mankind in general.

REFERENCES

1. Amor Y, Babiychuk E, Inzé D, Levine A. 1998. The involvement of poly(ADP-ribose) polymerase in the oxidative stress response in plants. *FEBS Lett.* 440: 1-7
2. Belkhadir Y, Sumramaniam R, Dangl JL. 2004. Plant disease resistance protein signaling: NBS-LRR proteins and their partners. *Curr. Opin. Plant Biol.* 7: 391–399
3. Bergelson J, Kreitman M, Stahl EA, Tian D. 2001. Evolutionary dynamics of plant *R*-genes. *Science* 292:2281–2285
4. Blumwald E, Aharon GS, Lam BCH. 1998. Early signal transduction pathways in plant–pathogen interactions. *Trends Plant Sci,* 3: 342-346
5. Bohlmann H, Vignutelli A, Hilpert B, Miersch O, Wasternack C, Apel K. 1998. Wounding and chemicals induce expression of the *Arabidopsis thaliana* gene Thi2.1, encoding a fungal defense thionin, via the

octadecanoid pathway. *FEBS Lett.* 437: 281-286

6. Boller T, Keen NT. 2000. Resistance and the perception and transduction of elicitor signals in host–pathogen interaction. In: *Mechanisms of Resistance to Plant Diseases* (ed. Slusarenko A, Fraser RSS and Van Loon LC). Kluwer Academic Publishers, Dordrecht, The Netherlands, pp. 189-229
7. Bousset L, Pons-Kuhnemann J. 2003. Effects of acibenzolar-S-methyl and ethirimol on the composition of a laboratory population of barley powdery mildew. *Phytopathology* 93: 305-315
8. Chandrashekhara, Niranjan-Raj S, Deepak S, Manjunatha G, Shekar Shetty H. 2010. Thionins (PR protein-13) mediate Pearl Millet Downy Mildew disease resistance. *Arch Phytopathology Plant Protect.* 43:1356-1366
9. Clarke A, Desikan R, Hurst RD, Hancock JT, Neill SJ. 2000. NO way back: nitric oxide and programmed cell death in *Arabidopsis thaliana* suspension cultures. *Plant J.* 24: 667-677
10. Cohn J, Sessa G, Martin GB. 2001. Innate immunity in plants. *Curr. Opin. Immunol.* 13: 55-62
11. Colilla FJ, Rocher A, Mendez E. 1990. Gamma-Purothionins: amino acid sequence of two polypeptides of a new family of thionins from wheat endosperm. *FEBS Lett.* 270: 191-194
12. Dangl J, Holub E. 1997. La dolce vita: a molecular feast in plant–pathogen interactions. *Cell* 91: 17-24
13. Dangl JL, Jones JD. 2001. Plant pathogens and integrated defence responses to infection. *Nature* 411: 826–833
14. Davis HA, Daniels MJ, Dow JW. 1997. Induction of extracellular matrix glycoproteins in Brassica petioles by wounding and in response to *Xanthomonas campestris. Mol. Plant Microbe In.* 10: 812–820
15. Deepak SA, Nirajan-Raj S, Umemura K, Kono T, Shetty HS. 2003. Cerebroside as an elicitor for incuding resistance against downy mildew disease of pearl millet. *Ann. Appl. Biol.* 143: 169-173
16. Deepak S, Shailashree S, Kini KR, Hause B, Shetty HS, Mithofer A. 2007a. Role of hydroxyproline-rich glycoprotein in resistance of pearl millet against downy mildew pathogen *Sclerospora graminicola. Planta* 226: 313-333
17. Deepak S, Shailashree S, Sujeeth N, Kini KR, Shetty

HS, Mithofer A. 2007b. Purfication and characteri-zation of praline/hydroxypro-line–rich glycoprotein from pearlmillet coleoptiles infected with downy mildew pathogen *Sclerospora graminicola*. *Phytochemistry* 68: 298-305

18. Deepak S, Niranjan-Raj S, Shailasree S, Kini KR, Boland W, Shetty HS, Mithofer A. 2007c. Induction of resistance against downy mildew pathogen in pearl millet by a synthetic jasmonate analogon. *Physiol. Mol. Plant Pathol.* 71:96-105
19. Deepak S, Manjunath G, Manjula S, Niranjan-Raj S, Geetha NP, Shetty HS. 2008. Involvement of silicon in pearl millet resistance to downy mildew disease and its interplay with cell wall proline/ hydroxyproline-rich glycoproteins. *Australas. Plant Path.* 37: 498-504
20. Durner J, Wendehenne D, Klessig DF. 1998. Defense gene induction in tobacco by nitric oxide, cyclic GMP, and cyclic ADP-ribose. *Proc. Natl. Acad. Sci. USA* 95: 10328-10333
21. Durrant WE, Dong X. 2004. Systemic acquired resistance. *Annu. Rev. Phytopathol.* 42:185–209
22. Edreva A. 2005. Pathogenesis-related proteins: Research progress in the last 15 years. *Gen. Appl. Plant Physiol.* 31:105-124
23. Gachomo EW, Shonukan OO, Kotchoni SO. 2003. The molecular initiation and subsequent acquisition of disease resistance in plants. *Afr. J. Biotechnol.* 2: 26-32
24. Geetha HM, Shetty HS. 2002a. Induction of resistance in pearl millet against downy mildew disease caused by *Sclerospora graminicola* using benzothiadiazole, calcium chloride and hydrogen peroxide-a comparative evaluation. *Crop Prot.* 21: 601-610
25. Geetha HM, Shetty HS. 2002b. Expression of oxidative burst in cultured cells of pearl millet cultivars against *Sclerospora graminicola* inoculation and elicitor treatment. *Plant Sci.* 163: 653-660
26. Giudici M, Poveda JA, Molina ML, de la Canal L, Gonzalez-Ros JM, Pfuller K, Pfuller U, Villalain J. 2006. Antifungal effects and mechanism of action of viscotoxin A(3). *FEBS J* 273:72–83
27. Hammond-Kosack KE, Jones JDG. 1997. Plant disease resistance genes. *Annu Rev Plant Physiol Plant Mol Biol* 48:575-607
28. Harder DE, Chong J. 1991. Rust haustoria. In: *Electron*

Microscopy of Plant Pathogens. (ed. Mendgen K and Lesemann DE) Springer-Verlag. Berlin pp. 235–250

29. Heath MC. 2000. Hypersensitive response-related death. *Plant Mol. Biol.* 44: 321-334
30. Heil M, Bostock RM. 2002. Induced Systemic Resistance (ISR) Against Pathogens in the Context of Induced Plant Defences. *Ann. Bot.* 89: 503-512
31. Hindumathy CK, Shailashree S, Kini KR, Shetty HS. 2005. Spore cell wall components of *A. niger* elicit downy mildew disease resistance against Pearl Millet. *Phytoparasitica* 34: 111-118
32. Jabs T, Slusarenko AL. 2000. The hypersensitive response. In: *Mechanisms of Resistance to Plant Diseases* (ed. Slusarenko A, Fraser RSS and Van Loon LC). Kluwer Academic Publishers, Dordrecht, The Netherlands, pp. 279-323
33. Janssens S, Beyaert R. 2003. Role of Toll-like receptors in pathogen recognition. *Clin. Microbiol. Rev.* 16: 637-646
34. Jones AL, Dangl JL. 1996. Logjam at the Styx: programmed cell death in plants. *Trends Plant Sci.* 1: 114-119
35. Kader JC. 1996. Lipid-transfer proteins in plants. *Annu. Rev. Plant Physiol. Plant Mol. Biol.* 47: 627-654
36. Keller B, Feuillet C, Messmer M. 2000. Genetics of disease resistance. In: *Mechanisms of Resistance to Plant Diseases* (ed. Slusarenko A, Fraser RSS and Van Loon LC). Kluwer Academic Publishers, Dordrecht, The Netherlands, pp. 101-160
37. Kini KR, Vasanthi UK, Shetty HS. 2000. Induction of ß-1,3-glucanase in seedlings of pearl millet in response to infection by *Sclerospora graminicola. Eur. J. Plant Pathol.* 106: 267-274
38. Kumudini BS, Vasanthi NS, Shetty HS. 2001. Hypersensitive response, cell death and histochemical localisation of hydrogen peroxide in the host and nonhost seedlings infected with downy mildew pathogen *Sclerospora grami-nicola. Ann. Appl. Biol.* 139:217-225
39. Kunkel B, Brooks D. 2002. Cross talk between signaling pathways in pathogen defense. *Curr. Opin. Plant Biol.* 5: 325-331
40. Lam E, Kato N, Lawton M. 2001. Programmed cell death, mitochondria and the plant hypersensitive response. *Nature* 411: 848-852
41. Lamb C, Dixon RA. 1997. The oxidative burst in plant

disease resistance. *Annu. Rev. Plant Physiol. Plant Mol. Biol.* 48: 251-275

42. Lay FT, Anderson MA. 2005. Defensins–components of innate immune system in plants. *Curr. Protein Pept. Sci.* 6: 85-101

43. Mallouk Y, Vayssier-Taussat M, Bonventre JV, Polla B. 1999 Heat shock protein 70 and ATP as partners in cell homeostasis. *Int. J. Mol. Med.* 4: 463-474

44. Manjunatha G, Niranjan-Raj S, Nandini P Shetty, Shekar Shetty H. 2008a. Nitric oxide donor seed priming enhances defense responses and induces resistance against pearl millet downy mildew disease. *Pestic. Biochem. Phys.* 91: 1-11

45. Manjunatha G, Roopa KS, Geetha NP, Shetty HS. 2008b. Chitosan enhances disease resistance in pearl millet against downy mildew caused by *Sclerospora graminicola* and defense-related enzyme activation. *Pest Manag. Sci.* 64: 1250-1257

46. Michelmore R. 2000. Genomic approaches to plant disease resistance. *Curr. Opin. Plant Biol.* 3: 125-131

47. Miller G, Shulaev V, Mittler R. 2008. Reactive oxygen signaling and abiotic stress. *Physiol. Plantarum* 133: 481-489

48. Mittler R, Simon L, Lam E. 1997. Pathogen-induced programmed cell death in tobacco. *J. Cell Sci.* 110: 1333-1344

49. Montesano M, Brader G, Palva ET. 2003. Pathogen derived elicitors: searching for receptors in plants. *Mol. Plant Pathol.* 4: 73-79

50. Nakashita H, Yoshioka K, Yasuda M, Nitta T, Arai Y, Yoshida S, Yamaguchi I. 2002. Probenazole induces systemic acquired resistance in tobacco through salicylic acid accumulation. *Physiol. Mol. Plant Pathol.* 61: 197–203

51. Nimchuk Z, Eulgem T, Holt BF, Dangl JL. 2003. Recognition and response in the plant immune system. *Annu. Rev. Genet.* 37:579-609

52. Niranjan-Raj S, Chaluvaraju G, Amruthesh KN, Shetty HS, Reddy MS, Kloepper JW. 2003. Induction of growth promotion and resistance against downy mildew on pearl millet (*Pennisetum glaucum*) by rhizobacteria. *Plant Dis.* 87:340-345

53. Niranjan- Raj S, Shetty NP, Shetty HS. 2004a. Seed biopriming with *Pseudomonas fluorescens* isolates enhances growth of pearl millet plants and induces resistance against downy mildew. *Int. J. Pest Manage.* 50:41-48

54. Niranjan-Raj S, Shetty NP, Shetty HS. 2004b. Proline a novel inducer of resistance in pearl millet against downy mildew disease caused by *Sclerospora graminicola*. *Phytoparasitica* 32: 523-525

55. Niranjan-Raj S. 2005. Plant growth promoting rhizobacteria mediated induction of resistance against pearl millet downy mildew – histological, biochemical and molecular analysis. Ph.D. Thesis submitted to the University of Mysore, Mysore, Karnataka, India.

56. Nürnberger T, Brunner F, Kemmerling B, Piater L. 2004. Innate immunity in plants and animals: striking similarities and obvious differences. *Immunol. Rev.* 198: 249-266

57. Obradovic A, Jones JB, Momol MT, Olson SM, Jackson LE, Balogh B, Guven K, Iriarte FB. 2005. Integration of biological control agents and systemic acquired resistance inducers against bacterial spot on tomato. *Plant Dis.* 89:712-716

58. Pushpalatha HG, Mythrashree SR, Shetty R, Geetha NP, Sharathchandra RG, Amruthesh KN, Shetty HS. 2007. Ability of vitamins to induce downy mildew disease resistance and growth promotion in pearl millet. *Crop Prot.* 26: 1674-1681

59. Sarosh BR, Sivaramakrishnan S, Shetty HS. 2005. Elicitation of defense related enzymes and resistance by L-methionine in pearl millet against downy mildew disease caused by *Sclerospora graminicola*. *Plant Physiol. Bioch.* 43:808-815

60. Sarosh BR. 2002. Induced systemic resistance in pearl millet against downy mildew and molecular genetics of disease resistance. Ph.D. Thesis submitted to the University of Mysore, Mysore,Karnataka, India.

61. Sasabe M, Takeuchi K, Kamoun S, Ichinose Y, Govers F, Toyoda K, Shiraishi T, Yamada T. 2000. Independent pathways leading to apoptotic cell death, oxidative burst and defense gene expression in response to elicitin in tobacco cell suspension cultures. *Eur. J. Biochem.* 267: 5005-5013

62. Shailasree Sekhar, Sarosh BR, Vasanthi NS, Shetty HS. 2001. Seed treatment with β-amino-n-butyric acid protects *Pennisetum glaucum* systemically from *Sclerospora graminicola*. *Pest Manag. Sci.* 57: 725-728

63. Shailashree S, Kini KR, Deepak S, Kumudini BS, Shetty HS. 2004. Accumu-

lation of Hydroxyproline rich glycoproteins in pearl millet seedlings in response to *Sclerospora graminicola* infection. *Plant Sci.* 167: 1227-1234

64. Sharada MS, Shetty SA, Shetty HS. 1995. Infection processes of *Sclerospora graminicola* in *Pennisetum glaucum* lines resistant and susceptible to downy mildew. *Mycol. Res.* 99:317-322

65. Sharathchandra RG, Geetha N, Amruthesh KN, Kini KR, Sarosh BR, Shetty NP, Shetty HS. 2006. Isolation of a protein elicitor from *Sclerospora graminicola* and elicitor mediated induction of defense responses in cultured cells of *Pennistem glaucum. Funct. Plant Biol.* 33: 267-278

66. Shirasu K, Schulze-Lefert P. 2000. Regulators of cell death in disease resistance. *Plant Mol. Biol.* 44: 371-385

67. Shivakumar PD, Shekar Shetty H. 2009. Purification of an infection-related acidic peroxidase from pearl millet seedlings. *Pestic. Biochem. Phys.* 94:119–126

68. Shivakumar PD, Geetha HM, Shetty HS. 2009. Induction of systemic resistance in pearl millet (*Pennisetum glaucum*) against downy mildew (*Sclerospora graminicola*) by *Datura metel* extract. *Crop Prot.* 28:783-791

69. Shivakumar PD, Vasanthi NS, Shetty HS, Smedegaard-Petersen V. 2000. Ribonucleases in the seedlings of pearl millet and their involvement in resistance against downy mildew disease. *Eur. J. Plant Pathol.* 106:825-836

70. Singh BP, Chauhan RS, Singhal LK. 2003. Toll-like receptors and their role in innate immunity. *Curr. Sci. India* 85: 1156-1164

71. Sreedhar AS, Csermely P. 2004. Heat shock proteins in the regulation of apoptosis: new strategies in tumor therapy. *Pharmacol. Therapeut.* 101: 227-257

72. Stec B, Markman O, Rao U, Heffron G, Henderson S, Vernon LP, Brumfeld V, Teeter MM. 2004. Proposal for molecular mechanism of thionins deduced from physico-chemical studies of plant toxins. *J. Pept. Res.* 64: 210–224

73. Szabo LJ, Bushnell WR. 2001. Hidden robbers: The role of fungal haustoria in parasitism of plants. *Proc. Natl. Acad. Sci. USA.* 98: 7654-7655

74. Tameling WIL, Elzinga SDJ, Darmin PS, Vossen JH, Takken FLW, Haring MA, BJC. 2002. The tomato R gene products I-2 and MI-1 are functional ATP binding proteins with ATPase activity. *Plant Cell* 14: 2929–2939

75. Terras FRG, Torrekens S, Van Leuven F, Osborn RW, Vanderleyden J, Cammue BPA, Broekaert WF. 1993. A new family *of* basic cysteine-rich plant antifungal proteins from Brassicaceae species. *FEBS Lett.* 316: 233-240

76. Van Loon LC, Van Strien EA. 1999. The families of pathogenesis-related proteins, their activities and comparative analysis of PR-1 type proteins. *Physiol. Mol. Plant Pathol.* 55:85-97

77. Van Loon LC. 1997. Induced resistance in plants and role of pathogenesis-related proteins. *Eur. J. Plant Pathol.* 103, 235-270

78. Velazhahan R, Radhajeyalakshmi R, Thangavelu R, Muthukrishnan. 2001. An antifungal protein purified from pearl millet seeds shows sequence homology to lipid transfer proteins. *Biol. Plantarum* 44: 417-421

79. Yoshioka K, Nakashita H, Klessig DF, Yamaguchi I. 2001. Probenazole induces systemic acquired resistance in *Arabidopsis* with a novel type of action. *Plant J.* 25:149-157

80. Young TE, Gallie DR. 2000. Programmed cell death during endosperm development. *Plant Mol. Biol.* 44: 283-301

81. Yulong G, Wangzhen G, Lei W, Tianzhen Z. 2006. Isolation and characterization of resistance and defense gene analogs in cotton (*Gossypium barbadense* L.) *Sci. China C. Life Sci.* 49 : 530—542

MOLECULAR IDENTIFICATION, RECOGNITION, CELL DEFENSE RESPONSES AND AGRICULTURAL IMPORTANCE OF AM FUNGI

B.N. Chakraborty* and U. Chakraborty

Immuno-Phytopathology Laboratory, Department of Botany,
University of North Bengal, Siliguri 734 013, West Bengal, India
[**E-mail: bncnbu@gmail.com*]

KEY WORDS : Arbuscular mycorrhizae, molecular markers, PCR, rDNA, biocontrol

ABSTRACT

Arbuscular Mycorrhizal fungi (AMF) are ubiquitous in terrestrial ecosystems and form mutualistic relationship with more than 80 percent of major group of vascular plants. Root colonization with AMF is a dynamic process, which is influenced by several edaphic factors. Mycorrhiza form critical link between the plant and soil structure and make a large direct contribution to soil fertility and quality through increased uptake of immobile nutrients, especially phosphorus that are mobilised by the fungus. However, there is increasing evidence that AMF have a range of other effects, for example, protection against plant parasite, water stress tolerance, alleviation of salt stress and in sustainable maintenance of plant health and soil fertility. Molecular analysis has provided the first opportunity to consistently identify the AMF taxa present in the plant roots. Molecular methods allow the identification of AMF taxa independently of morphological criteria and potentially at low levels of colonization. The successful establishment of mutualistic association between plants and AMF constitutes a strategy to improve the nutritional status of both partners. Multiple signals and differential induction of gene expression mediate the complex interaction between AMF and plant cells. The differential activation of defense genes reflect that some mechanism exists that regulate the suppression of the defense response to allow compatible Arbuscular mycorrhizal fungi to form a mutualistic symbiotic interaction with the plant. During symbiosis, specific elements of the plant and fungal genetic program are expressed in order that both symbionts can live together and complement each other. During colonization, AMF can also prevent root infections by reducing the access sites and stimulating host defense. In order to efficiently use mycorrhizal fungi in agriculture, it is necessary to understand their ecophysiology and metabolic functioning. Concerted efforts should be made to develop bioformulations (AMF-PGPR biocontrol consortium) which can be used as biofertizers and bioprotectors for improvement of plant health status. However, stringent quality control measures must be adopted so that farmers get quality products.

INTRODUCTION

Arbuscular Mycorrhizal (AM) fungi are ubiquitous in terrestrial ecosystems and form mutualistic relationship with more than 80 per cent of major group of vascular plants. An apparently low taxonomic diversity coupled with a broad geographic range had led to the view that AMF are a rather homogenous group, both functionally and morphologically. It is routinely possible, however, to find 10-30 spore types in the soil at a single site. This high local diversity against a background of low global diversity is paradoxical if all fungi are equally capable of colonising all plants. The apparent high genetic diversity of AMF is similarly paradoxical in relation to the low morphological diversity. Improved methods of isolation from soil, and of microscopy and molecular analysis have shown that there may be substantial diversity present in field soils, but the relationship between morphological and molecular diversity is still unclear.

Root colonization with AMF is a dynamic process, which influenced by several edaphic factors such as nutrient status of soil, seasons, arbuscular mycorrhizal (AM) strains, soil temperature, soil pH, host cultivar susceptibility to AM colonization and feeder root condition. There has been growing appreciation of the importance of plant and fungal interaction especially AMF on terrestrial ecosystem (58, 99). A mycorrhiza forms critical link between the plant and soil structure and make a large direct contribution to soil fertility and quality through contribution of soil organic matter. The assumed primary benefit to plants of the Mycorrhizal symbiosis is an increased uptake of immobile nutrients, especially phosphorus that are mobilised by the fungus. However, there is increasing evidence that AMF have a range of other effects, for example, protection against plant parasites (3, 13), water stress tolerance (91) alleviation of salt stress (49) and in sustainable maintenance of plant health and soil fertility (71,133). This evidence of multiple functions, host selectivity (or non-random distribution) and higher diversity than is apparent from cultures and morphology leads us to conclude that communities of AMF are much more diverse than previously thought. At the community level, recent experiments have shown that AMF are major players and may influence plant species diversity (75,120) and recruitment of plant species into the population (74). Colonisation of host plants by AMF exhibits significant spatial and temporal heterogeneity (66, 86). High spore diversity has been found in the field (11,48) and the number of spores produced by each AM fungus may be host-dependant (11,108). Traditionally, AMF community diversity has been measured using spore counts from field soils, but such counts are a measure of the sporulation activity of the fungi rather than a direct measurement of diversity, and may be highly variable and almost impossible to correlate among habitat types (89). Molecular genetic analysis provides a way around this obstacle as it has the potential to identify objectively the taxa present in the roots of the plants.

MOLECULAR IDENTIFICATION OF AM FUNGI

In the study of the functional ecology, biodiversity and phylogenetic relationships of AMF communities, it is necessary to distinguish individual taxa consistently with the group. Various morphological characters used for Taxonomy and identification of AM fungi have recently been critically reviewed (99).

Molecular analysis has provided the first opportunity to consistently identify the taxa present in the roots of plants. Molecular analysis of AMF has opened up possibilities that could only be dreamt of for studying AMF both in the laboratory and in the field (95,98,114). Earlier studies showed conclusively that the population of spores in the soil may not reflect root colonisation (39). Molecular methods allow the identification of AMF taxa independently of morphological criteria and potentially at low levels of colonisation. Analysis of AMF from field samples relies on the Polymerase Chain Reaction (PCR) that allows the DNA from small quantities of starting material to be bulked up into sufficient quantities for analysis. AMF can be targeted either within DNA of a plant root or from small amounts of template such as a single spore. The data from PCR depend upon the design of the primers used to initiate the reaction and the specificity of the primer determines the range of sequences that may be designed to amplify sequences from a single isolate up to those that recognise all AMF (6,22,67).

The majority of nucleic acid information derived from the Glomales is from the ribosomal RNA genes. In most organisms, the ribosomal RNA genes are present in multiple copies arranged in tandem arrays, with each repeat unit consisting of a small subunit (SSU or 18S) and a large subunit (LSU or 28S) ribosomal RNA gene separated by an internal transcribed spacer (ITS) that includes the 5.85 rRNA gene. The multiple copies are normally very similar in sequence due to the genetic processes of unequal crossing over and gene conversion. It is expected that the ribosomal genes of the Glomales are similarly arranged, but an unusual feature that has come to light is the frequent occurrence of multiple rRNA gene sequences, sometimes quite divergent, in a single spore or culture.

Primers developed

First molecular analyses of DNA isolated from AMF spores (*Glomus intraradices* and *Gigaspora margarita*) were done (113) and the SSU rRNA gene sequenced using PCR fragments generated with universal eukaryotic primers. It was possible to amplify sequences from several AMF directly from plant roots by using VANS1, a taxon-specific primer. A phylogenetic analysis of sequences from spores using VANS1 showed that the molecular phylogeny was congruent with the family classification, and that the sequence divergence was consistent with the fossil record (112). Group specific primers VAGLO (*Glomus*), VAACAU

(*Acaulospora*), VALETC (*G. etunicatum*) and VAGIGA (*Gigaspora* and *Scutellospora*) were subsequently designed (Table 1). However, using VANS1, it was found difficult to amplify products directly from roots (39,40,109,114).

Table 1. Primers used to amplify SSU rDNA and LSU rDNA sequences from Mycorrhizal roots

SSU rDNA	
VANS1,	5'GTCTAGTATAATCGTTATACAGG
NS21,	5'AATATACGOCTATTGGAGCTGG
VALETC,	5'ATCACCAAGGTTTAGTTGGTTGC
VAGLO,	5'CAAGGGAATCGGTTGCCCGAT
VAACAU,	5'TGATTCACCAATGGGAAACCCC
VAGIGA,	5'TCACCAAGGGAAACCCGAAGG
VANS22,	5'TAAACACTCTAATTTTTTCAA
VANS32,	5'AAGCTCGTAGTTGAATTTCGG
SS1492',	5'GCGGOCCGCTACGGMWACCTTGTTACGACTT
LSU rDNA	
LSU 0061[1,2]	5′ AGCATATCAATAAGCGGAGGA
LSU 3f[1]	5′ AGTTGTTTGGGATTGCAGC
LSU 4f[3]	5′ GGGAGGTAAATTTCTCCTAAGGC
LSU 6f[1]	5′ AAATTGTTGAAAGGGAAACG
LSU 9f[1]	5′ ATTCGTTAAGGATGTTGACG
LSU 5r[1]	5′ CCCTTTCAACAATTTCACG
LSU 7r[3]	5′ ATCGAAGCTACATTCCTCC
LSU 8r[1]	5′ GGGTATCCGTTGCAATCCTC
LSU 0599[2]	5′ TGGTCCGTGTTTCAAGACG
LSU 0805[1,2]	5′ CATAGTTCACCATCTTTCGG

[1] Primers used for sequencing.
[2] Primers used for primary PCR amplification.
[3] Primers used for the second nested-PCR amplification.

Based on a large amount of reference sequence information that was available to earlier studies, another primer AM1 was designed to amplify AMF sequences by excluding plant sequences (65). AM1 amplifies a section of the SSU gene in combination with the general primer NS31, that has a high degree of information sequence variation (111). However, drawback of this primer combination is that it also amplifies sequences from some other fungi of Pyrenomycetes. While the AMF that fall within the three long-accepted families can be detected successfully using this primer, it does not match the SSU sequences of the more divergent clades which are now designated as the new families Archaeosporaceae and Paraglomaceae. SSU primers that are specific for each of the five Glomalean families have been described (95,96).

Other studies have focused on the ITS region of the rRNA family. Universal sequences ITS1 and ITS4 have been used (132) to obtain PCR- restriction fragment length polymorphism (RFLP) and sequence information from spores collected from the field (107), and to study the relationships among AMF species (82,97). As high level of diversity is observed within species and even within spores, utility of ITS variation for field identification is limited. Other taxon-specific primers have been designed to amplify fragments (ca. 45 bp) of the large subunit (LSU) rRNA gene (123). The design of PCR primers for community analysis necessarily evolves as more information becomes available. The history of molecular analysis of AMF is one of the evolutions of more precise targeting (109). No single system will satisfy all purposes, and the regular evaluation of a system is necessary as more reference sequence information becomes available and a better understanding of the genetic diversity of AMF is required.

Family or Species specific Primers

The use of family or species-specific primer sets has demonst-rated the power of molecular approaches for positive identification of fungi (Table 2). Targets have been the SSU of the ribosomal gene complex (45,140), the LSU (123) and the ITS (88). Specific primers have also been generated from random amplified polym-orphic DNA (RAPD) analysis of spores (1,134) or characterization of highly repeated DNA sequences (83,136). Species- specific primers were used to investigate the influence of two host plants on the community structure of *Glomus mosseae* (123).

Table 2. Primers for amplification of fungal nuclear ribosomal RNA genes

rRNA	Gene primer[a]	Product size (bp)[b]
Nuclear small		
NS1 5′	GTAGTCATATGCTTGTCTC	555
NS2 5′	GGCTGCTGGCACCAGACTTGC	
NS3 5′	GCAAGTCTGGTGCCAGCAGCC	597
NS4 5′	GTTCCGTCAATTCCTTTAAG	
NS5 5′	AACTTAAAGGAATTGAGGGAAG	310
NS6 5′	GCATCACAGACCTGTTATTGCCCTC	
NS7 5′	GAGGCAATAACAGGTCTGTGATGC	377
NS8 5′	TCCGCAGGTTCACCTACGGA	
Nuclear ITS		
ITS1 5′	TCCGTAGGTGAACCTGCGG	290
ITS5 5′	GGAAGTAAAAGTCGTAACAAGG	315
ITS2 5′	GCTGCGTTCTTCATCGATGC	290
ITS3 5′	GCATCGATGAAGAACGCAGC	330
ITS4 5′	TCCTCCGCTTATTGATATGC	580
ITS1F 5′	CTTGGTCATTTAGAGGAAGTAA	700
ITS4B 5′	CAGGAGACTTGTACACGGTCCAG	
TW13 5′	GGTCCGTGTTTCAAGACG	1200

[a]All odd-numbered primers are 5′ primers; even numbers indicate 3′ primers. Sequence are written 5′–3′.

[b]Product sizes are approximate based on rRNA genes of *Saccharomyces cerevisiae.*

Gigasporaceae- specific primer - VANS1 (113, 114) was found to perform poorly for investigation of the structure of natural AMF communities, specially *Scutellospora* colonising the roots of bluebell (39,40), using the primer combinations SS38/VAGIGA and VANS1/VAGIGA. These primer sets were used to amplify sequences from bluebell roots and field spores. They demonstrated a surprisingly high level of SSU sequence variation with single spores as discussed earlier. Additionally, it was noted that several of the AMF sequences obtained from spores would not have been amplified using VANS1. This illustrates the limited value of using specific primers to study community diversity in any situation in which the starting material is not well characterised with regard to its inherent sequence variability. Specific primers are ideally suited for the experiments where the aim is to rapidly test for the presence /absence of fungi added from well characterised pot cultures, but they are not appropriate for field collected samples unless the aim is to focus on just one component of the community.

Primers Intended to Target all Glomalean Fungi

An allied approach is to use primers designed to amplify Glomalean templates but exclude other sequences likely to contaminate DNA extracts. Studies of this kind have all used cloning and PCR-RFLP to separate classes and sequencing to give insight into the identity of the clone. Interest has again concentrated on the ribosomal gene complex, in particular the SSU and ITS regions.

RECOGNITION

The relative unspecificity of fungi and hosts in all kinds of mycorrhiza comes as a surprise to those who might be led to expect that organisms which form permanent mutualistic associations would be more likely to be highly specific than pathogenic biotrophs whose association is temporary, ending perhaps with the premature death of the host. Indeed it might further be expected that ectomycorrhizal fungi, symbiotic with long-lived perennials, might be more specific than those of annual herbs, but that is not so either. The explanation lies in the selective advantage of the associated state and a consequent selection against variation towards incompatibility. In addition to the nutritional advantages mentioned, there is for the fungus a further selective advantage arising out of non-specificity. There will always be, in any soil or in any ecosystem, suitable habitat, i.e. suitable host roots, for a non-specific fungal symbiont. Mycorrhizal dependency may be defined as "the degree to which a plant is dependent on the mycorrhizal condition to produce maximum growth or yield at a given level of fertility" of soil. A great variation of dependency in *Citrus* seedlings using *Glomus fasciculatum* in artificial soil cultures of different fertility has been demonstrated. "Mycorrhizal dependency" is essentially a crop or host-oriented concept where interest centres on the growth or development of the host rather than of the fungus. "Mycorrhizal effectiveness" considers the effectiveness of the mycorrhiza as organ active in some physiological processes e.g. nutrient absorption, but it is often measured in similar terms, of dry weight increment per unit time compared to an uninfected control, as mycorrhizal dependency.

Very little experimental work has been performed on mutual recognition of mycorrhizal fungi and their hosts. Their specificity is not close; a single fungal isolate may form vesicular-arbuscular mycorrhiza with a wide range of species of host of all the phyla of land plants in laboratory experiments. Specificity seems to be closer in the competition of natural vegetation than in pure culture. The extent to which this impression is real is questionable. At all events, any mycorrhizal host is usually compatible with a wide range of fungi and each mycorrhizal fungus with a wide range of hosts. Moreover at a single time one individual plant may associate with several species of fungi. Mycorrhizal fungi have extensive compatibility with potential hosts which is perhaps only limited by the inhibitory properties in the host, which itself can be universal or selective in response to the fungi.

Roots of every plant are colonized by root-surface populations of microorganisms which also includes the mycorrhizal fungi. In the first stage of mycorrhizal infection these fungi colonize the surface in competition with other members of the population. The members of the root surface populations rely for growth particularly on the presence of essential nutrients including carbohydrates, organic and amino acids and a multiplicity of other compounds including vitamins and growth factors in the root exudates. The mycorrhizal fungi must possess the ability to penetrate the middle lamella region of the cortical tissues and they must have common property that allows them to do this, and that property differentiates them from the other denizens of the root surface which do not penetrate.

The available evidence suggests that penetration occurs in a region of the root where the cells of the epidermis and cortex are reaching or have reached their full size and are in the process of laying down their cell walls. The electron microscopic observations of many mycorrhizas show evidence of an "interfacial matrix", a "contact zone" or an "involving layer" derived from the wall of the host (and perhaps that of the fungus also) separating the hypha from plasmalemma of the host, both in intercellular and intracellular infections. This indicates an action of the fungus upon the process of condensation of polymers to form the cell walls of the host. Such an action could arise as proposed in the case of pathogens, by the proteins (enzymes) of the fungal walls complexing with the condensing enzymes forming the cell walls of the host and inhibiting them. It could also arise if the fungus produced on or from its surface an inhibitor of the host enzymes. The appearance of the interfacial matrix in endomycorrhizas gives the impression of the host plasmalemma continuing to secrete vesicles and to form carbohydrate fibres, but being unable to organize them as a coherent wall. The ability to form a wall is recovered when the fungus senesces, degenerates and becomes encapsulated within the cell. It is a corollary of the views that variations of the constitutions of the enzymes which bring about wall-building of the host, or of the proteins of the fungal surface, will provide a primary explanation of the limitation of ectomycorrhizal fungi to particular groups of host, ericoid mycorrhiza fungi to others, and vesicular-arbuscular mycorrhiza fungi, to a wide range of hosts.

This hypothesis clearly needs examination, especially into the nature of the enzymes or surface proteins or glycoproteins of the walls or developing walls of both host and fungus. It is implicit in it that wall-synthesizing enzymes of the host are affected in their activity by substances on the fungal wall. In addition, a special examination is needed of the exact location where infection occurs and the condition there of the host cells and their walls. The hypothesis, however, has the great merit of explaining how it is that fungi which do not secrete enzymes hydrolyzing carbon polymers such as pectins or cellulose penetrate the cortex and into the cells.

PLANT DEFENSE RESPONSES IN ARBUSCULAR MYCORRHIZAL SYMBOSIS

The endosymbioses formed between plants and micro-organ-isms play an important role in agriculture natural ecosystems. The most widespread mutualistic endosymbiotic interactions are formed between plant roots and AMF. The successful establishment of this mutualistic association constitutes a strategy to improve the nutritional status of both partners. The fungi receive fixed carbon compounds from the host plant, while the plant benefits from the association by the increased nutrient uptake of phosphorus, enhanced tolerance to abiotic stress, and resistance to pests and pathogens (13,116). Generally, AMF show little or no specificity and the factors that determine whether mycorrhiza are formed or not appear to depend on the genotype of the host plant (76). Evidence for this is provided by the existence of non-host plant species (59) and Myc mutant plants unable to form AM symbiosis (60).

Penetration into the root and intercellular growth of the AM fungi involves a complex sequence of biochemical and cytological events and intracellular modifications (20), which imply that the fungus must clearly be recognized by the host plant. The mechanisms controlling AM developments are largely unknown. However, evidence suggests that, as in plant-pathogen interactions, the induction/suppression of mechanisms associated with plant defence play a key role in AM fungal colonization and compatibility with its host.

Defense responses in plants during the early stages of development of arbuscular symbiosis

Recognition of a potential invader is a prerequisite for the initiation of an effective defence response by the plant which is achieved through the recognition of specific signal molecules known as elicitors. Exogenous elicitors are secreted from the microbe, whereas endogenous elicitors formed as a result of physical and/or chemical cleavage of the plant cell wall. After perception of an elicitor, a number of biochemical changes contribute to the early response in host cells. These processes include changes in the ion permeability of the plasma membrane, the activation of plasma membrane-bound enzymes, the activation of kinases, phosphatases, phospholipases, and the production of signal molecules, including active oxygen species. The result of these processes is the transcriptional activation of defence-related genes (117). Some events such as signal perception, signal transduction and defence gene activation, similar to those found in plant-pathogen interactions have also been found in the plant interaction with AMF. AM fungi secrete similar chitin elicitors, which could induce a defence response (102). For instance, the elicitor derived from an extract of extraradical mycelium of *Glomus intraradices* was able to induce phytoalexins (glyceollin) synthesis in soybean cotyledons (77). Chalcone synthase, the first enzyme in the metabolism of a flavonoid compound in *Medicago truncatula* was induced by *G. intraradices* (19). Furthermore, a more hypersensitive-like response, commonly observed when the plant is confronted with a pathogen, could be observed in compatible AM associations. An oxidative burst could be detected at sites where hyphal tips of *G. intraradices* attempted to penetrate a cortical root cell of *M. truncatula* (104). Moreover, necrosis and cell death have been observed at sites of *Gigaspora margarita* infection of *Medicago sativa* roots (46). The most evident effects were observed in incompatible associations between AM fungi and non-host plants or Myc mutant plants. In the case of non-host plants, the molecular bases for incompatibility with AM fungi remain unclear, but in some associations like resistance reaction seems to involve a hypersensitive defence response, characterized by the deposition of callose, PR-1 protein, and phenolics has been observed when pea Myc mutant plants were challenged with an AMF (60).

Transient increases of catalase and peroxidase activity were also observed in mycorrhizal plants, during appressoria formation and fungal penetration into the tobacco, onion and bean roots (17,77,118) which coincided with the accumulation of salicyclic acid (SA), a signal molecule involved in the signal transduction pathway activated in plant-pathogen reactions (30). A transient accumulation of SA during the early stages of infection also has also been observed in the interaction between rice and *G. mosseae* (18) which was correlated to an increase in the expression of genes encoding lipid transfer protection (LTP) and phenylalanine ammonia-lyase (PAL). Most of the defense-related genes which express during the early stages of AM fungal penetration are also activated by pathogen infection, treatment with elicitors, or by SA. In this context, the induction of defence gene expression could be considered to be a result of fungal elicitor recognition and signal transduction pathway activation. The weak and transient character of the plant defense response could be a consequence of the low capacity of the fungus to trigger such a response and/to induce a plant mechanism which suppresses an already activated defence response at several levels to allow for fungal growth within the plant tissue.

Defense responses in plant cells containing arbuscules

In the early stages of root colonization by AMF, the plant defense response is characterized by a weak and transient activa-tion, but at later stages, after arbuscule formation, this activation appears to become stronger, although only in those cells which contain fungal structures. For instance, the use of specific probes in studies of *in situ* expression revealed that some mRNAs of genes associated with the plant defense response specifically accumulated in plant cells containing arbuscules. Likewise, members of different classes of plant defense genes, including genes encoding hydroxyproline-rich glycoproteins (HRGP), enzymes involved in metabolism of phenlypropanoid, reactive oxygen species and plant hydrolase have been detected in plant cells containing arbuscules. An increase in the quantity of mRNA encoding HRGP was observed in the root cells that contained arbuscules (16).

Genes encoding enzymes that catalyse core reactions of the metabolism of phenlypropanoid have been expressed in cells with arbuscules. *In situ* localization of phenlyalanine ammonia lyase (PAL) and chalcone synthase (CHS) transcripts were observed in cells containing arbuscules. However, expression of other genes encoding enzymes such as chalcone isomerase (CHI) or isoflavone reductase was not significantly affected in mycorrhizal roots (63). Genes involved in the catabolism of reactive oxygen species, such as catalase and peroxidase have also been localized in bean and wheat root cells containing arbuscules (16) which play a role in the catabolism of hydrogen peroxidase and/or in cross-linking reactions between proteins and polysaccharides in the interface between the arbuscule and the plant cell plasma membrane. Corroborating the biochemical data, differential gene expression of acidic and basic forms of chitinase and β-1,3-glucanase has been observed during mycorrhiza formations in different plant-fungal combinations (25,78,103). Accumulation of basic forms of chitinase and β-1,3-glucanase transcripts have been observed in the intercellular region between cortical cells containing arbuscules (15,78), suggesting that these enzymes might be involved in the control of intraradical fungal growth. Interestingly, the accumulation of β-1,3-glucanase mRNA in cells containing arbuscules was modulated by phosphorus (P) concentration. The level of mRNA accumulation increased as concentration of phosphorus decreased. In comparison, the amount of basic chitinase transcripts did not change with mycorrhization or P concentration (78). The enzymatic activity and chitinases and β-1,3-glucanases could form part of the defence response by the plant to the invading fungus. A specific induction of a class III chitinase gene family in mature *M. truncatula* mycorrhizae is evident, which may be involved in suppression of plant defence reactions in the later stages of the AM development (103).

Results from the split root experimental system clearly demonstrated for the first time that arbuscular mycorrhiza formation by *G. mosseae* induces not only localized but also systemic resistance against *Phytophthora parasitica* in tomato roots (42). The localized resistance of mycorhizal tissues is seen in the rare occurrence of *P. parasitica* hyphae and their inability to invade arbuscules containing host cells. The ISR is characterized by large reductions in root damage and in *P. parasitica* development within nonmycorrhizal root systems, in comparison to roots of nonmycorrhizal plants. This bioprotection is directly linked to root colonization by the Arbuscular Mycorrhizal fungus since neither a microbial filtrate from *G. mosseae* inoculum nor low Mycorrhizal levels are sufficient to induce it. Cytomolecular investigations provide clear evidence that both the localized and systemic protective effects induced by Arbuscular mycorrhiza involve the accumulation of plant defense related molecules in association with the elicitation of wall reactions in the host roots.

Regulation of defense mechanisms in AM symbiosis

Multiple signals and differential induction of gene expression mediate the complex interaction between AMF and plant cells. The differential activation of defence genes reflect that some mechanism exists that regulated the suppression of the defence response to allow compatible AM fungi to form a mutualistic symbiotic interaction with the plant.

One possible mechanism to attenuate the plant defence response in the AM symbiosis might be the degradation of exogenous elicitor molecules produced by the AM fungi and/or the prevention of endogenous elicitor release from the plant cell wall. Theoretically, hydrolytic enzymes regulate both processes as plant hydrolases could hydrolyse fungal elicitor components while fungal enzymes could hydrolyse plant cell wall components. During AM development the expression of constitutive and mycorrhiza-specific chitinases, and β-1, 3-glucanse isoforms (potential hydrolases of fungal elicitors) are differently regulated, thereby pointing towards a central role for these enzymes in AM formation (103). Furthermore, the induction of a plant defense response by an elicitor released from an extract of extraradical mycelium of *G. intraradices* has been observed (77). A similar mechanism for elicitor degradation and plant defense attenuation in AM chitinase enzymes by the host plant was demonstrated (102, 103). It was proposed that constitutively expressed plant chitinase, in the early stages, and the mycorrhiza- specific isoforms in the later stages, were the enzymes responsible for elicitor degradation.

The production of endogenous elicitors derived from the degradation of the cell wall does not seem to be an effective mechanism for the induction of the plant defence response. Evidence supporting the hypothesis is that mycorrhizal fungi produce very little plant cell wall degrading enzymes (53). Furthermore, the cell wall degrading enzymes produced in mycorrhizal roots show similar electrophoretic and biochemical characteristics as plant enzymes produced in non-mycorrhizal roots (55) which suggest that the role of fungal enzymes is only to selectively and specifically break plant cell wall components for fungal penetration, but not to participate in an unspecific and uncoordinated plant cell wall degradation process, which could produce endogenous elicitors.

Nevertheless, the intracellular colonization by AM mycorrhizal fungi creates a new interface compartment composed of membranes from both partners separated by apoplastic material containing molecules common to the plant primary wall, such as cellulose, pectin, xyloglucane, and HRGP which are not assembled into a fully structured wall (20). The lytic activity of plant and/or fungal enzymes over these molecules could then putatively generate oligo-fragments which could act as elicitors for a localized defensive response at the arbuscular level.

Another possible mechanism to attenuate the plant defense response could be by blocking components of the signal transduction pathway that activate this response. Among these components, salicyclic acid and reactive oxygen species (ROS) have been implicated as second messengers in AM associations. Although levels of H_2O_2 and other ROS have not been measured in AM roots, indirect evidence exists that suggests that the levels of H_2O_2 in mycorrhizal associations are increased (104). Alterations in the pattern of anti-oxidative enzymes, such as catalase and peroxidase in mycorrhizal roots may indicate that oxidative compounds are produced during the colonization process (77). The increase in catalase and peroxidase activity could be due to their function as antioxidants for any active oxygen molecules generated during the initial stages of fungal penetration. Since H_2O_2 and other reactive oxygen species are involved in signal transduction cascades in plant-pathogen interactions, it is possible that degradation of H_2O_2 by catalase in AM could be a possible mechanism for avoiding the activation of defence response genes. Transient increases in catalase and peroxidase activities in AM tobacco roots occur at the same time as the transient enhancement of free SA (17). In this way there may be a synergistic relationship between SA and H_2O_2 in some plant defence responses. The role and significance of SA in AM interactions remains unclear. Nevertheless, the finding that transient increases of SA are detected in tobacco and rice AM roots and that the exogenous application of SA to rice roots did not affect appressoria formation, but did cause a transitory delay of mycorrhization of root (18) suggest that the regulation of defence response in plants against AM fungi may be through the SA pathway.

Alterations in the flux of nutrients and in the levels of plant hormones could generate cell-signaling pathways that regulate the plant defence response during AM development. Evidence for this is provided by the fact that phosphate levels in the plant are negatively, and carbohydrate levels are positively correlated with root colonization by AMF. The exact mechanism and molecular basis that could mediate the inhibition of AM colonization by P remain unknown. One possibility is that defense gene expression is mediated by a signaling mechanism that senses the level of P in the root, resulting in an upregulation of these genes in plants with high levels of P. A sensor system based on the flux of carbohydrates regulates defence gene expression in plant cells containing arbuscules. It has been postulated that an increase in the flux of sucrose, glucose, and fructose and the activation of genes associated with the catabolism of sucrose in the cells containing arbuscules constitute a mechanism responsible for the activation of defense genes (16). However, the mechanism that mediates defence regulation by nutritional factors remains unknown.

It is intriguing that plant hormone levels in mycorrhizae are altered, as the ethylene level in AM tobacco roots is low (125) and cytokinin levels are significantly increased in alfalfa and tobacco AM roots (57, 122). The significance and role of plant hormones in AM symbiosis have been reviewed (12), but the exact role of plant hormones in the regulation of gene expression during the plant – AMF colonization is not clear. Nevertheless, the effective role of plant hormones during AM is not clear and it is difficult to determine if changes in hormone levels in AM symbiosis are directly implicated in the process of mycorrhization.

An investigation of the signalling mechanisms underlying the activation and cross-talk processes that participate in the regulation of the AM symbiosis may contribute to our understanding of the formation and functioning of this symbiosis, that these mechanisms could have on the induction of the plant resistance against pathogens observed in mycorrhizal plants.

AM FUNGI IN AGRICULTURE

The excessive use of chemical inputs to control disease poses an important problem for present-day plant production systems. Research priorities for alternative management practices compatible with sustainable agriculture and the environment include the use of beneficial soil microorganisms. Antagonistic microorganisms like bacteria (29,31,34,35,80,121,130) and fungi (36,50) have been reported tc have protective effects against soil borne pathogens. AM fungi, which occur widely in the rhizosphere of all types of crops and trees are known to exert beneficial effects on plants by forming symbiotic associations with a wide range of plant species, are another interesting group of microorganisms that effectively reduce root disease being considered promising candidates for improving agricultural output (7,24,27,28,44,70, 81,100). Arbuscular mycorhiza occur on roots of most of the food, medicinal and aromatic, horticultural crops and tropical trees (101). It has also been estimated that about 95% of the world's present species of vascular plants belong to the families that are characteristically mycorrhizal, which demonstrate their usefulness in agriculture (3,13). Besides arbuscular mycorrhiza, other mycorrhizal types are also important in particular situations: ectomycorrhizas (ECM) for forest production, and in reafforestation programmes, ericoid (ERM) for fruit crops and orchid mycorrhizas for enhanced propagation particularly for conservation. This has led to the early recognition that different mycorrhizal symbioses may be manipulated to increase crop yields in different types of primary production systems. Though there are several examples where crop productivity is influenced by AM symbioses, examples where inoculation and management to increase AM colonization are carried out as part of normal commercial practice are few. Application of large amounts of pesticides and fertilizers is quite common in agricultural systems (62, 79, 87), which may reduce the potential of the fungal symbionts to have a net significant effect on plant nutrition and growth (116). Recent emphasis on organic farming for sustainable agriculture offers some hope for better utilization of such natural components of soil for increasing crop productivity. Other than their role in enhancing plant nutrition, in recent years, attention is also being laid to the possibility that arbuscular mycorrhizas may increase tolerance of plants to toxic elements, salinity and drought, or reduce soil borne diseases, root pathogens and pests (13,79,93,110).

In order to efficiently use mycorrhizal fungi in agriculture, it is necessary to understand their ecophysiology and metabolic functioning. It is now established that over a period of 400 million years of coevolution the plant and AM fungi have become very interdependent, both at the ecological and physiological level. During symbiosis, specific elements of the plant and fungal genetic program are expressed in order that both symbionts can live together and complement each other (92). Most important role of the myorrhizal fungi is to improve the host plant's uptake of mineral nutrients, mainly N and P.

Current trends in AM utilization point to their packaging and commercialization as inoculants and this has attracted interest of a number of biotechnology companies. However, being obligate symbionts, the process of producing inoculum is rather difficult and time consuming, making it more costly than other easily producible inoculants. Hence, their application would be rather limited to high value crops. Commercial available products (25) have focused more on increasing yields, with less attention being paid to establishment or maintenance of production systems that are 'sustainable' in preserving, or improving soil resources. In such environments, and in less developed systems with low inputs, naturally occurring AM populations may play important roles in crop nutrition that is not always appreciated. It has been demonstrated that even when there are no net increases in yield, AM pathway of phosphorus uptake operates in colonized plants in the field, indicating that AM fungi are integral to the function of root systems and must be taken into consideration in programmes aimed at developing sustainable agriculture. Wherever costs of production and application of fertilizers are important, or minimum input or organic culture is practiced, the contribution of biological processes and organisms, including AM fungi, in nutrient dynamics deserve to receive more attention (68). Another area of concern in agriculture is the knowledge that the huge deposits of rock phosphate may be depleted within few decades, which increases the urgency to harness plant adaptations that maximize the use of P accumulated in the soil. Since AM fungi have multiple roles to play, their inclusion in programmes like pest management, management of water and nutrient use etc. needs to be considered for full their maximum utilization.

Crop nutrition and Plant growth

The relationship between the development of arbuscular mycorrhizas and increased growth and nutrition of the host was recognized by Asai (5) as early as 1944. Pioneering work on the potential of mycorrhizas in plant nutrition was carried out by Mosse (90) on apples, Baylis (8,9) on *Griselinia* and other New Zealand plants and Gerdemann (56) on *Liquidambar* and maize. McGonigle (85) evaluated seventy-eight field trials with AM fungi and found that inoculation increased yield by about 37%. Though at that time, it was not clear whether the observed increase in yield could be correlated to increased nutrient uptake facilitated by AM fungi, later developments have shown that indeed, AM fungi increase yield through improved nutrition. Since roots of most plants are colonized by mycorrhizas, unequivocal proof that AM colonization increases nutrient uptake has been difficult to obtain. Nevertheless, the role of AM fungi in increasing nutrient uptake and plant growth has now been documented. It is now established that even when there is no increase in nutrient uptake efficiency in tissue concentrations or in total plant nutrient content, the fungal partner can make a significant contribution to nutrient uptake.

The development of research on roles of AM fungi in nutrient uptake has been closely linked to the knowledge of soil chemistry, particularly in relation to pools and availability of phosphorus (119). The amount and form of phosphorus in soil and the factors affecting its availability are important in determining the ways in which AM fungi influence uptake by plants (41). Phosphorus, one of the most essential elements of plants, is not easily available from soil because, in spite of being present in relatively large amounts, much of it is insoluble. Indirect evidence that AM roots can be more efficient in nutrient uptake than non-mycorrhizal roots came from the observation that responsive AM plants are both larger and contain higher concentrations of P in their tissues than uncolonized controls. The explanation first suggested was that AM colonization increased efficiency of absorption by roots. However, this is not the only possible explanation for the increases. Increased total root length or efficiency in AM plants would certainly contribute to increased total uptake, but would not necessarily lead to elevated tissue concentration. If growth keeps pace with P upake, tissue concentrations remain constant, for they are dependent on the relative rates of uptake and growth. It is not known exactly what mechanisms are involved by AM fungi in the P uptake of inorganic and organic P. External phosphatase can make organically bound P available for uptake. It is however not believed that a large amount of phosphatases is excreted by AM fungal extraradical mycelium (92). The uptake of P by AM fungi must be an active process since the difference in electrochemical potential between soil and extraradical mycelium is large and since the inorganic P concentration inside the mycelium is high compared with the soil solution. It is generally accepted that the uptake of P takes place partly via proton cotransport, which is driven by a membrane-bound proton-ATPase (10). Proton-ATPases were found to be expressed in the extraradical mycelium of two *Glomus* species, indicating that P uptake is mediated by a high-affinity transporter of which the functioning depends on the electrochemical gradient of proton generated by a plasma membrane proton-ATPase (51).

Elevated concentrations may also result from increased C use in AM plants, leading to C-limitation and 'luxury' accumulation of P (116). More direct evidence of increased efficiency of P absorption was obtained by expression uptake on the basis of amount of absorbing tissue. The first demonstration of increased inflow of P in AM roots was by Sanders and Tinker (105,106) in highly responsive *Allium cepa* colonized by *Glomus* sp. They calculated that AM fungi contributed about 70% of the P absorbed by the AM plants. Direct confirmation of the role of AM fungi in improved plant nutrition was obtained by Cavagnaro *et al.* (23), who, using mutant tomato with reduced AM colonization, showed that AM fungal populations not only contribute to nutrition, but also enhance food quality in terms of nutrient densities. Effects of AM fungi on nutritional value of crops deserve particular attention particularly in the light of concerns that use of highly purified fertilizers and other modern agricultural practices are reducing micronutrient densities below those required for human health (128,129). Another area of current investigation is to explore the probability of AM fungi increasing beneficial compounds in the plants like antioxidants.

Other than P, evidence is also now accumulating that the AM pathway makes considerable contribution to plant N uptake from soil, regardless of total N uptake and N responses. In most cases, improved nodulation and N_2 fixation in AM plants appears to be the result of relief from P stress and possibly uptake of some essential micronutrients, which result in both a general improvement in growth and indirect effects upon the N_2 fixing system. Experiments using ^{15}N have shown that hyphal N transfer occurs between AM fungi and host roots (64, 84). However, the translocation of N has not been correlated with increased plant N content or growth. There is also evidence now that the efficiency of Zn and Cu is increased in AM plants. Further, the interactions between AM colonization and accumulation of heavy metals and other toxic elements is an area of considerable interest in relation to both production of safe food and bioremediation programmes.

In order to enhance the sustainability of agro-ecosystems, there is a shift towards low-input instead of conventional high-input agricultural systems. In these low-input systems the Mycorrhizal symbiosis is regaining its importance. This can lead to better fungal development in soil, higher root colonization, and an enhanced nutrient uptake.

Plant protection

Plants and pathogens interact with a wide variety of orga-nisms throughout their lifecycle. These interactions can signifi-cantly affect plant health in various ways. The types of inter-actions were referred to as mutualism, protocooperation, commen-salism, neutralism, competition, amensalism, parasitism, and predation. While the terminology was developed for macroecology, examples of all of these types of interactions can be found in the natural world at both the macroscopic and microscopic level. And, because the development of plant diseases involves both plants and microbes, the interactions that lead to biological control take place at multiple levels of scale. From the plant's perspective, biological control can be considered a net positive result arising from a variety of specific and non-specific interactions. Mutualism is an association between two or more species where both species derive benefit. Sometimes, it is an obligatory lifelong interaction involving close physical and biochemical contact, such as those between plants and mycorrhizal fungi.

Different hypotheses have been proposed to explain bioprotection by AM fungi. These include (a) improvement of plant nutrition and root biomass in Mycorrhizal plants, which could contribute to an increased plant tolerance and compensate for root damage caused by a pathogen, (b) changes in root system morphology, (c) modification of antagonistic microbial populations in the mycorrhizosphere, and (d) competition between AM fungi and pathogenic fungi to colonize root tissues, with the possible induction of resistance mechanisms.

Many of the microbes isolated and classified as biocontrol agents (BCAs) can be considered facultative mutualists involved in protocooperation, because survival rarely depends on any specific host and disease suppression will vary depending on the prevailing environmental conditions. Further down the spectrum, commensalism is a symbiotic interaction between two living organisms, where one organism benefits and the other is neither harmed nor benefited. Most plant-associated microbes are assumed to be commensals with regards to the host plant, because their presence, individually or in total, rarely results in overtly positive or negative consequences to the plant. And, while their presence may present a variety of challenges to an infecting pathogen, an absence of measurable decrease in pathogen infection or disease severity is indicative of commensal interactions. Neutralism describes the biological interactions when the population density of one species has absolutely no effect whatsoever on the other. Related to biological control, an inability to associate the population dynamics of pathogen with that of another organism would indicate neutralism. In contrast, antagonism between organisms results in a negative outcome for one or both. Competition within and between species results in decreased growth, activity and/or fecundity of the interacting organisms. Biocontrol can occur when non-pathogens compete with pathogens for nutrients in and around the host plant. Direct interactions that benefit one population at the expense of another also affect our understanding of biological control. Parasitism is a symbiosis in which two phylogenetically unrelated organisms coexist over a prolonged period of time. In this type of association, one organism, usually the physically smaller of the two (called the parasite) benefits and the other (called the host) is harmed to some measurable extent. The activities of various hyperparasites, i.e., those agents that parasitize plant pathogens, can result in biocontrol. And, interestingly, host infection and parasitism by relatively avirulent pathogens may lead to biocontrol of more virulent pathogens through the stimulation of host defense systems. Lastly, predation refers to the hunting and killing of one organism by another for consumption and sustenance. While the term predator typically refer to animals that feed at higher trophic levels in the macroscopic world, it has also been applied to the actions of microbes, e.g. protists, and mesofauna, e.g. fungal feeding nematodes and microarthropods, that consume pathogen biomass for sustenance. Biological control can result in varying degrees from all of these types of interactions, depending on the environmental context within which they occur. Significant biological control, as defined above, most generally arises from manipulating mutualisms between microbes and their plant hosts or from manipulating antagonisms between microbes and pathogens.

The role of mycorrhizal fungi in control of various soil borne plant diseases has been reviewed by many workers (3,70,71,110) and a few cases have been tabulated here (Table 3). Besides, the association of AM fungi with plant nematodes and the beneficial effect of mycorrhizal symbiosis on plant growth had led to investigations into the potential of AM fungi to limit yield losses due to nematodes (13).

During colonization, AM fungi can prevent root infections by reducing the access sites and stimulating host defense. AM fungi have been found to reduce the incidence of root-knot nematode (13). Various mechanisms also allow AM fungi to increase a plant's stress tolerance. This includes the intricate network of fungal hyphae around the roots which block pathogen infections. Inoculation of apple-tree seedlings with the AM fungi *Glomus fasciculatum* and *G. macrocarpum* suppressed apple replant disease caused by phytotoxic myxomycetes. AM fungi protect the host plant against root-infecting pathogenic bacteria. The damage due to *Pseudomonas syringae* on tomato may be significantly reduced when the plants are well colonized by mycorrhizae (54). The mechanisms involved in these interactions include physical protection, chemical interactions and indirect effects. The other mechanisms employed by AM fungi to indirectly suppress plant pathogens include enhanced nutrition to plants; morphological changes in the root by increased lignification; changes in the chemical composition of the plant tissues like antifungal chitinase, isoflavonoids etc., alleviation of abiotic stress and changes in the microbial composition in the mycorrhizosphere (81).

Table 3. VAM Fungi in biocontrol of fungal pathogens

VAM fungi	Host Plant	Disease	Pathogen	Ref.No
G. fasciculatum	Tomato	Wilt	*Fusarium oxysporum* f.sp. *lycopersici*	21
G. fasciculatum	Green gram Black gram	Root rot	*Macrophomina phaseolina*	37,69,73
	Chick pea Sugar cane	Wilt	*Fusarium moniliforme* and *Cephalosporium sacchari*	
G. mosseae	Tomato	Blight	*Phytophthora parasitica*	42
G. mosseae	Tomato and Egg plant seedlings	Wilt	*Verticillium dahlia*	72
G. mosseae *G. intraradices* *G. clarum* *Gigaspora margarita*	Common bean plant	Rot	*Fusarium solani*	4
G. fasciculatum	Pea	Root rot	*Aphanomyces enteiches*	124

Gigaspora	Pigeon pea	Blight	*Phytophthora drechsleri* f.sp.*cajani*	14
G. mosseae	Chilli	Damping off	*Pythium aphanidermatum*	115
G. geosporum *G. mosseae*	*Dalbergia sissoo*	Wilt	*Fusarium solani*	115
G. mosseae	Soybean	Blight	*Phytophthora megasperma* var. *sojae*	61
G. fasciculatum	Poinsettia	Rot	*Rhizoctonia solani*	
		Damping off	*Pythium ultimum*	
G. mosseae *G. fasciculatum*	Tea	Brown root rot	*Fomes lamaoensis*	25,26
		Sclerotial blight	*Sclerotium rolfsii*	

In studies of colonization patterns of tomato roots by the mycorrhizal fungus *Glomus mosseae* and the pathogen *Phytophthora parasitica*, Cordier *et al.* (42) showed that proliferation of the pathogen is greatly reduced in mycorrhizal root system of tomato, compared with non mycorhhizal ones. Moreover, they observed that the host cells containing typical haustoria-like arbuscule structures of the mycorrhizal fungus were not infected by *P. parasitica* and that pathogen proliferation was reduced not only in mycorrhizal parts but also in nonmycorrhizal parts of Mycorrhizal root systems. Cytomolecular phenomena underlying bioprotection against *P. parasitica* in *G. mosseae*-colonized root systems of tomato has been demonstrated.

Effectiveness of arbuscular mycorrhizal fungi in the protection of common bean plant (*Phaseolus vulgaris* L.) against Fusarium root rot disease was investigated by Al-Askar and Rashad (4) under natural conditions in pot experiment. A mixture of arbuscular mycorrhizal fungi consisting of propagated units of *G. mosseae*, *G. intraradices*, *G. clarum*, *Gigaspora gigantea* and *Gi. margarita* in suspension form (10^6 unit L^{-1} in concentration) was used at dilution of 5 ml L^{-1} water. The obtained results demonstrated that, arbuscular mycorrhizal colonization significantly reduced the percentage of disease severity and incidence in infected bean plants. On the other hand, mycorrhizal colonization significantly increased the tested growth parameters and mineral nutrient concentrations. While, infection with *Fusarium* root rot disease negatively affected on the mycorrhizal colonization level in bean roots. Finally, mycorrhizal colonization led to a significant increase in the phenolic content and the activities of the investigated defense related enzymes (phenylalanine ammonia-lyase, polyphenol oxidase and peroxidase enzyme). From the obtained results, it can be concluded that the application of arbuscular mycorrhizal fungi as a biocontrol agent played an important role in plant resistance and exhibit greater potential to protect bean plants against the infection with *F. solani*.

The results indicated that, mycorrhizal colonization significantly reduced the percentage of disease severity and incidence in infected bean plants. These results are in agreement with that of Dar *et al.* (43), who found that inoculation of common bean plants with *G. mosseae* decreased root rot by 34 to 77%. Many authors have reported that the AM colonization can reduce root disease caused by several soil borne pathogens (2,38,135). Among the potential mechanisms involved in the resistance of mycorrhizal systems, the induction of plant defenses is the most controversial (127), where a number of biochemical and physiological changes have been associated with mycorrhizal colonization. Alteration in isoenzymatic patterns and biochemical properties of some defense-related enzymes such as chitinases, chitosanases and β-1,3-glucanases have previously been shown during mycorrhizal colonization (94). These hydrolytic enzymes are believed to have a role in defense against invading fungal pathogens because of their potential to hydrolyze fungal cell wall. Stimulating the host roots to produce and accumulate sufficient concentrations of metabolites which impart resistance to the host tissue against pathogen invasion have also been reported (47).

Direct (via interference competition, including chemical interactions) and indirect (via exploitation competition) interactions have been suggested as mechanisms by which AM fungi can reduce the abundance of pathogenic fungi in roots. These have generally been proposed in response to observations of negative correlations in the abundance of AM fungal structures and pathogenic microorganisms in roots (52). Presumably, pathogenic and AM fungi exploit common resources within the root, including infection sites, space and photosynthates within the root (131). Interference competition may also arise if carbon availability within intercellular spaces and the rhizosphere (61) or the number of infection loci within the root system (126) is reduced as a result of AM fungal colonization. Moreover, increasing the richness of AM fungal taxa colonizing the root system may result in more intense competition with a pathogenic fungus.

Another group of rhizosphere influencing root-microbe interaction are the plant growth promoting Rhizobacteria (PGPR) which are described as soil bacteria that colonise the roots of plants and enhance plant health. PGPR are known to participate in ecosystem processes including biocontrol of pathogens. *Pseudomonas* and *Bacillus* are the genera most commonly known as PGPR, but many other taxa also contain PGPR properties (32,33). Microbial populations in the rhizosphere where a typical response exerted by a kind of bacterial population called "mycorrhiza-helper bacteria" (MHB) known to stimulate mycelial growth/ formation of AMF and secretes compound responsible for enhanced root exudation. This in turn stimulates AMF mycelia in the rhizosphere. The bacteria have been found adhering to the AMF hyphae as well as embedded within the spore walls. Bacteria adhering to AMF mycelium may utilize hyphal exudates or use mycelium as vehicle for colonization of rhizosphere. Bacteria from genus *Paenibacillus,* which are antagonistic to a broad range of root pathogens are able to stimulate mycorrhizal colonization. They are found frequently to be associated with *G. intraradices* mycelium.

It should be mandatory to detect the cohesiveness of both AMF and PGPR participating in a particular rhizosphere while maintaining the healthy rhizosphere. The key step is to ascertain whether an antifungal biocontrol agent will negatively affect the AMF populations. There is need to exploit the possibilities of dual (AMF and PGPR) inoculation to provide plant defense against root pathogens. The success of these microbes will depend on our ability to manage the rhizosphere to enhance survival and competitiveness of these beneficial microorganisms. Rhizosphere management will require consideration of soil and crop cultural practices as well as inoculants formulation and delivery. It has also been established that combined uses of both AMF and PGPR increases the efficacy towards pathogen control (26,110). But the magnitude of pathogen suppression varies with rhizosphere to rhizosphere, which need to be deciphered and customized. Hence, there is need to modulate the mycorrhizosphere to maintain higher hyphosphere activity to manage resident AMF and PGPR for improving the plant and soil health and should be the key aim of the applied research in the future. With the growing interest in reducing chemical inputs, companies involved in the manufacturing and marketing of bioformulations (AMF-PGPR biocontrol consortium) should experience continued growth. However, stringent quality control measures must be adopted so that farmers get quality products. New, more effective and stable formulations also will need to be developed.

Literature Cited

1. Abbas, J.D, Hetrick, B.A.D., Jurgenson, J.E. 1996. Isolate specific detection of mycorrhizal fungi using genome specific primer pairs. *Mycologia* 88: 939-946.
2. Abdalla, M.E., Abdel-Fattah, G.M. 2000. Influence of the endomycorrhizal fungus *Glomus mosseae* on the development of peanut pod rot disease in Egypt. *Mycorrhiza* 10: 29-35.
3. Aggarwal, A., Prakash, V., Mehrotra, R.S. 2006. VAM fungi as biocontrol agents for Soil-borne plant pathogens. In: *Plant Protection in New Millennium* Vol II. Eds. A.V. Gadewar and B.P. Singh, Satish Serial Publishing House, New Delhi. p. 137-162.
4. Al-Askar, A.A., Rashad, Y.M. 2010. Arbuscular mycorrhizal fungi: A biocontrol agent against common bean *Fusarium* root rot disease. *Plant Pathol. J.* 9: 31-38.
5. Asai, T. 1944 Uber die Mykorrhizenbildung der leguminosen Pflanzen. *Jap.J. Bot.* 13 : 463-485.
6. Avise, J.C. 1994. Molecular markers, natural history and evolution.Chapman and Hall, London.
7. Barea, J.M., Jeffries, P. 1995. Arbuscular Mycorrhizas in sustainable soil plant systems. In: *Mycorrhizae structure and function, Molecular biology and Biotechnology.* Eds B. Hock and A. Varma. Springer-Verlag, Berlin. p. 521-559.
8. Baylis, G.T.S. 1959. Effect of vesicular-arbuscular mycorrhizas on growth of *Griselinia littoralis* Cornaceae. *New Phytol.* 58: 274-280.
9. Baylis, G.T.S. 1967. Experiments on the ecological significance phycomycetous mycorrhizas. *New Phytol.* 66: 231-243.

10. Becard, G., Kosuta, S., Tamasloukht, M., Sejalon-Delmas, N., Roux, C. 2004. Partner communication in the Arbuscular mycorrhizal interaction. *Can. J. Bot.* 82: 1186-1197.

11. Bever, J.D., Morton, J.B., Antonovics, J., Schultz, P.A. 1996. Host-dependent sporulation and species diversity of arbuscular mycorrhizal fungi in a mown grassland. *J.Ecol.* 84: 71-82.

12. Beyrle, H. 1995. The role of phytohormones in the function and biology of mycorrhizas. In : *Mycorrhiza structure, function, molecular biology and biotechnology*. Eds. A. Varma and B. Hock. Springer-Verlag, Berlin, 365-390.

13. Bhargava, S., Sharma, M.P., Pandey, R. Gour, H.N. 2008. Suppression of plant parasites including nematodes by AM fungi induced resistance in plants. *Rev.Plant Pathol.* 4: 421-466.

14. Bisht, V. S., Krishnan, K.R., Nene, Y.L. 1985. Interaction between vesicular arbuscular mycorrhiza and *Phytophthora drechsleri* f. sp. *cajani*. *Inter. Pigeon Pea Newsletter* 4: 63-64.

15. Blee, K.A., Anderson, A.J. 1996. Defense-related transcript accumulation in *Phaseolus vulgaris* L. colonized by the arbuscular mycorrhizal fungus *Glomus intraradices*, Schenk & Smith. *Plant Physiol.* 110: 675-688.

16. Blee, K.A., Anderson, A.J. 2000. Defense responses in plants to arbuscular mycorrhizal fungi. In: *Current advances in mycorrhizae research*, Eds. G.K. Podila and D.D. Douds. The American Phytopathological Society, Minnesota, USA, 27-44.

17. Blilou, I., Bueno, P., Ocampo, J.A., Garcia-Garrido, J.M. 2000a. Induction of catalase and ascorbate peroxidase activities in tobacco roots inoculated with the arbuscular mycorrhizal fungus *Glomus mosseae*. *Mycol. Res.* 104: 722-725.

18. Blilou, I., Ocampo, J.A., Garcia-Garrido, M. 2000b. Induction of *Ltp* (Lipid transfer protein) and *Pal* (phenylalanine ammonia lyase) gene expression in rice roots colonized by the arbu-scular mycorrhizal fungus *Glomus mosseae*. *J. Exp. Bot.* 51: 1969-1977.

19. Bonanomi, A., Oetiker, J.H., Guggenheim, R., Boller, T., Weimken, A., Vogeli-Lange, R. 2001. Arbuscular mycorrhizas in mini mycorrhizotrons: first contact of *Medicago truncatula* roots with *Glomus intrardices* induces chalcone synthase. *New Phytol.* 150: 573-582.

20. Bonfante, P. 2001. At the interface between mycorrhizal fungi and plants: the structural organization of cell wall, plasma membrane and cytoskeleton. In : *The Mycota*

IX. Eds. K. Hesser and B. Hock. Spinger-Verlag, Berlin, Heidelberg, 45-61.

21. Caron, M., Fortin, J. A., Richard, C. 1986. Effect of phosphorus concentration and *Glomus intraradices* on *Fusarium* crown and root rot of tomatoes. *Phytopathology* 76: 942-946.
22. Carvalho, G.R. 1998. Advances in molecular ecology (ed). NATO science series, Vol.306, IOS Press, Amsterdam, The Netherlands.
23. Cavagnaro, T.R., Jackson, L.E., Six, J. 2006 Arbuscular mycorrhizas, microbial communities, nutrient availability, and soil aggregates in organic tomato production. *Plant Soil* 282: 209-225.
24. Chakraborty, B.N. 2007. Mycorrhizal network: extent, function, recognition and importance in agricultural biotechnology. *Journal of Mycology and Plant Pathology*. 37:562-563
25. Chakraborty, B.N., Chakraborty, U., Barman, B.C., Bhutia, L., Ghosh, P.K.(2007). Induction of systemic resistance in tea plants against root rot pathogens upon field application of VAM, phosphate solubilizing fungus and bacterium. In: *Rhizosphere Biotechnology / Microbes in relation to plant health: Retrospects & Prospects*.Eds. A.K.Roy, B.N. Chakraborty, D.S. Mukadam and Rashmi, Scientific Publi-shers (India), Jodhpur, p.61-71.
26. Chakraborty, B.N., Chakraborty, U., De, U., Chakraborty, A.P., Rai, K. 2009. Dual application of *Glomus mosseae* and *Bacillus pumilus* to enhance growth of tea and suppress sclerotial blight disease. *Journal of Mycology and Plant Pathology*.39:565-566
27. Chakraborty, B.N., De, U.K. 2009. Screening of arbuscular mycorrhizal fungi associated with plantation crops and their evalution on plant health improvement. *Journal of Mycology and Plant Pathology*.39:195
28. Chakraborty, B.N., Sunar, K. 2009. Arbuscular mycorrhizal fungal association in rhizosphere of *Hevea brasiliensis NBU J. Plant Sci.* 3 : 67-70.
29. Chakraborty, U., Basnet, M., Roy Chowdhury, P., Chakraborty, B. N. 2004. Interaction of *Bacillus pumilus* with tea root rot pathogens. In: *Stress Biology*, Eds. U. Chakraborty and B. N. Chakraborty. Narosa Publishing House, New Delhi. p. 252-256.
30. Chakraborty, U., Chakraborty, B.N. 2008. Involvement of salicylic acid in plant defense against stresses. In : *Abiotic Stresses in Plants*. Eds. N.A. Khan & S. Singh. I.K. International Publishers, New Delhi, p 233-246.

31. Chakraborty, U., Chakraborty, B.N. and Basnet, M. 2006. Plant growth promotion and induction of resistance in *Camellia sinensis* by *Bacillus megaterium* *J. Basic Microbiol.* 46: 186-195.

32. Chakraborty U., Chakraborty B.N., Basnet M. and Chakraborty A.P. 2009. Evaluation of *Ochrobactrum anthropi* TRS-2 and its talc based formulation for enhancement of growth of tea plants and management of brown root rot disease. *J. Appl. Microbiol.* 107:625-634.

33. Chakraborty, U., Chakraborty, B.N., Chakraborty, A.P. 2010. Influcnco of *Serratia marcescens* TRS-1 on growth promotion and induction of resistance in *Camellia sinensis* against *Fomes lamaoensis*. *J. Plant Int.*. DOI: 10 1080/17429140903551738.

34. Chakraborty, U., Chakraborty B.N., Basnet, M., Bhutia, L.P. 2005. PGPR-mediated induction of resistance in tea plants for management of root rot diseases. In: *Emerging trends in Plant-microbe interactions,* Eds. S.S. Gnanamnickam, R. Balasubramanian and N. Anand. University of Madras, Chennai, p. 42-50.

35. Chakraborty, U., Chakraborty, B.N., Basnet, M., Deb, D. 2007. Plant growth promoting activity of *Serratia marcescens* isolated from tea rhizosphere and its application for induction of resistance against *Fomes lamaoensis.* In : *Rhizosphere Biotechnology/ Microbes in relation to plant health: Retrospects & Prospects*.Eds. A.K.Roy, B.N.Chakraborty, D.S.Mukadam and Rashmi, Scientific Publishers (India), Jodhpur, p. 135-146.

36. Chakraborty, U., Sarkar, B., Chakraborty, B. N. 2003. Protection of soybean root rot by *Bradyrhizobium japonicum* and *Trichoderma harzianum,* associated changes in enzyme activities and phytoalexins production. *J. Mycol. Plant Pathol.* 33: 21-25.

37. Chandra, S., Kehri, H.K., Dwivedi, R. 2000. Managing soil-borne diseases of horticulture cropsthrough VAM fungi. In: *Glimpses in plant sciences.* Eds. K.R. Aneja, M.U. Charaya, A. Aggarwal., D. K. Hans and S.A. Khan, Pragati Prakashan, Meerut, India pp. 82-94.

38. Chandanie, W.A., Kubota, M., Hyakumachi, M. 2009. Interactions between the arbuscular mycorrhizal fungus *Glomus mosseae* and plant growth-promoting fungi and their significance for enhancing plant growth and suppressing damping-off of cucumber (*Cucumis sativus* L.). *Appl. Soil Ecol.*41: 336-341.

39. Clapp, J.P., Young, J.P.W., Merryweather, J.W., Fitter, A.H. 1995. Diversity of fungal

symbionts in arbuscular mycorrhizas from a natural community. *New Phytol.* 130:259- 265.

40. Clapp, J.P., Young, J.P.W., Fitter, A.H. 1999. Ribosome small subunit sequence variation within spores of an arbuscular mycorrhizal fungus, *Scutellospora* sp. *Mol. Ecol.* 8: 915-921.

41. Comerford, N.B. 1998. Soil phosphorus bioavailability. In: *Phosphorus in Plant Biology*. Eds JP Lynch and J Deikman. American Society of Plant Physiologists, Rockville, Marryland USA. pp. 136-147.

42. Cordier, C, Pozo, M.J., Barea, J.M., Gianinazzi, S. And Gianinazzi-Pearson, V. 1998. Cell defense responses associated with localized and systemic resistance to *Phytophthora parasitica* induced in tomato by an arbuscular mycorrhizal fungus. *Mol. Plant-Microbe Inter.* 11: 1017-1028.

43. Dar, G.H., Zargar, M.Y. Beigh, G.M. 1997. Biocontrol of *Fusarium* root rot in the common bean (*Phaseolus vulgaris* L.) by using symbiotic *Glomus mosseae* and *Rhizobium leguminosarum*. *Microb. Ecol.* 34: 74-80.

44. Dehne, H.W., Schonbeck, F. 1979. The influence of endotrophic mycorrhiza on plant diseases.II. Phenol metabolism and lignifications. *Phytopathol. Z.* 95: 210-216.

45. Di Bonito, R., Elliott, M.L., Jardin, E.A.D. 1995. Detection of an arbuscular mycorrhizal fungus in roots of different plant species with the PCR. *Appl. Environ. Microbiol.*61: 2809-2810.

46. Douds, D.D., Galvez, L., Becard, G., Kapulnik, Y. 1998. Regulation of arbuscular mycorrhizal development by plant host and fungus species in alfalfa. *New Phytol.* 138: 27-35.

47. El-Khallal, S.M. 2007. Induction and modulation of resistance in tomato plants against *Fusarium* wilt disease by bioagent fungi (arbuscular mycorrhiza) and or hormonal elicitors (jasmonic acid and salicylic acid): 2-changes in the antioxidant enzymes, phenolic compounds and pathogen related-proteins. *Aust. J. Basic Applied Sci.*, 1: 717-732.

48. Eom, A., Hartnett, D.C., Wilson, G.W.T. 2000. Host plant species effects on arbuscular mycorrhizal fungal communities in tall grass prairie. *Oecologia* 122: 435-444.

49. Evelin, H., Kapoor, R., Giri, B. 2009. Arbuscular mycorrhizal fungi in alleviation of salt stress: a review. *Ann. Bot.* 104: 1263-1280.

50. Falk, S.P., Pearson, R.C., Gadoury, D.M., Seem, R.C., Sztejnberg, A. 1996. *Fusarium proliferatum* as a biocon-

trol agent against grape downy mildew. *Phytopathology* 86: 1010-1017.

51. Ferrol, N., Barea J.M., Azcon-Aguilar, C. 2002. Mechanism of nutrient transport across interfaces in Arbuscular Mycorrhizas. *Plant Soil* 244 : 231-237.

52. Filion, M., St-Arnaud, M. Jabaji-Hare, S.H. 2003. Quantification of *Fusarium solani* f. sp. *phaseoli* in mycorrhizal bean plants and surrounding mycorrhizosphere soil using realtime polymerase chain reaction and direct isolations on selective media. *Phytopathology,* 93: 229-235.

53. Garcia-Garrido, J.M., Garcia-Romera, I., Ocampo, J.A. 1992. Cellulase production by the vesicular-arbuscular mycorrhizal fungus *G. mosseae. New Phytol.* 121: 221-226.

54. Garcia-Garrido, J.M., Ocampo, J.A. 1989. Effect of VA Mycorrhizal infection of tomato on damage caused by *Pseudomonas syringae. Soil Biol Biochem.* 21 : 165-167.

55. Garcia-Garrido, J.M., Tribak, M., Rejon-Palomares, A., Ocampo, J.A., Garcia-Romera, I. 2000. Hydrolytic enzymes and ability of arbuscular mycorrhizal fungi to colonize roots. *J.Exp.Bot.* 51: 1443-1448.

56. Gerdemann, J.W. 1965. Vesicular-arbuscular mycorrhizae formed on maize and tuliptree by *Endogone fasciculata. Mycologia* 57: 562-575.

57. Ginzberg, I., David, R., Shaul, O., Elad, Y., Wininger, S., Ben-Dor, B., Badani, H., Fang, Y., van Rhijn, P., Li Y., Hirsch, A.M., Kapulnik, Y. 1998. *Glomus intraradices* colonization regulates gene expression in tobacco plants. *Symbiosis* 24: 145-157.

58. Giovannetti, M., Avio, L. Fortuna, P., Pellegrino, E., Sbrana, C. Strani, P. 2006. At the root of the wood wide web: self recognition and non-self incompatibility in mycorrhizal networks. *Plant Signal.Behav.* 1: 1-5.

59. Giovannetti, M., Sbrana, C. 1998. Meeting a non-host: the behaviour of AM fungi. *Mycorrhiza* 8: 123-130.

60. Gollotte, A., Gianinazzi, P.V., Giovannetti, M., Sbrana, C., Avio, L., Gianinazzi, Z. 1993. Cellular localization and cytochemical probing of resistance reactions to arbuscular mycorrhizal fungi in a 'locus a' mutant *Pisum sativum* L. *Planta* 191: 112- 122.

61. Graham, J.H., 2001. What do root pathogens see in mycorrhizas? *New Phytol.* 149: 357-359.

62. Hall, I.R. 1978 Potential for exploiting vesicular –arbuscular mycorrhizae in agriculture. In: *Biotechnology in Agriculture* Ed. R. Alan Liss Inc., New York, USA. pp. 141-174.

63. Harrison, M., Dixon, R. 1994. Spatial patterns of expression of flavonoid/isoflavonoid pathway genes during interactions between roots of *Medicago truncatula* and the mycorrhizal fungus *Glomus versiforme. The Plant J.* 6: 9-20.

64. Hawkins, H.J., Johansen, A., George, E. 2000. Uptake and transport of organic and organic nitrogen by arbuscular mycorrhizal fungi. *Plant and Soil* 226 : 275-285.

65. Helgason, T., Daniell, T.J., Husband, R., Fitter, A.H., Young, J.P.W. 1998. Ploughing up the wood-wide web? *Nature* 394:431.

66. Helgason, T., Fitter, A.H., Young, J.P.W. 1999. Molecular diversity of arbusclar mycorrhizal fungi colonizing *Hyacinthiodes non-scripta* (blue-bell) in a natural woodland. *Mol. Ecol.* 8: 659-666.

67. Innis, M.A., Gelfand, D.H., Sninsky, J.J., White, T.J. (eds). 1990. PCR protocols: a guide to methods and applications. Academic Press, New York.

68. Jakobsen, I., Chen, B.D., Munkvold, L., Lundsgaard, T., Zhu, Y.G. 2005a Contrasting phosphate acquisition of mycorrhizal fungi with that of root hairs using the non-mycorrhizal root hairless barley mutant. *Plant Cell Environ* 28 : 928-938.

69. Jalali, B.L., Chabra, M.L., Singh, R.P. 1990. Interaction between vesicular arbuscular mycorrhizal endophyte and *Macrophomina phaseolina* in mung bean. *Ind.Phytopath.* 43: 527-535.

70. Jalali, B.L., Jalali, I. 1991. Mycorrhiza in plant disease control. In: *Handbook of Applied Mycology.Vol I. Soil and Plants.* Eds. D.K.Arora, B. Rai, K.G. Mukherjee and D. Knuden, Marcel Dekker, New York. p. 131-154.

71. Jeffries, P., Gianinazzi, S., Perotto, S., Turnau, K., Barea, J.M. 2003. The contribution of arbuscular mycorrhizal fungi in sustainable maintenance of plant health and soil fertility. *Biol. Fertil.Soils*. 37: 1-16.

72. Karagiannidis, N., Bletsos, F., Stavropoulos, N. 2002. Effect of *Verticillium* wilt (*Verticillium dahliae* Kleb.) and mycorrhiza (*Glomus mosseae*) on root colonization, growth and nutrient uptake in tomato and eggplant seedlings *Sci. Hort.* 94:145-156.

73. Kehri, H.K., Chandra, S. 1990. Mycorrhizal association in crops under sewage farming. *J. Indian. Bot. Soc.* 69: 267-272.

74. Kiers, E.T., Lovelock, C.E., Krueger, E.L., Herre, E.A. 2000. Differential effects of tropical arbuscular mycorrhizal fungal inocula on root colonization and tree seedling growth: implications for

tropical forest diversity. *Ecol. Lett.* 3:106-1 I3.

75. Klironomos, J.N., McCune, J., Hart, M., Neville J. 2000. The influence of arbuscular mycorrhizae on the relationship between plant diversity and productivity. *Ecol. Letters* 3: 137-141.

76. Koide, R.T., Schreiner, R.P. 1992. Regulation of the vesicular arbuscular mycorrhizal symbiosis. *Ann. Rev. Plant Physiol. Plant Mol. Biol.* 43: 557-581.

77. Lambais, M.R. 2000. Regulation of plant defence-related genes in arbuscular mycorrhizae. In : *Current advances in mycorrhizae research*, Eds. G.K. Podila and D.D. Douds The American Phytopathological Society, Minnesota, USA, 45-59.

78. Lambais, M.R., Mehdy, M.C. 1998. Spatial distribution of chitinases and β-1, 3 glucanase transcripts in bean arbuscular mycorrhizal roots under low and high soil phosphate conditions. *New Phytol.* 140: 33-42.

79. Larsen, J., Ravnskov, S., Sorensen, J. 2007. Capturing the benefits of arbuscular mycorrhizae in horticulture. In *Mycorrhizae and Crop production*. Eds. C. Hamel and C. Plenchette the Haworth press, Inc., Binghamton, New York, USA. pp. 1232-1250.

80. Lemanceau, P., Alabouvette, C. 1993. Suppression of *Fusarium* wilt by fluorescent pseudomonas: Mechanisms and applications. *Biocont. Sci. Tech.* 3: 219-234.

81. Linderman, R.G. 1994. Role of VAM fungi in biocontrol. In: *Mycorrhizae and Plant Health.* Eds. F.L. Pfleger and R.G. Lindermann. American Phytopathological Society, St. Paul, MN p. 1-25

82. Lloyd-MacGilp, S.A., Chambers, S.M., Dodd, J.C., Fitter, A.H., Walker, C., Young, J.P.W. 1996. Diversity of the ribosomal internal transcribed spacers within and among isolates of *Glomus mosseae* and related fungi. *New Phytol.* 133: 103-111.

83. Longato, S., Bonfante, P. 1997 Molecular identification of mycorrhizal fungi by direct amplification of microsatellite regions. *Mycol. Res.* 101:425-432.

84. Mader, P., vierheilig, H., Streitwolf-Engel, R. 2000. Transport of ^{15}N from a soil compartment separated by a polytetrafluoroethylene membrane to plant roots via the hyphae of arbuscular mycorrhizal fungi. *New Phytol.* 146:155-161

85. McGonigle, T.P. 1988. A numerical analysis of published field trials of vesicular – arbuscular mycorrhizal fungi. *Funct. Ecol.* 2: 473-478..

86. Merryweather, J., Fitter, A.H. 1998b. The arbuscular mycorrhizal fungi of *Hyacinthiodes non-scripta*.II. Seasonal and spatial patterns of fungal populations. *New Phytol.* 138: 131-142.

87. Miller, J.C., Rajapakse, S., Garber, R.K. 1986 Vesicular – arbuscular mycorrhizae in vegetable crops. *Hort. Sci.* 21: 974-984.

88. Millner, P.D., Mulbry, W.W., Reynolds, S.L., Patterson, C.A. 1998. A taxon-specific oligonucleotide probe for temperate zone soil isolates of *Glomus mosseae*. *Mycorrhiza* 8:19-27.

89. Morton, J.B., Betivenga, S.P., Bever, J.D. 1995. Discovery, measurement and interpretation of of diversity in arbuscular mycorrhizal fungi (Glomales, Zygomycetes) *Can.J.Bot.* 73:S25-S32.

90. Mosse, B. 1957. Growth and chemical composition of mycorrhizal and non-mycorrhizal apples. *Nature* 179:922-924.

91. Newsham, K.K., Fitter, A.H., Watkinson, A.R. 1995. Multifunctionality and biodiversity in arbuscular mycorrhizas. *Trends. Ecol.Evol.* 10: 407-411.

92. Olsson, P.A., Van Aarle, I.M., Allaway, W.G., Ashford A.E., Rouhier, H. 2002. Phosphorus effects on metabolic processes in monoxenic Arbuscular mycorrhiza cultures. *Plant Physiol.* 130 : 1162-1171.

93. Plenchette, C., Clermont-Dauphin, C., Meynard, J.M., Fortin, J.A. 2005. Managing arbuscular mycorrhizal fungi in cropping systems. *Can. J. Plant Sci.* 85:31-40.

94. Pozo, M.J., Cordier, C., Dumas-Gaudot, E., Gianinazzi, S., Barea, J.M., Azcon-Aguilar, C. 2002. Localized versus systemic effect of arbuscular mycorr-hizal fungi on defence responses to *Phytophthora* infection in tomato plants. *J. Exp. Bot.* 53: 525-534.

95. Redecker, D. 2000. Specific PGR primers to identify arbuscular mycorrhizal fungi within colonized roots. *Mycorrhiza* 10: 73-80.

96. Redecker, D., Morton, J.B., Bruns, T.D. 2000. Ancestral lineages of arbuscular mycorrhizal fungi (Glomales). *Mol. Phyl. Evol.* 4:276-284.

97. Redecker, D., Thierfelder, H., Walker, C., Werner, D. 1997. Restriction analysis of PCR-amplified internal transcribed spacers of ribosomal DNA as a tool for species identification in different genera of the order Glomales. *Appl. Environ. Microbiol.* 63: 1756-1761.

98. Reddy, S.R., Pindi, P.K. and Reddy, S.M. 2005. Molecular methods for research on arbuscular mycorrhizal fungi in India : problems and prospects. *Curr.Sci.* 89: 1699- 1709.

99. Rodrigues, B.F. 2008. Taxonomy of Glomeromycota:

a review. *Rev.Plant Pathol.* 4: 193-216.

100. Rosendahl, S. 1985. Interactions between the vesicular-arbuscular Mycorrhizal fungus *Glomus fasciculatum* and *Aphanomyces euteiches* root rot of peas. *Phytopathol.Z.* 114: 31-40.
101. Roy, A. K., Kumar, R., Chakraborty, B. N., Chakraborty, U. 2002. VA Mycorrhiza in relation to growth of different tea varieties. *Mycorr. News* 14: 9-11.
102. Salzer, P. Boller, T. 2000. Elicitor- induced reactions in mycorrhizae and their suppression. In : *Current advances in mycorrhizae research*, Eds. G.K. Podila and D.D. Douds, The American Phytopathological Society, Minnesota, USA, 1-10.
103. Salzer, P., Bonanomi, A., Beyer, K., Vogeli-Lange, R., Aeschbacher, R.A., Lang, J., Wiemken, A., Kim, D., Cook, D.R., Boller, T. 2000. Differential expression of eight chitinase genes in *Medicago truncatula* roots during mycorrhizal formation, nodulation and pathogen infection. *Mol. Plant-Microbe Int.* 13: 763-777.
104. Salzer, P., Corbiere, H., Boller, T. 1999. Hydrogen peroxide accumulation in *Medicago truncatula* roots colonized by the arbuscular mycorrhizal-forming fungus *Glomus mosseae. Planta* 208: 319-325.
105. Sanders, F.E., Tinker, P.B. 1971. Mechanism of absorption of phosphate from soil by Endogone mycorrhizas. *Nature* 233: 278-279.
106. Sanders, F.E., Tinker, P.B. 1973. Phosphate flow into mycorrhizal roots. *Pest. Sci.* 4: 385-395.
107. Sanders, I.R., Alt, M., Groppe, K., Boller, T., Wiemken, A. 1995. Identification of ribosomal DNA polymorphisms among and within spores of the Glomale: application to studies on the genetic diversity of arbuscular mycorrhizal fungal communities. *New Phytol.* 130: 419-427.
108. Sanders, I.R., Fitter, A.H. 1992. Evidence for differential responses between host-fungus combinations of vesicular arbuscular mycorrhizas from a grassland. *Mycol.Res.* 96:415- 419.
109. Schubler, A., Gehrig, H., Schwarzott, D., Walker, C. 2001. Analysis of partial Glomales SSU rRNA gene sequences: implications for primer design and phylogeny. *Mycol. Res.* 105: 5-15.
110. Sharma, M.P., Sharma, S.K., Gupta, G.K. 2009. Arbuscular mycorrhizal fungi and plant growth promoting rhizobacteria in the control of soil borne diseases of selected oilseed crops. In: *Agriculturally Important Microorga-*

nisms, vol II. eds. G.G. Khachatourians, D.K. Arora, T.P. Rajendran & A.K. Srivastava, Academic World International, p.45-74.

111. Simon, L. 1996. Phylogeny of the Glomales: deciphering the past to understand the present. *New Phytol.* 133: 95-101.

112. Simon, L., Bousquet, J., Levesque, R.C., Lalonde, M. 1993. Origin and diversification of endomycorrhizal fungi and coincidence with vascular land plants. *Nature* 363: 67-69.

113. Simon, L., Lalonde, M., Bruns, T.D. 1992. Specific amplification of 18S fungal ribosomal genes from VA endomycorrhizal fungi colonizing roots. *Appl. Environ. Microbiol.* 58: 291-295.

114. Simon, L., Levesque, R.C., Lalonde, M. 1993. Identification of endo-mycorrhizal fungi colonizing roots by fluorescent single-strand conformation polymorphism-polymerase chain reaction. *Appl. Environ. Microbiol.* 59: 4211-4215.

115. Singh, R., Adholeya, A. And Mukerji, K.G. 2000. Mycorrhiza in control of soil borne pathogens. In: *Proceedings of Mycorrhizal Biology*. Ed. K.G. Mukerji. Kluwar Academic Publishers, New York.p. 173-196

116. Smith, S.E., Read, D.J. 2008. *Mycorrhizal symbiosis,* 3rd edn. Academic Press Ltd, London, UK.

117. Somssich, I.E., Hahlbrock, K. 1998. Pathogen defence in plants: a paradigm of biological complexity. *Trends Plant Sci.* 3: 86-90.

118. Spanu, P., Bonfante-Fasolo, P. 1998. Cell wall-bound peroxidase activity in roots of mycorrhizal *Allium porrum* : regulation and localization. *Planta* 177:447-455.

119. Van Aarle, I.M. 2009. Mycorrhizal fungi in agriculture: ecophysiology and phosphorus metabolism. In: *Agriculturally Important Microorganisms,* vol II. eds. G.G. Khachatourians, D.K. Arora, T.P. Rajendran & A.K. Srivastava, Academic World International, p. 25-43

120. van der Heijden, M.G.A., Klironomos, I.N., Ursic, M., Moutoglis, P., Streitwolf-Engel, R., Boller, T., Wiemken, A., Sanders, I.R. 1998. Mycorrhizal fungal diversity determines plant biodiversity ecosystem variability and productivity. *Nature* 396:69- 72.

121. Van Peer, R., Niemann, G.J., Schippers, B. 1991. Induced resistance and phytoalexin accumulation in biological control of *Fusarium* wilt of carnantion by *Pseudomonas* sp. strain WCS 417r. *Phytopathology* 81: 728-734.

122. van Rhijn, P., Fang, Y., Galili, S., Shaul, O., Atzmon, N., Wininger, S., Eshed, Y., Lum,

M., Li, Y., To, V., Kapulnik, Y., Hirch, A.M. 1997. Expression of early nodulin genes in alfalfa mycorrhizae indicates that signal transduction pathways used in forming arbuscular mycorrhizae and *Rhizobium*-induced nodules may be conserved. *PNAS*, USA 94: 5467-5472.

123. Van Tuinen, D., Jacquot, E., Zhao, B., Gollotte, A., Gianinazzi, P.V. 1998. Characteriasation of root colonization profiles by a microcosm community of arbuscular mycorrhizal fungi using 25S rDNA – targeted nested PCR. *Mol. Ecol.* 7: 879- 887.

124. Vidyasekharan, P. 1990. Current trends in Mycorrhizal research. In: *Proceedings of National Conference on Mycorrhiza*. Eds. B.L.Jalali & H. Chand, HAU, Hisar.p. 91-93.

125. Vierhlig, H., Alt, M., Mohr, U., Boller, T., Wiemken, A. 1994. Ethylene biosynthesis and activities of chitinase and β -1, 3 glucanase in the roots of host and non-host plants of vesicular-arbuscular mycorrhizal fungi after inoculation with *Glomus mosseae. J. Plant Physiol.* 143: 337-343.

126. Vigo, C., Norman, J.R., Hooker, J.E. 2000. Biocontrol of the pathogen *Phytophthora parasitica* by arbuscular mycorrhizal fungi is a consequence of effects on infection loci. *Plant Pathol.*, 49: 509-514.

127. Wehner, J., Antunes, P.M., Powell, J.R., Mazukatow, J. Rillig, M.C. 2009. Plant pathogen protection by arbuscular mycorrhizas: A role for fungal diversity?. *Pedobiologia.* 10. 1016/j.pedobi. 2009. 10.002.

128. Welch, R.M. 2002. The impact of mineral nutrients in food crops on global human health. *Plant Soil* 245:83-90.

129. Welch, R.M., Graham, R.D. 2002. Breeding crops for enhanced micronutrient content. *Plant Soil* 245:205-214.

130. Weller, D.M. 1988. Biological control of soilborne plant pathogens in the rhizosphere with bacteria. *Ann.Rev. Phytopathol.* 26 : 379-407.

131. Whipps, J.M., 2004. Prospects and limitations for mycorrhizas in biocontrol of root pathogens. *Can. J. Bot.*, 82: 1198-1227.

132. White, T.J., Bruns, T., Lee, S., Taylor, J. 1990. Amplification and direct sequencing of fungal ribosomal genes for phylogenies. In:. *PCR Protocols: a guide to methods and applications*. Eds. Innis, M.A., Gelfand, D.H., Sninsky, J.J., White, T.J. Academic Press, New York, 315-322.

133. Wright, S.F., Upadhyaya, A. 1998. A survey of soils for aggregate stability and glomalin, a glycoprotein produced by hyphae of arbuscular mycorrhizal fungi. *Plant Soil* 198: 97-107.

134. Wyss, P., Bonfante, P. 1993. Amplification of genomic DNA of arbuscular mycorrhizal (AM) fungi by PCR using short arbitrary primers. *Mycol. Res.* 97: I 351-1357.

135. Yao, M.K., Tweddell, R.J., Desilets, H. 2002. Effect of two vesicular-arbuscular mycorrhizal fungi on the growth of micropropagated potato plantlets and on the extent of disease caused by *Rhizoctonia solani*. *Mycor-rhiza*, 12: 235-242.

136. Zeze, A., Hosny, M., Gianinazzi, P.V, Dulieu, H. 1996. Characteris-ation of a highly repeated DNA sequence (5C1) from the arbuscular mycorrhizal fungus *Scutellospora castanea* and its detection *in planta*. *Appl. Environ. Microbiol.* 62:2443-2448

CONTRIBUTIONS TO OUR UNDERSTANDING OF PATHOGEN DIVERSITY AND MANAGEMENT OF RICE DISEASES IN INDIA

Samuel S. Gnanamanickam

Former Professor of Botany (Plant Pathology), Centre for Advanced Studies in Botany, University of Madras, India;
Senior Scientist, R & D, Novozymes Biologicals, Inc, 5400 Corporate Circle, Salem, VA 24153, USA. E-mail: SSGn@novozymes.com.

KEY WORDS: Transgenic, pathogen diversity, backcross breeding, biocontrol, DNA fingerprint, plasmid

ABSTRACT

In this article I would be reviewing research carried out by us at the University of Madras during 1978-2006 in the belief that it does provide a blue-print for the right research approaches made to rice disease management in India. We demonstrated for the first time that a combination of Pi-1+Pi-2 genes for blast resistance can be quite durable against blast caused by *Magnaporthe grisea* in southern India. Efforts were made to incorporate a combination of 4 resistance genes, *Xa4+xa5+xa13+Xa21* of bacterial blight pathogen, *Xanthomonas oryzae* pv. *oryzae* to improve the genetic resistance of elite indica rice cultivars, IR50 and Jyothi through backcross breeding in combination with molecular marker-assisted selection. We also identified R-gene *Xa-23* in native accessions of wild rice *Oryza rufipogon* as a new and valuable source of resistance to Indian strains of *X. oryzae* pv. *oryzae*. Bioassay of 20 lines of transgenic plants showed enhanced resistance to blast and bacterial blight. *Pseudomonas fluorescens* was introduced to suppress several diseases of banana, citrus, peanut and rice by seed treatments and foliar spray applications, and it enhanced yields in rice and groundnut. For the first time from India we reported production of 2,4-diacetylphloroglucinol [DAPG] by Indian strains of *P. fluorescens* and its implication in the biological suppression of rice bacterial blight.

In this review I would be unfolding our own research carried out at the University of Madras for two decades and a half (1978-2006) in the belief that it does provide a blue-print for the right research approaches made to rice disease management in India. It took a lot of hard thinking as to what would be a most appropriate treatise. These approaches include, 1. Careful analyses of pathogen diversity, 2. Backcross breeding and molecular marker-assisted selection (gene pyramiding), 3. Generation of transgenic indica rices, and 4. Biological control of rice diseases. While it is hard to imagine that all these kinds of research was carried out in one laboratory, the network of collaborations with so many laboratories around the world and the financial assistance available, were the key elements. In each of the above research category, are methods used that were pioneering, traditional and advanced in nature. I hope that the readers of the Annual Review of Plant Pathology volume will benefit in gaining a comprehensive over-view of the rice pathogens, their diversity in the Indian sub-continent, and disease management options that have been developed.

I. Rice Pathogens: Analysis of Pathogen Populations and Their Virulence Characteristics

A. Blast pathogen: *Magnaporthe grisea*

1. Genetics of rice blast fungus, *Magnaporthe grisea*

Laboratory production of sexual structures of rice blast fungus, *Magnaporthe grisea* was not achieved. Although several scientists had intensively studied the blast fungus in the laboratory, the sexual structures of *M. grisea* (perithecium, ascus and ascospores) could not be produced in the laboratory until 1995. Our laboratory was one of the two laboratories in the country to successfully produce these structures (67, 68) and demonstrate cross compatibility. The ability to produce the sexual stage of the blast pathogen in the laboratory led to further studies on the genetics of *M. grisea*.

2. Mating type distribution of *M. grisea*

Studies by Viji and Gnanamanickam (69) and Dayakar et al (14) demonstrated the prevalence of *Mat*1-1 and *Mat*1-2 in field populations of *M. grisea* and identified *Mat*1-1 as the dominant and functional mating type.

3. DNA fingerprint groups or lineages of *M. grisea*

By the use of MGR (*Magnaporthe grisea* repeat element) 586, more than 600 isolates of *M. grisea* assembled from the southern states of Tamil Nadu, Kerala, Karnataka and Andhra Pradesh along with isolates also from Orissa and Maharashtra were analyzed by the DNA fingerprinting and were grouped into 29 (named, A to Z) major genetic families or "lineages" (54, 55). In subsequent analyses, MGR DNA fingerprinting of isolates from northern states of India revealed the presence of 23 lineages. Isolates of the blast pathogen from the state of Meghalaya showed maximum genetic diversity (14). In our laboratory, the DNA polymorphisms of *M. grisea* isolates that are pathogenic to finger millet and rice was analyzed by Viji et al (70) by RFLP. In this analysis, DNA probes such as *MGR* 586, *grasshopper* and probes for the *Avr*-genes were used to distinguish the two pathogenic populations. The results revealed that the pathogen populations were distinct and did not show any shared RFLPS among field isolates of the pathogen that infected rice and finger millet. DNA fingerprinting analyses were carried out in collaboration with Morris Levy, Purdue University, USA. Molecular tools were also developed to detect the pathogen in rice and other millet leaves (5, 35).

4. Lineage-exclusion tests for *M. grisea* and gene pyramiding for durable blast Resistance (Molecular marker-assisted backcross breeding)

To identify blast resistance genes (R-genes) which would afford durable blast-resistance in southern India, representative strains from the above MGR lineages were inoculated on to the near-isogenic rice lines of CO-39 developed at IRRI each of which contained a single major R-gene for blast resistance (30-32). These tests showed for the first time that a combination of Pi-1+Pi-2 genes for blast resistance can be quite durable against blast in southern India (54-56). This information was very valuable not only for India but also for the rest of Asia and China. When a traditional backcross breeding method was used to pyramid the Pi-1+Pi-2 genes for blast resistance (also in combination with the use of molecular marker-assisted selection, MAS), the rice pyramids thus constructed in elite rice cultivars, CO39 and IR50 showed high levels of blast resistance. These have been tested in hot-spot locations for blast incidence in southern India (Table 1) and in IRRI, Philippines (4-6, 42-44, 56).

Table 1. Incidence of rice blast observed on rice cultivar CO39 (blast-susceptible) and its near-isogenic lines carrying blast-resistance genes, Pi-1 and Pi-2 and pyramid rice lines carrying both genes planted in hot-spot locations. Field experiment, South India (56)

Rice cultivar line/pyramid	Experimental location[a] and year		
	Pattambi, Kerala		Ponnampet, Karnataka
	1996	1997	1996
CO39	S	S	S
C101LAC (Pi-1)	S	S	R
C101A51 (Pi-2)	R	R	R
Pyramid lines of Pi-1+Pi-2 (seven lines)	R	R	R

[a]Proven hot-spot locations for natural incidence of rice blast.

Disease reactions to blast: R = resistance with severity scores of 0-3; S = susceptibility with severity scores of 4-9 in the SES system for rice

B. Bacterial Blight (BB) Pathogen, *Xanthomonas oryzae* pv. *oryzae*

Rice bacterial blight (BB) has been intensively studied in our laboratory for several years (1994-2004). Our major interest was to understand the genetic and pathogenic variability of the pathogen.

Detection of the seedborne inoculum of the BB pathogen has been a debated issue. Whether the pathogen is successfully transmitted through the rice seed to initiate fresh incidence of BB in rice has not been answered satisfactorily (25). Available evidences did not demonstrate evidence for seed transmission of the BB pathogen.

1. Use of monoclonal antibodies for grouping Indian strains of *X. oryzae* pv. *oryzae*

During 1994, in collaboration with Dr Anne Alvarez at the University of Hawaii, USA, we initiated a study with pathovar-specific monoclonal antibodies of the BB pathogen which improved the detection of X. *oryzae* pv. *oryzae* in infected rice seeds and also aided in making a serological classification of the Indian strains (1, 2, 28, 29).

2. DNA fingerprinting of Indian strains of *X. oryzae* pv. *oryzae*

In collaboration with Dr P. K. Ranjekar, National Chemical Laboratory, Pune, we fingerprinted a large number of Indian strains of the rice BB pathogen (34, 47, 48). In this study we used a set of hypervariable probes which included microsatellites, a human minisatellite probe, *avr* gene probe (*avrXa10*) and a repeat clone pBS101. Five major clusters/ groups were identified at 56% similarity thus suggesting high level of possible genetic polymorphisms within the field populations of the BB pathogen.

3. Identification of R-gene *Xa-23* in native accessions of wild rice *Oryza rufipogon* as a new and valuable source of resistance to Indian strains of *X. oryzae* pv. *oryzae* (7-9)

4. Gene pyramiding of BB-resistance genes (Marker-Assisted Backcross Breeding)

Efforts were made to incorporate a combination of 4 BB resistance genes, *Xa4+xa5+xa13+Xa21* to improve the genetic resistance of elite indica rice cultivars, IR50 and Jyothi through backcross breeding in combination with molecular marker- assisted selection. There were reasons why we resorted to the use of a 4-gene pyramid. We identified sub-populations of *X. oryzae* pv. *oryzae* in Kerala that could overcome the resistance offered by the most useful BB resistance gene, *Xa21* (66). This virulent sub-population was not able to break down the resistance in the rice line NH56 [*Xa4+xa5+xa13+Xa21*]. Plants of IR50 and Jyothi improved for their resistance to rice BB were obtained through systematic backcross breeding combined with the use of molecular markers linked to *xa5, xa13* and *Xa21* (10, 12).

C. Sheath Blight Pathogen: *Rhizoctonia solani*

Sheath blight used to be a minor disease but has recently become one of the most important production constraints for rice production both in India and many other rice-growing regions of the world. *R. solani* [teleomorph: *Thanatephorus cucumeris* (A.B. Frank) Donk] is a fungus representing a collective species which has been divided into 13 anastomosis groups (AG1 to AG13). We have carried out detailed analysis on the Indian isolates that belong to the anastomosis group pathogenic to rice, AG1 (12, 13). Recent molecular analyses carried out by Taheri *et al* (57) on the populations of sheath blight pathogen shows that *R. oryzae sativae* is associated with sheath blight as a minor pathogen.

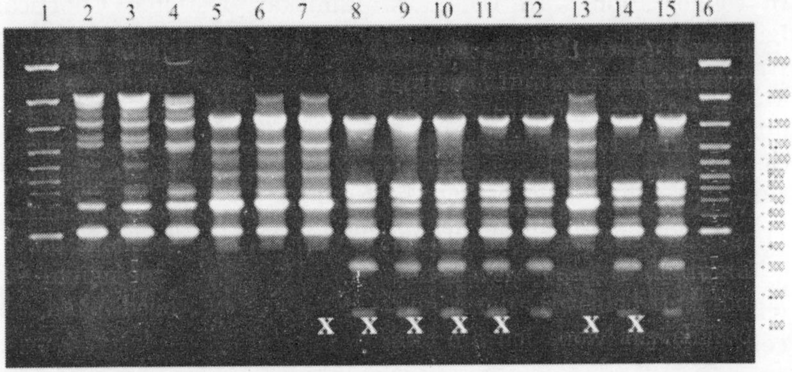

Fig. 1a. Rep-PCR fingerprints of *Rhizoctonia solani* isolates from India (41).

In collaboration with Bruce McDonald (ETH-Zurich), we evaluated the population structure of 96 Indian isolates of *R. solani* AG1 1A using seven RFLP loci. The field population of Indian *R. solani* AG1 IA has shown characteristics that are consistent to their high levels of gene flow and frequent sexual reproduction (as indicated by their Hardy-Weinburg equilibrium) among populations (Fig. 1a) (41). Their RFLP profiles and a phenogram showing them in 2 major clusters with 3 sub-clusters under each major cluster which also compares them to a Texas (USA) population of *R. solani* are shown in Fig. 1 b.

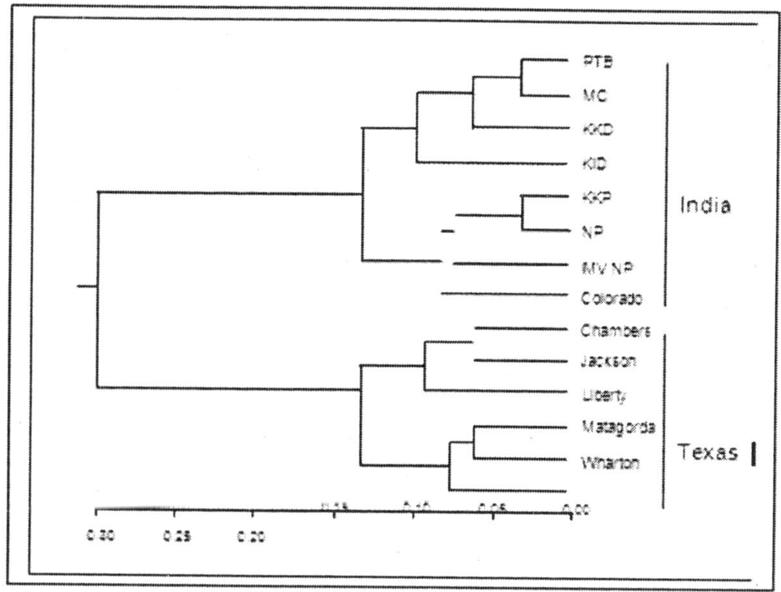

Fig. 1b. Phenogram derived by UPGMA based on Nei's genetic distance between Texas and Indian populations of *R. solani* AG1 IA (41).

In more recent studies on the population structure, Taheri et al (57) observed that *Rhizoctonia* spp. associated with sheath blight in India also had a number of (11/110 isolates) *R. oryzae-sativae*. Amplified fragment length polymorphism (AFLP) analyses carried out separated isolates of the three intraspecific groups (96/99 isolates of AG1-1A, 1/99 of 1B and 2/99 of 1C) of *R. solani* AG1 and differentiated *R. solani* from *R. oryzae-sativae.* They also observed that AG1-IA and AG1-IB isolates were more virulent to rice (cv. Zenith) than AG1-IC and *R. oryzae- sativae* (57).

II. Development of Transgenic Rices with Resistance to both Blast and Bacterial Blight

In our laboratory, research towards the improvement of elie indica rice cultivars, IR50 and CO39 or blast resistance and cv. Jyothi and IR50 for bacterial blight resistance was carried out through pyramiding of R-genes and marker-assisted selection of resistant plants (as described in the previous section). In order to create multiple resistances in these rice's, blast-resistant pyramids of IR50 and CO39 available in our laboratory (listed in Table 1) were transformed with *Xa21* gene. The bacterial blight resistance gene *Xa21* is known to confer resistance to all known races of *X. oryzae* pv. *oryzae* in India and the Philippines. This work was carried out in collaboration with Dr Swapan Datta of IRRI and was supported by the Rockefeller Foundation (42,45,46).

A. Starting materials for transformation

Target rice pyramids (carrying blast resistant gene/genes) for the transformation were, IR50 (*Piz-5*) and CO39 (*Pi-1+Piz-5*). They were used as the starting material for the rice tissue culture and transformation with *Xa21* gene using particle bombardment. Immature embryos and mature seeds were used as explants in both the varieties for tissue culture process.

B. Construct used for bombardment

The plasmid pC822 that contained *Xa21* coding sequence was kindly supplied by Dr. PC Ronald, University of California, Davis, USA. The primer U1 and I1, developed to amplify a 1.4-Kb DNA fragment of *Xa21* that was polymorphic to fragments amplified from other *Xa* genes were used for quick genetic analysis of the transgenic progeny. Plasmid pROB5 contained the selectable marker, the hph-coding region, flanked by the cauliflower mosaic virus (CaMv) 35S promoter and polyadenylation signals (Poly A). This plasmid provided a selectable marker that confers resistance to hygromycin for cotransformation with the pC822 plasmid.

C. Particle bombardment and selection

Immature embryos (IEs), immature embryo derived callus (IECs) and mature seed derived callus (MCs) were arranged and bombarded with the *Xa-21* + pROB5 plasmid (Purified by CsCl/EtBr method) by the particle gun PDC-1000/He system (BIO-RAD, Hercules, CA) following manufacturer's instructions. After bombardment, the explants were left in the dark overnight in the same medium. In the morning, the cultures were transferred to MS + 30g/l maltose + 7.5g/l agar medium supplemented with 50mg/l hygromycin B for selection of transformed calli and incubated in dark at 27°C. The newly developing hygromycin-resistant calli were subcultured in fresh media under continued selection pressure at every fortnight interval for 5 cycles.

D. Plant regeneration

The embryogenic calli were carefully selected and transferred to 50 ml flasks containing plant regeneration medium (For IR50, MS medium with 2mg kinetin/L and 1 mg NAA/L and for CO39, MS medium with 5mg kinetin/L and 1 mg NAA/L). The cultures were incubated for 2-3 week in 16h photoperiod of 3000-lux intensity at 26°C. Three to four week old plantlets were transferred to MS basal medium (MSo) for rooting. These plantlets with vigorous roots were transferred to styrofoam boards with holes in a plastic tray containing Yoshida's culture solution. The regenerants in the culture solution were allowed to grow for 2 weeks for hardening and then were transferred to soil directly in the transgenic green house with a day/night temperature regime of 29/23°C.

E. Molecular assays

These included hygromycin phosphotransferase (hpt) assay, PCR analysis and Southern blotting. The protocols have been described adequately elsewhere by Narayanan et al. (44-46).

F. Bioassays

The transgenic T_1 plants were grown in an IRRI containment greenhouse under the following conditions: 29°C and 85% humidity during the day and 25°C and 90% humidity at night. Since the plants contained both blast and bacterial blight resistance genes, bioassays were done for both the pathogens individually and in combinations to see the effect of multiple genes in the genotypes. Therefore, the bioassays were done in the following three ways:

1. Bioassay for Blast resistance

Twenty lines from each of the transgenic test cultivar, CO39, IR50 (susceptible), and C101A51 (*Piz-5*), were sown in plastic trays. Inoculum of *M. grisea* containing 50,000 conidia/ml of inoculum was spray-inoculated onto the rice seedlings at 4th leaf stage. Inoculated seedlings were incubated in a moist chamber for 24 hr and then transferred to the dew misty chamber at 22-25°C for 7 days till scoring. Blast disease severity was scored 10 day after inoculation using a 0-9 scoring method.

Leaves of IR50 and CO39 (transgenic T1) plants showed high levels of resistance to blast. This observation is also supported by data on percent diseased leaf area (DLA) presented in Tables 2 and 3.

Table 2. Screening of IR50 transgenic plants (T1) against the rice blast fungus *Magnaporthe grisea* (45)

Genotype	Mean size of lesion (mm)	DLA%[a]	Score[b]
CO39	12 ± 0.8	65	9
IR50	8 ± 0.7	60	9
C101A51	1 ± 0.2	0.5	0-1
13-T	2 ± 0.4	0.3	0-1
14-T	2 ± 0.7	0.6	0-1
15-T	1 ± 0.8	0.8	0-1

[a]DLA %, Percentage diseased leaf area
[b]Score based on a scale of 0-9 (SES system, IRRI).

Table 3. Screening of CO39 transgenic plants (T1) against the rice blast fungus *Magnaporthe grisea* (46)

Genotype	Mean size of lesion (mm)	DLA%[a]	Score[b]
CO39	11 ± 0.6	65	9
IR50	7 ± 0.6	60	9
C101A51	1 ± 0.3	0.5	0-1
C101LA C	1 ± 0.4	0.5	0-1
18-T	2 ± 0.4	0.3	0-1
19-T	2 ± 0.7	0.6	0-1

a. DLA %, Percentage diseased leaf area
b. Score based on a scale of 0-9 (SES system, IRRI).

2. Bioassay for bacterial blight resistance (45, 6)

The set of test cultivars IR50, IR24 (susceptible controls), IRBB21 (near isogenic line for *Xa-21*), IRBB4 (near isogenic line for *Xa-4*), and 20 lines from each transgenic line were sown in plastic trays. Plants were tested against three different races of *X. oryzae* pv. *oryzae* (*Xoo*) to differentiate the genes *Xa-21* and endogenous *Xa-4* such as *PXO*86 (race 2), *PXO*99 (race 6) and *PXO*341 (race 10). The inoculum of each strain was prepared by incubating the bacteria on Wakimoto's medium (Medium composition: Modified Wakimoto's medium (MF-P): sucrose- 30 g, bacteriological peptone- 5g, calcium nitrate- 0.5g, sodium phosphate (dibasic)- 0.82g, ferrous sulphate- 0.05g, agar- 15-17g, pH- 6.0) for 72h at 30°C, then suspending each pure culture in sterile distilled water and adjusting the inoculum to about 10^9 cells per milliliter.

At the maximum tillering stage, each plant was inoculated with the above three strains of *Xoo* using the leaf clipping method at the transgenic greenhouse, IRRI. Reaction of rice plants to each race of *Xoo* was scored 14 days after inoculation.

Resistance to bacterial blight was observed in T_1 plants of IR50 and CO39 (Table 4). Bacterial blight lesions of <2.0 cm length observed in transgenic plants were characteristic resistance reactions. The non-transformed BB-susceptible parent plants showed bacterial blight lesions of >10.0 cm length.

Table 4. Resistance response to bacterial blight in transgenic plants

Genotype	Race 2 (PX086)				Race 6 (PX099)			
	7d		14d		7d		14d	
	MLL (cm)	Reaction	MLL (cm)	Reaction	MLL (cm)	Reaction	MLL (cm)	Reaction
IR50 (C)	5.3	S	7.8	S	5.8	S	9.4	S
IR24 (C)	13.5	S	15.8	S	14.5	S	17.8	S
IRBB21 (C)	1.3	R	2.4	R	3.4	R	4.7	R
IRBB4(C)	4.6	R	6.8	S	6.4	S	7.9	S
T13 R-1	2.6	R	2.6	R	2.8	R	3.8	R
T13 S-1	5.6	S	7.8	S	12.8	S	15.3	S

1. Average of 15/20 T: progenies showing resistance to BB pathogen
2. Average of 5/20 T: progenies showing susceptible to BB pathogen.

C - non-transformed parental lines; R-Resistant; S-Susceptible; MLL-Mean Lesion Length.

3. Bioassyas for blast and bacterial blight resistance

All 20 lines of transgenic plants showed enhanced resistance to blast and bacterial blight (Fig. 2) similar to the results obtained in separate bioassays for blast and bacterial blight.

Fig. 2. Reactions of transgenic T_1 CO39 plants with *Xa21* introgressed with *Piz-5* for blast resistance when inoculated with IK81-3. Leaves 1 to 4 represent CO39, C101LAC, C101A51 (*Piz-5*), and the traand transgenic CO39 (*Pi-1+ Piz-5+Xa21*), respectively (45).

These transgenic plants have to be evaluated in the rice fields in southern India under strict biosafety precautions and their field performances have to be monitored carefully. This has not happened yet and efforts are in progress.

III. Development of Bacterial Biocontrol agents for Rice Pathogens

Pioneering discoveries made by our team relate to the development of biological control of crop diseases, and in particular to biocontrol of rice diseases. The important features of this development are listed below:

1. Introduction of *Pseudomonas fluorescens* to the nation as a potential biocontrol agent that suppresses plant pathogens (60)

From this very first introduction, several diseases of banana, citrus, groundnut (peanut) and rice were suppressed by seed treatments and foliar spray applications of *Pseudomonas fluorescens* (3, 17, 21, 23, 49-51, 53). The bacteria were reported to cause enhanced yields in rice and groundnut (51, 52).

2. Biological control of rice blast by seed treatment of rice with *Pseudomonas fluorescens* (24)

Blast (leaf and neck blasts) suppression was mediated by the production of an antifungal antibiotic [tentatively identified as phenazine-1-carboxylic acid (PCA)] by the biocontrol agent.

3. Genetic analysis of antifungal antibiotic production by *Pseudomonas fluorescens* Pf7-14 (11, 61) (Fig. 1)

This led to the generation of *Tn*-5 mutants of the bacterium that were altered in afa (antifungal antibiotic) production, cloning of ant+ DNA, and evaluation of the role of antibiotic production in the about 80% control of blast (leaf and neck blasts) (Fig. 3) and sheath blight of rice.

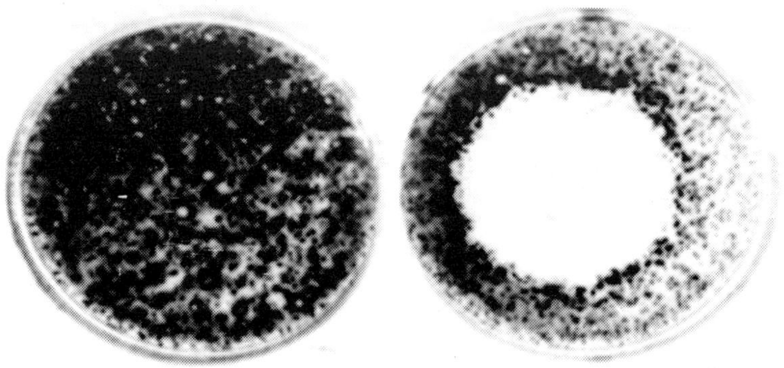

Fig. 3. A. Plast control by *Pseudomonas fluorescens* strain *Pf*7-14 (24, 61). A. Laboratory assay shows inhibition of *Pyricularia grisea* (rice blast fungus) by *Pf*7-14 (24).

Fig. 3B. Rice panicles from a field experiments show suppression of neck blast due to treatment with *Pf*7-14 (plants on the right) and lack of protection against neck blast in plants treated with its afa- mutant (plants on the left).

1. Development of *Pseudomonas fluorescens* and *P. putida* strains for the suppression of different rice diseases

This research carried out over a number of years is summarized in Table 5. More than 5500 strains of plant-associated bacteria have been isolated, assembled and studied in the authors' laboratory at University of Madras during 1983-2003.

Table 5. Bacterial biocontrol agents developed for major diseases of rice at Center for Advanced Studies in Botany, University of Madras, India (18, 19, 26, 27, 33).

Disease	Pathogen (causal agent)	Biocontrol agent developed	Reference
Blast (Bl)	*Pyricularia grisea* (Teleomorph: *Magnaporthe grisea*)	*Pseudomonas fluorescens*, *Bacillus* spp., *B. polymyxa*, *B. pumulus*, *B. coagulans*, *Enterobacter agglomerans*	24, 36, 61
Sheath blight (ShB)	*Rhizoctonia solani* (Teleomorph: *Thanetophorus sasaki)*	*P. fluorescens*, *P. putida*, *Bacillus megaterium*, *B. polymyxa*, *B. pumulus*, *B. coagulans*, *Enterobacter agglomerans*	36-40, 58, 59, 62

Sheath-rot (Sh-R)	*Sarocladium oryzae*	*P. fluorescens*	22, 50, 51
Stem rot (St-R)	*Sclerotium oryzae/ Helminthosporium sigmoideum* (Teleomorph: *Magnaporthe salvinii*	*P. fluorescens*	15
Bacterial Blight (BB)(Fig.2)	*Xanthomonas oryzae* pv. *oryzae*	*Bacillus* spp., *B. lentus, B. cereus, B. circulans* and *P. fluorescens*	63-65, 71, 73, 74
Tungro disease	Tungro virus (RTV)	*P. fluorescens* that kill the leaf-hopper vector of RTV	16

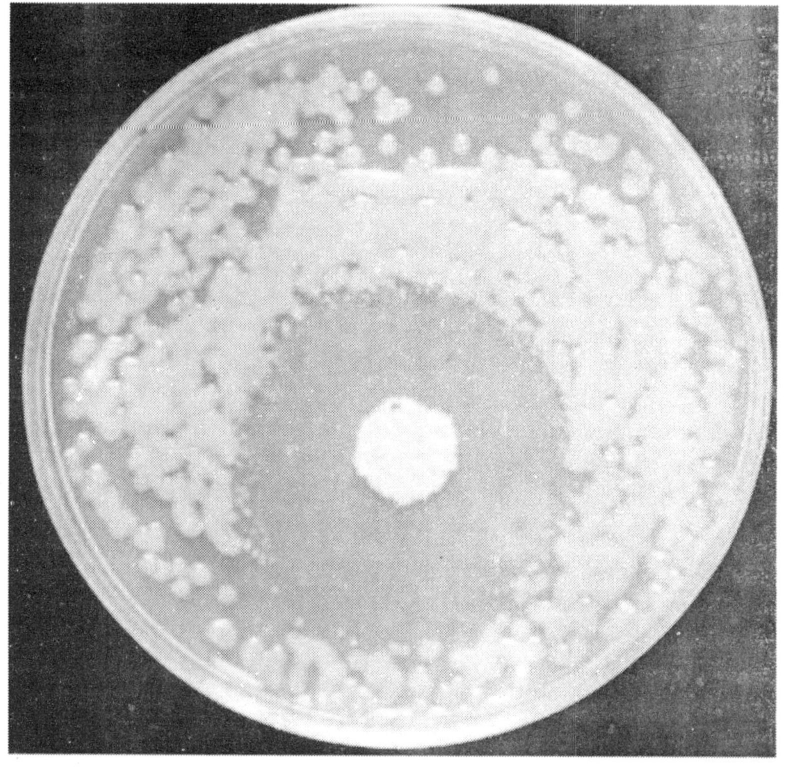

Fig. 4. Inhibition of *Xanthomonas oryzae* pv. *oryzae* by *Bacillus lentus* ALP18 in laboratory assays (63).

5. Identification of mechanisms involved in the biological control of rice diseases by *Pseudomonas fluorescens*

As a laboratory carried out basic research, we were always interested in elucidating the mechanisms involved in biological disease suppression in rice. The following is a summary of many years of research (Table 6).

Table. 6. A summary of mechanisms identified in the biological suppression of rice diseases (64)

Rice disease studied	Bacteria used	Mechanism(s) identified
Blast	*Pseudomonas fluorescens Pf*7-14	Antifungal antibiotics (phenazines, antifungal proteins), induced systemic resistance (ISR).
Sheath blight	*Pf*7-14, *Bacillus* spp.	Chitinase, antibiotics, ISR
Bacterial blight	*P. fluorescens, Bacillus* spp. (*B. lentus, B. circulans, B. coagulans, Bacillus* sp.)	Antibacterial antibiotics, kanosomine (Fig.2) and 2,4-diacetylphloroglucinol
Rice tungro virus	*P. fluorescens*	Antibiotics that kill the insect vector

6. A new and exciting discovery is the production of 2,4-diacetylphloroglucinol [DAPG] by Indian strains of *Pseudomonas fluorescens* (first report from India by Velusamy and Gnanamanickam (71) and its implication in the biological suppression of rice bacterial bacterial blight (20, 72, 73, 74).

Before 2004, production of DAPG has been implicated in the control of important diseases of tobacco and wheat and became increasingly well known as the agent that caused "take-all decline' in wheat soils. Our report (first report for India) showed DAPG production in the tropical soils of India and its role in the suppression of this major bacterial plant pathogen of rice, *Xanthomonas oryzae* pv. *oryzae* (71) Velusamy *et al* (72-74) (Table 3).

Table 7. Plant-associated fluorescent pseudomonads which inhibited the growth of *Xanthomonas oryzae* pv. *oryzae* (*Xoo*) and produced 2,4- diacetylphloroglucinol (DAPG) (71)

S.N.	Place of collection	Code number of strain	Diameter (in cm) of inhibition zone of *Xoo*	Production of 2,4-DAPG[1]	Percent (%) suppression of rice BB	
					Net-house[2]	Field[2]
1	Karnataka	KAD7	1.8	+	56.87	53.93
2	Karnataka	IMV14	2.8	+	58.68	56.74
3	Karnataka	IMV2	2.8	+	36.78	33.18
4	Karnataka	BGR19	0.5	+	27.48	42.71
5	Kerala	PTB9	2.1	+	58.78	64.46
6	Kerala	MON1	3.3	+	19.70	37.13
7	Tamil Nadu	TVM8	1.2	+	16.32	23.51
8	Tamil Nadu	VEL17	2.3	+	23.33	54.08
9	Tamil Nadu	VEL10	0.7	+	17.03	17.52
10	Tamil Nadu	GDY4	2.1	+	30.73	41.44
11	Tamil Nadu	GDY7	2.3	+	51.57	45.30
12	Tamil Nadu	TRP5	1.7	+	33.68	34.54
13	Tamil Nadu	TRP18	2.3	+	53.15	48.34
14	Tamil Nadu	MDR9	2.1	+	30.68	23.29
15	Tamil Nadu	MDR7	1.3	+	57.11	54.43
16	Tamil Nadu	STR7	1.8	+	19.51	36.04
17	Tamil Nadu	VGP13	2.5	+	55.39	51.62
18	Tamil Nadu	MDR16	1.4	+	37.60	46.12
19	Tamil Nadu	PDY5	1.8	+	58.83	56.88
20	Tamil Nadu	VLB7	1.7	+	33.64	9.08
21	Tamil Nadu	KVR5	2.5	+	55.92	49.75
22	Tamil Nadu	TNI13	2.2	+	26.05	28.19

23	Tamil Nadu	KOV8	3.0	+	38.69	9.71
24	Tamil Nadu	RJP31	2.4	+	10.31	0.50
25	Tamil Nadu	KOV3	1.8	+	21.18	49.21
26	Tamil Nadu	PDU1	2.9	+	8.16	11.80
27	Tamil Nadu	PDU9	2.7	+	16.94	2.04
28	Control	-	-	-	0.00	0.00

[1]Production of DAPG was identified through a PCR-based screening procedure which amplified a 745 bp DNA fragment in these strains.

[2]Results of a replicated net-house/field experiment (RBD) conducted at the Regional Agricultural Research Station, Pattambi, Kerala. Each figure is a mean of 3 replications.

Concluding remarks

The work of contributions from our laboratory during 1978-2007 is documented in this retrospective writing. These contributions fall in four major areas of rice disease management. They are, 1. Assessment of rice pathogen diversity in India with the use of both traditional and molecular tools; 2. Pyramiding of R-genes on the basis of the molecular information on pathogen diversity and their virulence characteristics, 3. Generation of blast and bacterial blight-resistant transgenic indica rice's (IR50, CO39) and 4. Development of bacterial biocontrol agents for major rice diseases. They have led to a whole new perspective on rice disease management. I have provided here a mere summary listing of the research developments. For more detailed information, readers may refer to original journal papers listed in the references, published in international journals of great reputation.

Acknowledgements

I acknowledge the commitment, innovation and hard work of all my Ph.D. students in my laboratory. It is their work that helped me weave a design for rice disease management. I also thank the Director, Centre for Advanced Studies in Botany, University of Madras for providing laboratory facilities and encouragement.

My contributions to agricultural research were made possible by a number of research collaborations developed within India and outside India. I am deeply thankful to Swapan Datta (IRRI), T.W. Mew (IRRI), P. Ranjekar (NCL), A. Alvarez (U. of Hawaii), M. Levy (Purdue U), B. A. McDonald and G. Defago (ETH-Zentrum, Zurich), Linda Thomashow (USDA-ARS, Pullman, WA) and M. Hofte (Ghent, Belgium).

I deeply appreciate the hospitality of Kerala Agricultural University for the greenhouse and field facilities made available by the Associate Director, Regional Agricultural Research Station, Pattambi, Kerala.

I acknowledge the financial support I received from the Rockefeller Foundation, The International Rice Research Institute (IRRI), Asian Development Bank, Bayer Crop Science (Belgium) and the national funding agencies, DST, DBT and UGC.

Literature cited

1. Alvarez, A. M. and Gnanamanickam, S. S. 1993. Monoclonal antibodies. Pages 137-148, In Genetic Engineering, Molecular Biology and Tissue Culture for Crop Pests and Disease Management. P. Vidhyasekaran, ed, Daya Publishing House, New Delhi.
2. Alvarez, A. M., Benedict, A. A, and Gnanamanickam, S. S. 1993. Identification of seed-borne bacterial pathogens of rice with monoclonal antibodies. Pages 3-8, In *Proc. Int'l Seed Testing Assoc.* Symp. On Seed Health Testing, Ottawa, Canada, August 9-11, 1993.
3. Anuratha, C. S. and Gnanamanickam, S. S. 1990. Biological control of bacterial wilts caused by *Pseudomonas solanacearum* in India with antagonistic bacteria. *Plant Soil* 122: 121-128.
4. Babujee, L and Gnanamanickam, S. S. 2001. Blast Disease of Rice. Chapter 12, Pages 299-320 In *IPM in Agriculture,* Vol. 8 (Eds: R.K. Upadhyay, D. K. Arora and O.P. Dubey), Aditya Books Pvt Ltd, New Delhi.
5. Babujee, L, and Gnanamanickam, S. S. 2000. Molecular tools for the characterization of rice blast pathogen (*Magnaporthe grisea*) population and molecular marker-assisted breeding for disease resistance. *Curr. Sci.* 78: 248-257.
6. Babujee, L., Venkatesan, B., Kavitha, S., Gnanamanickam, S. S., Leenakumar, S., McCouch, S. R, and Leong, S.A. 2003. Marker-assisted selection for transferring resistance to blast in high-yielding but susceptible Jyothi. Pages 253-255 In: *Advances in Rice Genetics*, eds. G. S. Khush, D. S. Brar and B.

Hardy, The International Rice Research Institute, Los Banos, Philippines.

7. Bharathkumar, S., and Gnanamanickam, S. S. 2005. Identification of a new source of resistance in wild rice, *Oryza rufipogon* to bacterial blight of rice caused by Indian strains of *Xanthomonas oryzae* pv. *oryzae*. *Curr. Sci.* 88: 1229-1231.

8. Bharathkumar Srinivasan and Gnanamanickam, S. S. 2004. Identification of *Oryza rufipogon* as a new source of bacterial blight resistance for India. *Proc. 1st International Conference on Bacterial Blight of Rice*, Tsukuba, Japan, March 17-19, 2004.

9. Bharathkumar Srinivsan, David Paulraj, R. S., Brindha, P. V., Kavitha, S, and Gnanamanickam, S.S. 2007. Improvement of bacterial blight resistance in rice cultivars Jyothi and IR50 via marker-assisted backcross breeding. *J. Crop Improvement* 19: in press.

10. Brindha Priyadarisini, V., Leenakumari, S., Raji, P., Rosemary Francis., P. V. and Gnanamanickam, S. S. 2003. Molecular analysis of *Xanthomonas oryzae* pv. *oryzae* populations in Kerala and pyramiding of genes for bacterial blight resistance in rice. Pages 146-172 In *Priorities and Strategies for Rice Research in High Rainfall Tropics*, Kerala Agricultural University, Trichur, India.

11. Chatterjee, A., Valasubramanian, R., Vachani, A., Mau, W-L., Gnanamanickam, S. S. and Chatterjee, A. K. 1996. Biological control of rice diseases with *Pseudomonas fluorescens* (Pf) 7-14: isolation of ant mutants altered in antibiotic production, cloning of ant+ DNA and an evaluation of a role for antibiotic production in the control of blast and sheath blight. *Biological Control* 7: 185-195.

12. David PaulRaj, R.S. 2003. *Molecular strategies for management of sheath blight and bacterial blight diseases of rice through pathogen population analyses and breeding for disease resistance*. Ph.D Thesis, University of Madras, 120 p.

13. David Paul Raj, R. S., Velusamy, P, and S. S. Gnanamanickam 2001. Sheath blight disease of rice. Chapter 13, Pages 329-355 *In IPM in Agriculture*, Vol. 8 (Eds. R.K. Upadhyay, D. K. Arora, & O.P. Dubey), Aditya Books Pvt Ltd, New Delhi.

14. Dayakar, B. V., Narayanan, N. N, and Gnanamanickam, S. S. 2000. Cross-compatibility and distribution of mating type alleles of rice blast fungus *Magnaporthe grisea* in India. *Plant Disease* 84: 700-704.

15. Elangovan C. and Gnanamanickam, S.S. 1992. Incidence of *Pseudomonas fluorescens* in rhizosphere of rice and their antagonism towards *Sclerotium oryzae*. *Indian Phytopath*. 43: 358-361.

16. Ganesan, P. 1999. Biological control of *Fusarium* wilts of cotton and tomato with bacterial strains of *Pseudomonas fluorescens* and *P. putida*. Ph.D. Thesis, University of Madras, 81 p.

17. Ganesan, P. and Gnanamanickam, S. S. 1987. Biological control of *Sclerotium rolfsii* Sacc. in peanut by inoculation with *Pseudomonas fluorescens*. *Soil Biol. Biochem*. 19: 35-38.

18. Gnanamanickam, S. S. 2002. *Biological Control of Crop Diseases*. Marcel Dekker Publishers, New York, 450 p.

19. Gnanamanickam, S. S. 2006. *Plant-Associated Bacteria*. Springer, The Netherlands, 712p.

20. Gnanamanickam, S. S. 2004. Biological control of bacterial blight of rice: insights into mechanisms. *Proc. 1st International Conference on Bacterial Blight of Rice, Tsukuba, Japan*, March 17-19, 2004.

21. Gnanamanickam, S. S. and Mew, T. W. 1990. Biological control of rice diseases (blast and sheath blight) with bacterial antagonists: an alternative strategy for disease management. Pages 87-110, In *Pest Management in Rice*, B. Grayson, Ed, Elsevior Science Publishers, London, UK.

22. Gnanamanickam, S. S. and Mew. T. W. 1991. Interactions between *Sarocladium oryzae* and stem-attacking fungal pathogens of rice. *Plant and Soil* 138: 213-219.

23. Gnanamanickam, S. S. and Anuratha, C. S. 1991. Moko disease of banana. Pages 283-299 In *Plant Diseases of International Importance: Diseases of Fruit Crops, Vol. III*, J. Kumar, H. S. Chaube, U.S. Singh, and A. Mukhopadhyay, eds, Prentice Hall, Inc, New Jersey.

24. Gnanamanickam, S. S. and Mew, T. W. 1992. Biological control of blast disease of rice (*Oryza sativa* L.) with antagonistic bacteria and its mediation by a *Pseudomonas* antibiotic. *Ann. Phytopathol. Soc. Japan* 58: 380-385.

25. Gnanamanickam, S. S, and Sakthivel, N. 1996. Seed transmission of bacterial blight pathogen of rice. Pages 155-175. In '*Proc. National Seminar on Seed Pathology*. H. Shekar Shetty and H. S. Prakash, eds, University of Mysore Press.

26. Gnanamanickam, S. S, and Velusamy, P. 2003. Plant-Associated bacteria for biological control of rice diseases. *Souvenir, 6th PGPR Int'l*

Workshop, Oct, 2003, Calicut, India.

27. Gnanamanickam, S. S., and A. R. Podile. 2003. Overview of PGPR work in India. Pages 43-45 In *Proc. 6th International PGPR Workshop*, October, 2003, Calicut, India.
28. Gnanamanickam, S. S., Shigaki, T., Medalla, E. S., Mew, T. W, and Alvarez, A. M. 1994. Problems in detection of *Xanthomonas oryzae* pv, *oryzae* in rice seed and potential for improvement using monoclonal antibodies. *Plant Disease* 78: 173-178.
29. Gnanamanickam, S. S., Sakthivel, N., Alvarez, A.M., Benedict, A. A. and Leach, J. E. 1996. Serological and molecular probes for detection of seedborne bacterial pathogens of rice and lack of demonstrable evidence for seed transmission of *Xanthomonas oryzae* pv. *oryzae*. *Proc. IX Int'l Conf. On Plant Pathogenic Bacteria*, Aug., 1995, Chennai, India.
30. Gnanamanickam, S. S., Lavanya, B., Viji, G., Krishnamurthy, K., Dayakar, B. V., Brindha Priyadarisini, V., Sivaraj, R, and Levy, M. 2000. Lineage-exclusion resistance breeding strategy and pyramiding of genes for blast-resistance for the management of rice blast in India. Pages 195-198 *In Proc. Int'l Conf. on Integrated Plant Dis. Mgmt for Sustainable Agriculture*, Golden Jubilee, Indian Phytopathol. Soc, New Delhi.
31. Gnanamancikam, S. S., Lavanya Babujee, Sivaraj, R, and Levy, M. 2000. Genetic diversity of *Magnaporthe grisea* in India and pyramiding of genes for blast-resistance for the management of rice blast. Pages 28-35, *In Molecular Approaches in Plant Disease Management*, G. S. Shekawat, R. A. Singh and B. P. Singh, eds, Malhotra Publ. House, New Delhi.
32. Gnanamanickam, S. S., Lavanya Babujee, Brindha V. Priyadarisini, Dayakar, B. V., Leenakumari, S., Sivaraj, R., Levy, M and Leong, S. A. 2000. Lineage-exclusion resistance breeding: pyramiding of blast-resistance genes for management of rice blast in India. Pages 172-179 In *Advances in Rice Blast Research*, D. Tharreau, M. H. Lebrun, N. J. Talbot, and J-L. Notteghem, eds, Kluwer Academic Publ, Dordrecht, NL.
33. Gnanamanickam, S. S., Preeti Vasudevan, Reddy, M. S., Defago, G, and Kloepper, J. 2002. Principles of biological control. Pages 1-9, In *Biological Control of Crop Diseases*, S. S. Gnanamanickam, ed, Marcel Dekker Publishers, New York, 450 p.
34. Gupta V. S, Rajebhosale, M. D., Sodhi, M., Sukhwindersingh, Gnanamanickam, S. S., Dhaliwal, H. S. and Ranjekar, P. K. 2001. Assessment of

genetic variability and strain identification of *Xanthomonas oryzae* pv. *oryzae* using RAPD and IS112- based PCR. *Curr. Sci.* 80: 1043-1049.

35. Karthikeyan, V., and Gnanamanickam, S. S. 2005. A rapid and specific detection method for blast infection casued by *Setaria italica*. *Curr. Sci.* 88: 349-250.

36. Kavitha, S. 2002. *Strategies for management of rice blast and sheath blight with bacterial biocontrol agents in combination with major gene for disease resistance.* Ph.D. Thesis, University of Madras, 125 p.

37. Krishnamurthy, K. and Gnanamanickam, S. S. 1997. Biological control of sheath blight of rice: induction of systemic resistance by plant-associated *Pseudomonas* spp. *Curr. Sci.* 72: 331-334.

38. Krishnamurthy, K and S. S. Gnanamanickam. 1998. Biological control of rice blast by *Pseudomonas fluorescens* strain *Pf*7-14: Evaluation of a marker gene and formulations. *Biological Control* 13: 158-165.

39. Krishnamurthy, K and S. S. Gnanamanickam. 1998. Induction of systemic resistance and salicylic acid accumulation in *Oryza sativa*, L. in the biological suppression of rice blast caused by treatments with *Pseudomonas* spp. *World J. Microbiol. Biotechnol.* 14: 135-137.

40. Krishnamurthy, K and Gnanamanickam, S. S. 1998. Biocontrol of rice sheath blight with formulated *Pseudomonas putida*. *Indian Phytopath.* 51: 233-236.

41. Linde, C., Zala, M., David Paulraj, R. S., McDonald, B. A, and Gnanamanickam, S. S. 2005. Population structure of the rice sheath blight pathogen *Rhizoctonia solani* AG1-IA from India. *Euro. J. Plant Pathol.* 112:113-121.

42. Narayanan, N.N. 2001. *Molecular breeding for the development of transgenic rice resistant to blast and bacterial blight diseases.* Ph.D. Thesis, University of Madras, 108 p.

43. Narayanan, N., Dayakar, D. V, and Gnanamanickam, S. S. 2000. Evidence for possible genetic recombination in isolates of *Magnaporthe grisea* from Himachal Pradesh by DNA fingerprinting and plant inoculation assays. *Indian Phytopath.* 53: 480-482.

44. Narayanan, N., Baisakh, N., Oliva, N. P., Vera Cruz, C. M., Datta, K., Gnanamanickam, S. S, and Datta, S. K. 2002. Gene pyramiding through marker- assisted selection and transformation for blast and bacterial blight resistance in rice. (abs.) *3rd Int'l Rice Blast Conference*, 11-14 Sept, 2002, Tsukuba, Japan.

45. Narayanan, N. N., Baisakh, N., VeraCruz, C. M., Gnanamanickam, S. S., Datta, K, and Datta, S. K. 2002. Molecular breeding for the development of blast and bacterial blight resistance in rice cv. IR50. *Crop Science* 42: 2072-2079.

46. Narayanan, N. N., Baisakh, N., VeraCruz, C. M., Gnanamanickam, S. S., Datta, K, and Datta, S. K. 2004. Gene stacking strategy for developing transgenic indica rice [CO39] for blast and bacterial blight resistance. *Molecular Breeding* 14: 61-71.

47. Rajebhosale, M. D., Chowdari, K. V., Ramakrishna, W., Tamhankar, S., Gupta, V., Gnanamanickam, S. S. and Ranjekar, P. K. 1997. DNA fingerprinting of Indian strains of *Xanthomonas oryzae* pv. *oryzae*. *Theor. Appl. Genet.* 95: 103-111.

48. Rajebhosale, M. D., Gupta, V. S., Tamhangar, S. A., Gnanamanickam, S. S, and Ranjekar, P. K. 1998. A new range of DNA markers to evaluate genetic diversity in Indian isolates of *Xanthomonas oryzae* pv. *oryzae* and their potential in pathogen diagnostics. Pages 44-46, In *Proc. 4th Asia-Pacific Conference in Agricultural Biotechnology*, P. A. Larkin, ed, Darwin, Australia, July, 1998.

49. Sakthivel, N and Gnanamanickam, S. S. 1986. Isolation and assay for cerulenin produced by rice sheath-rot pathogen, *Sarocladium oryzae* (Saw) Gams. *Curr. Sci.* 55: 988-989.

50. Sakthivel, N. and Gnanamanickam, S. S. 1987. Evaluation of *Pseudomonas fluorescens* strains for suppression of sheath-rot disease and for enhancement of grain yields in rice, *Oryza sativa*, L. *Appl. Environ. Microbiol.* 53: 2056-2059.

51. Sakthivel, N. and Gnanamanickam, S. S. 1989. Incidence of different biovars of *Pseudomonas fluorescens* in flooded rice rhizospheres of India. *Agriculture, Ecosystem and Environment* 25: 287-295.

52. Savithiry, S. and Gnanamanickam, S. S. 1987. Bacterization of peanut with *Pseudomonas fluorescens* for biological control of *Rhizoctonia solani* and for enhanced yield. *Plant and Soil* 102: 11-15.

53. Sivamani, E. and Gnanamanickam, S. S. 1988. Biological control *of Fusarium oxysporum* f.sp. *cubense* in banana by inoculation with *Pseudomonas fluorescens*. *Plant and Soil* 107: 3-9.

54. Sivaraj, R. 1995. *Genetic structure and pathotype organization of rice blast fungus Pyricularia grisea Sacc in Southern India*. Ph.D. Thesis, University of Madras, 121 p.

55. Sivaraj, R., Gnanamanickam, S. S. and Levy, M. 1996. Studies on the genetic diversity of *Pyricularia grisea*: a

molecular approach for management of rice blast. Pages 958-962, In *Rice Genetics III* (Proc. 3rd Int'l Rice Genetics Symp., October 1995), IRRI publication, Manila.

56. Sivaraj, R., Gnanamanickam, S. S, and Levy, M. 2000. Lineage-exclusion tests for blast resistance in southern India. Pages 154-161, *In Advances in Rice Blast Research*, D. Tharreau, M. H. Lebrun, N. J. Talbot, and J-L. Notteghem, eds, Kluwer Academic Publ, Dordrecht, NL.

57. Taheri, P., Gnanamanickam, S. and Hofte, M. 2007. Characterization, genetic structure, and pathogenicity of *Rhizoctonia* spp. associated with rice sheath diseases in India. *Phytopathology* 97: 373-383.

58. Thara, K. V. 1994. *Biological control of rice sheath blight by bacterial antagonists: Mechanisms of disease suppression.* Ph.D. Thesis, University of Madras, 108 p.

59. Thara, K. V. and Gnanamanickam, S. S. 1994. Biological control of rice sheath blight in India: lack of correlation between chitinase production by bacterial antagonists and sheath blight suppression. *Plant and Soil* 160: 277-280.

60. Unnamalai, N. and Gnanamanickam, S. S. 1983. *Pseudomonas fluorescens* is an antagonist to *Xanthomonas citri* (Hasse) Dye, the incitant of citrus canker. *Curr. Sci.* 50: 703-704.

61. Valasubramanian, R. 1994. *Biological control of rice blast with Pseudomonas fluorescens Migula: Role of antifungal antibiotic in disease suppression.* Ph.D. Thesis, University of Madras, 111 p.

62. Vasantha Devi, T., Malarvizhi, R., Sakthivel, N. and Gnanamanickam, S. S. 1989. Biological control of sheath blight of rice in India with bacterial antagonists. *Plant and Soil* 119: 121-128.

63. Vasudevan, P. 2002. *Isolation and characterization of Bacillus spp. from the rice rhizosphere and their role in biological control of bacterial blight of rice caused by Xanthomonas oryzae* pv. *oryzae.* Ph.D Thesis, University of Madras, 129 p.

64. Vasudevan, P., Kavitha, S., Brindha Priyadrisini, S., Lavanya Babujee, and Gnanamanickam, S. S. 2002. Biological control of rice diseases. Pages 11-32, *In Biological Control of Crop Diseases*, S. S. Gnanamanickam, ed, Marcel Dekker Publishers, New York, 450 p.

65. Vasudevan, P, and Gnanamanickam, S. S. 2002. Suppression of rice bacterial blight and enhancement of plant growth by *Bacillus* spp. (abs.) *Int'l Rice Congress,* 16-20 Sept., 2002, Beijing, China.

66. Venkatesan, Brindha and Gnanamanickam, S.S. 1999. Occurrence of a sub-population of *Xanthomonas oryzae* pv. *oryzae* with virulence to rice cultivar IRBB21 (*Xa21*) in southern India. *Plant Dis.* 83: 781.

67. Viji, G. and Gnanamanickam, S. S. 1996. Biological control of blast disease of finger millet (*Eleusine coracana* (L.)) and an analysis of fertility of *Magnaporthe grisea. Curr. Sci.* 71: 144-147.

68. Viji, G, and Gnanamanickam, S. S. 1996 (1999). Effect of cultural factors on the formation of perithecium in *Magnaporthe grisea. Kavaka* 24: 51-56.

69. Viji, G. and Gnanamanickam, S. S. 1998. Mating type distribution and fertility status of *Magnaporthe grisea* from different hosts in India. *Plant Disease* 82: 36-40.

70. Viji, G., S. S. Gnanamanickam, and M. Levy. 2000. DNA polymorphisms of isolates of *Magnaporthe grisea* from India pathogenic to finger millet and rice. *Mycological Res.* 104: 161-167.

71. Velusamy, P. and Gnanamanickam, S. S. 2003. Identification of 2,4-diacetylphloroglucinol production by plant-associated bacteria and its role in suppression of rice bacterial blight in India. *Curr. Sci.* 85: 1270-1273.

72. Velusamy, P. and Gnanamanickam, S. S. 2003. Plant-associated bacteria, 2,4-diacetylphloroglucinol (DAPG) production and suppression of rice bacterial blight in India. *National conference on 'Biological Control of Crop Diseases: Present Perspectives and Future Challenges'* at Centre for Advanced Studies in Botany, University of Madras, Guindy Campus, Chennai, India. March 5-6, 2003 (abst.).

73. Velusamy, P., Defago, G., Thomashow, L.S. and Gnanamanickam, S. S. 2004. Role of 2,4-diacetylphloroglucinol (DAPG) for plant disease control: its importance to rice bacterial blight suppression in India. Pages 182-191 In *Biotechnological Approaches to the Integrated Management of Crop Diseases*, C. D. Mayee et al (eds), Daya Publishing House, New Delhi.

74. Velusamy, P., Ebenezar Immanuel, J., Gnanamanickam, S.S and Thomashow, L.S. 2006. Biological control of rice bacterial blight by plant associated bacteria producing 2,4-diacetylphloroglucinol. *Can. J. Microbiol*. 52: 56-65

MANAGING PLANT DISEASES IN ORGANIC FARMING

Anju Bala, T.S. Thind, S. Gowswami and S. Raheja

Department of Plant Pathology, Punjab Agriculture University, Ludhiana
E-mail: anjuchani@yahoo.co.in

KEY WORDS: Plant diseases, organic farming, sustainability

ABSTRACT

Organic farming is gaining wide attention among farmers, entrepreneurs, policy makers and agricultural scientists. Disease management in organic farming begins by taking right decision at right time such as growing crops that are naturally resistant to diseases or choosing sowing times that prevent disease outbreak. Use of resistant and tolerant varieties, selection of pathogen free site and application of biocontrol agents can help to reduce diseases caused by pathogens like *Armillaria, Fusarium, Sclerotium, Rhizoctonia, Verticillium* etc. Seed borne pathogens like *Xanthomonas compestries* pv. *oryzae, Ustilago segetum* var. *tritici*, MLOs etc. can be controlled through the use of heat therapy. Decomposition products of cover crops like brocoli, mustard, sudan grass etc. help to control many soil borne diseases. Excessive irrigation should be avoided to prevent damping off of seedlings. Soil solarization helps to kill many fungal, bacterial and nematode propagules as well as weed seeds. Field sanitation and use of net house, poly house, green house and destruction of infected plant debris help to exclude the pathogen. Safer chemicals like, some sulphur and copper based compounds can be used when we expect a disease to appear. Organically approved bicarbonate salts and peroxide materials are also available for suppression of diseases on plant surface. Plant oils, extracts, bio-regulators are being used to control many plant diseases. Correct disease diagnosis at the right time and record keeping of field history help to manage disease problems in organic systems.

The green revolution technologies involving greater use of synthetic agrochemicals such as fertilizers and pesticides with adoption of nutrient responsive, high – yielding varieties of crops have boosted the production output per hectare in most of the cases. However, priorities in agriculture research are gradually moving from a focus on individual crop performance to total system productivity with due attention on product quality and environment safety. Environmental and health problems associated with agriculture have been increasingly well documented. As a result more and more emphasis is being given towards returning to nature and adoption of organic agriculture. Organic farming is not new to Indian farming community. Several forms of organic farming are being successfully practiced in diverse climate, particularly in rain fed, tribal, mountains and hilly areas of the country. Much of the forest produce of economic importance like herbs, medicinal plants, etc., by default come under this category. Among all farming systems, organic farming is gaining wide attention among farmers, entrepreneurs, policy makers and agricultural scientists for varied reasons such as it minimizes the dependence on chemical inputs (fertilizers; pesticides; herbicides and other agro-chemicals) thus safeguards/improves quality of resources, and environment, it is labour intensive and provides an opportunity to increase rural employment and achieve long term improvements in the quality of resource base. Organic production systems are based on specific standards precisely formulated for food production and aim at achieving agro ecosystems, which are socially and ecologically sustainable. Many definitions have been proposed for organic agriculture with primary focus on ecological principles as the basis for crop production and animal husbandry.

Codex Alimentarius Commission, a joint body of FAO/WHO defines "organic agriculture as holistic food production management systems, which promotes and enhances agro-ecosystem health, including biodiversity, biological cycles and soil biological activity. It emphasizes the use of management practices in preference to the use of off-farm inputs, taking into account that regional conditions require locally adapted systems. This is accomplished by using, where possible, agronomic, biological and mechanical methods, as opposed to using synthetic materials, to fulfil any specific function within the system". In its simplistic form organic agriculture may be defined as a kind of diversified agriculture wherein crops and livestock are managed through use of integrated technologies with preference to depend on resources available either at farm or locally.

The popularity of organic farming is gradually increasing and now organic agriculture is practiced in almost all countries of the world, and its share of agricultural land and farms is growing. As per a recent report of International Federation of Organic Agriculture Movements (IFOAM) the total organically managed area is more than 24 million hectares world-wide. Organic farming is practiced in approximately 130 countries of the world and the area under organic management is continually growing. The market for organic products is growing, not only in Europe and North America but also in many other countries. The demand for organic food is steadily increasing both in developed and developing countries, with annual average growth rate of 20-25% .The countries with the largest areas of organic farmland are: Australia, Argentina, Italy, Canada and USA. Some countries have reached a substantial proportion (close to or more than 10%) of organic land; these include Sweden, Austria, Switzerland, Finland and Italy (3).

Disease management in organic cultures

Disease management in organic farming begins by taking right decision at right time such as growing crops that are naturally resistant to diseases or choosing sowing times that prevent disease outbreaks. Other methods generally employed for the management of diseases are clean cultivation, improving soil health to resist soil pathogens and promote plant growth, encouraging natural biological agents for control of diseases. All such methods come under preventive crop protection methods. Some of these methods are listed here:-

- Use of adapted and resistant varieties
- Site selection and appropriate soil cultivation

- Use of pathogen free seed and planting material
- Selection of suitable cropping pattern
- Input of organic matter
- Efficient water management
- Sanitation measures
- Organically approved protectant chemicals
- Use of oils and plant extracts
- Use of bio agents

The features and benefits of the above mentioned methods in managing disease problems in organic farming are discussed below:

Resistant and tolerant varieties

Resistance is plant hosts ability to suppress the activity and progress of the pathogen, which results in the absence/reduction of symptoms whereas tolerant plants can endure severe disease without suffering significant losses in quality or yield. However these tolerant plants do not significantly inhibit the pathogen's activity and disease symptoms may be clearly evident. Such selections are completely non-disruptive to the environment and their use enables growers to reduce and in some cases eliminate the application of chemicals used for pathogen control (12).

The greatest shortcoming is that resistance is not available for all diseases on all crops e.g. tomato late blight and white rot of Alliums, acceptable no resistant cultivars are yet available. In addition, commercial seed companies and plant breeders rarely invest in efforts to develop resistant cultivars for specialty and minor crops. Sometimes long development time and high market demand make the seed expensive and that will affect farmer budgets. Another shortcoming is that some resistant cultivars lack adequate horticultural characteristics in regard to appearance, quality, colour, yield, and other important criteria. Celery resistant to *Fusarium oxysporum* f. sp. *apii* is unacceptably ribby, short, or low yielding. A final disadvantage to resistance is that, depending on host pathogen systems, resistance is not long - lasting and new strains of the pathogen readily develop, making the crop susceptible once again. Despite the challenges of developing resistant cultivars and setbacks of resistance breakdown, resistant plants remain among the most important weapons for disease control in organic systems. Thus organic growers are encouraged to actively and thoroughly investigate which resistant cultivars are available and to test to determine which cultivars perform best under their particular growing conditions.

Selection of pathogen free site

A grower can incur significant losses if he plants susceptible crops in a field known to be infested with persistant soil borne pathogens. Plant pathogenic fungi such as *Armillaria, Fusarium, Plasmodiophora, Sclerotium, Rhizoctonia* and *Verticillium* are true soil inhabitants and will persist in soil for many years, even in the absence of a plant host. A grower is advised to select a planting site away from such fields. Pastures, foothills, river-banks, grasslands and areas supporting weeds and natural vegetation often are reservoirs of pathogens that cause virus or virus like diseases. For example the aster yellows phytoplasma and its leafhopper vector can be found in weedy grasslands.

Consider pertinent environmental factors while selecting a planting site. Crops planted near river or other water bodies tend to be more at risk from downy mildew diseases as a result of the increased and persistent humidity. An understanding of soil factors is critical in avoiding some root and crown diseases. A site that has well drained sandy soils reduces the risk of damping off and root rot for sensitive crops like spinach. As a grower one should keep notes on previous soil borne disease problems associated with certain fields, position of fields in relation to other key areas (weed reservoirs), the environmental characteristics of importance for each location and the nature of soil, water, and other physical features of each site.

Pathogen free seed and planting material

For some diseases, seed borne pathogens are a primary means of pathogen dissemination. Grower should purchase seed that has been tested and certified to be below a certain threshold infest-ation levels. If a grower produces or purchases transplants, they too should be as free as possible of pathogen contamination. Seed should be disinfected with heat treatments. Seedlings/ transplants should be inspected for disease symptoms before planting in field and control disease, if appears, in seedlings in nursery instead of waiting till they are in field. Some of the seed borne diseases and their control suggestions are listed in Table 1.

Table 1. Important seed borne diseases and their possible control measures

Crop	Disease	Pathogen	Control measure	Reference
Rice	Bacterial blight	*Xanthomonas campestris* pv. *oryzae*	Soaking seed at room temp. for 12 hours followed by hot water treatment at 53°C for 30 min.	(30)
Wheat	Loose smut	*Ustilago segetum* var. *tritici*	Solar heat treatment	(23)
Bajra	Ergot	*Claviceps fusiformis*	Soaking seed in 1% common salt solution	(27)
Sugarcane	Grassy shoot and mosaic	MLOs and Virus	Hot air treatment	(24)
Grapevine	Petri diseases	*Phaeoacremonium* sp.	Hot water treatment at 51 °C	(8, 11)

Adoption of cropping system

Selected cropping system should provide genetic diversity. Diverse crops, cover crops and fallow period reduce inoculum level. Decomposition products of cover crops like Broccoli, mustard, sudan grass releases natural chemicals that reduces number of *Verticillium* microsclerotia. Avoiding disease buildup is probably the most widely emphasized benefit of crop rotation. Many diseases build up in the soil when the same crop is grown in the same field year after year. Rotation to a non-susceptible crop can help break this cycle by reducing pathogen levels. To be effective, rotations must be carefully planned. Since diseases usually attack plants related to each other, it is helpful to group vegetable rotations by family e.g., nightshades, alliums, cole crops, cucurbits. The susceptible crop, related plants, and alternate host plants for the disease must be kept out of the field during the rotation period. Since plant pathogens persist in the soil for different lengths of time, the length of the rotation will vary with the disease being managed. To effectively plan a crop rotation, it is essential to know what crops are affected by what disease organisms (31). Rotation periods to control some of the vegetable diseases have been summarized in Table 2.

Table 2. Rotation period to reduce vegetable soil borne diseases

Vegetable	**Disease**	**Years**
Asparagus	Fusarium rot	8
Beans	Root rots	3-4
Cabbage, Radish	Clubroot	7
Cabbage	Blackleg	3-4
Cabbage	Blackrot	2-3
Muskmelon, Peas	Fusarium wilt	5
Parsnip	Root canker	2
Peas	Root rots	3-4
Pumpkin	Black rot	2

Input of organic matter

Organic soil amendments and conditions traditionally used by organic growers as part of soil fertility programmes also have potential against a variety of disease problems (21). Organic amendments containing high nitrogen such as poultry manure, peat, bone meal and soymeal significantly reduce populations of a wide spectrum of soil borne plant pathogens because of the increased production of ammonia and (or) nitrous acid (20). Management of macro and microelement nutrition plays a significant role in reducing soil–borne pathogens in organic production systems (15) Compost has been used effectively in the nursery industry, in high value crops, and in potting soil mixtures for control of root rot diseases. It increases soil microflora diversity and population and gives the benefit of fertility and tilth. Some of the successful examples of disease suppression by composting are listed in Table 3.

While compost and farmyard manure are added in decomposed form, the materials in green manure and special organic amendments are added in undecomposed form and decomposition takes place in the field soil where pathogens survive. Organic amendments are one way of promoting naturally occurring biological control of pathogens in the soil. The reduction in common scab by green manuring with soybean through prevention of build up of inoculum was the first report, in 1926 providing experimental evidence of benefits of organic amendments. Partial list of diseases in which organic amendments have caused suppression of the pathogen are listed in Table 4 (29).

Table 3. Disease management through compost treatment

Crop	Pathogen/Disease	Treatment	Comments/Reference
Alfalfa	"Clover tiredness"	Four years of treating fields with high quality compost	Stand thickness and yield doubled (22)
Wheat/ Barley	Powdery mildew (*Erysiphe graminis*)	Compost added to soil	Disease incidence suppressed 95% when1:1 soil:compost mix used (34)
Beans	*Rhizoctonia* sp.	Compost added to soil at varying rates (36-72 tons/ acre)	Disease reduced 80% in areas with highest compost rates, 40% where intermediate rates were used (14)
Cucumber	*Sphaerotheca* sp./ Powdery mildew	Young cucum-ber plants grown in soil/ compost mix of variable rates.	1:1 soil: compost mix decreased PM by 20% over control, 1:3 mix decreased infection by 40% (34)
Pea	*Pythium* sp./ Damping off	Seed treatment; seeds soaked in dilute compost extract, dried before sowing	(34)
Peppers, Soybean	*Phytophthora* sp.	40 tons of compost per acre	Reduced *Phytophthora* incidence (14)

Table 4. Pathogens suppressed by organic amendments of soil

Pathogen	Host/ Disease	Amendment
Aphanomyces euteiches	Pea root rot	Crucifer, green manure
Fusarium oxysporum f.sp .*cubense*	Banana wilt	Sugarcane bagasse
Phytophthora cinnamomi	Avacado root rot	Lucern, alfalfa meal
Fusarium oxysporum f.sp.*corianderi*	Coriander wilt	Oil-cake

Gaeumannomyces graminis	Wheat take all	Rape, Pea green manure
Thielaviopsis basicola	Bean root rot	Oat straw, Corn stover, Lucerne hay
Phymatotrichum omnivorum	Cotton root rot	Green manure with pea
Macrophomina phaseolina	Cowpea charcoal rot, Cotton root rot	Margosa oil cake, barley straw, Lucerne hay
Rhizoctonia solani	Potato black scurf	Sawdust, oil cakes Wheat straw
Streptomyces scabies	Potato scab	Soybean, Green manure
Verticillium albo- atrum	Potato wilt	Barley straw
Meloidogyne spp.	Root knot	Oil cakes, Saw dust, green leaves

Water and irrigation management

The aim of irrigation should be to wet the soil only to the extent that roots easily get water and mineral nutrients. Timing and duration of irrigation should satisfy crop water requirement without allowing excess water. The disadvantage of excess irrigation and prolonged stay of water around root zone are many. The juvenile stage of seedlings is lengthened making them susceptible to damping off. Frequent but low quantity of watering is therefore recommended for avoiding damping off. Sprinkler irrigation should be avoided to prevent foliar diseases (29). Water draining from fields should be directed away from greenhouse and irrigation pond.

Soil management

Shaping top soil into ridges or beds is a part of tillage and is a well known and useful means of keeping dry the soil that is in direct contact with collar and root crown. Ridges are effective in reducing incidence of *Sclerotium rolfsii* on many vegetables and on groundnut, Pythium fruit rot of cucurbits and *Sclerotinia sclerotiorum* in some vegetables. Deep ploughing to turn the top soil and exposure to sun is one way of killing propagules of many pathogens present in the surface soil. Tillage should reduce plant residues left from previous crops. Good tilth is helpful for seed emergence and better plant stand where as poor soil preparation results in stressed plants and increase damping off problems.

Soil solarization is a novel technique for treatment of soil by solar heat. It is a hydrothermal treatment where moistened soil is covered under transparent polythene film during summer to prevent dissipation of the trapped solar heat, the temperature of the soil raises, sometimes 13°C higher than the temperature outside, reaching 50-55°C at 8 cm depth. This kills most of the fungal, bacterial and nematode propagules as well as weed seeds. It has been found very helpful to destroy soil borne pathogens like *Fusarium, Rhizoctonia, Verticillium, Sclerotium* etc (4).

Exclusion of pathogen through sanitation

These methods are aimed at preventing new pathogens and diseases from reaching an uninfected area and avoiding contact between the pathogen and the crop of field. Following points should be considered for this purpose:

1. Grow crop in enclosures e.g. in net house, polyhouse, green/ glass house.
2. Use new pots or treat old pots with steam, bleach or other disinfectant.
3. Do not recycle old potting mixture.
4. Avoid movement of contaminated soil from one field to other through shoes, wind and equipment.
5. Power wash the equipment while moving between fields.
6. Burn old plants and crop residue immediately after harvest to limit over-wintering of the pathogen.
7. Rouge off infected plants.

Use of safer chemicals

When we expect a disease to appear as when late blight is in the neighbourhood, or if scouting turns up the first signs of a disease like powdery mildew, then it may be necessary to apply a protectant material to prevent crop loss. Quite a few copper and sulfur-based compounds are allowed for organic use and they are relatively inexpensive and pose minimum environmental risk. Copper based compounds have some activity against a wide range of fungi and bacteria, but their effectiveness is limited and under high disease pressure probably will not provide excellent control. Sulfur products also provide some control of many pathogens, but they usually are excellent for managing only some diseases such as powdery mildews. Moreover both copper and sulfur compounds can burn sensitive crops under warm climates.

Organically – approved bicarbonate salts have recently become available. They have demonstrated good activity against powdery mildews and a few other diseases. Peroxide type materials are also available for general suppression of diseases on plant surface (36).

Plant oils, extracts and bio-regulators are the other natural products that can be used to control diseases in organic crops. Gibberellic acid (GA_3) @ 5 ppm have been reported to reduce tomato leaf curl incidence by 26% (18). Some of the successful examples of plant oils and extracts used in disease control are listed in Tables 5 and 6. Some of the commercial formulations of plant-based products used in diseases control are listed in Table 7 (32).

Table 5. Use of plant oils in disease control

Oils	Pathogen/Disease	Reference
Emulsified rape oil	Apple powdery mildew	(28)
Rape oil derivatives	Grape powdery mildew	(2)
Cinnamomum zeylanicum	*Aspergillus niger, A. flavus*	(33)
Tagetes erecta	*Pythiym aphanidermatum*	(11)
Eucalyptus oil	*Candida albicans*	(16)
Clove oil	*A. niger, Fusarium chlamydosporum, Alternaria alternata*	(7)

Table 6. Use of plant extracts in disease control

Plant extract	Pathogen	Reference
Datura innoxia	*Colletotrichum capsici*	(5)
Achyranthes	*Alternaria pisi*	(19)
Vinca rosea	*Pyricularia oryzae* *Helminthosporium oryzae*	(9)
Garlic	*A. flavus* *A. parasiticus*	(10)
Azadirachta indica	*Alternaria sesame*	(13)
Cabbage, Garlic, Alfalfa	*Phytophthora capsici*	(6)

Table 7. Commercial formulations of plant based products.

Formulation	Plant Base	Effective against	Manufacturers
Nimin	Neem	Club root	Agaricare, Bangalore
Agaricare	---do---	Black arm	Agaricare, Bangalore
Plantolyte	Neem	Black arm	Agaricare, Bangalore
Neemark	Neem	Black arm	West coast Rasayan Int. Pvt, Bombay
RD-9	Neem	Black arm	ITC Ltd, Hyderabad
NO-60 EC	Neem	Bacterial blight of French beans	TNAU, Coimbatore
Triact 70	Neem	Powdery mildew and Rust	Thermo triology, US
Milsana	Giant knot weed	PM of Cucumber	_
Funga -stop	Mint based	Soil borne pathogens	_

Table 8. Commercial formulations of bioagents used in plant disease control

Formulation	Bio-agent	Target disease/ pathogen
Gliogard	*Gliocladium virens*	Seedling diseases
F- Stop	*Trichoderma harzianum*	Soil borne fungi
BINAB-T	*T. harzianum T. polysporum*	Wood decay fungi
Gallex	*Agrobacterium radiobacter* K-84	*A. tumifaciens*
Galltrol	*Agrobacterium radiobacter* K-84	*A. tumifaciens*
Dagger G	*Pseudomonas fluorescens*	*Pythium* and *Rhizoctonia*
Kodiak	*Bacillus subtilis*	Seed borne diseases

Use of bio-agents

Other organic option for disease suppression are products that contain microbes or their by–products, including species of *Trichoderma, Gliocladium virens*, *Pseudomonas*, *Bacillus*, *Coniothyrium minitans* and non pathogenic strains of *Fusarium* and *Pythium* etc. Several examples of biocontrol of plant pathogens can be given. Common scab of potato grown in sterilized soil and inoculated with *Streptomyces scabies* can be controlled through simultaneous inoculation of soil with *S. praecox,* a vigorous saprophytic species (25). Common saprophytic fungus *Trichoderma viride* is able to parasitize the mycelia of other fungi (35). Some of the important books and review articles (1, 26) contains detailed information of use of bio-agents for disease control. A partial list of formulations of biocontrol agents available for disease control are given in Table 8 and 9.

Table 9: Commercial bio-formulations available in India

Product	Bioagent	Developing agency
Trichogaurd	*Trichoderma viride*	Anu Biotech. International Ltd., Faridabad.
Funginil	*T. viride*	Crop health Biproduct research centre Gaziabad.
Ecofit	*T. viride*	Hoechest and Schering Agr Evo Ltd., Mumbai.
Bioderma	*T. viride* + *T. harzianum*	Biotech. International Ltd., New Delhi
Ecoderma	*T .viride* + *T . harzianum*	Margo Biocontrol Pvt. Ltd., Bangalore.
Defence-SF	*T. viride*	Wockhardt Life Science Ltd., Mumbai.
Tricho-X	*T. viride*	Excel Industries Ltd. Mumbai.
Biogaurd	*T. viride*	Krishi Rasayan export pvt. Ltd., Solan.
Biocure-F	*T. viride*	T. Sanes Company Ltd., Coimbatore.
Biotok	*Bacillus subtilis*	Tocklai experimental station, Tea Research Association Jorhat, Assam.
Pant Biocontrol Agent-3	*Pseudomonas fluorescence*	G.B.Pant Univ. of Agric. & Tech., Pantnagar.
Biocure-B	*P. fluorescence*	T. Sanes Company Ltd., Coimbatore.
Bioshield	*P. fluorescence*	Anu Biotech.International Ltd., Faridabad.

CONCLUSION

To feed the future generations without degrading the resource base, agriculture must become economically viable and ecologically sustainable. Through organic farming, sustainability in production could be achieved. Disease management in organic farming is usually preventive rather than curative. Organic growers need to take appropriate and timely decision in order to ensure better management of different diseases. Apart from agronomically suitable disease tolerant varieties, seed treatment with safe chemicals and nursery bed management through integration of solarization and bioagents need to be ensured for getting proper germination and seedling stand. Disease management in the standing crop raised for organic farming can be achieved by following integrated approach through use of cultural and biological methods along with application of certain organically approved protectant chemicals. Correct disease diagnosis at right time and record keeping of field history help to reduce disease problems in organic systems.

REFERENCES

1. Adams P. 1990. The potential of mycoparasites for biological control of plant diseases. *Ann Rev Phytopath* 28 :59-72.
2. Azam MGN, Gaur GM, Magarey PA. 1998. Efficacy of compound based on canola oil as a fungicide for control of powdery mildew caused by *Uncinula necator*. *Aust Plant Pathol* 27: 116.
3. Bhattacharyya P. 2005. Salient features of organic farming. Organic Agriculture. *www.dacnet.nic.in.*
4. Chauhan YS. 1988. Effect of soil solarization on pigeonpea and chickpea wilt, *ICRISAT Res Bull* : 11 pp 16.
5. Chitra H and Kannabiran B. 2001. Antifungal effect of *Datura innoxia* on the anthracnose fungus *Colletotrichum capsici* in *in vitro*. *Adv Plant Sci* 14: 317-320.
6. Demirci F and Dolar FS. 2006. Effects of some plant materials on Phytophthora blight (*Phytophthora capsici* Leon.) of Pepper. *Turk J Agric* 30: 247-252.
7. Farrukh A, Beg AZ, Iqwal A. 2001. *In vitro* toxicity of plant essential oils against soil fungi. Proc Natl seminar on the Frontiers of research and development in medicinal plants, Lucknow, India 16-18 September 2000. *J Medicinal Aromatic Plant Sci* 22 : 177-181.
8. Fourie PH and Halleen F. 2004. Proactive control of Petri diseases of grapevine through treatment of propagation material. *Pl Dis* 88 (11): 1241-1245.
9. Ganguly L K. 1994. Fungi toxic effects of certain plant extracts against rice blast and brown spot pathogen. *Environ Ecol* 12: 731-733.
10. Garcia RP and Garcia ML. 1988. Laboratory evaluation of plant extracts for the control of *Aspergillus* growth

and aflatoxin production. *Proc Japanese Assoc Mycotoxicology* suppl no 1 pp 190-193.

11. Gramaje D, Garcia JJ and Armengol J. 2008. Sensitivity of Petri disease pathogens to hot water treatments *in vitro Annals of Applied Biology* 153 (1): 95-103.
12. Grubinger V. 2005. Organic vegetable disease management. *www.uvm.edu/disease management. html.*
13. Gularia S and Kumar A. 2006. *Azadirachta indica* leaf extract induces resistance in sesame against *Alternaria* leaf spot disease. *J Cell Mol Biol* 5: 81-86.
14. Hudson B D .1994. Soil organic matter and available water capacity. *J Soil Water Cons* 30 189-194.
15. Jones JP, Engelhard AW and Woltz SS. 1989 Management of fusarium wilt of vegetables and ornamentals by macro- and microelement nutrition. *In: A.W. Engelhard (ed.). Soilborne Plant Pathogens: Management of Diseases with Macro- and Microelements. Am Phytopathological Soc* pp. 18–32.
16. Khan M and Zubairy HN. 2001. Mycoses III: antimycotic activity of camphor eucalyptus oil and thymol. *Hamdard - Medicus* 44:17.
17. Kishore N and Dwivedi RS. 1991. Fungitoxicity of essential oils of *Tegetes erecta* L. against *Pythium aphanidermatum* Fitz. the damping off pathogen. *Flavour and Fragrance* 6: 291-294.
18. Kumar K, Shabnaz E, Razdan V K and Tewari A K. 2008. Effect of bio-regulators and plant extracts on the yield attributes and incidence of tomato leaf curl. *J. Mycol Pl Pathol* 38 (2): 175-177.
19. Lal HC, Upadhyay JP and Ojha KL. 1998. Management of Alternaria leaf blight of pigeonpea through plant extracts and chemicals of plant origin. *J. Appl Biol* 8: 80-85.
20. Lazarovits G. 2001. Management of soil borne plant pathogens with organic soil amendements. a disease control strategy salvaged from the past. *Can. J. Pl Pathol* 23 : 1-7.
21. Lazarovits G, Tenuta M, Conn K.L. 2001. Organic amendments as a disease control strategy for soil borne diseases of high value agricultural crops. *Australasian Pl Pathol* 30: 111-117.
22. Logsdon G. 1995. Using compost for plant disease control. *Farm Scale Composting*. JG Press, Inc. Emmaus, PA.
23. Luthra JC and Sattar A. 1934. Some experiments on the control of loose smut of wheat. *Indian J Agric Sci* 4:117-199.
24. Mehrotra RS and Aggarwal A. 2003. *Plant Pathology*, Second edition. Tata McGraw-Hill Publishing Company Ltd. New Delhi.

25. Millard WA and Taylor CB. 1927. Antagonism of micro-organisms as the controlling factor in inhibition of scab by green manuring. *Ann Appl Biol* 14: 202-215.
26. Mukhopadhyay AN. 1994. Biological control of soil borne fungal plant pathogens, current status, future prospects and potential limitations. *Indian Phytopath* 47:119-126.
27. Nene YL and Singh SD. 1976. Downy mildew and ergot of pearl millet. PANS. 22: 366-385.
28. Northover J and Schneider KE. 1993. Activity of plant oils on diseases caused by *Podosphaera leucotricha, Venturia ineaqualis* and *Albugo occidentalis. Plant Dis* 77: 152.
29. Singh RS. 2002. Introduction *to Principles of Plant Pathology.* Fourth edition, Oxford & IBH Publishing Co. Pvt. Ltd. New Delhi, pp 320.
30. Sinha SK and Nene YL. 1967. Eradication of seed borne inoculum of *Xanthomonas oryzae* by water treatment of paddy seeds. *Pl Dis Reptr* 51: 882-884.
31. Sullivan P. 2004. Sustainable management of soil borne plant diseases. *www. attra.ncat.org.*
32. Thind TS, Dhaliwal HJS, Arora JK and Mohan C. 2005. Role of plant products in eco-friendly management o plant diseases. In: *Diversity and Biotechnology* (M.K. Rai and Deshmukh S.K, (Eds). Scientific Publishers, Jodhpur, pp 404-434.
33. Tiwari R. 1997. Fungitoxicity of volatile constituents of some higher plants against predominant storage fungi. *Indian Phytopathol.* 50: 548-551.
34. Trankner A. 1992. Use of agricultural and municipal organic wastes to develop suppressiveness to plant pathogens. p. 35-42. In: *Biological Control of Plant Diseases: Progress and Challenges for the Future.* EC Tjamos, GC Papavizas, and RJ Cook (Ed). NATO ASI Series No. 230. Plenum Press, New York.
35. Weindling R. 1932. *Trichoderma lignorum* as a parasite of other fungi. *Phytopathol.* 22: 837-845.
36. Wszelaki AL and Miller SA. 2005. Determining the efficacy of disease management products in organically produced tomatoes. *Plant Health Progress* 10: 1094-1095.

PEARL MILLET DOWNY MILDEW: CHALLENGES AND PROSPECTS

Arun Kumar[1], *V.K. Manga*[1], *H.N. Gour*[2] *and A.K. Purohit*[3]

[1]Central Arid Zone Research Institute, Jodhpur - 342 003, India
[2]Former Head, Department of Plant Pathology, MPUAT, RCA, Udaipur-3130 01, India
[3]Transcience Transactions, Jodhpur, Rajasthan 342 001, India

KEYWORDS: Pearl millet, downy mildew, *Sclerospora graminicola*, pathogen, epidemiology, disease resistance, disease management

ABSTRACT

Downy mildew or 'green ear' is a very destructive disease of pearl millet in Asia and Africa wherever pearl millet is grown as a food and fodder crop. The disease is soil borne in nature and caused by a fungus *Sclerospora graminicola*. In the present paper current status of knowledge about the symptoms of disease, pathogen variability, epidemiology and disease management along with disease resistance has been discussed in detail. Recent approaches of disease management through biological control of the disease have been discussed at length with particular emphasis on the use of BABA, cerebrosides, enzymes, vitamins, botanicals and other biocontrol agents like milk and *Gliocladium virens*. Advances made in the field of developing disease resistant varieties using various breeding strategies and molecular approaches have also been discussed in detail.

INTRODUCTION

Pearl millet [*Pennisetum glaucum* (L.) R. Br.] is an important cereal and forage crop of arid and subtropical regions of the Indian subcontinent as well as several African regions. It is believed that it was domesticated for not less than 3000 years ago as a forage or cereal crop in Africa. India is considered to be the secondary center of diversity (2). Pearl millet is predominantly a rainy season crop able to thrive well in the rainfall as low as 250 mm on relatively poor soils. Being most tolerant to drought and salinity, the crop is by and large grown in different countries of the world such as India, Nigeria, Chad, Tanzania, Mali, Niger, Ethiopia, China and Russia (16, 19, 56, 73, 120) As a warm season crop it has a great potential as a food source in the temperate zones also.

The crop is also known by several different common names, including bajra in Hindi, bulrush, spiked millet and cattail in English, mil chandelles in Arabic and dukhn in French. Areas planted with pearl millet are estimated at 15 million hectares annually in Africa and 14 million hectares in Asia. India alone accounts for 10.3 million ha with a total annual production of 7 million tonnes (87).

Downy mildew (DM) or 'green ear' disease caused by *Sclerospora graminicola* (Sacc.) Schroet. occurs most destructively in Asia and Africa (8, 57, 61, 87). The disease was first reported in India as the disease of ill-drained lands where it developed into epidemics of severity (17). Other published reports include detection in the temperate and tropical areas of the world, especially in Europe, Australia and various other countries of the world (Table 1). However, there is no report of its occurrence in the western hemisphere (47, 89). The disease was considered of minor importance till 1970, as its incidence was sporadic on local cultivars. The first epidemic of downy mildew occurred in 1971 on the first popular pearl millet hybrid, HB 3, resulting in severe grain loss of about 4.6 million metric tonnes (87, 89).

Because of continued large-scale cultivation of the susceptible hybrids, the disease caused serious epidemics during 1974, 1984, 1987 and 1988. Availability of data on grain loss is incomplete due to greater variability from field to field, farmer to farmer, and from season to season.

Table 1. Geographical distribution of *Sclerospora graminicola* on pearl millet

Continent	Country	Reference
Africa	Chad, Egypt, Gambia, Malawi, Mozambique, Niger, Nigeria, Senegal, South Africa, Zimbabwe, Burkina Faso, Ghana, Ivory Coast, Kenya, Mali, Sudan, Tanzania, Togo, Uganda, Zambia	Nene and Singh, 1976; Singh *et al.*, 1993, Singh, 1995, Wilson, 1999.
Asia	India, Pakistan, Israel	

On the basis of a few localized estimates in India the average annual yield losses can reach up to 40%, whereas 10-50% losses have been reported from Nigeria. However, worldwide annual grain yield losses may not exceed 20% (38, 57, 89, 119). Information on extent of losses caused by DM in pearl millet is rather scanty. Severe and systemic infection reduces fresh weight of the shoot along with the number of basal and nodal tillers (32).

In India, high yielding single cross F_1 hybrid cultivars based on A_1 cytoplasmic-nuclear male-sterility (CMS) with good tillering ability and a large number of compact, well filled ear heads were introduced in the mid-sixties but the desired production potential could not be attained due to the downy mildew pathogen (8, 87, 108). In a few states of India, yield losses caused by re-occurrence of DM in farmers' fields in the last 40 years led to the withdrawal of even promising cultivars (Fig. 1) that succumbed to the disease (86, 92). Since all the earlier hybrids had A_1 cytoplasm as source of male sterility, it was assumed that this cytoplasm is the cause of downy mildew susceptibility. However, it has been confirmed now that cytoplasm alone is not the cause, as different male sterile (ms) lines bred in same cytoplasm have been found to be resistant, hence ms lines with different genetic make up and resistance to downy mildew can be bred in the A_1 cytoplasm. Hybrids affected with more than 20% mean DM incidence have been proposed to be removed from production in recently concluded on-farm DM surveys conducted during 1994-2004 in four states of India (72, 113). It has been further observed that in spite of a progressive increase in the release of pearl millet hybrids no widespread epidemics of the disease have occurred in the past 10 years.

Information on downy mildew is available in the form of published reviews (8, 57, 87, 89, 94, 119). The present paper thus outlines the recent advances made after 1995 on this disease.

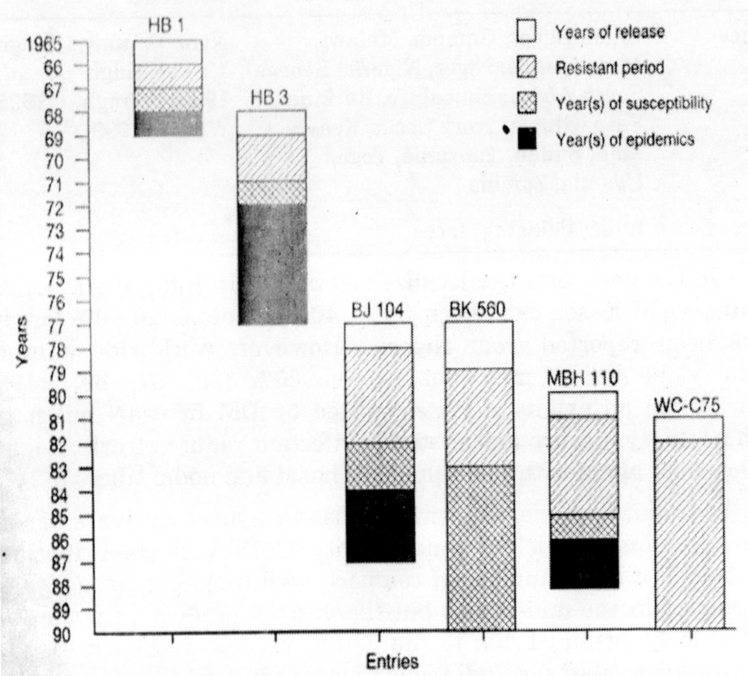

Fig. 1. Graphical presentation of six popular cultivars of pearl millet grown in India showing relative susceptibility and resistance periods, 1965-90 (Source: Singh *et al.*, 1997).

DISEASE SYMPTOMS

In this disease two types of symptoms are produced viz: 'downy mildew' and 'green ear'. Symptoms often vary according to host, time of their expression and ambient conditions (46). The variation in symptoms may occur due to systemic infection. Downy mildew infected plants develop severe disease syndrome from the very beginning and succumb even before reaching maturity. Reductions in plant height, number of leaves and nodes are commonly observed in susceptible cultivars resulting in reduced grain and fodder yields.

The disease normally appears in the form of chlorosis at the base of infected leaves followed by production of sporulation on the lower side of leaves known as the 'half-leaf' symptom. Abundant white asexual sporulation on the lower leaf surface of an infected chlorotic leaf produce white downy growth (Fig. 2) under high relative humidity (>95%) and moderate temperature (20-22°C). Subsequently, the leaves turn reddish brown and dry due to oospore production. In acute cases, shredding of leaves occurs. Severely infected plants (up to 60%) are generally stunted because of root retardation and do not produce panicles.

Fig. 2. Downy mildew infected plant showing asexual sporulation as half-leaf symptom.

At the time of panicle emergence, green ear symptoms become visible (Fig. 3). Symptoms appear on ear head with all possible degrees of proliferations and malformations. In malformation, the florets are converted into leafy structures of diverse appearance (sometimes also referred as virescence). The leafy structures are chlorotic and sometimes produce sporulation. Primary source of infection are from soil and from systemically infected seeds. The invasion of the fungus in floral primordia plays a crucial role in deciding the extent of malformation. Generally four types of malformations have been observed: a) the entire inflorescence is transformed into a green leafy tuft, which is symbolically referred as 'green ear'; b) the lower half of the ear-head is proliferated and the upper half contains normal flowers; c) only bristles become long and no malformed leafy structures are formed, and d) leafy tufts or frenching occurs where the shoots remain stunted and produce leafy tufts at the top (Fig. 4). In latent infections, green ear is the only manifestation of the disease.

Fig. 3. Green ear symptom at panicle emergence

Fig. 4. Green ear heads showing types of malformations with a healthy ear head (extreme left)

In all the diseased plants, as a rule, normal ears when outgrow the disease at primary shoots produce green ear followed by tufting and yellowing of leaves on the secondary shoots in successive downward order. This shows that leaf yellowing is an expression of the highest susceptibility, followed by tufting and green ear (Arun Kumar and Bhansali, unpublished).

Manifestation of changes in the infected flowers

The bristles subtending the spikelets and all its parts: the glumes, palea, pistil and stamens are affected due to the disease and present varied pattern of suppression, proliferation and malformation. As a result of these changes the ears appear green and conspicuous. The stamens and pistil are generally transformed into leafy structures although sometimes all parts may be transformed. The pistil is most commonly affected and is hypertrophied into a long (5.0 to 7.5 cm), thick, curled (on drying) structure, while the stamens remain normal. The stamens are also transformed into varied proportions and turn leafy in structure. Occasionally the glumes are hypertrophied and grow larger than other floral parts.

PATHOGEN DESCRIPTION

There is no disagreement on the taxonomy of the downy mildew causing fungus. *Sclerospora* is now monotypic, the sole representative of Ito's eusclerospora group. The very name *Sclerospora* is drawn from the thick-walled oospore with its dark-walled exosporium and adherent oogonial envelope. It appears that there are remote chances of a change in the name of *Sclerospora graminicola* (57, 100). Phylogenetic relationship among sequences from Indian and African samples of *S. graminicola* was examined using a nested PCR technology. All the samples from India, Mali, Nigeria and Niger, with the exception of Niger 4, formed a monophyletic group with the oomycetes (118). Diagnosis of the fungus is based on two types of spores produced by *S. graminicola*. Asexual spores (sporangia, zoospores) are thin walled, hyaline, ellipsoid or elliptic and papillate measuring 15-22 x 12-21 µm. Under natural conditions sporangia are produced in abundance during the night (03.00 h). Normal temperature (20 to 25°C) and high relative humidity (95-100% RH) favour sporangial production (34, 89). Sporangia are actively ejected and germinate immediately, producing zoospores, or else die within a few hours. Under suitable conditions sporangia form in great abundance on the under surface of the diseased leaf (and, when conditions are favourable, also on the upper surface) forming a conspicuous and characteristic white 'downy' growth (26).

Oospores are the sexual spores, which can survive for 14 years under laboratory conditions, but their viability is reduced after 4 years of storage. Oospores germinate in multiple ways; by vesicle-like structures, by both vesicles and germ tubes, by typical irregular structures different from germ tubes and vesicles, by germ tubes and by extrusion of small round bodies/sporangia-like structures (50). The optimum temperature for oospore germination is 28±2°C. Highest germination is observed in sterile distilled water after 24 h. Dry oospores are more resistant to steam sterilization than wet oospores. To ensure the total destruction of oospores from soil, sterilization of wet soil for more than 2 h is needed (90). Oospore density in the soil is highly correlated with disease incidence at 90 days after sowing, indicating the importance of oospores in disease epidemiology (28). The oospores play an important role as far as transmission of the disease is concerned. Unsterilized seeds produce 23.4% disease incidence when sown in sterilized soil. This suggests that seed borne spores are important in inciting downy mildew (55, 80). However, there is no significant correlation between the mean incidence of downy mildew and oospore production (71).

PATHOGEN VARIABILITY

S. graminicola reproduces asexually by means of sporangia and sexually through oospores. The fungus is largely heterothallic with two mating types. These characteristics of the fungus make it highly variable and adaptable to diverse environmental conditions. Similarly, its host pearl millet is a highly outcrossing crop species. The information generated for the past few years on genetics of resistance, availability of host differentials, and development of molecular techniques, has made it easier to understand *S. graminicola*- pearl millet interactions.

Single cross F_1 hybrids have greatly contributed to increasing productivity of pearl millet. Early maturity, uniform crop stand and high harvest index of these hybrids have made them popular among farmers. As a result hybrid cultivars cover about 55% of the total 10 million ha area under pearl millet in India with the cultivation of around 50 hybrids (105, 108). Several races or pathotypes of *S. graminicola* have now evolved in India. Stability of resistance in pearl millet lines developed at ICRISAT was studied through a collaborative International Pearl Millet Downy Mildew Virulence Nursery (IPMDMVN). The reactions to downy mildew of 11 pearl millet lines at 17 locations in India, Burkina Faso, Mali, Niger, and Nigeria from 1995 to 1999 were recorded (42, 107, 108, 111, 112). Seven pearl millet lines (IP 18292, IP 18293, 700651, P310-17, P7-4, MBH 110 and 852B) provided differential reactions permitting classification of the 23 populations into 15 putative pathotypes at the global level (Table 2). The existence of a highly virulent pathotype of *S. graminicola* is reported from Jodhpur, India (111).

Table 2. Populations of S. graminicola from 23 locations in seven countries identified into 15 putative pathotypes on seven host differential lines, based on field testing during 1992 to 1999.

Location/ pathogen group	Host differential line							
	IP 18292	IP 18293	700651	P310-17	P7-4	MBH110	852B	Putative patho-type
Bambey	R	R	R	R	R	R	R	1
Jamnagar	R	R	R	R	R	R	R	
Gwalior	R	R	R	R	R	R	R	
Bengou	R	R	R	R	S	S	S	2
Ludhiana	R	R	R	R	S	S	S	
Kamboinse	S	S	R	S	S	S	S	3
Durgapura	S	S	R	S	S	S	S	
Jodhpur	S	S	R	S	S	S	S	
Mysore	R	R	S	S	S	R	S	4
Pune	R	R	S	S	S	R	S	
Dhule	R	R	S	S	S	R	S	
Bagauda	S	S	S	S	S	S	S	5
Cinzana	S	S	S	S	S	S	S	
Patancheru	R	R	S	R	S	R	R	6
Hisar	R	R	R	S	R	R	S	7
Jalna	R	R	S	R	R	S	R	8
Coimbatore	R	R	R	R	S	S	R	9
Aurangabad	R	R	S	R	R	R	S	10
Nioro-du-Rip	S	R	R	R	R	R	S	11

Samanko	R	S	R	S	R	S	S	12
Sadore	S	R	R	S	S	S	S	13
Mandor	S	R	S	S	S	S	S	14
El-Obeid	R	R	R	R	R	S	S	15

In a recent study on the inheritance of avirulence in *S. graminicola* by hybridizing the isolates Sg 139-4 (Mat A), highly virulent, and Sg 110-9 (Mat B), avirulent, on a pearl millet genotype IP 18292 containing a downy mildew resistance gene Rsg1., it was observed that avirulence was dominant over virulence and a single gene pair (AA/Aa) controlled avirulence in isolate Sg 110-9 to a corresponding resistance gene Rsg1 in IP 18292. The pattern of segregation of virulence : avirulence suggested the presence of a gene-for-gene interaction between *S. graminicola* and *P. glaucum* (59).

Molecular aspects of pathogen variability

Molecular methods have been used to study the mechanism of virulence and variability in DM fungus. Pathogenic and genetic diversities among 15 isolates of *S. graminicola* on various differential lines of pearl millet cultivars in India have been demonstrated. Fifteen isolates have been classified into five major virulence groups based on disease incidence data. Using a specific set of primers designed for the conserved region the presence of retrotransposable elements in the genome has been detected by amplification of part of the reverse transcriptase gene in a PCR reaction. A characteristic of retrotransposon activity responsible for mutation of the organism's genes has also been identified (99). The DM fungus actively expresses the P5-like sequences in the susceptible host whereas the element is suppressed in a resistant host. A genomic library has been prepared of the Path-6, a virulent isolate with P5 retroelements (a member of the copia-like retrotransposon family). The library clones, designated as SgP5-1, SgP5-2, SgP5-3 and SgP5-4 are distributed invariantly among genomes of five host-genotype-specific downy mildew pathotypes. A genomic library of this pathotype has been constructed in the *lambda* gt11 vector. Based on colony hybridization of total *S. graminicola* DNA with a genomic library, repetitive DNA content has been estimated. Digestion with isoschizomeric methylation-sensitive restriction enzymes reveals partial methylation of GATC and CCGG sequences in the genome. The presence of retrotransposable elements in the genome can be detected by amplification in a PCR reaction, using a specific set of primers designed for the conserved region (76). Genetic variability in host genotype-specific pathotypes of S. *graminicola* can be studied at the molecular level. Micro-satellites containing (GAA) 61, (GACA) 4, and especially (GATA) 4 are quite informative and show high levels of polymorphism among the pathotypes. Six pathotypes could be classified into five groups based on the cluster analysis of their genetic similarities, thereby confirming the existence of distinct host genotype-specific virulence in *S. graminicola* pathotypes. AFLP techniques have also been used to detect genetic variation in mildew isolates to screen breeding material for identification of resistant millet, and for monitoring changes in *S. graminicola* in relation to changes in host for effective disease management. Fourteen AFLP primer combinations produced 184 polymorphic bands. An unweighted pair-group method of averages cluster analysis represented by dendrogram and principal coordinate analysis separated the mildew collections into four distinct groups (98).

ENVIRONMENT AND AGRONOMY

Temperature, humidity, rainfall, cloudiness and intensity of radiation are among the important environmental factors that influence the development and spread of the disease. Apart from environment, edaphic and biotic factors considerably determine the course of DM development. Low pH (7.5), higher soil bulk density (2.12 g cm^{-1}), and increased soil moisture content (96.8% ASM) with addition of organic matter (farm yard manure) reduce the disease development. Adding *Rhizobium*, *Azospirillum* or *Azotobacter* inocula as combined seed and soil treatment also reduce the disease (33). The response of DM is also influenced by the direction of sowing. In an arid environment the NW-SE sown crop showed less disease in comparison to the N-S sown crop (31). Another important observation in relation to oospores and the disease development is that oosporic powder, when placed over the seed in the furrow gives better results than the present practice of pre-sowing oospores below the seed for field screening (32). The amount of oospores is positively correlated with disease incidence and negatively with the time taken to express disease symptoms. A highly negative correlation is established between the age of seedlings at the time of sporangial inoculation and disease incidence. Barring sparse production of sporangia in the daytime, maximum production is observed at 03.00 h. The sporulation period in infected plants varies from 3 to 20 days (34). Recently it has been indicated that intra-row spacing had no significant effect on the incidence and severity of the disease but significantly increased plant height, panicle length and weight, 1000-grain weight and grain yield (122). Certain anatomical features of the host have also been worked out elucidating the defense mechanism in pearl millet against DM. Accordingly the cuticular and epidermal thickness of leaves of pearl millet were correlated with resistance to DM. However, the highly resistant cultivars had a significantly lower stomatal index and higher epicuticular wax content when compared with the highly susceptible cultivars.

Downy mildew and Tissue culture

There are many potential advantages in using tissue culture techniques to understand the obligate nature of *S. graminicola*. The inciting fungus can be cultured together with its host in a controlled chemical and physical environment. The technique provides possibilities of a simplified experimental system for investigating the structure and physiology of host-parasite interactions. It also allows close scrutiny of pathogen products with an insight to the mechanism of resistance at the cellular level.

Dual culture of *S. graminicola*

Downy mildew fungus has been successfully cultured on medium with no evident loss of fructifications. *S. graminicola* was first grown on host callus tissue and subsequently on a modification of White's basal medium containing casein- hydrolysate, 2,4-dichlorophenoxyacetic acid (2,4-D) and kinetin (114, 115). Dual cultures of host and fungal biotrophs provide a controlled system to maintain isolates of pathogen and to study the interaction between host and pathogen. A few attempts have been made to establish dual cultures using infected shoot tip and inflorescence axes or artificially inoculating healthy calli obtained from pearl millet seeds with. *S. graminicola* sporangia (9, 15, 117). Dual cultures can easily be established from malformed florets of pearl millet infected with *S. graminicola*. Malformed floral tissues usually produce more callus with fungus than shoot tips. Fungal mycelium covers the entire surface of the callus within 30 days of placement of explants on the Murashige and Skoog's (MS) medium with 2 mg l^{-1} of 2,4-D. The infected calli also differentiate and produce plantlets when transferred to MS medium devoid of 2, 4-D (117).

Disease resistance through somaclonal variation

Pearl millet has considerable genetic variability useful for the improvement of germplasm. Tissue culture methods have been used for inducing desirable characters, genetic transformation and regeneration through embryogenesis (58, 101, 102). Many workers have demonstrated successful regeneration from protoplasts. Somaclonal variation, therefore, has shown great promise in enhancing the production and productivity of pearl millet. In tissue cultures, large population of callus cultures is required for the induction and selection of disease resistant lines. In the case of *S. graminicola,* three different ways have been used 1) a tissue culture system for studying the host-pathogen interaction; 2) somaclonal variation as a source of disease resistance, and 3) selection, including the use of culture filtrate and toxic metabolites (83, 85). Isolation of pathogen specific SG-toxin and development of toxin resistant pearl millet plants through callus cultures have been reported to be the most efficient method for inducing DM resistant cell lines. The adverse effect of SG-toxin has been observed on callus initiation and its further new growth and regeneration of plantlets from callus and seedling growth in many susceptible pearl millet lines. Callus initiation decreases with increase in concentration of SG-toxin. Toxin resistant callus lines have been regenerated via somatic embryogenesis in pearl millet cvs. PNMS 6B line, ICTP 8203, ICMS 9555 and HB3 at 10% SG-toxin. Such regenerated plants exhibited resistance throughout the growth period when tested for resistance to DM under field conditions (12, 83, 85).

DISEASE MANAGEMENT

Chemical control

Like other oomycetes, *S. graminicola* has a complex disease cycle where developmental forms differ in physiology and anatomy along with different impacts on host-parasite interactions, survival and distribution. The disease can efficiently be managed with metalaxyl, the systemic fungicide. Seed dressing with Apron 35-SD (6 kg t^{-1} seeds) and foliar application of Ridomil MZ-72 (2 kg ha^{-1}) for seed crops have been recommended (88). However, looking to examples of acquired tolerance to metalaxyl in the recent past, a search for new systemic fungicides has been conducted. Consequently, it has been inferred that the fungicides of acylanilide series exhibit the highest efficacy of DM control. However, their benefit: cost ratio generally proves to be uneconomical (20). Seed treatment along with foliar application of three strobilurin fungicides viz. azoxystrobin, kresoxim-methyl and trifloxystrobin have reportedly shown enhanced protection against the DM. Disease curative activity of azoxystrobin was found to be higher with partial systemic activity (103).

Biological control

With increasing environmental awareness the focus of managing plant diseases has been shifted towards viable and sustainable alternatives (4). The importance of biological control for managing downy mildew disease has been discussed. The lack of durable resistance, existence of pathogenic variability, and concerns about fungicide resistance opened the way for searching alternative and eco-friendly methods of DM control by activating the plant's own defense mechanisms with specific biotic or abiotic elicitors. The scientific basis of induced resistance has emerged with the identification and characterization of genes and biochemical pathways governing resistance in plants against invading pathogens. In this regard, operability of induced systemic resistance is gaining importance, as some biotic and abiotic agents have conferred protection against the disease (3, 6, 68, 69, 70, 82). The efficacy of cerebrosides (glycosphingolipids extracted from various plant pathogens), as resistance elicitors against DM of pearl millet has been reported (22). β-Aminobutyric acid (BABA) treatment of pearl millet seeds influenced seedling vigour and protected the seedlings from downy mildew disease. Seeds treated with 50 mM for 6 h resulted in the maximum of 1428 seedling vigour and showed 23% disease incidence in comparison with the control, which recorded seedling vigour of 1260, and 98% disease incidence i.e. 75% protection from disease (77). A combination of raw cow's milk as seed treatment and *Gliocladium virens* as seed and soil treatments resulted in the lowest disease incidence and highest number of tillers of the crop (6). The protection by raw cow's milk and *G. virens* in managing DM is ascribed to induced systemic resistance based on an increase in resistance related enzymes (PPO, CA and PO) and metabolites (phenols, ODP) in the treated plants (7). Recently, polyphenol oxidase (PPO) activity was analyzed in seedlings of 10 resistant and susceptible pearl millet cultivars with or without inoculation of *S. graminicola*. Seedlings of resistant varieties had greater PPO activity than susceptible seedlings, and inoculated seedlings had significantly higher PPO levels than un-inoculated seedlings. PPO activity was positively correlated with levels of downy mildew resistance in different pearl millet cultivars under field conditions. Native PAGE staining showed four isoforms of PPO, which were differentially induced in relation to the time of appearance and intensities in the un-inoculated seedlings, whereas a fifth PPO isoform appeared after inoculation with *S. graminicola*. PPO activity was significantly higher in the shoot and leaves of pearl millet than in the root. The enzyme was predominantly expressed after pathogen inoculation and was localized in the epidermal and vascular regions (66). The importance of integrating indigenous knowledge of using raw cow's milk with biocontrol agents has been emphasized as a logical strategy to induce resistance in crops of low economic value largely grown by resource-poor semi-arid tropical (SAT) farmers (5, 30). A commercial preparation, Elexa, (an aqueous chitosan formulation) has been tested under field conditions. A combination of seed treatment and foliar spray to 7-day-old seedlings recorded 69% protection. The nature of disease control is ascribed to induction of systemic resistance (78). Methanolic extracts of plant species such as *Clematis gouriana, Evolvulus alsinoides, Mimusops elengi, Allium sativum* and *Piper nigrum*, commonly growing across India were reported to be having antisporulant activity against *S. graminicola*. The extracts of 11 species (*Agave americana, Artemisia pallens, Citrus sinensis, Dalbergia latifolia, Helianthus annuus, Murraya koenigii, Ocimum basilicum, Parthenium hysterophorus, Tagetes erecta, Thuja occidentalis and Zingiber officinale*) also exhibited remarkable anti-sporulant effect even after 10-fold dilution of the crude extracts and were found commensurable to that of the marketed botanical fungicides (21).

In an attempt to induce resistance in pearl millet for DM management certain biochemicals have been tried. Seed treatment with the amino acid proline (50 mM for 3 h) provided significant (58% protection under greenhouse and 67% protection under field conditions) enhancement of seed germination and seedling vigour besides protecting the pearl millet plants from downy mildew. The benefit was sustained throughout the life of the plants (67). Treating pearl millet seed with amino acids showed that under field conditions, Serine, DL-Tryptophan, DL-Leucine and DL-Isoleucine (25mM for 6h) reduced DM disease from 51 to 57%. In this case activation of the phenylpropanoid pathway and accumulation of phenolic compounds correlated with amino acid-mediated induced systemic resistance (74, 84). Defense related enzymes comparable to the resistant cultivar are elicitated by L-methionine application through activation in mRNA levels of genes for Pr-1a, beta-1, 3-glucanase, chitinase, peroxidase, lipoxygenases and chalcone synthase in pearl millet against downy mildew (74). Likewise, vitamin menadione sodium bisulphite (MSB) (6 h seed-soak treatment at 20 mM) enhanced germination and seedling vigour significantly (73%) and also induced DM resistance followed by niacin (63%) and riboflavin (62%), respectively. Seeds treated with vitamins induced maximum resistance in the seedlings by the fourth day after pathogen inoculation, and the resistance persisted till the end of the growth period of the crop. The vitamin treatments had a growth promotional effect and considerably increased the yield (60).

Both seed as well as foliar treatments with *Pseudomonas fluorescens* controlled downy mildew, but efficacy was significantly higher when seed treatment was followed by a foliar application (116). Under laboratory conditions, trichoshield (a talc formulation consisting of spores of *Trichoderma harzianum*, *Trichoderma lignorum*, *Gliocladium virens* and *Bacillus subtilis*) seed treatment enhanced seed germination and seedling vigour of pearl millet significantly over the control; under greenhouse conditions vegetative growth parameters like height, fresh and dry weight, leaf area and number of tillers were significantly enhanced over the control. Trichoshield formulation provided better protection (52 to 71%) under field conditions against downy mildew in comparison with individual strains of *T. harzianum*, *T. lignorum*, *G. virens* and *B. subtilis*. Among the methods of application, foliar spray was found to be a more efficient delivery method than seed or slurry treatment. Under field conditions, trichoshield treatment enhanced yield significantly (28%) over the control. Days required for 50% flowering was reduced by 4 days compared to the control (68). Nitric oxide (NO) donors Nitroso-R-Salt, 2-Nitroso-1-Naphthol and Sodium Nitro Prusside (SNP) were evaluated for their effectiveness in protecting pearl millet plants against DM. Aqueous SNP seed treatment with or without polyethylene glycol (PEG) priming was the most effective in inducing the host resistance against DM both under greenhouse and field conditions. Spatio-temporal studies corroborated that the protection offered by NO donor treatment was systemic in nature and a minimum of 3-day time gap between the inducer treatment and subsequent pathogen inoculation was necessary for maximum resistance development. Expression of primary defense responses like hypersensitive response, lignin deposition and defense related enzyme phenylalanine ammonialyase (PAL) were enhanced by NO donor treatments (52).

Host-plant resistance

Host plant resistance is the only reliable, environment-friendly and economical measure against the disease. Over the past thirty years considerable progress has been made in the field of host resistance. Research to improve screening systems to identify and use plant resistance has largely been successful. Efforts were directed to search other sources of cytoplasm to avoid any situation as was faced in case of maize. Presently, several sources are available, but still A_1 cytoplasm is being agronomically better than other A_2 to A_5 cytoplasms is used in most of the male sterile (ms) lines. Efforts made at different national research centers in India led to the development of genetically different ms lines for use in the hybrid development programmes.

Genetic diversification of the restorers was also made at different centers and private seed agencies, and now a large number of genetically diverse restorers are available. Combining inbred lines from ICRISAT, ARS Durgapura and Mandor, and CAZRI, Jodhpur, Mandor Restorer Composite (MRC) and Mandor Early Restorer Composite (MERC) was developed. Seed of this composite was distributed by ICRISAT to NARS scientists and new resistant restorer lines suitable for use in arid regions have been developed. Nineteen inbred restorer lines have been developed from MERC at CAZRI, Jodhpur itself, and are being used in the hybrid breeding. These efforts have now resulted in a broader genetic base of hybrids in India than was available a decade ago. Some of the new CMS systems (A4 and A5) in pearl millet, for which a large amount of the germplasm is in the maintainer class than in the A_1 system, offer further opportunity to accelerate diversification of the genetic base of pearl millet seed parents (35, 64, 65). These two new CMS systems also appear to have more stable male sterility than the A_1 CMS system (63). Recently, it has been observed that high resolution and reproducibility of ISSR-PCR assay shall provide a simple, rapid and cost-effective system for the characterization of huge collection of germplasm and testing of purity of genetic stocks/hybrids. The study also demonstrates the need for selecting/generating genetically diverse restorer as well as CMS lines for the breeding of genetically diverse hybrids in pearl millet (121).

A new antiserum in pearl millet screening for disease resistance has been developed after raising purified proline/ hydroxyproline-rich glycoprotein (p/hrgp) cell walls and polyclonal antibodies (pab-p/hrgp) from pearl millet in Rabbit. Based on this antiserum, an enzyme immunoassay was developed that displayed a linearity detection range from 0. 01 to 10 μg P/hrgp. The specificity of the antiserum was confirmed by Western blot analysis and immuno-cytochemical studies on P/hrgp localization in epidermal peelings or in suspension-cultured cells (23).

Sources of resistance

The successful development and operation of field and greenhouse screening techniques resulted in the identification of a large number of germplasm and breeding lines with high levels of downy mildew resistance (92). About 3500 germplasm accessions from almost all pearl millet growing countries have been evaluated at ICRISAT for their reaction to downy mildew. These downy mildew resistant sources [P7 (ICML-12), SDN 503 (ICML-13), 700251 (ICML 14), 700516 (ICML 15), 700651 (ICML 16)] have shown stability across sites and years. Other sources of resistance are IP 16438, IP 16762, P310-17 and P 1449-3. Out of these P7, 700651, 700516 and P 310-17 have been utilized in breeding programmes (40). Donor parents like 863 B (IP 22303),

P1449-2 (IP21168), ICMB 90111 (IP 22319), ICMP 451 (IP 22442) and IP 18293 were also identified as important sources of resistant genes against different pathotypes of downy mildew in India. Similarly, from 534 wild accessions of pearl millet, 220 were found to be resistant to downy mildew (119).

Across the sites and years ICMP 423, ICMB 90111 and ICMA 90111 have been found resistant to DM. However, their resistance has been difficult to manipulate with pedigree selection, suggesting its complex inheritance. ICMP 85410 has been found downy mildew resistant as it derives its resistance from SC 14(M)-1 (39, 105). Downy mildew resistance of this line has been found to be stable across the sites in western Africa. Some male sterile F_1 lines were highly resistant to 4 pathotypes (from Patancheru, Durgapura, Jalna and Jamnagar) of downy mildew in field in Andhra Pradesh, whereas seven ms lines of ICRISAT origin were found resistant to moderately resistant in western India (51, 62).

Reselected Sources of resistance

Selection of variability within susceptible parents led to the development of resistant types e.g. ICML 22 (7042 DMR) a downy mildew resistant version of susceptible 7042 (IP 2696) a landrace from Chad (95). The line ICML 22 is being used at ICRISAT Asia Center for breeding of pollinators. Recent study on effectiveness within progeny selection for DM resistance in pearl millet revealed significant difference between the selected and unselected lines in 6 S3 progenies, with the DM incidence levels in the selected lines reduced from half to one-third in the unselected lines (62). However, the DM incidence in the six selected progenies varied from 14 to 46%, indicating that one-stage selection during the inbreeding process was not effective in improving the resistance of the crop to acceptable levels.

Recovery resistance

Recovery resistance is a phenomenon in which systemically infected plants outgrow the disease to produce healthy panicles. Here, pathogen and host co-exist, apparently without adversely affecting the yield. The phenomenon brings heritable changes in the recovered plants. Plants with this type of resistance quickly recover from the disease and subsequently behave like conventional resistant genotypes. Marker assisted selection would allow this type of resistance to be pyramided with more conventional resistances. Sources of recovery resistance are P1449, SDN 503 and ICMB 841 (91, 96).

Sources of complete resistance

Accessions that show 100% resistance in greenhouse inoculation tests against major pathotypes in India and a large number of field populations of *S. graminicola* in field disease nurseries in India and western Africa have been identified in the past. Five accessions viz. IP 18292, IP 18293, IP 18294, IP 18295 and IP 18298 were initially found to have zero disease incidences regardless of inoculum level (40, 87).

BREEDING METHODS

Since host plant resistance is the only reliable, environment-friendly and economical measure against the disease, breeding for resistance to downy mildew has always remained an integral part of the breeding programmes. This is mainly because, a large number of resistance sources are available and also suitable screening techniques have been developed (92) and also inheritance of resistance is better understood, resistance being dominant and exhibiting continuous variation in segregating populations (87).

Conventional Breeding Methods

Conventional breeding procedures utilize greenhouse or field screening methods to incorporate adequate levels of downy mildew resistance (DMR) into breeding populations, parental lines and experimental open pollinated varieties that have superior agronomic performance and product quality. The following methods are being used in breeding for resistance to DM with varying degree of success: Pure line selection (type of selection within partially inbred lines), Pedigree selection and Recurrent selection (the most widely and successfully used methods for improving downy mildew resistance) and Backcrossing (conventional backcrossing procedures have seldom been used in breeding for resistance to pearl millet downy mildew. The only reported success using backcrossing is breeding of DMR seed parents MS 5054A and MS 5141A in the elite background of Tift 23 by IARI, New Delhi, India). Induced mutations (these have been used to produce

MS 5071 B from Tift 23 B, and MS 5071 A was used in producing NHB series of hybrids with downy mildew resistance).

Selection for within line variability

Selection for DMR is done within a susceptible line using pedigree procedures. Large population of a susceptible line is grown in the disease nursery. Disease free plants are selfed and selfed progenies are screened panicle-to-row against the disease. A notable success is the development of ICMA 841A from susceptible 5141A and 5141B. This method provides an opportunity to revive once popular cultivars that have become susceptible to downy mildew.

Inheritance of Resistance

There have been conflicting reports about the nature of inheritance of downy mildew resistance mainly both the pathogen and the host is allogamous and highly variable (109) and segregation for host plant resistance generally shows continuous variation. Regional variability in pathogen population used and difficulty in maintaining high and uniform disease pressure have led to conflicting conclusions, even for the same source (43).

It is clear that A_1 cytoplasm is not associated with susceptibility or resistance to downy mildew (1). However, there is ample evidence that genes in the nucleus control host plant resistance to this disease. Downy mildew resistance has been found to be dominant and variation in segregating population is continuous (89). The picture of mode of inheritance largely remains unclear and incomplete due to the variable pathogen population used in all these studies. Quantitative inheritance of downy mildew resistance in pearl millet has been more successful in identifying parental materials with the ability to transmit high levels of resistance. It has been concluded by a number of workers that non-additive gene action is responsible for much of the heritable variability (24, 45). Such non-additive gene action can contribute substantially to general combining ability (GCA), since parents having dominant resistance can be expected to have high GCA for this trait when compared with more susceptible parents.

Regional variability of resistance causes additional hindrance to breeding (14, 29, 81, 93, 106). This regional variability has been found to be mainly due to genetic variability between pathogen populations rather than environmental differences between locations (11). Consequently, the testing of breeding material under multilocational trials has become expensive and time consuming but necessary (97). Selection for multilocational resistance could, therefore, theoretically be carried out by simultaneously screening breeding material in one location against a range of pathogen populations.

Marker-Assisted Selection (MAS)

MAS uses highly heritable 'tags' as selection criteria that are genetically linked to the portion of genome controlling characters of interest. These markers can be morphological traits and proteins (including isozymes or DNA markers as RFLPs, RAPD and others). MAS is an excellent alternative to time consuming testcrosses, otherwise required to pyramid DMR genes. MAS and graphical maps based on markers can help breeders; more rapidly recover recurrent parent genotype in backcrossing programme, and more efficiently identifying desired recombinants in pedigree programmes (25, 41, 48). In order to use MAS in improvement of downy mildew resistance it is necessary to have a:

- Donor parent with identified linked markers for some of its resistance genes.
- An elite parent with inadequate resistance, which has different marker alleles than the donor parents and lacks these resistance genes (at least the specific alleles linked to markers in the donor genome).
- A system for efficiently determining genotype (or at least phenotype) at the marker loci in individual segregating plants derived from crosses between the resistant source and the elite parent.

MAS has considerable potential as a tool for pearl millet improvement. A saturated genetic linkage map based on molecular markers has been developed (43, 44). This map was initially based on RFLP markers, since these can reveal an almost unlimited number of polymorphism and can be used directly to tag loci controlling traits of interest for MAS. Further, a skeleton map based on sequence tagged sites (STS – markers based on the polymerase chain reaction with less stringent requirements for large quantities of good quality DNA) has also been developed that should reduce the cost of transferring this map to other breeding populations (54). Recent inheritance studies based on Molecular Marker Genetic Linkage Maps are yielding interesting results that will facilitate genetic manipulation of disease resistance (36, 43, 44). To date markers have been identified for at least sixteen different putative DM resistant quantitative trait loci (QTLs).

Development of bulk segregant analysis (53), similar in many ways to the concept of near isogenic populations (18), has dramatically reduced the time and expense required to identify molecular markers for new resistance genes with large effect. However, it can miss genes of small effect. Molecular markers are also being used to characterize populations of pearl millet pathogens (75). Overall it looks like molecular markers and MAS will become important tools for pearl millet breeders in the decade ahead. Marker assisted selection offers breeders a new array of disease screening method with maximized heritabilities. Although not a universal solution to all breeding problems, MAS offers tremendous opportunities to breed for important characters that are otherwise difficult and expensive to assess.

Molecular markers linked to resistance genes would allow resistances to different pathogen populations to be selected for a single location in the absence of pathogen. Linkage drag and confounding effects of environmental variation associated with conventional breeding would also be taken care of. The ability to map genes contributing towards variation in complex traits with enough accuracy has only recently been made possible through the development of comprehensive molecular marker maps.

The first molecular-marker map for pearl millet has recently been constructed so that QTL analysis is now possible (49). Following QTL analysis, Jones *et al.* (43) found independent inheritance of resistance to pathogen populations from India, Senegal and populations from Niger and Nigeria. These results showed the existence of virulence differences in the pathogen populations from within Africa and between Africa and India. QTLs of large effect, contributing towards a large proportion of variation in resistance were consistently detected in repeated screens. No QTLs that were effective against all four-pathogen populations were observed, demonstrating that pathotype-specific resistance is a major mechanism of downy mildew resistance in this cross. For all but one of the QTLs, resistance was inherited from the resistant parent and the inheritance of resistance tended to be the result of dominance or over-dominance.

Ultimately the goal of identified linked markers is their use as indirect selection tool in the breeding programme. In addition to the identified linked markers, it is necessary to have a low cost, high throughput marker service facility. At present SSR markers are considered appropriate for most species. It is predicted that SNP markers may become attractive in the coming years. A major limitation in MAS is the cost to select for the recurrent genome in the backcrossing programmes (normally, only markers linked to the trait/gene of interest are used).

Apart from this, molecular mapping has also expanded knowledge of downy mildew resistance. It has shown that:

- Many genes contribute to downy mildew resistance
- These genes are scattered throughout the host genome
- Pathogen strain specificity is the rule for each of these genes, and
- A large portion of resistance to a given pathogen population can be accounted for by relatively few genes.

Use of MAS for downy mildew resistant hybrid

Downy mildew surveys conducted in the pearl millet hybrid-intensive states of Gujarat, Haryana, Maharashtra and Rajasthan in India during 1994–2004 forewarned about the performance of hybrids and their possible replacement in certain areas (110). The surveys indicated the likely resistance breakdown of cv. HHB 67, the most popular hybrid in Haryana and parts of Rajasthan, cultivated on at least 400,000 hectares in these states. Being very early in maturity (within 65 days), it escapes terminal drought and also provides an opportunity for double cropping.

Since HHB 67 is highly preferred by the farmers, and with no other hybrid in this maturity group being available, attempts were made to improve the parental lines of HHB 67 for DM resistance. The parental lines (843 A and H 77/833-2) of the original hybrid were improved for DM resistance through marker-assisted as well as conventional backcross breeding programmes at the ICRISAT campus at Patancheru. A gene for downy mildew resistance was added to the male parent, H 77/833-2, through marker-assisted breeding using ICRISAT elite parent ICMP 451 as the resistance gene donor. The gene for DM resistance was added to the female parent, 843A/B, from ICRISAT line ICML 22 through conventional backcross breeding. Field-testing of the new hybrid was done at various locations through All India Coordinated Pearl Millet Improvement Project (AICPMIP). The advantage of the marker-assisted selection was that male parent for HHB 67-2 could be developed in one-third of the time required for developing the female parent by conventional selection methods. By identifying and marking the gene responsible for DM resistance in ICMP 451, it could be checked whether the gene had transferred to the next generation in the progeny of crosses between ICMP 451 and the male parent of HHB 67. By using molecular marker technology the presence of the gene can be tested even while the next generation is a seedling, saving precious breeding time. This is the first example of development of a DM-resistant version of hybrid HHB 67, using both conventional and DNA marker technologies to produce a DM resistant hybrid in pearl millet. It was recommended for release in the state of Haryana in the year 2005 (37, 104).

Physiology of host resistance

Recently, investigations have been made to determine the physiological attributes of a DM resistant pearl millet plant. The activity of a disease resistance related enzyme phenylalanine ammonia lyase (PAL) has been studied in pearl millet cultivars having different levels of DM resistance (27). The PAL activity increases in resistant host cultivar and decreases in the susceptible cultivars following infection of DM pathogen. The enzyme activation varies between cultivars and is correlated with the degree of resistance to the disease. An assay has been developed for identifying DM resistant and susceptible pearl millet genotypes at the seedling stage using arachidonic acid (AA). The acid induces a hypersensitive response (HR) comparable with HR induced by *S. graminicola* on coleoptile/root regions of 2-day-old pearl millet seedlings. A time gap in the appearance of cell necrosis among genotypes is related to the degree of resistance to downy mildew. The appearance of hypersensitive cell necrosis is rapid in genotypes having high resistance to downy mildew and relatively slows in genotypes with high susceptibility. This simple method of screening various genotypes in the absence of the pathogen can assist in the identification of DM resistant/ susceptible host cultivars without the risk of introducing the virulent race of the pathogen. Similarly, a rapid method of detecting incipient infection of DM fungus in pearl millet has been developed (13). Dormant nodal buds isolated from systemically infected plants show incipient infection of DM. Microscopic examination revealed that 80% of the meristems and bud scales showed DM fungus when stained with 0.1% Trypan blue. The fungus colonizes all embryonic leaves and grows inside the developing leaves. Examination of systemically sprouted nodal buds also confirms that the mycelium detected in meristems and bud scales of nodal, apical and floral buds belong to *S. graminicola*. This method of detection may prove to be useful in forecasting the presence of infection at an early stage of plant to avert the outbreak of DM in field grown pearl millet plants.

Differential induction of superoxide dismutase (SOD) in DM-resistant and -susceptible genotypes of pearl millet is reported (10). Accordingly, the SOD activity increases by 2-3 fold in resistant seedlings upon inoculation. Roots show highest activity of SOD. This study also revealed increased intensity of all four isozymes of SOD in the resistant genotype upon inoculation. The investigations conducted with an enzyme lipoxygenase (LOX) indicated a good correlation between enzyme activity and downy mildew reaction in field. Seeds of highly resistant genotypes show maximum activity. Seeds obtained from plants recovered from the disease exhibited more LOX activity of the original parent seeds. Thus, LOX activity can be used as a biochemical marker for screening different genotypes of pearl millet for downy mildew. (10).

Stomatal behaviour of pearl millet plants has been observed during DM pathogenesis (79). The infected plants showed lower leaf conductance and transpiration rate compared with healthy plants, indicating that infection elicited stomatal closure. Downy mildew infection inhibited stomatal opening at 12.00 h and pronounced stomatal closure was observed at 16.00 h, which resulted in decreased transpiration rate. Stomatal closure was more pronounced at the boot stage.

FUTURE PROSPECTS

Downy mildew is one of the important diseases of pearl millet, mainly because of frequent recent epidemics. Host resistance against DM has been exploited to a large extent and the farmers' now use resistant varieties. However, the shift in virulence pattern of the pathogen, especially in endemic areas is causing concern. Breeding for disease resistance appears to be the best approach to reduce the disease. The on-going programmes utilize resistant stocks to developing varieties having better agronomic features with DM resistance. Apart from breeding disease resistant varieties, significant advances have been made in the past decade towards searching for biological control agents. This includes elicitors of host defense, including biochemical and natural products. In this direction a significant progress has been made showing elicitation of resistance and defense related enzymes by raw cow milk and amino acids against downy mildew disease of pearl millet (7, 104). The success of biocontrol agents is dependant upon the complex interaction that these beneficial products establish with the pathogen and plants in the soil ecosystem. Better understanding of biocontrol mechanism will pave a way for applying safer and less expensive methods to protect pearl millet plants from DM and increase its yield. Identification of biochemical components of resistance would also assist in breeding programmes. Tissue culture techniques may be followed for isolating desirable somaclonal variants arising from cultured cells of well-adapted varieties. The successful extraction and purification of SG toxin from *Sclerospora graminicola* will be useful in screening downy mildew tolerant pearl millet lines. Molecular tagging of disease resistant genes may be of great help in the development of resistant varieties through marker-assisted breeding. Transferring and pyramiding resistant genes in agronomically elite parental lines would be useful in developing hybrids with lasting resistance to downy mildew. Further identification of downy mildew resistant wild germplasm and its utilization in developing disease resistant useful material needs proper attention. *P. polystachion* a wild species of pearl millet has been found to be a very important source for downy mildew resistant genes.

However, it is clear that different approaches such as genetic, molecular, biochemical and ecological (including indigenous knowledge), should be integrated to conduct future course of studies on downy mildew of pearl millet.

LITERATURE CITED

1. Anand Kumar K, Jain RP, Singh SD. 1983. Downy mildew reaction in pearl millet lines with and without cytoplasmic male sterility. *Plant Disease* 67: 663-665.

2. Appa Rao S, de Wet JMJ. 1999. Taxonomy and evolution. In: *Pearl Millet Breeding*. pp. 29-47 (Khairwal IS, Rai KN, Andrews DJ, Harinrayana G. eds.). Oxford & IBH Publishing Co., New Delhi, India.

3. Arun-Kumar, Verma, SK. 2005. Milk in the management of plant diseases. In: *Bridging gap between ancient and modern technologies to increase agricultural productivity* pp. 67-73. (Chaudhary SL, Saxena RC, Nene YL eds.), CAZRI, Jodhpur, India.

4. Arun-Kumar 2008. Biocontrol of plant diseases: Need to tap the options. *Journal of Arid Legumes* 5: 99-108.

5. Arun-Kumar, Bhansali RR, Gajja, BL. 2005. Validating people's knowledge: The role of formal sciences. *Honey Bee* 16: 4-5.

6. Arun-Kumar, Bhansali RR, Mali PC. 2004. Raw cow's milk and *Gliocladium virens* induced protection against downy mildew of pearl millet. *International Sorghum and Millets Newsletter* 45: 64-65.

7. Arun Kumar and Purohit AK. 2012. The Role of Indigenous Knowledge in Biological Control of Plant Pathogens: Logistics of New Research Initiatives. Chapter 7, In: *Plant Defence: Biological Control*, J.M. Mérillon and K.G. Ramawat, eds.), Progress in Biological Control, Springer Science, pp. 161-194.

8. Arya HC, Kumar A. 1976. Diseases of bajra- A serious problem of Rajasthan desert economy. *Transactions of Indian Society of Desert Technology & University Center of Desert Studies* 1: 177-182.

9. Arya HC. Tiwari MM. 1969. Growth of *Sclerospora graminicola* on callus tissues of *Pennisetum typhoides* and in culture. *Indian Phytopathology* 22: 446-452.

10. Babitha MP, Bhat SG, Prakash HS, Shetty HS. 2002. Differential induction of superoxide dismutase in downy mildew-resistant and susceptible genotypes of pearl millet. *Plant Pathology* 51: 480–486.

11. Ball SL, Pike D J. 1984. Intercontinental variation of *Sclerospora graminicola*. *Annals of Applied Biology*, 104: 41-51.

12. Bhansali RR, Arun-Kumar 2004. Induction of SG toxin (*Sclerospora graminicola*) tolerant callus lines of pearl millet. *Proceedings of National Seminar on Millets Research and Development: Future Policy Options in India*, March 11-12, 2004, RAU, Mandor, Jodhpur, India, pp.13-14.

13. Bhansali RR, Kumar A, Kaul, RK. 1997. Rapid detection of incipient infection of downy mildew in pearl millet (Abstract). *Proceedings of International Conference on Integrated Plant Disease Management for Sustainable Agriculture*, pp. 185, Nov. 10-15, IARI, New Delhi, India.

14. Bhat SS. 1973. Investigations on the biology and control of Sclerospora graminicola on bajra. *Ph.D thesis*, Department of Botany, University of Mysore, India, pp.165.

15. Bhat SS, Safeeulla KM, Shaw CG. 1980. Growth of *Sclerospora graminicola* in host tissue culture. *Transactions of British Mycological Society* 75: 303-309.

16. Bouriquet G. 1963. Les principaux ennemis et maladies du sorgho et du mil en Afrique tropicale. (In Fr. Summaries in En, Es.) *Agronomie Tropicale* 18: 132-136.

17. Butler E J. 1907. Some diseases of cereals caused by *Sclerospora graminicola. Memoirs of the Department of Agriculture in India, Botanical Series* 2: 1-24.

18. Burton GW, Wells HD. 1981. Use of near isogenic host populations to estimate the effect of three foliage diseases on pearl millet forage yield. *Phytopathology* 71: 311-333.

19. Chevaugeon J. 1952. Maladies des plantes cultivees en Moyenne-Casamance et dans le Delta Central Nigerien. (In Fr.) *Revue de Pathologie Vegetale et d'Entomologie Agricole de France* 31: 3-51.

20. Deepak SA, Chaluvaraju G, Basavaraju P, Amutesh KN, Shetty HS, Oros G. 2005. Response of pearl millet downy mildew (*Sclerospora graminicola*) to diverse fungicides. *International Journal of Pest Management* 51: 7-16.

21. Deepak SA, Oros G, Sathyanarayana SG, Shetty NP, Shetty HS, Sheena S. 2005. Antisporulant activity of leaf extracts of Indian plants against *Sclerospora graminicola* causing downy mildew disease of pearl millet. *Arch-*

ives of Phytopathology and Plant Protection 38: 31-35.

22. Deepak SA, Raj SN, Umemura K, Kono T, Shetty HS. 2003. Cerebroside as an elicitor for induced resistance against the downy mildew pathogen in pearl millet. *Annals of Applied Biology* 143: 169-173.

23. Deepak SA, Shailasree S, Sujeeth N, Kini RK, Mithöfer A, Shetty HS. 2008. Serodiagnosis of pearl millet resistance to downy mildew by quantitating cell wall P/hrgp using polyclonal antiserum Pab-P/HRGP. *European Journal of Plant Pathology* 121: 77-85.

24. Deswal DP, Govila OP. 1994. Genetics of resistance to downy mildew (*Sclerospora graminicola*) in pearl millet (*Pennisetum glaucum*). *Indian Journal of Agricultural Sciences*, 64: 661-663.

25. Fahr S, Messmer MM, Melchinger AE, Lee M, Woodman WL. 1993. Graphical genotype of maize inbred B86 revealed by RFLPs. *Plant Breeding*, 110: 29-34.

26. Francis SM, Williams RJ. 1998. *Sclerospora graminicola*. IMI Descriptions of Fungi and Bacteria, 77, Sheet 770.

27. Geetha NP, Amruthesh KN, Sharathchandra RG, Shetty, H S. 2005. Resistance to downy mildew in pearl millet is associated with increased phenylalanine ammonia lyase activity. *Functional Plant Biology* 32: 267-275.

28. Gilijamse E, Frinking HD, Jeger MJ. 1997. Occurrence and epidemiology of pearl millet downy mildew, *Sclerospora graminicola*, in south-west Niger. *International Journal of Pest Management* 43: 279-283.

29. Girard JC. 1975. Downy mildew of pearl millet in Senegal. In: *Proc. of the consultants group meetings on downy mildew and ergot of pearl millet* (Williams RJ ed.). 1-3 October, ICRISAT (International Crops Research Institute for the Semi-Arid Tropics), Patancheru, India, pp. 59-73.

30. Gupta AK. 2006. From sink to source: The honey bee network documents indigenous knowledge and innovations in India. *Innovations* (summer): 49-66.

31. Gupta GK, Singh D. 1996a. Effectiveness of water absorbents, FYM application and sowing direction in downy mildew reduction in pearl millet under arid

environment. *Indian Phytopathology* 49: 41-47.

32. Gupta GK, Singh D. 1996b. Studies on the influence of downy mildew infection on yield and yield-contributing plant characters of pearl millet in India. *International Journal of Pest Management* 42: 89-93.

33. Gupta GK, Singh D. 1999. Role of edaphic factors in the development of downy mildew (*Sclerospora graminicola*) in pearl millet. *Journal of Agricultural Science* 133: 61-68.

34. Gupta GK, Singh D. 2000. Epidemiological studies on downy mildew of pearl millet (*Pennisetum glaucum*). *International Journal of Tropical Plant Diseases*18: 101-115.

35. Hanna WW. 1989. Characteristics and stability of new cytoplasmic-nuclear male sterile source in pearl millet. *Crop Science* 29: 1457-1459.

36. Hash CT, Cavan G, Bidinger FR, Howarth CJ, Singh SD. 1995. Downy mildew resistance QTLs from a seedling heat tolerance mapping population. *International Sorghum and Millets Newsletter* 36: 66-67.

37. Hash CT, Sharma A, Kolesnikova-Allen M., Serraj R, Thakur R, Bidinger FR, Bhasker Raj RG, Rizvi SMH, Beniwal CR, Yadav HP, Yadav YP, Srikant, Bhatnagar SK, Yadav RS, Howarth GJ, Breese WA, Witcombe JR. 2003. Marker-assisted breeding to improve pearl millet hybrid HHB 67: Lab to Field. *In the Wake of the Double Helix: From the Green Revolution to the Gene Revolution*, 27-31 May 2003, Bologna, Italy (Abstract.). pp. 82.

38. Hash CT, Singh SD, Thakur RP, Talukdar BS. 1999. Breeding for disease resistance. In: *Pearl Millet Breeding*. pp.337-379. (Khairwal IS, Rai KN, Andrews DJ, Harinrayana G. eds.). Oxford & IBH Publishing Co., New Delhi, India.

39. Hash CT, Witcombe JR. 1994. Pearl millet mapping populations at ICRISAT. In: *Use of molecular markers in sorghum and pearl millet breeding for developing countries*. pp. 69-75 (Witcombe JR, Duncan RR eds.). Overseas Development Administration, London, U.K.

40. Hash CT, Witcombe JR., Thakur RP, Bhatnagar S, Singh SD, Wilson JP. 1996. Breeding for pearl millet disease resistance. In: *Proc. of the International Conference on Genetic Improvement of Sorghum and Pearl Millet,*

September 22-27, 1996, Lubbock, Texas, pp. 337-372.

41. Hospital F, Chevalet C, Mulsant P. 1992. Using markers in gene introgression breeding programs. *Genetics* 132: 1199-1210.

42. International Pearl Millet Downy Mildew Virulence Nursery (IPMDMVN)- A Summary Report 1992-1998. ICRISAT, Patancheru, India, pp. 33.

43. Jones ES, Liu CJ, Gale MD, Hash CT, Witcombe JR 1995. Mapping quantitative trait loci for downy mildew resistance in pearl millet. *Theoretical Applied Genetics* 91: 448-456.

44. Jones ES, Witcombe JR, Hash CT, Singh SD, Gale MD, Shaw DS, Liu CJ. 1994. Mapping QTLs controlling resistance to downy mildew in pearl millet and their application in plant breeding programmes. In: *Use of molecular markers in sorghum and pearl millet breeding for developing countries*. pp. 76-79. (Witcombe JR, Duncan RR. eds.). Overseas Development Administration, London, U.K.

45. Kataria R P, Yadav HP, Beniwal CR, Narwal MS. 1994. Genetics of incidence of downy mildew (*Sclerospora graminicola*) in pearl millet (*Pennisetum glaucum*). *Indian Journal of Agricultural Sciences* 64: 664-666.

46. Kenneth R. 1981. Downy mildew of cereals. In: *The Downy Mildews*, (Spencer D M. ed.). Academic Press, London and New York, pp. 367-394.

47. Kenneth R. 1998. *Sclerospora graminicola*. (Descriptions of Fungi and Bacteria). CAB International: IMI Descriptions of Fungi and Bacteria, 46: Sheet 452.

48. Lande R, Thompson R. 1990. Efficiency of marker assisted selection in the improvement of quantitative traits. *Genetics* 124: 734-736.

49. Liu CJ, Witcombe JR, Pittaway TS, Nash M, Hash CT, Busso C S, Gale MD. 1994. An RFLP-based genetic map of pearl millet (*Pennisetum glaucum*). *Theoretical and Applied Genetics* 89: 481-487.

50. Lukose C, Dave HR. 1995. Germination of oospores of *Sclerospora graminicola*. *Indian Phytopathology* 48: 74-76.

51. Manga VK, Arun-Kumar, Dubey LK. 2004. Reaction of pearl millet male-sterile lines against downy mildew in arid environment. *Journal of Mycology and Plant Pathology* 34: 340-341.

52. Manjunatha G, Niranjan Raj S, Shetty NP, Shetty HS. 2008. Nitric oxide donor seed priming enhances defense responses and induces resistance against pearl millet downy mildew disease. *Pesticide Biochemistry and Physiology* 91: 1-11.

53. Michelmore RW, Paran, Kesseli RV. 1991. Identification of markers linked to disease resistance genes by bulked segregant analysis: A rapid method to detect markers in specific genomic regions by using segregating populations. *Proc. of the National Academy of Science* 88: 9828-9832.

54. Money TL, Liu CL, Gale MD. 1994. Conversion of RFLP markers for downy mildew resistance in pearl millet to sequence-tagged sites. In: *Use of molecular markers in sorghum and pearl millet breeding for developing countries*. pp. 65-68 (Witcombe JR, Duncan RR. eds.). Overseas Development Administration, London, U.K.

55. Nagaraja A, Siddiqui MR. 1994. Importance of oospores in the transmission of *Sclerospora graminicola* (Sacc.) Schroet. In: *Pennisetum glaucum* (L.) R.Br. *Seed Research* 22: 81-82.

56. N'Doye M, Gahukar RT, Carson AG, Selvaraj CJ, Mbaye DF, Diallo S. 1986. Les problemes phytosanitaires du mil dans le Sahel. (In Fr.) In: *Compte rendu du Seminaire International du Project CILSS de Lutte Integree*, 6-13 December 1984. Niamey, Niger. Ouagadougou, Burkina Faso: CILSS (Comite Permanent Inter-Etats de Lutte contre la secheresse dans le Sahel), pp.79-94.

57. Nene YL, Singh SD. 1976. Downy mildew and ergot of pearl millet. *PANS* 22: 366-385.

58. O'Kennedy MM, Smith G, Botha FC. 2004. Improved regeneration efficiency of a pearl millet (*Pennisetum glaucum* [L.] R.Br.) breeding line. *South African Journal of Botany*, 70: 502-508.

59. Pushpavathi B, Thakur RP, Rao KC. 2006. Inheritance of avirulence in *Sclerospora graminicola*, the pearl millet downy mildew pathogen. *Plant Pathology Journal Faisalabad* 5: 54-59.

60. Pushpalatha HG, Mythrashree SR, Shetty R, Geetha NP, Sharathchandra RG, Amruthesh KN, Shetty H S. 2007. Ability of vitamins to induce downy mildew disease resistance and growth

promotion in pearl millet.*Crop Protection* 26: 1674-1681.

61. Rachie KO, Majmudar JV. 1980. *Pearl millet*. University Park, Pennsylvania, USA, Pennsylvania State University Press. pp. 307.

62. Rai KN, Kulkarni VN, Singh AK 2004. Performance of a male-sterile F_1 and its inbred parental lines in pearl millet. *International Sorghum and Millets Newsletter* 45: 43-44.

63. Rai KN, Murty DS, Andrew DJ, Bramel-Cox P J. 1999. Genetic enhancement of pearl millet and sorghum for the semi-arid tropics of Asia and Africa. *Genome* 42: 617-628.

64. Rai KN, Rao A S. 1996. A new CMS system and its restorers in pearl millet. *Interntional Sorghum and Millets Newsletter* 36: 49-50.

65. Rai, KN, Virk DS, Harinrayana G, Rao AS. 1996. Stability of male-sterile sources and fertility restoration of their hybrids in pearl millet. *Plant Breeding* 115: 494-500.

66. Raj SN, Sarosh BR, Shetty HS. 2006. Induction and accumulation of polyphenol oxidase activities as implicated in development of resistance against pearl millet downy mildew disease. *Functional Plant Biology* 33: 563-571.

67. Raj SN, Shetty NP, Shetty HS. 2004. Proline - an inducer of resistance against pearl millet downy mildew disease caused by *Sclerospora graminicola*. *Phytoparasitica* 32: 523-527.

68. Raj SN, Shetty NP, Shetty HS. 2005. Synergistic effects of Trichoshield on enhancement of growth and resistance to downy mildew in pearl millet. *Bio Control* 50: 493-509

69. Raj SN, Chaluvaraju G, Amruthesh KN, Shetty HS, Reddy M S, Kloepper JW. 2003a. Induction of growth promotion and resistance against downy mildew on pearl millet (*Pennisetum glaucum*) by rhizobacteria. *Plant Disease* 87: 380-384.

70. Raj SN, Deepak SA, Basavaraju P, Shetty HS, Reddy MS, Kloepper JW. 2003b. Comparative performance of formulations of plant growth promoting rhizobacteria in growth promotion and suppression of downy mildew in pearl millet. *Crop Protection* 22: 579-588.

71. Rao VP, Thakur RP. 2004. Downy mildew incidence and oospore production by *Sclerospora graminicola* in pearl millet hybrids in Maharashtra and Rajasthan. *International Sorghum and Millets Newsletter* 45: 57-61.

72. Rao VP, Thakur RP, Rai KN, Sharma YK. 2005. Downy mildew incidence on pearl millet cultivars and pathogenic variability among isolates of *Sclerospora graminicola* in Rajasthan. *International Sorghum and Millets Newsletter* 46: 107-110.

73. Saccas AM. 1954. Les champignons parasites des sorghos et des penicillaires en Afrique Equatoriale Francaise. *Agronomie Tropicale* 9: 135-173, 263-301 and 647-686.

74. Sarosh BR, Subramoniam S, Shetty HS. 2005. Elicitation of defense related enzymes and resistance by L-methionine in pearl millet against downy mildew disease caused by *Sclerospora graminicola. Plant Physiology and Biochemistry* 43: 808-815.

75. Sastry JG, Ramakrishna W., Sivaramakrishnan S, Thakur RP, Gupta VS, Ranjekar PK. 1995. DNA finger printing detects genetic variability in the pearl millet downy mildew pathogen (*Sclerospora graminicola*). *Theoretical and Applied Genetics* 91: 856-861.

76. Sastry JG, Sivaramakrishnan S, Thakur RP, Gupta VS, Ranjekar PK. 1997. Characterization of the genome of *Sclerospora graminicola*, the causal fungus of downy mildew of pearl millet. *World Journal of Microbiology and Biotechnology* 13: 693-697.

77. Shailasree S, Ramchandra KK,. Shetty HS. 2007. ß - Amino butyric acid-induced resistance in pearl millet to downy mildew is associated with accumulation of defence-related proteins. *Australasian Plant Pathology* 36: 204-211.

78. Sharathchandra RG, Raj SN, Shetty NP, Amruthesh KN, Shetty HS. 2004. A chitosan formulation ElexaTM induces downy mildew disease resistance and growth promotion in pearl millet. *Crop Protection* 23: 881-888.

79. Sharma YK, Bora KK. 2003. Effect of downy mildew on stomatal behaviour, transpiration and leaf temperature in pearl millet. *Journal of Eco-Physiology* 6: 127-130.

80. Sheoran RK, Govilla OP, Kaushik CD, Chandra N. 2000. Studies on mode of transmission of downy mildew (*Sclerospora graminicola* Sacc. Schroet.) in pearl millet. *International Journal of Tropical Agriculture* 18: 165-168.

81. Shetty HS, Ahmad R. 1981. Physiologic specialization in *Sclerospora graminicola. Indian Phytopathology* 34: 307-309.

82. Shetty H S, Kumar VU. 2000. Biological control of pearl millet downy mildew: present status and future prospects. In: *Biocontrol potential and its exploitation in sustainable agriculture*. pp. 251-265. Volume I. (Upadhyay R, Mukerji KG, Chamola BP. eds.). Springer Verlag, Germany.

83. Shetty HS, Nagarathna K C, Shetty SA, Vidhyasekaran P. 1993. Tissue culture for management of downy mildew disease in pearl millet. *Genetic engineering, molecular biology and tissue culture for crop pest and disease management*, pp. 283-293.

84. Shetty R, Pushpalatha HG, Geetha NP, Shetty HS. 2005. Comparative studies on pearl millet seed treatment with amino acids and associated difference in activation of phenylpropanoid pathway and induced systemic resistance. (Abstract), *Proc. Global Conference II 'Plant Health-Global Wealth'*, November 25-29, ISMPP, MPUAT, Udaipur, India, p.73.

85. Shetty SH, Nagarathna KC, Umesha S, Shetty SA. 1999. Selection of downy mildew resistant somaclones, from a susceptible B line of pearl millet. *Plant Cell, Tissue and Organ Culture* 58: 159-162.

86. Singh SD. 1994. Recycling of pearl millet cultivars for the control of downy mildew. *Indian Journal of Plant Protection* 22: 164-169.

87. Singh SD. 1995. Downy mildew of pearl millet. *Plant Disease* 79: 545-550.

88. Singh SD, Shetty HS. 1990. Efficacy of systemic fungicide metalaxyl for the control of downy mildew (*Sclerospora graminicola*) of pearl millet (*Pennisetum glaucum*). *Indian Journal of Agricultural Sciences* 60: 575-581.

89. Singh SD, King SB, Werder J. 1993. Downy mildew disease of pearl millet. *Information Bulletin no. 37*, ICRISAT, Patancheru, A. P., India, pp.36.

90. Singh SD, Navi SS. 1996. Factors affecting germination of oospores of *Sclerospora graminicola*. *Indian Journal of Mycology and Plant Pathology* 26: 271-277.

91. Singh SD, Talukdar BS. 1996. Recovery resistance to downy mildew in pearl millet potential line ICMA 1 and ICMB 1. *Crop Science* 36: 201-203.

92. Singh SD, Wilson JP, Navi SS, Talukdar BS, Hess DE, Reddy KN. 1997. Screening techniques and sources of resistance to downy mildew and rust in pearl millet.

Information Bulletin no. 48. Patancheru, Andhra Pradesh, India: ICRISAT, pp.36.

93. Singh SD, Singh G. 1987. Resistance to downy mildew in pearl millet hybrid NHB-3. *Indian Phytopathology* 40: 178-180.

94. Singh SD, King SB, Werder J. 1993. Downy mildew disease of pearl millet. *Information Bulletin no. 37.* International Crops Research Institute for the Semi-Arid Tropics, Patancheru, India, pp.30.

95. Singh SD, Singh P, Andrew DJ, Talukdar BS, King SB. 1992. Reselection of a pearl millet cultivar utilizing residual variability for downy mildew reaction. *Plant Breeding* 109: 54-59.

96. Singh SD, Talukdar BS. 1996. Recovery resistance to downy mildew in pearl millet parental lines ICMA 1and ICMB 1. *Crop Science* 36: 201-203.

97. Singh SD, Wilson JP, Navi SS, Talukdar BS, Hess DE, Reddy KN. 1997. Screening techniques & sources of resistance to downy mildew and rust in pearl millet. International Crops Research Instt. for the Semi-arid Tropics, *Information Bulletin No. 48.* ICRISAT, Patancheru, India, pp. 36.

98. Singru R, Gupta VS, Ranjekar PK. 2003. Identification of a P5 retrotransposon-like element in *Sclerospora graminicola* and contemplation of its role. *World Journal of Microbiology and Biotechnology* 19: 387-390.

99. Singru R, Sivaramakrishnan S, Thakur RP, Gupta VS, Ranjekar PK. 2003. Detection of genetic variability in pearl millet downy mildew (*Sclerospora graminicola*) by AFLP. *Biochemical Genetics* 41: 361-374.

100. Spencer DM. 1981. *The Downy Mildews*, Academic Press, London and New York, pp. 636.

101. Srivastava S, Kothari SL. 2003. Assessment of somaclonal variations in two lines of pearl millet [*Pennisetum glaucum* (L.) R. Br.]. *Indian Journal of Genetics and Plant Breeding* 63: 295-298.

102. Subba Rao MV, Mythali MK, Sreeram Murthy J, Saideswara Rao Y, Manga V. 1999. Biotechnology, In: *Pearl Millet Breeding*, pp. 417-424 (Khairwal IS, Rai, KN, Andrews DJ, Harinrayana eds.). Oxford & IBH, New Delhi, India.

103. Sudisha J, Amruthesh KN, Deepak SA, Shetty NP, Sarosh BR, Shetty HS. 2005. Comparative efficacy of strobilurin fungicides against

downy mildew disease of pearl millet. *Pesticide Biochemistry and Physiology* 81: 188-197.

104. Sudisha, J., Arun-Kumar, Amruthesh, KN, Niranjana, SR and Shetty, HS. 2011. Elicitation of resistance and defense related enzymes by raw cow milk and amino acids in pearl millet against downy mildew disease caused by *Sclerospora graminicola*. *Crop Protection* 30: 794-801.

105. Talukdar BS, Khairwal IS, Singh R. 1999. Hybrid Breeding. In: *Pearl Millet Breeding* pp.269-301. (Khairwal IS, Rai, KN, Andrews DJ, Harinrayana eds.). Oxford & IBH Publishing Co., New Delhi, India.

106. Talukdar BS, Prakash PP, Rao AM, Ramakrishna C, Witcombe JR, King SB, Hash CT. 1998. Registration of ICMP 85410 dwarf, downy mildew resistant, restorer germplasm line of pearl millet. *Crop Science* 38: 904-905.

107. Thakur DP. 1987. Studies on downy mildew epiphytotic in pearl millet hybrid BJ 104 in Haryana. *Haryana Agriculture University Journal of Research* 17: 68-74.

108. Thakur RP, Rao VP. 1997. Variation in virulence and aggressiveness among pathotypes of Sclerospora graminicola on pearl millet. *Indian Phytopathology* 50: 41-47.

109. Thakur RP, Rai KN, Rao VP, Rao AS. 2001. Genetic resistance of pearl millet malesterile lines to diverse Indian pathotypes of *Sclerospora graminicola*. *Plant Disease* 85: 621-626.

110. Thakur RP, Shetty KG, King S B. 1992. Selection for host-specific virulence in asexual populations of *Sclerospora graminicola*. *Plant Pathology* 41: 626-632.

111. Thakur RP, Rao VP, Amruthesh KN, Shetty HS, Datar VV. 2003. Field surveys of pearl millet downy mildew–Effects of hybrids, fungicide and cropping sequence. *Journal of Mycology and Plant Pathology*, 33: 387-394.

112. Thakur RP, Rao VP, Hash CT. 1998. A highly virulent pathotype of *Sclerospora graminicola* from Jodhpur, Rajasthan, India. International Sorghum and Millets Newsletter 39: 140-142.

113. Thakur RP, Rao VP, Wu BM, Subbarao KV, Shetty HS, Singh G, Lukose C, Panwar MS, Sereme P, Hess DE, Gupta SC, Dattar VV, Panicker S, Pawar NB, Bhangale GT, Panchbhai SD. 2004. Host resistance stability to downy mildew in pearl millet and pathogenic variab-

ility in *Sclerospora graminicola. Crop Protection* 23: 901-908.

114. Thakur RP, Shetty HS, Khairwal IS. 2006. Pearl millet downy mildew research in India. *International Sorghum and Millets Newsletter* 47: 125-130.

115. Tiwari MM, Arya HC. 1967. Growth of normal and diseased *Pennisetum typhoides* tissue infected with *Sclerospora graminicola* in tissue culture. *Indian Phytopathology* 20: 336-338.

116. Tiwari MM, Arya HC. 1969. *Sclerospora graminicola* axenic culture. *Science* 163: 291-293.

117. Umesha S, Dharmesh SM, Shetty SA, Krishnappa M, Shetty HS. 1998. Biocontrol of downy mildew disease of pearl millet using *Pseudomonas fluorescens. Crop Protection* 17: 387-392.

118. Upadhyaya G, Shivanna MB, Prakash HS, Shetty HS. 1992. A novel approach to the establishment of dual cultures of pearl millet and *Sclerospora graminicola. Plant Cell, Tissue and Organ Culture* 31: 203-206.

119. Upadhyaya, H.D., Reddy, K.N. and Gowda, C.L.L. 2007. Pearl millet germplasm at ICRISAT gene bank – Status and Impact. *SAT ejournal.* 3: 1-5

120. Viswanathan A. 2005. Phylogenetic analysis of *Sclerospora graminicola* using internal transcribed spacer region-2. *M.Sc. Thesis*, Texas A&M University, Texas, USA.

121. Williams RJ. 1984. Downy mildews of tropical cereals. *Advances in Plant Pathology* 2: 1-103.

122. Wilson JP. 1999. Pearl millet diseases: A compilation of information on the known pathogens of pearl millet, *Pennisetum glaucum*, U. S. Department of Agriculture, Agriculture Research Service, *Agriculture Handbook No. 716*, pp.50.

123. Yadav S, Jain S, Jain, V, Jain RK. 2007. Genetic analysis of CMS, restorer, hybrid and open-pollinated genotypes of Indian pearl millet (*Pennisetum glaucum* (L.) R. Br.) using ISSR markers. *Indian Journal of Biotechnology* 6: 349-353.

124. Zarafi AB, Emechebe AM. 2006. Effect of intra row spacing on the incidence and severity of pearl millet downy mildew and grain yield. *Archives of Phytopathology and Plant Protection* 39: 9-14.

INCIDENCE AND TOXICOLOGY OF DEOXYNIVALENOL

Shilpa Penugonda, S. Girisham* and S.M. Reddy

Department of Microbiology, Kakatiya University, Warangal 506 009
*E-mail: sivasrigirisham@gmail.com

KEY WORDS: Deoxynivalenol, nivalenol, *Fusarium* species, natural incidence, biological activity, vomitoxin, pharmacokinetics, management

ABSTRACT

Deoxynivalenol (DON), elaborated by different species of *Fusarium* proved to be common contaminant in different foods and feeds all over the world. Domestic animals are reported to be primarily affected by this toxin. The main effect of DON at low dietary doses appears to be decreased growth and anorexia, while higher doses result in emesis immunotoxicity and bring neurochemical changes in brain. Pigs proved to be more sensitive than mice, poultry and ruminants to DON. *In vivo* DON suppresses normal immune response and concurrently induces auto immune-like effects which are similar to human immunoglobulin A (IgA) nephropathy. Mechanism of action and pharmacokinetics to contain DON are discussed. DON is partially decomposed by heating and dry milling of food grains. Since DON is elaborated in surface layer of grain, it is eliminated to the extent of 80-90% during the milling. The fate of DON in food grains during post-harvest practices is also discussed.

Finally DON contaminations in certain managerial process for future are discussed.

INTRODUCTION

Deoxynivalenol (DON) is one of the trichothecenes, produced by species of *Fusarium.* It was first isolated and characterized from moldy barley in Japan (1976) and subsequently found to be identical to the emetic factor from corn in the USA (167). Deoxynivalenol or Vomitoxin is a frequently occurring mycotoxin elaborated by species of *Fusarium* which infect maize and other food grains (20,42,112). DON at high concentration inhibits protein synthesis (47) and is responsible for numerous human and animal disorders (51). Lower doses of DON have been shown to provoke elevation of serum IgA level and are known to affect cell-mediated and humoral immunity in several animal species (48). In 1993, IARC placed DON in group 3, not classifiable as to its carcinogenicity to humans (46,77). At cellular level, the main toxic effect of DON is immunosuppressant and immunostimulation depending upon the dose and duration of exposure (61,205). Consumption of contaminated feeds by livestock has been associated with a variety of adverse health effects including gastrointestinal system, feed refusal, reduced weight, diarrhoea and emesis (92,95). Rocha et al (168), Pestka and Smolinski (154) and Faixova et al (63) have excellently reviewed different aspects of incidence and toxicity of DON.

DON is shown to interfere with the metabolism of membrane phospholipids and to increase liver lipid per oxidation. DON is known to cause animal intoxications and adversely affect the growth and reproduction in animals causing serious economic losses (32,65). It is relatively stable to cooking and processing, therefore, food preparation procedures cannot be expected to remove DON to safe level (28).

STRUCTURE AND PHYSICO-CHEMICAL PROPERTIES

DON is a type B trichothecene, and soluble even in water and in polar solvents such as aqueous methanol, aqueous acetonitrile, ethyl acetate and aqueous alkali. Deoxynivalenol has a 12, 13-epoxy group, three OH functions, and α, β- unsaturated keto group. Its chemical name is therefore, 12-13-epoxy-3 α, 7 α, 15-trihydroxy trichothec-9-ene-8 one. Its molecular formula, C_{15} H_{20}

O_6, shows that its relative molecular mass is 296.3. The structure of DON was first elucidated by Godtfredsen et al (69). DON crystallizes as colorless needles, with a melting point of 151-153°C. Its specific rotation has been determined to be $[\alpha]^{20}$ D= +6.35. The α, β unsaturated keto causes in absorption of ultraviolet (UV) radiation of short wavelength. DON is stable in organic solvents (182), but ethyl acetate and acetonitrile are the most suitable solvents, particularly for long-term storage. It is stable at 120 °C and is not decomposed under mild acidic conditions. It exhibits blue florescence when excited by long wave UV light (360nm), solubility in water is about 20 μg/ml. It is also stable at neutral and acidic pH consequently it is not hydrolyzed in the stomach after ingestion (168). They can be deactivated under strong acidic and alkaline conditions (185). The effect of solvent, temperature, moisture, pH and heat on the stability of DON was well documented by Widestrand and Pettersson (204) and Lauren and Smith (102). Structure of DON and its derivatives and some synthetic analogues of DON are depicted in Fig 1.

ORGANISMS

The organisms reported to produce DON are *Fusarium roseum* (130), *F. graminearum* (81), *F. culmorum* and *F. tricinctum* (199), and *F. nivale* (17). Ramakrishna et al. (164) reported that *F. moniliforme* is also capable of producing DON. DON primarily occurs in *F. graminearum* and *F. culmorum* belonging to section discolor (80,134). A recent study by Clear and Patrick (41) ranked *F. culmorum* as the most dominant Fusaria in cereals in Canada. However, under suitable environmental conditions, *F. culmorum* is also capable of causing severe seedling disease and seed deterioration (97).

The success of infection depends on many factors including climatic conditions, such as temperature (22,122,176) and humidity (139), cultivar resistance level (13,45,111) and nitrogen fertilization (53) are prominent among others. However, it is the availability of moisture which is the deciding factor (97). More recent work containing a larger collection of strains and DNA sequence comparisons from a total of 11 genes at six independent genetic loci support the existence of nine distinct, cryptic species within the *F. graminearum* complex (125,142,203). *F. gramine-*

arum contains four chromosomes that can be visualized by cytological methods (190).

Deoxynivalenol (DON)

Iso Deoxynivalenol

Nivalenol (NIV)

De-epoxy Deoxynivalenol

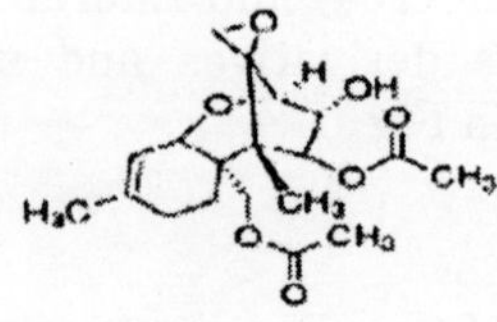

Neosolaniol

nor Deoxynivalenol A

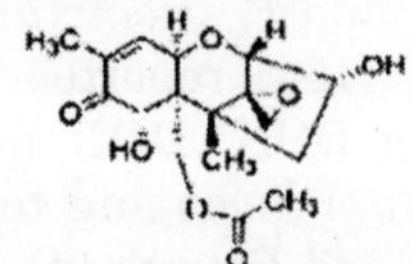

15- acetyl Deoxynivalenol

nor Deoxynivalenol B

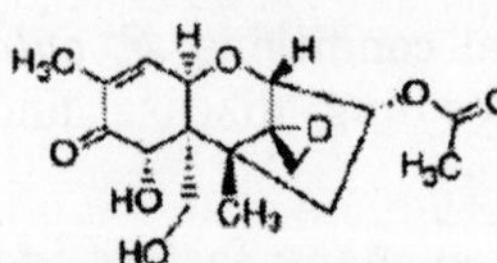

3-Acetyl Deoxynivalenol

nor Deoxynivalenol C

3-keto Deoxynivalenol

Deoxynivalenol -3-glucoside

Fig. 1. Structures of Deoxynivalenol and its derivatives

Nucleic acid sequencing will provide more objective information on identification of genera and species than that provided by conventional techniques. This technique can best be used for the identification of organisms that cannot be satisfactorily identified by their morphological features, particularly those that do not sporulate within 48 hrs. of subculture. Nucleic acid sequencing using the D2 large subunit rDNA sequencing kit offers promise of being an accurate identification system (108). The r DNA repeat unit contains genic and non non-genic or spacer regions. Each repeat unit consists of a copy of 18s- 5.8s and 28s-like rDNA and two spacers, the internal transcribed spacer (ITS) and intergenic spacer (IGS) (105,141). The rDNA genes have been employed to analyze major evolutionary events because it is highly conserved, where as the rDNA internal transcribed spacer (ITS1 and ITS2) are more variable (29).

Molecular phylogenetics and phylogenomics have greatly revised and enriched the fungal systematic in the last two decades. It is based on the comparison of more than one gene and resolves the taxonomic status of some morphologically mis identified species (31).

DETECTION OF DON

The detection of DON has been reviewed by Trucksess (195) and Lawrence and Scott (104). DON and some of its derivatives fluorescence blue-green when excited by UV radiation (228nm). This property has been exploited to detect DON on TLC plates (206). Krska et al (92) discussed different methods of analysis of DON in natural contaminated seeds. Quantitation with TLC followed by an aluminium chloride spray and visualization or densitometry under UV light (57,133,180,197). Detection of DON on TLC plate is reported by Williams (206) and confirmation of DON by some chemical methods by Scott et al (177). Bio assay of DON has also been reported by Koshinsky and Khachatourians (90).

DON is usually extracted with chloroform/ethanol, methanol, or acetonitrile in combination with water. Cleanup is accomplished with column chromatography using various packing's and often in combination with liquid-liquid partition. The amount of DON can be determined by using the formula (11).

DON (µg/gm) = Sx (C/X) x (V/W)

Where,

S= µl standard equal to test spot

C= Concentration of standard solution (20µg/ml)

X= µl test solution that has fluorescence intensity equal to standard spot.

V= Final volume of test solution (µl)

W= amount of test portion represented by final test solution.

Besides the spray reagents which are specific for the 12, 13-epoxy group of the trichothecenes, Al_2Cl_3 is the most useful reagent for visualizing DON (57). Some of the analytical techniques for detection of DON are précised in Table- 1.

Table 1. Analytical techniques for detection of Deoxynivalenol from various commodities

Extraction	Clean up column	Derivatization	Determination	Detection limit (µg/g)	Commodities	Reference
Acetonitrile/ Water	Column chromatography	None	TLC Al $_2Cl_3$ spray	30	Wheat	
	Charcoal alumina celite column	None	TLC Densitometry under long-wave UV light	0.3	Wheat	57, 19
Toluene/ Acctone	Preparative minicolumn chromatography	None	Minicolumn	500	Corn and wheat	70
Acetonitrile/ Water	Charcoal alumina celite column	Methanolic aluminium chloride	HPTLC	40	Wheat	
Methanol/ water	Silica-gel chromatography	None	HPTLC	5	Cereals	64,144
Chloroform/ Methanol	Silica-gel chromatography	None	HPLC HPLC	0.5	Wheat flour bran	10

	Silica-gel chromato-graphy	None	C_{18} Reverse phase	5	Alcoholic drinks (Wine, beer)	146, 10
	Silica-gel chromato-graphy	Heptaflurob utyic acid & 4-di methyl amino pyri-dine catalyst	HPLC	12	Corn, wheat and barley	35, 89
	Charcoal alumina celite column	None	HPLC (MS)	40	Milk and dairy products	88
Chloroform/ Methanol	Silica-gel chr-omatography	Methanolic Al_2cl_3	GC	40	Wheat	19,68
Methanol/ Water	Liquid-liquid partition	Silyation with TBT	GC electron capture	10	Cereals	126
Ethyl acet-ate/ acetoni-trile/Water	Liquid-liquid partition	Mass spect-roscopy with tri methyl silyation	GC	0.25	Corn	82
Acetonitrile/ Water	Liquid-liquid partition	HFBI hepta fluro butyry-limidazole	GC-MS	0.50	Grains	91
	Florisil column	TMS	GC-ECD GC-MS	10	Corn and their products	91
Methanol/ Water	Silica gel column	TMS HFB	GC-ECD GC-MS	20 50-200	Wheat, corn meal, rye flour	91
Acetonitrile/ Water	Immuno affi-nity chromat-ography	Mass spectr-oscopy	ELISA	100-500	Eggs	188
Methanol/ Water	Solid-phase	Electron capture	ELISA-HPLC, HPTLC	200	Rice	49
Chloroform/ Water Acetonitrile/ Water	Charcoal alumina celite column	None	ELISA,LC-MS,GC-HPTLC	0.3-0.8	Wheat	3, 4, 173

A rapid method has been introduced in which Mycosep is used for clean-up, with fluorimetric detection of DON derivitized with Zirconylnitrate and ethylene diamine in methanol (114). Several suitable HPLC methods have been reported for the determination of DON in foods and feeds. Separation is usually achieved on a C_{18} reversed phase column with methanol-water mixtures as the mobile phase (177). Reverse-phase HPLC is also used for quantitation with UV detection (101) or fluorescence detection following post column derivitization (171). Mini column screening method is also described by Gordon and Gordon (70).

DON and nivalenol (NIV) can be detected even at nanograms per gram, by employing LC-MS instruments, particularly those with atmospheric pressure chemical ionization interfaces (166). However, recent developments in high-resolution MS coupled with computer search capabilities have led to new MS methods for both identification and quantification of DON.

To overcome the difficulties encountered with the chemical and biological methods, new immunochemical methods have been developed (38,128). Consequently, many types of immunoassays, including Radioimmunoassay (RIA) and Enzyme linked immunosorbent assay (ELISA) as well as several novel immunochemical screening tests, were developed for DON analysis. Most of these methods are very sensitive, specific and simple to operate.

RIA procedure involves incubation of a specific antibody simultaneously with a solution of unknown sample or known standard and a constant amount of DON. The sensitivity of RIA for DON detection in corn and wheat are 50-5,000 pg/tube and detection limit of DON are 30 µg/kg.

Enzyme immunoassay (EIA) involves use of an enzyme as a marker for the detection of DON due to immunocomplex formation. The general principle of EIA is similar to that of RIA, there is an amplification system present in this assay, and thus, EIA is more sensitive. Enzyme labeling can be done by conjugation of enzyme to antigen (Ag) or antibody (Ab) via the periodate oxidation and subsequent reductive alkylation method or by cross-linking using glutaraldehyde. The sensitivity of ELISA for DON in corn and wheat is 1,000 µg/kg.

Competitive ELISA (C-ELISA) and non-competitive sandwich ELISA is used most frequently for DON analysis (212). In Direct competitive ELISA (dc-ELISA) specific antibodies against DON are coated on the ELISA plate and can be completed in 0.5-2hr. Limits of detection and range of quantization of DON in competition ELISA of DON and NIV is 80ppb (212). The indirect competitive ELISA or idc-ELISA or Double Antibody ELISA, a DON protein or polypeptide conjugate is first prepared on the plate and then incubated with specific rabbit antibody in the presence or absence of the DON. In general, ELISA is approximately 10 to 100 times more sensitive than RIA when purified DON is used.

Immunoscreening is a method, in which the incubation time, adjusting the antibody and enzyme concentration is modified in ELISA assay. By this method it is possible to screen and determine the DON. For screening, it is necessary to introduce a fluorophore to the materials eluted from the immunoaffinity column (IAC). The application of various immuno screening of DON has been summarized by Pestka (153) and Chu (39). Most of the screening tests are commercially available as kits.

Immunoaffinity columns (IACs) were made by conjugating antibodies to a solid-phase matrix. These columns are then used either in a screening test or for subsequent chemical analysis (38). Although earlier application of this technique was primarily aimed at biological fluids, but now it is also used as a cleanup tool and help to concentrate DON in large amount of sample. Thus it lowers the detection limits, sometimes as low as parts per trillion of DON can be measured. With the availability of sensitive ELISA methods, the combination with HPLC and TLC has proved effective as a post column monitoring system for HPLC.

Lateral flow immuno chromatography (LTF) is also called rapid one step assay of mycotoxins. It has become the most popular method for the rapid identification of mycotoxins and rough estimation of their concentration. The sensitivity of LTF is very high and is comparable to those of sophisticated methodologies such as LC-MS. The limit of detection and range of quantization of DON assayed by LTF is 0.5-2.0 ppb (198). Fluorescent polarization immunoassay is a very rapid test. Three minute extraction and 2 min assay were sufficient to detect 2 ppm of DON (110). Zöllner and Mayer-Helm, (216) have discussed in detail detection of DON with liquid chromatography mass spectrometry (LCMS).

Biosensor assay are designed to perform the simultaneous assay of many mycotoxins. The limits of detection of assays based on fluorescent antibodies against DON are at the ppb level (136,137). The assays are regenerable (135).

Molecular Identification: Real-time PCR amplifications

For each analyzed sample, DNA was amplified with either primer designed to track DON- producing fungi. Once primer pair- (Forward sequence: 5'-GACCCTAAGCGACTACAG-3'; reverse sequence;'-GTGCTACGGATAAGGTTC- 3' with accession numbers respectively) was designed to amplify a 287 bp fragment (Tm: 58°C) from the trichodiene synthase gene (Tri 5) of DON producing fungi (138) and to be more efficient in real time PCR. The primer specificity was assessed using genomic DNA from *Fusarium* species which are DON producers and from non-toxin producers.

BIOSYNTHESIS

Proctor et al (163) and Brown et al (26) have elegantly discussed the biosynthesis of DON. According to them the entire DON molecule can be synthesized from a single Farnesyl-pyrophosphate molecule (Fig. 2) as demonstrated by Lee et al (106) and Matsumoto et al (116).

Several genes are involved in the biosynthesis of DON which is localized in a gene cluster. The Tri 5 gene encodes the trichodiene synthase, which catalyzes the first step in the biosynthesis of DON. The nucleotide sequence of the Tri 5 gene has been characterized in several species of *Fusarium* (73,74). The Tri 6 gene encodes a protein that regulates the DON biosynthesis genes and has been sequenced in *F. sporotrichioides* (116) and *Giberella zeae* (26,194). Trichodiene undergoes a series of oxygenations, cyclisation, isomerisations and esterifications to yield bioactive trichothecenes such as T-2 toxin DON and acetylated DON. Like other microbial antibiotic biosynthetic pathways, the genes responsible for this are located in a gene cluster (83). The gene cluster in *F. graminearum* contains 10 functional genes (25,26). The larger number of functional genes in the *F. sporotrichioides* cluster are involved in the addition of oxygen with subsequent acetylation (25,26,107) found in T-2 toxin.

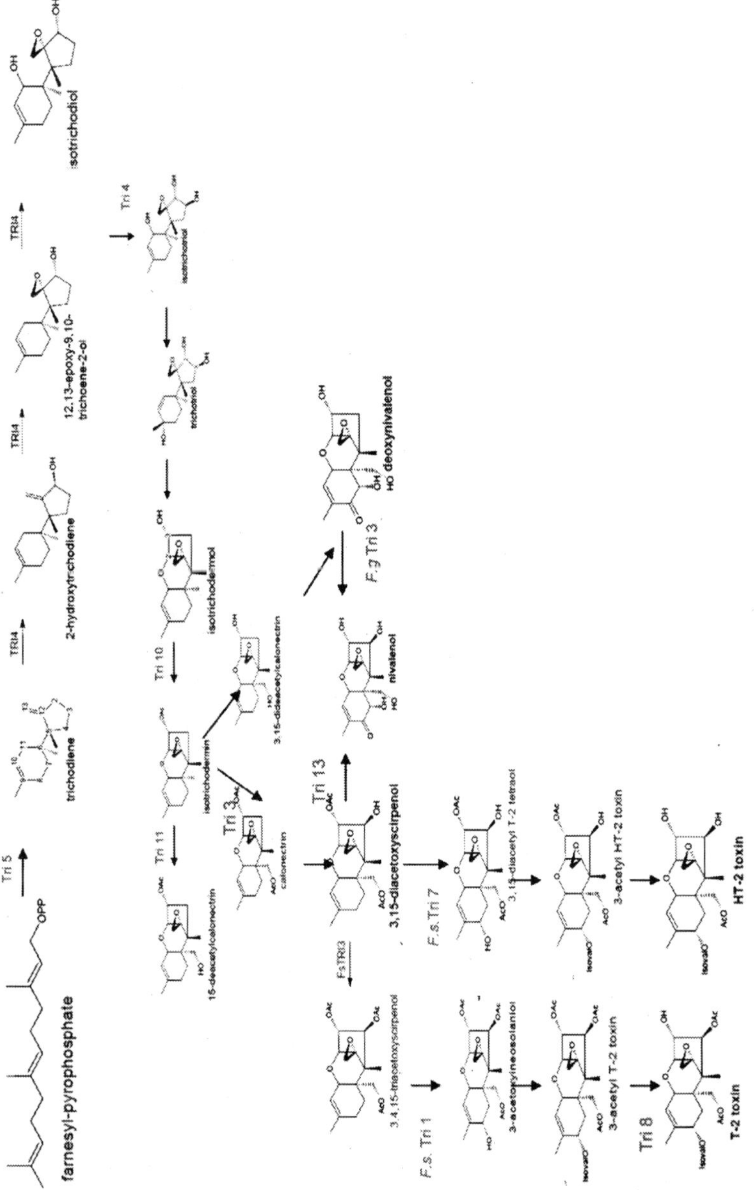

Fig. 2. Biosynthetic pathway of trichothecene. Tri-gene products are catalyzed in the steps of the pathways

The genes required for trichothecene biosynthesis and regulation in *F. sporotrichioides* are located at five separate loci within the genome (9,25,121), Tri 4 (118), Tri 11 (9), and Tri 13 (107), Five genes are involved with acylation / deacylation Tri 3 (120), Tri 101 (87,120), Tri 8 (117), Tri 7 (25), and Tri 16 (149), while 3 genes are involved with regulation Tri 6 (162), Tri 10 (189) and Tri 15 (9) while Tri 12 gene is involved with trichothecene transport (8), Tri 14 gene is involved with virulence (54), TOX_1 gene is involved with quantity of toxin (119) and Tri 9 gene whose function is unknown but appears to be co-regulated with the other biosynthetic genes (149).

NATURAL INCIDENCE

A worldwide survey for incidence of DON and NIV in 500 agricultural samples from 19 countries revealed the presence of DON approximately 40-50% with an average concentration of 292 μg/g DON and 267μg/g NIV. In particular DON was detected in maize (69%), followed by mixed feed (65%) and wheat (60%) (191). The simultaneous occurrence of DON with other *Fusarium* mycotoxins, mainly type B trichothecenes and zearalenone has been reported from variety of agricultural commoditities (174).

DON appears predominantly in wheat, corn, rye and barley (18). The natural occurrence of DON in barely has been reported from different parts of the World (127,150,191). *Fusarium graminearum* has also been more common on wheat grain in Germany (143) where high toxic levels of DON have been detected (158). This trend could be indicative of a change to warmer weather, a genetic change in the *F. graminearum* population or changes in cropping practices (14). Abnormally high concentration of nivalenol has been recorded in Japanese barley and in maize of New Zealand (158). Grains that have not been significantly dried because of wet weather at harvest or in grains that were stored wet (moisture >22%) were infested heavily by species of *Fusarium*. Worldwide occurrence of DON along with Zearalenone has been reported by Tanaka et al (191) from all over the world in Fig. 3 and Table 2.

The incidence of *Fusarium* in head blights is most closely related to moisture at the time of flowering (anthesis), and the timing of rainfall rather than the amount of rain (132). A direct relationship between the incidence of ear blight and the DON contamination of wheat has been established (52,184). As common contaminant of cereal crop, DON has been detected in foods consumed as part of the daily diet; snack foods, breakfast cereals and corn-barley-oat-and rice based products (5, 27,195).

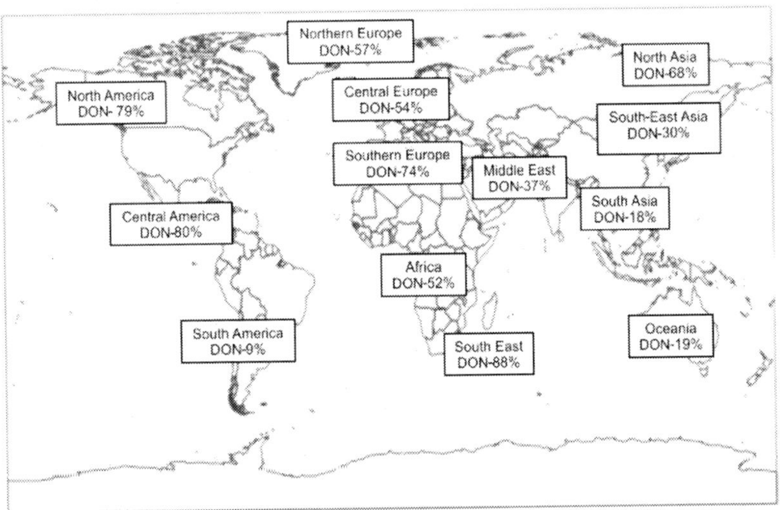

Fig. 3. Detection of DON in different parts of the world (from Tanaka et al, 191)

Table 2. Natural Incidence of DON and NIV from different parts of the World

Foods	Range		Countries	References
	DON (μg/g)	NIV (μg/g)		
Corn	60	40	Brazil	123
	0.065	0.13	Poland	93
	204-745	--	Nigeria	7
	5-2060	12-2440	Italy	34
	45-3430	200	North Italy	34
	68-967	--	Italy	34
	834.4	--	Argentina	24
	26.1-131.7	51.1-106.5	Spain	33
Wheat	15-1379	25-40	Germany	174
	10-2591	10-234	Denmark	165

	0.01	--	Poland	93
	642	--	Lithuania	115
	350	--	Norway	99
Barley	0.025	0.04	Poland	93
	372	--	Lithuania	115
	500-10,000	--	Uruguay	145
	1440	50	Norway	99
	40-2340	40-2340	Ethiopia	12
	980	--	Canada	113
Rye	0.023	0.002	Poland	93
	691	--	Lithuania	115
	10-257	48	Denmark	165
	0.027	0.033	Poland	151
	204	--	Lithuania	115
	849	211	Norway	99
	90	--	Canada	113
Rice	105-159	182-462	Korea	147
Cereals	0.007-0.930	--	Italy	40
	16-450	10-6935	Chine	109
	5-111	10-20	Finland	62
	15-1670	25-231	Germany	175
Maize	116-170	23-113	India	78, 195

BIOLOGICAL ACTIVITY

The toxicity of DON has been assessed by WHO (208), IARC (76) and Rotter et al (169). Risk assessments and toxicological reviews of DON have been published for Canada (94,103), the Nordic Council (58), the Netherlands (157) and the European Union (43). The acute symptoms of DON poisoning are charac-terized by skin irritation, feed refusal, vomiting, diarrhoea, haemorrhage, neural disturbance, abortion and death. The LD_{50} values for DON and its 3-acetyl and 15-acetyl metabolites admini-stered orally for different animals are given in Table 3. Transm-ission of DON into eggs has been reported by Prelusky et al (161). Low concentration of free and conjugated DON was found in milk of cows given a single oral dose of 920 mg/kg of DON in *Fusarium* contaminated maize (160). DON is a ubiquitous toxin and it was recently the subject of an International meeting (77). Guidance levels for DON toxicity were issued in 1982. Acceptable DON level was 1 ppm/kg for finished wheat products meant for human consumption and 4 ppm/kg for wheat and wheat milling by-products used in animal feed (207). The FDA advisory levels of DON are tabulated in Table 4. Dietary studies have been conducted in rats (131,196), rabbits (86), pigs (210) and mice (85). The DON given intraperitoneal (ip) to pregnant mice exhibited symptoms similar to that of Kashin-Beck disease (50). General signs of DON toxicity in animals include weight loss, decreased feed conversion, feed refusal, vomiting, bloody diarrhoea, severe dermatitis, hemorrhage and decreased egg production. Based on experimental information available on rats, it is concluded that other domestic animals also respond in a similar fashion (169). However, they differ with regard to the absorption, distribution, metabolism, and elimination of DON (32,154). Experimental studies reveal that it affects the immune system (155,193). Neuroendocrine toxic effects of DON on human and animal have been reviewed in detail by Pestka and Smolinski (154).

Table 3. Oral Toxicity of Deoxynivalenol

Species	Route	LD50 (mg/ KG bw)	Effects	Reference
Mouse	Oral	46	Maternal toxicity or embryotoxicity	211
Mouse	Oral	78	Adverse effects on epididymides	66
Mouse	Intra-peritoneal	43	Intestinal necrosis, renal glomerular lesions	192
Mouse	Oral	34	Decreased kidney, spleen weights	211
Mouse	Oral	56	Reduced feed intake, decreased thymus weight, changes in lymphnodes, gut	66
Chicken	Oral	140	Haematological, serum and histological parameters	75
Duck	Sub-cutaneous	27	Cardiac lesions	211

Table 4. FDA Advisory levels for DON

Class of Animal	Portion of Diet	Maximum DON level (in ppm)
Humans	Finished wheat products (flour, bran and germ)	1
Beef and cattle	Grain and grain by-products Not to exceed 50% of diet	10
Chickens	Grain and grain by-products Not to exceed 50% of diet	10
Swine	Grain and grain by-products Not to exceed 20% of diet	5
All other animals	Grain and grain by-products Not to exceed 40% of diet	5

From acute toxicity studies with animals, it is clear that DON might produce similar effects in humans (154) in Asia. Illness in humans such as vomiting, nausea, dizziness and headache can be attributed to the consumption of cereals contaminated with DON. Poor consumption of feed, depressed milk production, diarrhoea and total reproductive failure of cattle is attributed to feed contaminating DON at 500ppb (44) and is responsible for numerous human and animal disorders (51). DON caused multi-organ effect including emesis, nervous disorders, immunosuppr-ession and bone marrow damage in different animals (156,202).

MECHANISM OF ACTION

DON inhibits the synthesis of DNA, RNA and protein at the ribosomal level. The toxin has a hemolytic effect on erythrocytes. The effects are thought to be mediated by affecting the seroton-ergic activity in central nervous system or via peripheral actions on serotonin receptors (58,59,169). The molecular mode of action of DON involves disruption of normal cell structure and function by inhibiting protein synthesis via binding the ribosome and by activating critical cellular kinases involved in signal transduction related to proliferation, differentiation and apoptosis (37,154).

The mechanism of these effects has not been fully understood but alters in the serotonergic activity of both the peripherous and central nervous system may be involved (60). The ingestion of feed containing DON at concentrations that occur in contaminated food and feed results in impairment of intestinal transfer and uptake of nutrients such as glucose and 5-methyl tetrahydrofolic acid (76).

Most DON molecules inhibit protein synthesis, their potency depending on structural substituent's and requiring an unsaturated bond at the C9-C10 position and integrity of the 12, 13-epoxy ring. They bind to the 60s subunit of eukaryotic ribosome's and interfere with the activity of peptidyl transferase. DON, which lacks a substituent at C4, inhibits chain elongation (16,55). Inhibition of protein synthesis is considered to be the primary toxic effect of DON. The ID50 for inhibition of protein synthesis in rabbit reticulocytes was 2 µg/ml (172). Owing to differences in lipophilicity and other possible effects, the toxicity of DON in vivo is greater than would be expected from its effects in vitro (172,186,192). The minimum effective concentration of DON for inhibition of protein and DNA synthesis was 1000 ng/ml (67).

PHARMACOKINETICS

Studies on pharmacokinetics and metabolism indicates that the DON is fairly, rapidly adsorbed by following oral administr-ation and can be metabolized by intestinal tissue in pigs and possibly in humans. DON was responsible for emesis and feed refusal in swine (124) and it has been shown to be a strong emetic agent in dogs and ducklings (129). DON is known to isomerise and decompose under some conditions and form isomers or derivatives. Table 5 summarizes the DON degradation products or biological metabolites of DON. During baking, isomer of DON was known to be formed (21,71). Heating DON under alkaline conditions gave DON degradation products such as the norDON series, which are less cytotoxic (23). Deoxidized DON was not only found in milk and serum of cows fed with DON contaminated feed, but also in microbial inocula from rumen fluid, soil, and contents of the large intestines of chicken (CLIC) and swine (72,213). De-epoxy DON was reported to be about ten times less toxic than DON (61). DON-3- glucronide, a metabolite of DON in animals, was also less toxic than DON (209). The other metabolites of DON identified were 3-keto DON and DON-3-glucoside (DON-3G). Biliary excretion which enters hepatic circulation occurs in rats and mice, while urinary excretion predominates in rabbits. Urinary excretion is also the main route of excretion in pigs in spite of the demonstrated inter-hepatic circulation of DON, owing to high degree of re-absorption in the gut. Increase in post natal mortality was observed in mice (84). Mitogen activated protein kinase (MAPK) are important transducers of signaling events related to immune response and apoptosis and have been linked to DON mediated induction of pro-inflammatory cytokines which result in immunomodulation in mice (152).

Table 5. Toxicity of DON and its derivatives

Compound	Toxicity	Explanation	Reference
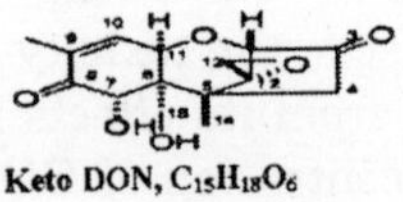 **Keto DON, $C_{15}H_{18}O_6$**	-----	The transformed product of DON under aerobic conditions	**183,200**
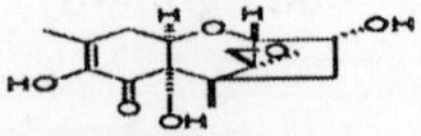 **IsoDON $C_{15}H_{20}O_6$**	------	The transformed product of DON found in bread	**21,71**
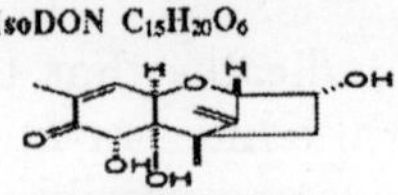 **DeepoxyDON,$C_{15}H_{20}O_5$**	IC50 for inhibition of DNA synthesis in mouse.	The transformed product found in milk and serum of cows fed with contaminated feed.	**61,72,213**
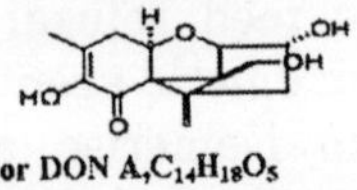 **nor DON A,$C_{14}H_{18}O_5$**	Cytotoxic upto 100mM in Human kidney epithelial IHKE cells	The transformed product found during heating under alkaline condtions	**23**
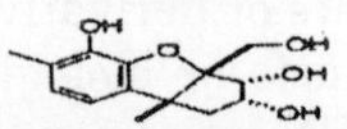 **nor DON B,$C_{14}H_{18}O_5$**	Cytotoxic upto 100mM in Human kidney epithelial IHKE cells	The transformed product found during heating under alkaline condtions	**23**
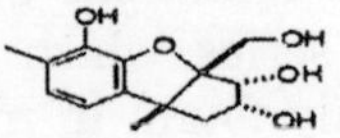 **nor DON C,$C_{14}H_{18}O_5$**	Cytotoxic upto 100mM in Human kidney epithelial IHKE cells	The transformed product found during heating under alkaline condtions	**23**
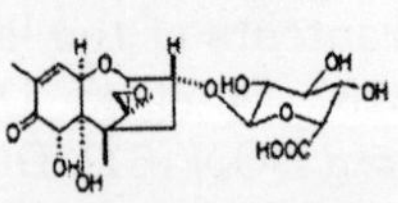 **DON-3-glucuronide,$C_{21}H_{28}O_{12}$**	270mM >can effect on human K 562 erythroleukamia and siginficant effect in immuno toxicity	The transformed product formed during detoxification in plasma or urine	**209**
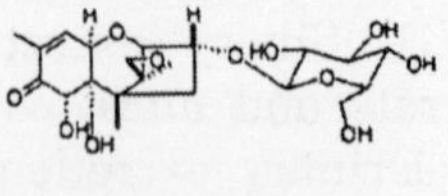 **DON-3-glucoside,$C_{21}H_{30}O_{11}$**	------	The transfomed product found in contaminated cereals	**15,181**

From numerous studies on laboratory animals and cell lines it was concluded that low dose exposure regulates expression of cytokines, chemokines and inflammatory genes with concurrent immune stimulation, whereas high dose exposure promotes leukocyte apoptosis with concomitant immune suppression (155).

MANAGEMENT

(1) Usual strategies both at pre-harvest and post-harvest stage in the management of mycotoxins also apply in true spirit in the management of DON. Methods aimed at preventing or reducing level of DON contamination can be broadly classified as preventive or curative which include:

(i) Prevention of the initial growth of mycotoxigenic fungus and subsequent production of DON.

(ii) Detection of DON in food and selective removal of contaminated portion which, however, reduces the toxin only to negligible level.

(iii) Inactivation or destruction of the DON by chemical, physical and biological means.

(iv) Development of effective and economic methods for controlling the mould growth and subsequent production of DON and use of some absorbents like charcoal, aluminium silicate, minerals, resins, microorganisms, β-glycans, bismuth carbonate, chlorophyllin and leguminous plants are useful in reducing DON (100)

(v) Utilization of genetic resource.

Measures taken to control or minimize *Fusarium* infection will also reduce the formation of DON. Such measures include culture techniques, growing resistant cultivars and use of fungicides or biological antagonists (148). Culture control techniques include suitable crop rotation, appropriate use of fertilizers, irrigation and weed control.

The effectiveness of milling practices for reducing the concentration of DON in flour depends to a large extent on the degree of fungal penetration of the endosperm. General cleaning was partially effective for removal of DON in different varieties of wheat (30,61,159,215). Nowicki et al (140) reported that scouring was effective because the toxin was unevenly distributed on the surface of kernels and DON- producing fungus was also removed from the surface. Numerous studies have documented that DON is heat-stable and very stable during baking at the temperature of 170-350°C, with no reduction of DON concentration (56,98,178). Since DON is water soluble, no reduction was observed in frying in oil (170). Some evidence indicates that DON levels may be reduced during the processing in basic condition such as boiling (140). Accerbi et al (6) reported 62% reduction of DON in wheat during extrusion in the presence of sodium bi sulfate. DON contaminated wheat is a great concern to animal and human health.

Prevention of *Fusarium* toxins including DON has been intensively studied by plant physiologists and plant pathologists all over the world because *Fusarium* is a typical field fungus infecting plants in the field, which results not only in reduction of yield but also accumulation of DON (215). The study on the retention of DON during primary processing (milling) and secondary processing (cooking) is important for the risk assessment and management for the people who live on the wheat-based food products. Heating infested grains is reported to reduce DON only to some extent (210)

DON could be removed on each process prior to milling to some extent. Scott et al (179) showed that cleaning was effective because higher concentration of DON was found in dockage (154). Because of economic feasibility and effective controlling of DON contamination of food and feed by traditional agricultural methods, efforts are being made to understand the molecular biology of the DON biosynthesis which helps to:

(1) Eliminate the toxin from the food chain.

(2) Understand the regulation and evaluation of the DON pathway.

(3) Understand the biological significance of DON to the producing fungus.

Since the Fusaria are opportunistic pathogens of mammals, insects and plants, furthering our understanding of the regulation of gene expression, development and secondary metabolism may provide important clues into their ecology and biology which helps not only in the effective control of DON but also to the control of growth of species of *Fusarium.* Genetic engineering may be useful in the elimination of DON from food and feed and increases host-plant resistance to fungal infection (Fig. 4). Factors associated with the inhibition of DON production in the field are complex and genetics associated with such properties would be equally more complex, increased understanding of the nature, origin, formation of DON may contribute to its control of contamination of foods and feeds.

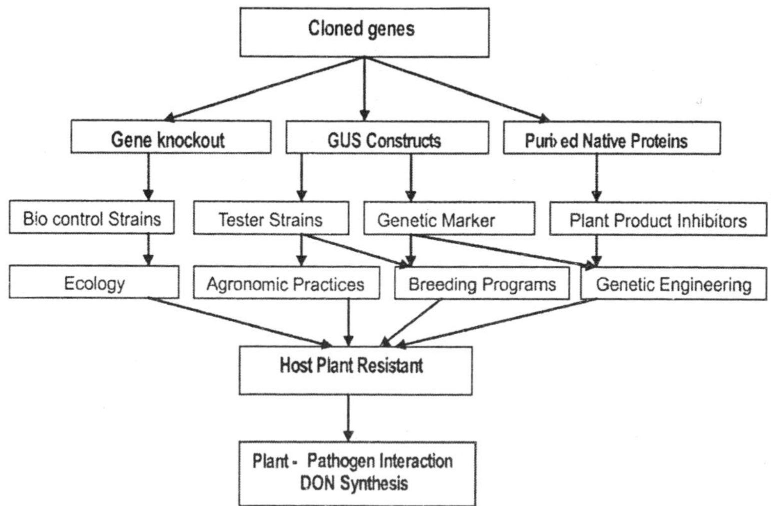

Fig. 4. Tools required for genetic engineering for plant resistance

An application of a new technology, superheated steam, decreased DON significantly in naturally *Fusarium*-infested wheat (36). Reduction of DON to the extent of 48% was achieved with a steam temperature of 185°C during processing time of 6 min. Decrease of DON during heating or process is debatable (1,2,214). Sugita-Konishi et al (187) reported the decrease in cytotoxicity level significantly after baking. Wageha et al (201) reported that the microorganisms present in the intestine of ruminants are responsible for detoxification of DON.

FUTURE PROSPECTS

In spite of intensive studies, a large gap still exists in the understanding of DON, its biological effects and management. Therefore, concentrative efforts are needed in the following areas.

1. Comparative studies of toxicity and toxicokinetics would help to clarify species differences in sensitivity to DON.

2. Additional studies are needed to determine the hormonal effect in pre-pubertal pigs.
3. Most of the studies deal with individual trichothecenes toxin, but interaction of DON and other trichothecenes may elucidate the real picture in nature.
4. In view of the widespread human exposure to DON, more detailed studies are needed to determine genotoxicity as well as its carcinogenicity.
5. Epidemiological studies of human exposure and health hazards should be conducted in those areas of the world, where the presence of scabby wheat or mouldy maize is a cyclic and endemic event.
6. Need for comparative studies on species difference including use of human cells in the metabolism of DON to elucidate long term carcinogenicity.
7. In view of its widespread occurrence, simpler and cheap but precise methods of analysis are to be developed.
8. Additional data on the natural incidence is desirable for reaching decisive conclusions in containing DON.
9. Development of visual methods to distinguish healthy food grains from Fusaria infested grains will avoid consumption of DON.
10. Development of analytical methods which facilitate finding the source of DON contamination where more than one organism is involved in food poisoning.
11. Assessment of DON and NIV in different foods may be assessed with more than one technique such as Gas chromatography coupled with mass spectrometry (GC-MS), and HPLC coupled with tandem mass-spectrometry (LC-MS) which are not only precise but also specific reveal facts of its incidence and possible health hazards (96).

ACKNOWLEDMENTS

Thanks are due to the Head, Department of Microbiology, Kakatiya University, Warangal, UGC, New Delhi and Jawaharlal Nehru Memorial Fund, New Delhi for the financial assistance.

LITERATURE CITED

1. Abbas HK, Mirocha CJ, Pawlosky RJ, and Pusch DJ. 1985. Effect of cleaning, milling, and baking on deoxynivalenol in wheat. *Appl Environ Microbiol* 50: 482-486.
2. Abbas HK, Mirocha CJ, Rosiles R, and Carvajal CJ. 1988. Decomposition of zearalenone and deoxynivalenol in the process of making tortillas from corn. *Cereal Chem* 65: 15-19.
3. Abdel-Aal ESM, Miah K, Young JC, and Rabalski I. 2007. Comparison of four c18 columns for the liquid chromatographic determination of deoxynivalenol in naturally contaminated wheat. *J AOAC Intern* 90: 995–999.
4. Abramovic B, Jajic I, Abramovic B, Cosic J, Juric V. 2007. Detection of deoxynivalenol in wheat by Fourier transforms infrared spectroscopy. *Acta Chem Slov* 54: 859–867.
5. Abouzied MM, Azcona JI, Braselton WE and Pestka JJ. 1991. Immunochemical assessment of mycotoxins in 1989 grain foods: Evidence for deoxynivalenol (vomitoxin) contamination. *Appl Environ Microbiol* 57:672–677.
6. Accerbi M, Rinaldi VE, Ng PK. 1999. Utilization of highly deoxynivalenol contaminated wheat via extrusion processing. *J Food Prot* 62: 1485-1487.
7. Adejumo TO, Hettwer U, Karlovsky, P. 2007. Occurrence of *Fusarium* species and trichothecenes in Nigerian maize. *Int J Food Microbiol* 116: 350-357
8. Alexander NJ, McCormick SP and Ziegenhorn SL. 1999. Phytotoxicity of selected trichothecenes using *Chlamydomonas reinhardtii* as a model system. *Nat Toxins* 7: 265-269.
9. Alexander NJ, Hohn TM, McCormick SP. 1998. The TRI11 gene of *Fusarium sporotrichioides* encodes a cytochrome P-450 monooxygenase required for C-15 hydroxylation in trichothe-cene biosynthesis. *Appl Environ Microbiol* 64: 221-225.
10. Anselme M, Tangni EK, Pussemier L, Motte JC, Van Hove F, Schneider YJ, Van Peteghem C, Larondelle Y. 2006. Comparison of ochratoxin A and deoxynivalenol in organically and conventionally produced beers sold on the Belgian market. *Food Addit Contam* 23: 910–918.
11. AOAC 17th edn, 2000. Official method 986, 17 Deoxyniv-

alenol in wheat Thin layer chromatographic method.

12. Ayalew AH, Fehrmann J, Lepschy R, Beck D, Abate. 2006. Natural occurrence of mycotoxins in staple cereals from Ethiopia. *Mycopathologia* 162: 57-63.

13. Bai GH and Shaner G. 1999. Scab of wheat. Prospects for control. *Plant Dis* 78: 760-766.

14. Bateman GL. 2005. The contribution of ground-level inoculm of *Fusarium culmorum* to ear blight of winter wheat. *Plant Pathol* 54: 299-307.

15. Berthiller FC, Dall Asta R, Schuhmacher M, Lemmens G, Adam R, Krska R. 2005. Masked mycotoxins: determination of a Deoxynivalenol glucoside in artificially and naturally contaminated wheat by liquid chromatography-tandem mass spectrometry. *J Agric Food Chem* 53: 3421-3425.

16. Betina V. 1989. Structure-activity relations among mycotoxins. *Chem Biol Interactions* 71: 105-146.

17. Betina V. (ed.). 1984. *Mycotoxins: production, isolation, separation and purification.* Elsevier/North-Holland Publishing Co., Amsterdam. p. 155.

18. Birzele B, Prange A, and Kramer J. 2000. Deoxynivalenol and ochratoxin A in German wheat and changes of level in relation to storage parameters. *Food Additives Contaminants* 17: 1027-1035.

19. Boehm C, Cichna-Markl M, Brenn-Struckhofova Z and Razzazi-Fazeli E. 2008. Development of a selective sample clean-up method based on immuno-ultrafiltration for the determination of deoxynivalenol in maize. *J Chromatogr A* 1202: 111–117.

20. Bottalico A and Perrone G. 2002. Toxigenic *Fusarium* species and Mycotoxins associated with head blight in small-grain cereals in Europe. *Eur J Plant Pathol* 108: 611-624.

21. Boyacioglu D, Hettiarachchy NS and D'appolonia BL. 1993. Additives affect deoxynivalenol (vomitoxin) flour during bread baking. *J Food Sci* 58: 416-418.

22. Brennan JM, Egan D, Cooke BM and Doohan FM. 2005. Effect of temperature on head blight of wheat caused by *Fusarium culmorum* and *F. graminearum*. *Plant Pathol* 54: 156-160.

23. Bretz M, Beyer M, Cramer B, Knecht A and Humpf HU. 2006. Thermal degradation of the *Fusarium* mycotoxin

deoxynivalenol. *J Agric Food Chem* 54: 6445-6451.

24. Broggi LE, Pacin AM, Gasparovic AC, Sacchi A, Rothermel A, Gallay S and Resnik. 2007. Natural occurrence of aflatoxins, deoxynivalenol, fumonisins and zearalenone in maize from Entre Rios Province, Argentina. *Myco Res* 23: 59-64.

25. Brown DW, McCormick SP, Alexander NJ, Proctor RH and Desjardins AE. 2002. Inactivation of a cytochrome P-450 is a determinant of trichothecene diversity in *Fusarium* species. *Fungal Genet Biol* 36: 224–233.

26. Brown DW, McCormick SP, Alexander NJ, Proctor RH and Desjardins AE. 2001. A genetic and biochemical approach to study trichothecene diversity in *Fusarium sporotrichioides* and *Fusarium graminearum*. *Fungal Genet Biol* 32: 121-133

27. Brumley WC, Trucksess MW, Alder SH, Cohen CK, White KD and Sphon JA. 1985. Negative ion chemical ionization mass spectrometry of deoxynivalenol (DON): application to identification of DON in grains and snack foods after quantitation/isolation by thin layer chromatography. *J Agric Food Chem* 33: 326-339.

28. Bullerman LB. 2002. Processing effects on Mycotoxins: introduction. In: *Mycotoxins and Food safety. Advances in Experimental Medicine and Biology*, vol. 504. Devries JW, Trucksess MW, Jackson LS. (Eds.). Kluwer Academic/ Plenum publishers, New York.

29. Bruns TD, White TJ and Taylor JW. 1991. Fungal molecular systematics. *Ann Rev Ecol Syst* 22: 525-564.

30. Canady RA, Coker RD, Egan SK, Krska R, Kuiper-Goodman T, Olsen M, Pestka J, Resnik SJ and Schlatter. 2001. Deoxynivalenol. In: *WHO/IPCS safety evaluation of certain mycotoxins in food. WHO Food Addit. Ser.* 47: 419-555.

31. Carbone, I and Kohn LM. 2004. Inferring process from pattern in fungal population genetics. In: *Fungal Genomics,* Arora, DK and Khachatourians, GG (Eds.). New York, Elsevier.

32. CAST. 2003. *Mycotoxins: risks in plant, animal and human systems.* Task Force Report no. 139. Council for Agricultural Science and Technol, Ames, IA, pp. 1-191.

33. Castillo MÁ, Montes R, Navarro A, Segarra R, Cuesta G, Hernández E. 2008. Occurrence of deoxynivalenol

and nivalenol in Spanish corn-based food products. *J Food Compost Anal* 21: 423-427.

34. Cavaliere C, D'Ascenzo G, Foglia P, Pastorini E, Samperi R and Laganà A. 2005. Determination of type b trichothecenes and macrocyclic lactone mycotoxins in field contaminated maize. *Food Chem* 92: 559-568.

35. Cavaliere C, Foglia P, Guarino C, Motto M, Nazzari M, Samperi R, Laganà A, and Berardo N. 2007. Mycotoxins produced by *Fusarium* genus in maize: Determination by screening and confirmatory methods based on liquid chromatography tandem mass spectrometry. *Food Chem* 105: 700–710.

36. Cenkowski SC, Pronyk D, Zmidzinska WE, Muir. 2007. Decontamination of food products with superheated steam. *J Food Eng* 83: 68-75.

37. Chayama Bonaziz, Martel C, Ossama SD, Salwa Abid-Essefi Catherine Brenner, Christophe L and Haseen B. 2009. Fusarial Toxin-induced toxicity in cultured cells and in isolated mitochondria involves PTPC- Dependent activation of mitochondrial pathway of Apoptosis. *Toxicological Sci* 110: 363-375.

38. Chu FS. 1996. Recent studies on immunoassays for mycotoxins. In: *Immunoassays for residue Analysis.* Beier RC, Stanker LH. (Eds.), Food Safety. ACS Symposium Series 621. American Chemical Society, Washington, DC, pp. 294-313.

39. Chu FS. 1986. Immunoassay for Mycotoxins. Current state of the art, commercial and epidemiological applications. *Vet Human Toxicol* 32 (suppl.): 42.

40. Cirillo T, Ritieni A, Galvano F, Cocchieri RA. 2003. Natural co-occurrence of deoxynivalenol and fumonisins B1 and B2 in Italian marketed foodstuffs. *Food Addit Contam* 20: 566-571.

41. Clear R and Patrick S. 2006. *Fusarium* head blight in Western Canada. Canadian Grain Commission. < http://grainscanada.gc.ca>.

42. Cole RJ and Cox RH. 1981. *Handbook of toxic fungal metabolites*; Academic Press New York, USA pp. 152-263.

43. Commission of the European Union. 1999. *Opinion on Fusarium toxins. Part 1 Deoxynvivalenol (DON).* Scientific Committee on Food Annex VI to the Minutes o the 119th Plenary Meeting, p 9.

44. Coppeck RW, Mostrom MS, Sparling CG, Jocobsen B and Ross SC. 1990. Apparent zearalenone intoxication in a dairy herd from feeding spoiled acid-treated corn. *Vet Hum Toxicol* 32: 246.

45. Cowger, C. 2005. Effects of moisture, wheat cultivar and infection timing on FHB severity and DON in wheat. In: Proc Nat Fusarium Head blight forum, 11-13 December 2005. Canty SM, Lewis J, Siler L and Ward RW (Eds). Hilton Milwaukee city center, Milwauke, Wisconsin, USA. Pp. 115-116.

46. Creppy EE. 2002 Update of survey, regulation and toxic effects of mycotoxins in Europe. *Toxicol Lett*, 127: 19-28.

47. Cundliffe E, Cannon M and Davies J. 1974. Mechanism of inhibition of eukaryotic protein synthesis by trichothecene fungal toxins. *Proc Nat Acad Sci* USA, 71: 30-34.

48. Danicke S, Valenta H, Klobasa F, Doll S, Ganter M, Flachowsky G. 2004. Effect of graded levels of *Fusarium* toxin contaminated wheat in diets for fattening pigs on growth performance, nutrient digestibility, deoxynivalenol balance and clinical serum characteristics. *Arch Anim Nutr* 58: 1-17.

49. Dawlatana M, Coker RD, Nagler MJ, Gibbs J and Blunden G. 1999. An HPTLC method for the quantitative determination of T-2 toxin and deoxynivalenol in rice. *Chromatographia* 49: 547-551.

50. Debouck CE, Haubruge P, Bollaerts D, Van Bignoot Y, Brostaux A, Werry and Rooze M. 2001. Skeletal deformities induced by the intraperitoneal administration of deoxynivalenol (vomitoxin) in mice. *Int Orthop* 25: 194-198.

51. Desjardins AE. 2006. *Fusarium Mycotoxins: Chemistry, Genetics, and Biology*. St, Paul, MN, USA, APS Press.

52. Dill-Macky R and Jones RK. 2000. The effect of previous crop residues and tillage on *Fusarium* head blight of wheat. *Pl Dis* 84: 71-76.

53. Doohan FM, Brennan JM, Cooke BM. 2003. Influence of climatic factors on *Fusarium* species pathogenic to cereals. *Eur J Plant Pathol* 109: 755-768.

54. DyerRex B, Ronald D, Plattner F, David Kendra and Daren Brown W. 2005. *Fusarium graminearum* TRI14 is Required for High Virulence and DON Production on Wheat but not for DON

Synthesis *in Vitro. J Agric Food Chem* 53: 9281-9287.

55. Ehrlich K and Daigle KW. 1987. Protein synthesis inhibition by 8-oxo-12, 13-epoxytrichothecenes. *Biochem Biophys Acta* 923: 206-213.

56. El-Banna AA, Lau PY and Scott PM. 1983. Fate of mycotoxins during processing of foodstuffs II- Deoxynivalenol (vomitoxin) during making of Egyptian bread. *J Food Prot* 46: 484-486.

57. Eppley RM, Trucksess MW, Nesheim S, Thorpe CW and Pohland AE. 1986. Thin layer chromatographic method for determination of deoxynivalenol in wheat: Collaborative study. *J Assoc Off Anal Chem* 69: 37–40.

58. Eriksen GS and Alexander J. (eds). 1998. *Fusarium Toxins in Cereals — A Risk Assessment.* TemaNord -Report 502. Copenhagen, Nordic Council of Ministers, pp. 10-165.

59. Eriksen GS. 2003. *Metabolism and Toxicity of Trichothecenes*, Doctoral thesis, Uppsala, Sweden

60. Eriksen GS and Pettersson H. 2004. Toxicological evaluation of trichothecenes in animal feed. *Anim Feed Sci Technol* 114: 205-239.

61. Eriksen Sundstol G, Pettersson H and Lundh T. 2004. Comparative cytotoxicity of deoxynivalenol, nivalenol, their acetylated derivatives and de-epoxy metabolites. *Food Chem Toxicol* 42: 619-624.

62. Eskola M, Parikka P and Rizzo A. 2001. Trichothecenes, ochratoxin A and zearalenone contamination and *Fusarium* infection in Finish cereal samples in 1998. *Food Addit Contam* 18: 707-718.

63. Faixova Z, Stefan F, Radka B and Leng Lubomir 2007. Effect of different doses of deoxynivalenol on metabolism in broiler chickens. *Bull Vet Inst Pulawy* 51: 421-424.

64. Fernandez C, Stack ME and Musser SM. 1994. Determination of deoxynivalenol in 1991 U.S. winter and spring wheat by high-performance thin layer chromatography. *J AOAC Int* 77: 628–630.

65. Fink GJ. 1999. Mycotoxins: their implications for human and animal health. *Vet Q* 21: 115-120.

66. Forsell JH, Jensen R, Tai JH, Witt M, Lin WS and Pestka JJ. 1987. Comparison of acute toxicities of deoxynivalenol (vomitoxin) and 15-acetyldeoxynivalenol in the B6C3F1 mouse. *Food Chem Toxicol* 25: 155–162.

67. Friedman L, Gaines DW, Eppley R, Smith MC, Chi RK and Braunberg RC. 1996. Effects of T-2 toxin (T2), deoxynivalenol (DON), zearalenone (Z) and DON+Z on macromolecular synthesis by male rat spleen slices. *FASEB J.*, 10, A458.

68. Gentili AF, Caretti G, D'Ascenzo LM, Rocca S, Marchese S, Materazzi D and Perret. 2007. Simultaneous determination of trichothecenes A, B, and D in maize food products by LC-MSMS. *Chromatographia* 66: 669-676.

69. Godtfredsen WO, Grove JF and Tamm CH. 1967. Trichothecenes. *Hev Chim Acta* 50: 1666–1668.

70. Gordon WC and Gordon LJ. 1990. Rapid screening method for deoxynivalenol in agricultural commodities by fluorescent minicolumn. *J Assoc Off Anl Chem* 73: 266.

71. Greenhalgh R, Gilbert J, King R, Blackwell LJ, Startin BA, Shepherd MJ. 1984. Synthesis, characterization and occurrence in bread and cereal products of an isomer of 4-deoxynivalenol (vomitoxin). *J Agric Food Chem* 32: 1416-1420.

72. He P, Young LG and Forsberg C. 1992. Microbial transformation of deoxynivalenol (vomitoxin). *Appl Environ Microbiol* 58: 3857-3863.

73. Hohn TM and Desjardins AE. 1992. Isolation and gene disruption of the *Tox5* gene encoding trichodiene synthase in *Gibberella pulicaris*. *Mol. Plant-Microbe Interact* 5: 249-256.

74. Hohn TM, Beremand PD. 1989. Isolation and nucleotide sequence of a sesquiterpene cyclase gene from the trichothecene-producing fungus *Fusarium sporotrichioides*. *Gene*, 79: 131-138.

75. Huff WE, Doerr JA, Hamilton PB and Vesonder RF. 1981. Acute toxicity of vomitoxin (deoxynivalenol) in broiler chickens. *Poultry Sci* 60: 1412–1414.

76. Hunder GK, Schumann G, Strugala J, Gropp B, Fichtl and Forth W. 1991. Influence of subchronic exposure to low dietary deoxynivalenol, a trichothecene mycotoxin, on intestinal absorption of nutrients in mice. *Food Chem Toxicol* 29: 809–814.

77. IARC (International Agency for Research on Cancer. 1993. Toxins derivated from *Fusarium graminearum, F. culmorum* and *F. crookwellense*: zearalenone, deoxynivalenol and fusarenone X. *IARC Monogr Eval Carcinog Risks Hum* 56: 397-444.

78. ICSI (International life Sciences Institute), 2004. Trichothecenes with a special Focus on DON. Summary report of a Workshop Held in September 2003. ILS Europe. Brussels, Belgium, pp. 1-36.

79. Janardhana GR, Raveesha KA and Shetty HS. 1999. Mycotoxin contamination of maize grains grown in Karnataka (India). *Food Chem Toxicol* 37: 863–868.

80. JECFA. 2001. Deoxynivalenol. (Joint FAO/WHO Expert Committee on Food Additives). Fifty sixth meeting. http://www.inchem.org/ documents/jecfa/jecmono/v47jeo5.htm.

81. Kamimura H, Nishizima M, Yasuda K, Saito K, Ibe A, Nagayama T, Ushiyama N. and Naoi Y. 1981. Simultaneous detection of several *Fusarium* Mycotoxins in cereal grains and food stuffs. *J Assoc Off Anal Chem*. 64: 1067-1073.

82. Kang Z, and Buchenauer H. 1999. Immunocytochemical localization of *Fusarium* toxins in infected wheat spikes by *Fusarium culmorum*. *Physiological and Molecular Plant Pathology* 55: 275-288.

83. Keller NP and Hohn TM. 1997. Metabolic pathway gene clusters in filamentous fungi. *Fungal Genet Biol* 21: 17–29

84. Khera KS, Arnold DL, Whalen C, Angers G and Scott PM. 1984. Vomitoxin (4-deoxynivalenol): Effects on reproduction of mice and rats. *Toxicol Appl Pharmacol* 74: 345–356.

85. Khera KS, Whalen C, Angers G, Vesonder RF and Kuiper-Goodman T. 1982. Embryotoxicity of 4-deoxynivalenol (vomitoxin) in mice. *Bull Environ contami Toxicol* 29: 487-491.

86. Khera KS, Whalen C and Angers G. 1986. A teratology study on vomitoxin (4-deoxynivalenol) in rabbits. *Food Chem Toxicol* 24: 421-424.

87. Kimura M, Kaneko I, Komiyama M, Takatsuki A, Koshino H, Yoneyama K, and Yamaguchi I. 1998. Trichothecene 3-O-acetyl transferase protects both the producing organism and transformed yeast from related mycotoxins. Cloning and characterization of Tri101. *J Biol Chem* 273: 1654-1661.

88. Kokkonen M, Jestoi M and Rizzo A. 2005. Determination of selected mycotoxins in mould cheeses with liquid chromatography coupled to tandem with mass spectrometry. *Food Addit Contam* 22: 449–456.

89. Kos G, Lohninger H, Mizaikoff B, Krska R. 2007. Optimisation of a sample preparation procedure for the screening of fungal infection and assessment of deoxynivalenol content in maize using mid-infrared attenuated total reflection spectroscopy. *Food Addit Contam* 24: 721–729.

90. Koshinsky HA and Khachatourians GG. 1992. Bioassay for deoxynivalenol based on the interaction of T-2 toxin with trichothecene Mycotoxins. *Bull Environ Contam Toxicol* 49: 246-51.

91. Krishnamurthy N, Mathew AG, Nambudiri ES, Shivashankar VS, Lewis VC and Natarajan CP. 1976. Oil and oleoresin of turmeric. *Tropical Sci* 18(1): 37-45.

92. Krska R, Baumgartner S and Josephs R. 2001. The state-of-the-art in the analysis of type-A and B trichothecenes Mycotoxins in cereals. Fresenius *J Anal Chem* 371: 285-299.

93. Krysińska-Traczyk E, Perkowski J, and Dutkiewicz J. 2007. Levels of fungi and mycotoxins in the samples of grain and grain dust collected from five various cereal crops in Eastern Poland. *Ann Agric Environ Med* 14: 159-167.

94. Kuiper-Goodman T. 1985. Potential human health hazards and regulatory aspects. In: *Mycotoxins: A Canadian Perspective.* Scott PM, Trenholm HL & Sutton MD (Eds.), (NRCC No. 22848), National Research Council of Canada, Ottawa, pp. 103–111.

95. Kuiper-Goodman T. 2002. Recent developments in the risk assessment of deoxynivalenol. In session 2: *Toxicology, quality and impact on industry.* Second Canadian Workshop on *Fusarium* head blight. Ottawa: http://res2.agr.gc.ca/ecorc/fusarium 01/session2.pdf.

96. Kushiro M, Tanaka K, Miyazaki S and Nagata T. 2006. Advances of liquid chromatographic determination of fumonisins; potential mycotoxins for humans. *Curr Pharm Anal* 2: 289-297.

97. Lacey J, Bateman GL, and Mirocha CJ. 1999. Effects of infection time and moisture on development of ear blight and deoxynivalenol production by *Fusarium* spp. in Wheat. *Ann Appl Biol* 134: 277-283.

98. Lancova K, Hajslova J, Kostelanska M, Kohoutkova J, Nedelnik J, Moravcova H and Vanova M. 2008. Fate of trichothecene mycotoxins

during the processing: milling and baking. *Food Addit Contam* 25: 650-659.

99. Langseth W, and Rundberget T. 1999. The occurrence of HT-2 toxin and other trichothecenes in Norwegian cereals. *Mycopathologia* 147: 157-165.

100. Larsen JC, Hunt J, Perin I, and Ruckenbauer P. 2004. Workshop on trichothecenes with a focus on DON: Summary report. *Toxicol Lett* 153: 1-22.

101. Lauren DR and Greenhalgh R. 1987. Simultaneous analysis of nivalenol and deoxynivalenol in cereals by liquid chromatography. *J. Assoc Off Anal Chem* 70: 479-483.

102. Lauren DR and Smith WA. 2001. Stability of the *Fusarium* mycotoxins nivalenol, deoxynivalenol and zearalenone in ground maize under typical cooking environments. *Food Addit Contam* 18: 1011-1016.

103. Lauret B, Videmann M, Mazallah M and Lecoeur S. 2010. Assessment of DON adsorbents and characterization *in vitro* tests. *Food Addi Contam* 27: 43-53.

104. Lawrence JF and Scott PM. 2000. Determination of mycotoxins and phycotoxins. In: *Sample Handling and Trace Analysis: Techniques, Applications and Quality Assurance*, Barceló D (Ed.), Amsterdam, Elsevier, pp. 413–456.

105. Lee YoungMi, Yong-Keel Choi and Byung-Re Min. 2000. PCR-RFLP and sequence analysis of the rDNA ITS region in the *Fusarium* spp. *J Microbiol* 38: 66-73.

106. Lee T, Oh D, Kim W, Lee HS, Kim J, Yun SH and Lee YW. 2001. Identification of deoxynivalenol- and nivalenol-producing chemotypes of *Gibberella zeae* by using PCR. *Appl Environ Microbiol* 67: 2966–2972.

107. Lee T, Han YK, Kim KH, Yun S H and Lee YW. 2002. Tri13 and Tri7 determine Deoxynivalenol and nivalenol-producing chemotypes of *Gibberella zeae*. *Appl Environ Microbiol* 68: 2148-2154.

108. Leslie Hall S, Wohlfiel and Roberts GD. 2004. Experience with the MicroSeq D2 large-subunit ribosomal DNA sequencing kit for the identification of filamentous fungi encountered in the clinical laboratory. *J Clin Microbiol* 42: 622-626.

109. Li FQ, Luo XY and Yoshizawa T. 1999. Mycotoxins (trichothecenes, zearalenone and fumonisins

in cereals associated with human red-mold intoxications stored since 1989 and 1991 in China. *Nat Toxins* 7: 93-97.

110. Lip polis V, Pascale M and Visconti A. 2006. Optimization of a fluorescence polarization immunoassay for rapid quantification of deoxynivalenol in durum wheat-based products. *J Food Prot* 69: 2712-2719.

111. Llorens A, Mateo R, Hinojo M, Valle-algarra FM and Jimenez M. 2004. Influence of environmental factors on the biosynthesis of type B trichothecenes by isolates of *Fusarium* spp. from Spanish crops. *Int J Food Microbiol* 94: 43-54.

112. Logrieco A, Mule G and Bottalico A. 2002. Toxigenic *Fusarium* species and mycotoxins associated with maize ear rot in Europe. *Eur J Plant Pathol* 108: 597-609.

113. Lombaert GA, Pellaers P, Roscoe V, Mankotia M, Neil R and Scott PM. 2003. Mycotoxins in infant cereal foods from the Canadian retail market. *Food Addit Contam* 20: 494-504.

114. Malone BR, Humphrey CW, Romer TR and Richard JL. 1998 One-step solid phase extraction clean-up and fluorometric analysis of deoxynivalenol in grains. *J AOAC Int* 81: 448–452.

115. Mankevičienė A, Butkutė B, Dabkevičius Z and Suproniené S. 2007. *Fusarium* mycotoxins in Lithuanian cereals from the 2004-2005 harvests. *Ann Agric Environ Med* 14: 103-107.

116. Matsumato G, Wuchiyama J, Shingu Y, Kimura M, Yoneyama K and Yamaguchi I. 1999. The trichothecene biosynthesis regulating gene from the type B producer *Fusarium* strains: sequence of *Tri6* and its expression in *Escherichia coli*. *Biosci Biotechnol Biochem* 63: 2001-2004.

117. McCormick SP and Alexander NJ. 2002. *Fusarium* Tri8 encodes a trichothecene C-3 esterase. *Appl Environ Microbiol*, 68: 2959-2964.

118. McCormick SP, Alexander NJ and Proctor RH. 2006. *Fusarium* Tri4 encodes a multifunctional oxygenase required for trichothecene biosynthesis. *Can J Microbiol* 52: 636-642.

119. McCormick SP, Alexander NJ, Trapp SE and Hohn TM. 1999. Disruption of TRI101, the gene encoding trichothecene 3-O-acetyltransferase, from *Fusarium sporotrich-*

ioides. Appl Environ Microbiol 65: 5252-5256.

120. McCormick SP, Hohn TM and Desjardins AE. 1996. Isolation and characterization of Tri3, a gene encoding 15-O-acetyltransferase from *Fusarium sporotrichioides. Appl Environ Microbiol* 62: 353-359.

121. Meek IB, Peplow AW, Ake C, Jr. Phillips TD and Beremand MN. 2003. Tri1 encodes the cytochrome P450 monooxygenase for C-8 hydroxylation during trichothecene biosynthesis in *Fusarium sporotrichioides* and resides upstream of another new Tri gene. *Appl Environ Microbiol* 69: 1607-1613.

122. Menetewab A, Rezanoor HN, Gosman N, Worland AJ and Nicholson P. 2000. Chromosomal location of Fusarium head blight resistance genes and analysis of the relationship between resistance head blight and brown foot rot. *Plant Breeding,* 119: 15-20.

123. Milanez TV, Valente-Soares LM and Baptista, GG. 2005. Occurrence of trichothecene mycotoxins in Brazilian corn-based food products. *Food Control* 17: 293-298.

124. Mirocha CJ, Pathre SV, Schauerhamer B and Christensen CM. 1976. Natural occurrence of *Fusarium* toxins in feedstuff. *Appl Environ Microbiol* 32: 553-556.

125. Mohmed S, Khalil Mohmed A, Abdel-Sattar, Ibrahim N, Aly Kamel A, Abd-Elsalam and Joseph A Verreet. 2003. Genetic affinities of *Fusarium* spp. and their correlation with origin and pathogenicity. *Afr J Biotechnol* 2: 109-113.

126. Moller TE and Gustavsson HF. 1992. Determination of type A and B trichothecenes in cereals by gas chromatography with electron capture detection. *J AOAC Internat* 75: 1049–1052

127. Molto G, Samar MM, Resnik S, Martinez EJ and Pacin A. 2000. Occurrence of trichothecenes in Argentinean beer. A preliminary exposure assessment. *Food Addit Contam* 17: 809–813.

128. Morgan MRA and Lee HA. 1990. Mycotoxins and natural food toxicants. In: *Development and application of immunoassay for food analysis.* JH Rittenburg (Ed.). Elsevier Appl Sci, London.

129. Morgavi DP and Riley RT. 2007. An historical overview of field disease outbreaks known or suspected to be caused by consumption of feeds contaminated with

Fusarium toxins. *Anim Feed Sci Technol* 137: 201-212.

130. Morooka N, Uratsuji N, Yoshizawa T and Yamamoto H. 1972. Studies on the toxic substances in barley infected with *Fusarium* spp. *J Food Hyg Soc Jpn* 13: 368–375.

131. Morrissey RE and Vesonder RF. 1985. Effect of deoxynivalenol (vomitoxin) on fertility, pregnancy, and postnatal development of Sprague-Dawley rats. *Appl Environ Microbiol* 49: 1062-1066.

132. Muthomi JW, Schutze A, Dehne HW, Mutitu EW and Oerke EC. 2000. Characterization of *Fusarium culmorum* isolates by Mycotoxins production and aggressiveness to winter wheat. *J Plant Dis Prot* 107: 113-123.

133. Nakajima M, Terada H, Hisada K, Tsubouchi H, Yamamoto K, Uda T, Itoh Y, Kawamura O and Ueno Y. 1990. Determination of ochratoxin A in coffee beans and coffee products by monoclonal antibody affinity chromatography. *Food Agric Immunol* 2: 189.

134. Nelson PE, Toussoun TA and Marasas WFO 1983. *Fusarium species: An illustrated Manual for identification.* Penn State Univ Press, Univ Park, Pennsylvania.

135. Ngundi MM, Qadri SA, Wallace EV, Moore MH, Lassman ME, Shriver-Lake LC, Ligler FS and Taitt CR. 2006. Detection of deoxynivalenol in foods and indoor air using an array biosensor. *Environ Sci Technol* 40: 2352-2356.

136. Ngundi MM, Shriver-Lake LC, Moore MH, Lassman ME, Ligler FS and Taitt CR. 2005. Array biosensor for detection of ochratoxin A in cereals and beverages. *Anal Chem* 77: 148-154.

137. Ngundi MM, Shriver-Lake LC, Moore MH, Ligler FS and Taitt C.R. 2006 Multiplexed detection of mycotoxins in foods with a regenerable array. *J Food Prot* 69: 3047-3051.

138. Niessen ML and Vogel RF. 1998. Group specific PCR-detection of potential trichothecene producing *Fusarium* species in pure cultures and cereal samples. *Systematic Appl Microbiol* 21: 618-631.

139. Nita M, Tilley K, De Wolf E and Kuldau G. 2005. Effects of moisture during and after anthesis on the development of Fusarium head blight of wheat and Mycotoxins production. In *Proc 2005 National Fusarium Head blight Forum,* Canty SM, Lewis J, Siler L and Ward

RW (Eds.), 11-13 December 2005, Hilton.

140. Nowicki TW, Gaba DG, Dexter JE, Matsuo RR and Clear RM. 1988. Retention of the *Fusarium* mycotoxin deoxynivalenol in wheat during processing and cooking of spaghetti and noodles. *J Cereal Sci* 8: 189-202.

141. O'Donnell K. 1992. Ribosomal DNA internal transcribed spacers are highly divergent in the phytopathogenic ascomycete *Fusarium sambucinum* (*Gibberella pulicaris*). *Curr Genetics* 22: 213-220.

142. O'Donnell K, Ward TJ, Geiser DM, Kistler HC and Aoki T. 2004. Genealogical concordance between mating type locus and seven other nuclear Genes support formal recognition of nine phylogenetically distinct species within the *Fusarium graminearum* clade. *Fungal Genet Biol* 41: 600–623.

143. Obst A and Fuchs H. 2000. The contamination of winter and spring sown cereals with *Fusarium* - results from random seed samples produced in Bavaria 1987-99. *Bayer Lande Bodenk Pflanzen Munch* 4: 21-25.

144. Ostry V and Skarkova J. 2000. Development of an HPTLC method for the determination of deoxynivalenol in cereal products. *J Pl Chrom – Modern TLC,* 13: 443–446.

145. Pan D, Bonsignore F, Rivas F, Perera G and Bettucci L. 2007. Deoxynivalenol in barley samples from Uruguay. *Int J Food Microbiol* 114: 149-152.

146. Papadopoulou-Bouraoui A, Vrabcheva T, Valzacchi S, Stroka J and Anklam E. 2004. Screening survey of deoxynivalenol in beer from the European market by an enzyme-linked immunosorbent assay. *Food Addit Contam* 21: 607–617.

147. Park JW, Choi SY, Hwang HJ, Kim YB. 2005. Fungal mycoflora and mycotoxins in Korean polished rice destined for humans. *Int J Food Microbiol* 103: 305-314.

148. Parry DW, Jenkinson P and McLeod L. 1995. *Fusarium* ear blight (scab) in small grain cereals - a review. *Plant Pathol* 44: 207-238.

149. Peplow AW, Tag AG, Garifullina GF and Beremand MN. 2003. Identification of new genes positively regulated by *Tri10* and a regulatory network for trichothecene mycotoxin production. *Appl Environ Microbiol* 69: 2731-2736.

150. Perkowski J, Jelen H, Kiecana I and Golinski P. 1997. Natural contamination of spring barley with groups A trichothecene mycotoxins in south-eastern Poland. *Food Addit Contam* 14: 321–325.

151. Perkowski J and Basiński T. 2002. Natural contamination of oat with group A trichothecene Mycotoxins in Poland. *Food Addit Contam* 19: 478-482.

152. Pestka JJ. 2008. Mechanisms of deoxynivalenol induced gene expression and apoptosis. *Food Addit Contam* 24: 1–13.

153. Pestka JJ, Yan D and King LE. 1994. Flow cytometric analysis of the effects of in vitro exposure to vomitoxin (deoxynivalenol) on apoptosis in murine T, B and IgA+ cells. *Food Chem Toxicol* 32: 1125-1136.

154. Pestka JJ and Smolinski AT. 2005. Deoxynivalenol: Toxicology and potential effects on humans. *J Environ Sci Health B*, 8: 39-69.

155. Pestka JJ, Zhou H, Moon Y and Chung YJ. 2004. Cellular and molecular mechanisms for immune modulation by deoxynivalenol and other trichothecenes: unravelling a paradox. *Toxicol Lett* 153: 61-73.

156. Pestka JJ, Tai JH, Witt MF, Dixon DE and Forsell JH. 1987. Suppression of immune response in the B6c3f mouse after dietary exposure to the *Fusarium* mycotoxin deoxynivalenol (vomitoxin) and Zearalenone. *Food Chem Toxicol* 25: 297-304.

157. Pieters MN, Fiolet DCM and Baars AJ. 1999. Deoxynivalenol: *Derivation of concentration limits in wheat and wheat containing food products* (RIVM Natural Toxins Project Report 388802 018). Bilthoven: National Institute of Public Health and the Environment, p. 32.

158. Placinta CM, D'Mello JBF and Macdonald AMC. 1999. A review of world contamination of cereal grains and animal feed with *Fusarium* mycotoxins. *Anim Feed Sci Techno* 78: 21-37

159. Poapolathep A, Ohtsuka R, Kiatipattanasakul W, Ishigami N, Nakayama H and Doi K 2002. Nivalenol-induced apoptosis in thymus, spleen and Peyer's patches of mice. *Exp Toxicol Pathol* 53: 441-446.

160. Prelusky DB, Trenholm HL, Lawrence GA and Scott PM. 1984. Nontransmission of deoxynivalenol (vomitoxin) to milk following oral administration to dairy cows. *J*

Environ Sci Health B, 19: 593–609.

161. Prelusky DB, Veira DM, Trenholm HL and Foster BC. 1987. Metabolic fate and elimination in milk, urine and bile of Deoxynivalenol following administration to lactating sheep. *J Environ Sci Health B*, 22, 125–148.

162. Proctor RH, Hohn TM, McCormick SP and Desjardins AE. 1995. *Tri6* encodes an unusual zinc finger protein involved in regulation of trichothecene biosynthesis in *Fusarium sporotrichioides*. *Appl Environ Microbiol* 61: 1923-1930.

163. Proctor RH, Hohn TM and McCormick SP. 1995. Reduced virulence of *Gibberella zeae* caused by disruption of a trichothecene toxin biosynthetic gene. *Mol Plant-Microbe Interact* 8: 593-601.

164. Ramakrishna Y, Bhat RV and Ravindranath V. 1989. Production of deoxynivalenol by *Fusarium* isolates from samples of Wheat associated with a human mycotoxicosis outbreak and from sorghum cultivars. *Appl Environ Microbiol* 34: 2619-2620.

165. Rasmussen P, Ghorbani H and Berg T. 2003. Deoxynivalenol and other *Fusarium* toxins in wheat and rye flours on the Danish market. *Food Addit Contam* 20: 396-404.

166. Razzazi-Fazeli E, Böhm J and Luf W. 1999. Determination of nivalenol and deoxynivalenol in wheat using liquid chromatography–mass spectrometry with negative ion atmospheric pressure chemical ionization. *J Chromatogr A*, 854: 45–55.

167. Richard JL, Bennett GA, Ross PF and Nelson PE. 1993. Analysis of naturally occurring Mycotoxins in feed stuffs and food. *J Animal Sci* 71: 2563-2574.

168. Rocha O, Ansari K and Doohan FM. 2005. Effects of trichothecene mycotoxins on eukaryotic cells. *Food Addit Contam* 22: 369-378.

169. Rotter BA, Prelusky DB and Pestka JJ. 1996. Toxicology of deoxynivalenol (vomitoxin). *J Toxicol Environ Health* 48: 1-34.

170. Samar MM, Resnik SL, Gonzalez HHL, Pacin AM and Castillo MD.2007. Deoxynivalenol reduction during the frying process of turnover pie covers. *Food Control* 18: 1295-1299.

171. Sano A, Satoshi M, Masao S and Shoji T. 1987. High-performance liquid chromato-

graphic method for determining trichothecene mycotoxins by post-column fluorescence derivatization. *J Chromatogr* 410: 427–436.

172. Sato N and Ueno Y. 1977. Comparative toxicities of trichothecenes. In *Mycotoxins in Animal and Human Health* Rodricks JV, Hesseltine CW & Mehlman MA, (Eds.),, Park Forest South, Illinois: Pathodox.

173. Schnerr H, Vogel RF, Niessen L. 2002. A biosensor-based immunoassay for rapid screening of deoxynivalenol contamination in wheat. *Food Agric Immuno,* 14: 313–321.

174. Schollenberger M, Jara HT, Suchy S, Drochner W and Müler HM. 2002 *Fusarium* toxins in wheat flour collected in an area in southwest Germany. *Int J Food Microbiol* 72: 85-89.

175. Schollenberger M, Suchy S, Jara HT, Drochner W and Müller HM. 1999. A survey of *Fusarium* toxins in cereal-based foods marketed in an area of southwest Germany. *Mycopathologia* 147: 49-57.

176. Schrödter R. 2004. Influence of harvest and storage conditions on trichothecene levels in various cereals. *Toxicol Lett* 153: 47-49.

177. Scott PM. 1993. Gas chromatography of mycotoxins. In *Chromatography of Mycotoxins — Techniques and Applications,* Betina V, (Ed.), Amsterdam: Elsevier, pp. 373–425.

178. Scott PM, Kanhere SR, Lau PY, Dexter JE and Greenhalgh R. 1983. Effects of Experimental flour milling and bread baking on retention of deoxynivalenol (vomitoxin) in hard red spring wheat. *Cereal Chem* 60: 421-424.

179. Scott PM, Kanhere SR, Dexter E, Brennan PW and Trenholm HL. 1984. Distribution of the trichothecene mycotoxin deoxynivalenol (vomitoxin) during the milling of naturally contaminated hard red spring wheat and its fate in baked products. *Food Addit Contam* 1: 313-323.

180. Scott PM. 1995. Mycotoxin methodology. *Food Addit Contam* 12: 395-403.

181. Sewald N, Lepschy von Gleissenthall J, Schuster M, Muller G and Aplin RT. 1992. Structure elucidation of a plant metabolite of 4-desoxynivalenol. *Tetrahedron Asymmetr* 3: 953–960.

182. Shepherd MJ and Gilbert J. 1988. Long-term storage stability of deoxynivalenol standard reference solutions.

J Agric Food Chem 36: 305-308.

183. Shima J, Takase S, Takahashi Y, Iwai Y, Fujimoto H, Yamazaki M and Ochi K. 1997. Novel detoxification of the trichothecene mycotoxin deoxynivalenol by a soil bacterium isolated by enrichment culture. *Appl Environ Microbiol* 63: 3825-3830.

184. Snijders CHA and Perkowski J. 1990. Effects of head blight caused by *Fusarium culmorum* on toxin content and weight of kernels. *Phytopathology* 80: 566–570.

185. Sudakin DL 2003. Trichothecenes in the environment: Relevance to human health. *Toxicol Lett* 143: 97-107.

186. Sugiama KI, Muro M, Tonamoto KI, Nishijima M and Sugita Konishi Y. 2010. DON or NIV inhibited lipopolysaccharide induced expression of interferon b (IFN-B) which both play an important role for host-protection against invading pathogen. *Toxicol Lett* 192: 150-154.

187. Sugita-Konishi Y, Park BJ, Kobayashi-Hattori K, Tanaka T, Chonan T, Yoshikawa K and Kumagai S. 2006. Effect of cooking process on the deoxynivalenol content and its subsequent cytotoxicity in wheat products. *Biosci Biotechnol Biochem* 70: 1764-1768.

188. Sypecka Z, Kelly M and Brereton P. 2004. Deoxynivalenol and zearalenone residues in eggs of laying hens fed with a naturally contaminated diet: Effects on egg production and estimation of transmission rates from feed to eggs. *J Agric Food Chem* 52: 5463–5471.

189. Tag AG, Garifullina GF, Peplow AW, Ake C, Jr. Phillips TD, Hohn TM and Beremand MN. 2001. A novel regulatory gene, Tri10, controls trichothecene toxin production and gene expression. *Appl Environ Microbiol* 67: 5294-5302.

190. Taga M, Waalwijk C, Flier WG and Kema GHJ. 2003. Cytological karyotyping of somatic chromosomes from *Phytophthora infestans, Mycosphaerella graminicola* and *Fusarium* spp. *Fungal Genet Newsl* 50 (suppl): 468.

191. Tanaka T, Hasegawa A, Yamamoto S, Lee US, Sugiura Y and Ueno Y. 1988. Worldwide contamination of cereals by the *Fusarium* mycotoxin nivalenol, deoxynivalenol and zearalenone. 1. Survey of 19 countries. *J Agric Food Chem* 36: 980-983.

192. Thompson WL and Wannemacher RW Jr. 1986. Struct-

ure–function relationships of 12, 13-epoxytrichothecene mycotoxins in cell culture: Comparison to whole animal lethality. *Toxicon* 24: 985-994.

193. Thuvander A, Wikman C and Gadhasson I. 1999 In vitro exposure of human lymphocytes to trichothecenes: Individual variation in sensitivity and effects of combined exposure on lymphocyte function. *Food Chem Toxicol* 37: 639–648.

194. Trail F, Gaffoor I and Vogel S. 2005. Ejection mechanics and trajectory of the ascospores of *Gibberella zeae* (anamorph *Fusarium graminearum*). *Fungal Genetics Biol* 42: 528-533.

195. Trucksess MW, Thomas F, Young K, Stack ME, Fulgueras WJ and Page SW. 1995. Survey of deoxynivalenol in US 1993 wheat and barley crops by enzyme-linked immunosorbent assay (ELISA). *J AOAC Int* 78: 631–636.

196. Tutelian VA, Krinitskaia NA, Avreneva LI, Kuzmina EE Levitskaia AB and Kravchenko LV. 1991. [Embryotoxic effects of deoxynivalenol mycotoxin (vomitoxin) in rats]. *Gig Sanit* 10: 49–51 (in Russian).

197. Ueno Y. 1989. Trichothecene Mycotoxins: mycology, chemistry, and toxicology, *Adv Nutr Res* 3: 301-353.

198. Van der Gaag B, Spath S, Dietrich H, Stigter E, Boonzaaijer G, van Osenbruggen T, and Koopal K. 2003. Biosensors and multiple mycotoxin analysis. *Food Control* 24: 252–254.

199. Vesonder RF, Ellis JJ and Rohwedder K. 1981. Elaboration of vomitoxin and zearalenone by *Fusarium* isolates and the biological activity of *Fusarium*-produced toxins. *Appl Environ Microbiol* 42: 1132-1134.

200. Völkl A, Vogler B, Schollenberger M and Karlovsky P. 2004. Microbial detoxification of mycotoxin deoxynivalenol. *J Basic Microbiol* 44: 147-156.

201. Wageha A, Awad Khaled Ghareeb, Joseph Bohm and Jugen Zentek. 2010 Decontamination and detoxification strategies for the *Fusarium* mycotoxin deoxynivalenol in animal feed and the effectiveness of microbial biodegradation. *Food Addit Contam* 27: 510-520.

202. Wannemacher Jr., RW, Brunner, DL and Neufeld, HA. 1991. Toxicity of trichothecenes and other Mycotoxins in laboratory animals. In *Mycotoxins and Animal foods.* Smith JE, Henderson, RS

(Eds.). CRC press, Boca Raton. pp. 499-552.

203. Ward TJ, Bielawski JP, Kistler HC, Sullivan E and O'Donnell K. 2002. Ancestral polymorphism and adaptive evolution in the trichothecene mycotoxin gene cluster of phytopathogenic *Fusarium*. *Proc Natl Acad Sci USA* 99: 9278–9283.

204. Widestrand J and Pettersson H. 2001. Effect of time, temperature and solvent on the stability of T-2 toxin, HT-2 toxin, deoxynivalenol and nivalenol calibrants. *Food Addit Contam* 18: 987-992.

205. Wijnands LM and Van Leusden FM. 2000. *An overview of adverse health effects caused by Mycotoxins and bioassays for their detection* RIVM report 257852 004, Bilthoven.

206. Williams SS. 1984. *Official methods of analysise of the association of official analytical chemists.* AOAC, Arlington, VA.

207. Wood GE and Carter L. 1989. Limited survey of deoxynivalenol in wheat and corn in the US. *J Asso Off Anal Chem.* 72: 38-40.

208. World Health Organisation, IPCS, 1990. Trichothecenes: In: *Selected Mycotoxins: Ochratoxins, Trichothecenes, Ergot.* World Healt Organisation, Geneva.

209. Wu X, Murphy P, Cunnick and Hendrich S. 2007 Synthesis and characteri zation of deoxynivalenc glucuronide: Its comparativ immunotoxicity wit deoxynivalenol. *Food Chen Toxicol* 45: 1846-1855.

210. Yumbe Guevana BE, Imoto and Yoshizawa T. 2003. Effec of heating procedures o DON, NIV and ZEA levels i naturally contaminate barley and wheat. *Foo Contam* 20: 1132-1140.

211. Yoshizawa T and Morooka N 1974. Studies on the toxi substances in infected cereals acute toxicities of new trich othecene mycotoxins: deoxyn valenol and its monoacetat *J Food Hyg Soc Jpn* 1 261-269.

212. Yoshizawa T, Kohno H, Iked K, Shinoda T, Yokohama H Morita K, Kusada O an Kobayashi Y. 2004. A pra tical method for measurin deoxynivalenol, nivaleno and T-2 + HT-2 toxin in foo by an enzyme-linked imm nosorbent assay using mon clonal antibodies. *Bios Biotechnol Biochem* 6 2076-2085.

213. Yoshizawa T, Cote S Swanson WB and Buck. 198 Confirmation of DOM-1, a d

epoxidation metabolite of deoxynivalenol, in biological fluids of lactating cows. *Agric Biol Chem* 50: 227-229.

214. Young JC, Fulcher RG, Hayhoe JH, Scott PM and Dexter JE. 1984. Effect of milling and baking on deoxynivalenol (vomitoxin) content of eastern Canadian wheats. *J Agric Food Chem* 32: 659-664.

215. Yuen GY and Schoneweis SD. 2007. Strategies for managing *Fusarium* head blight and deoxynivalenol accumulation in wheat. *Int J Food Microbiol* 119: 126-130.

216. Zollner P and Mayer-Helm B. 2006. Trace mycotoxin analysis in complex biological and food matrices by liquid chromatography-atmospheric pressure ionization mass spectrometry. *J Chromatogr A*, 1136: 123-169.

BIODIVERSITY, TAXONOMY AND SIGNIFICANCE OF AQUATIC CONIDIAL FUNGI

C. Manoharachary and I.K. Kunwar

Department of Botany, Osmania University,
Hyderabad-500 007, (A.P.), India

KEY WORDS: Conidial fungi, decomposition, foam, hyphomycetes, stream

ABSTRACT

Aquatic fungi predominantly live and multiply in fresh water and are significantly involved in the energy transfer and nutrient cycling through decomposition in aquatic habitats. They are mainly zoosporic or conidial fungi having branched, stauroform, sigmoid or other shaped spores. The shape and sizes are of biological significance. The conidia are the asexual phases of Ascomycotina and Basidiomycotina whose perfect states are either unknown or lacking. The foam, submerged dark brown leaves and water harbor such fungi. The present paper deals with the biodiversity, taxonomy and significance of aquatic conidial fungi.

Fungi colonize, survive, sporulate and multiply in/on diversified habitats such as water, foam, soil, air, plants, fruits, litter, dead organic matter, submerged dark brown to skeletonized leaves, keratin substrates, insects, fishes, animals and animal excreta etc. They vary in their morphology, reproduction, phylogeny, distribution, community composition, activity and occupy varied ecological niches. The types of fungi that occur on different habitats and extremophilic habitats have greater potentialities and biotechnological importance. Ecotypes, strains, biotypes and variants do occur. Further, about 29,000 fungi are reported from India and water-borne conidial fungi form a small fraction only (49). In recent times this field is getting some attention in India.

Biodiversity

Biodiversity encompasses the spectrum of variability among all types of living forms in nature and as selected by human intervention. Microbial diversity is an invisible resource but it deserves great attention. Already several billion dollar products have been discovered by exploration of vast diversity of fungi. There are fungi that tolerate or require extreme environmental conditions and are often endowed with unique metabolic machinery that has been at the central core of biotechnology industry. About 1.5 million fungi are estimated to exist on diversified habitats (37). Only 98,613 fungi are reported globally (44), out of which less than 5% prevalent fungal population is maintained as pure cultures in germplasm banks. Several genera and species have not been kept conserved in pure culture for undertaking further physiological and biotechnological studies.

Water as a medium

Water is an important resource besides being an ecological niche for a number of fungi. Water moulds/aquatic fungi have water as medium of life and are adapted to this environment. These fungi are of predominant occurrence and/or prevalent not only in pools ponds, lakes, rivers, streams and marshes but also in marginal and terrestrial habitats. Some forms of aquatic fungi are also reported in seawater, salt pools, logs, mangroves, tree holes, and thermal springs. Majority of these fungi live as saprobes and some being biotrophic on algae, fish, eggs, protozoa, aquatic plants and others. The distinct substrates in water which differ in their nutrient status and physicochemical setup support characteristic fungal communities. Fungi are known to colonize, multiply, sporulate and live in water besides completing their life cycle in water. The most prevalent groups of fungi that live in water are the zoosporic fungi, conidial fungi, trichomyctes, marine fungi and yeasts etc.

Aquatic fungi live in water, especially freshwater in contrast to marine fungi. Submerged natural substrates in aquatic habitats harbor diverse fungi viz. chytridiomycetes, meiosporic and mitosporic ascomycetes and basidiomycetes, and non-fungal saprolegniales of the oomycetes. The chief zoosporic fungi of freshwater are from Chytridiomycota and Oomycota especially Chytridiales and Saprolegniales which are commonly known as water molds. Chytrids parasitic on algae are also found both in fresh and marine water habitats. Some water molds are parasitic on fishes.

Fungal species reported from freshwater aquatic habitats can be indwellers/residents or immigrants/transients. "Indwellers" are completely adapted to aquatic environments and are able to grow and sporulate in water and are habitually modified to dispersal in water, while "immigrants" should frequently immigrate from other habitats to sustain their population in water (59). About the occurrence of fungi in aquatic habitats, it is often difficult to precisely characterize where species fall along this gradient of adaptation to aquatic habitats (83).

Though the water-borne conidial fungi were first discovered by de Wildeman (21) more than a century ago, but the existence of such fungi was appreciated only after the pioneering work of Ingold (40). These fungi are known only in the asexual or mycelial state and their perfect states are either unknown or may be lacking. Thus the so called aquatic conidial fungi are an artificial group of fungi frequently having branched or sigmoid spores. They are bound together by their common attribute of growing on submerged leaves/wood and the conidia of such fungi are readily trapped in foam besides sporulating under submerged conditions. These fungi represent the asexual phases of primarily Ascomycotina and also few Basidiomycotina.

In the review article by Shearer et al (83) based on presence/absence of data for a large number and variety of aquatic habitats approximately 3047 taxa have been reported from aquatic habitats thus far (Table 1), out of which 138 species belong to Saprolegniales and 576 to Chytridiomycota. The largest taxonomic group of fungi in aquatic habitats is comprised of meiosporic and mitosporic *Ascomycota,* followed by the *Chytriodiomycota.*

Table 1. The number of species in various taxonomic groups commonly found in aquatic habitats [from Shearer et al 2007 (83)]

Taxonomic group	No. of species
Chytridiomycota	576
Freshwater meiosporic ascomycetes	450
Mangrove meiosporic ascomycetes	612
Marine meiosporic and mitosporic ascomycetes	465
Ingoldian mitosporic fungi	290
Aeroaquatic mitosporic fungi	90
Miscellaneous mitosporic fungi	405
Saprolegniales	138
Basidiomycetes from freshwater habitats	11
Basidiomycetes from brackish and marine habitats	10
Total number of taxa	3047

Aquatic conidial fungi

Leaves falling into clean and well aerated streams and other water-bodies quickly get replaced by a distinguishing mycoflora of fungi-imperfecti, the aquatic hyphomycetes. These fungi develop on all sorts of leaves of deciduous trees and shrubs including needles of conifers. They also occur in foam or scum formed at barriers to water flow, thermal springs, tree canopies and as endophytes. Aquatic conidial fungi reported from various habitats in India are presented in Table 2. Meticulous extensive quantitative and qualitative investigations are necessary for the suitable understanding of the taxonomy, morphology, ecology and physiology of aquatic/water-borne conidial fungi. Hence mycologists have to wet their feet to make significant contribution to the aquatic mycology of India.

Aquatic hyphomycetes are broadly defined as anamorphic fungi typically with relatively large branched (often stauroform) and scolecoform conidia occurring mainly in lotic waters (83). They generally colonize submerged decaying tree leaves and woody debris. The widespread stauroform shape of conidia is assumed to facilitate colonization of substrates by contacting the surface simultaneously on three points by arm tips (99).

Table 2. Aquatic conidial fungi reported from India from various habitats viz, streams (26,64-66,71-73,79,80,94), foam (14,47,48,67,88,93), thermal springs (15), endophytes (58,74,75,76,78) and tree holes (42)

S. No.	Fungal species	Reference
1.	*Acaulopage tetraceros* Drechsler	75,6
2.	*Actinospora megalospora* Ingold	64-66,71,72,87,88
3.	*Alatospora acuminata* Ingold	15, 26,47,64,73,75,76,78, 87, 88
4.	*Alatospora flagellata* (J. Gönczöl) Marvanová	80
5.	*Alatospora pulchella* Marvanová	73,75,76,80
6.	*Anguillospora aquatica* Nilsson	64
7.	*Anguillospora crassa* Ingold	14,42,64,71-73,75,76,78
8.	*Anguillospora curvula* Iqbal	65,66
9.	*Anguillospora filiformis* Greath.	65,66
10.	*Anguillospora furtiva* Descals	80,94
11.	*Anguillospora longissima* (Sacc. & Sydow) Ingold	15,26,42,64-66,72,73, 75, 76,78,87, 88,98
12.	*Anguillospora gigantea* Ranzoni	65
13.	*Anguillospora* sp.	14
14.	*Arborispora palma* Ando	42
15.	*Articulospora moniliforma* Ranzoni	65,66
16.	*Articulospora tetracladia* Ingold	64,65,71,93
17.	*Bacillispora aquatica* Nilsson	64
18.	*Brachiosphaera tropicalis* Nawawi	87,88
19.	*Beltrania rhombica* Penzig	15,64-66,93

20.	*Beltrania* sp.	47
21.	*Blodgettia indica* Subram.	64
22.	*Camposporium antennatum* Harkness	25,64
23.	*Camposporium pellucidum* (Grove) Hughes	15,26,64,73,93
24.	*Campylospora chaetocladia* Ranzoni	42,64-66,71-73,75,76, 79,87,88
25.	*Campylospora filicladia* Nawawi	42,64-66,93
26.	*Campylospora parvula* Kuzuha	15,64,65,73,75,76,87,88
27.	*Chaetendophragmia triangularia* Matsushima	64
28.	*Chaetospermum chaetosporium* Smith & Ramsbottom	64,93
29.	*Chaetospermum elasticae* Koorders	47,67
30.	*Clavariana aquatica* Nawawi	26,65,66
31.	*Clavariopsis aquatica* de Wild	14,64,71,73,75,76,78,79, 93
32.	*Clavariopsis azlanii* Nawawi	64
33.	*Clavariopsis brachycladia* Tubaki	26,72
34.	*Clavariopsis bulbosa* Anastasiou	47,64
35.	*Clavatospora filiformis* Nawawi	72
36.	*Clavatospora longibrachiata* (Ingold) Sv. Nilsson ex Marvanová & Sv. Nilsson	48
37.	*Clavatospora tentacula* Sv. Nilsson	42,65,66,72,87,88
38.	*Condylospora spumigena* Nawawi	64,93
39.	*Cylindrocarpon aquaticum* (Nils.) Maranová & Descals	75
40.	*Cylindrocarpon destructans* Wollenweber	64
41.	*Cylindrocarpon* sp.	42
42.	*Dactylella aquatica* (Ingold) Ranzoni	14,64
43.	*Dactylella oviparasitica* Stirling & Mankau	64

44.	*Dactylella submersa* (Ingold) Nilsson	64
45.	*Dendromyces lobatum* Plakidas & Edgerton	64
46.	*Dendrospora erecta* Ingold	42,64
47.	*Diplocladiella longibrachiata* Nawawi & Kuthub.	73
48.	*Diplocladiella scalaroides* G Arnaud ex MB Ellis	26,47,64-67,72,76, 93
49.	*Dimorphospora foliicola* Tubaki	73,80
50.	*Dwayaangam dichotoma* Nawawi	80
51.	*Entomophthora* sp.	48
52.	*Flabellocladia gigantea* Nawawi	80
53.	*Flabellospora crassa* Alasoadura	42,64,87,88
54.	*Flabellospora curvula* Ingold	87, 88
55.	*Flabellospora multiradiata* Nawawi	15,42,64-66,87,88
56.	*Flabellospora* sp.	48
57.	*Flabellospora penicilloides* Ingold	48,72
58.	*Flabellospora verticillata* Alasoadura	15,26,42,64-66,72
59.	*Flabellospora tetracladia* Nawawi	87,88
60.	*Flagellospora curvula* Ingold	15,42,64
61.	*Flagellospora pencillioides* Ingold	15,26,42,64,73,79,87,88
62.	*Flagellospora saccata* Marvanová & Bärlocher	64
63.	*Geniculospora inflata* (Ingold) Sv. Nilsson ex Marvanová & Sv. Nilsson	48
64.	*Guilicidospora* sp.	72
65.	*Helicomyces colligatus* Moore	65,66
66.	*Helicomyces roseus* Link	64,76
67.	*Helicomyces torquatus* LC Lane & Shearer	65
68.	*Helicomyces* sp.	15,93

69.	*Helicosporium vegetum* Nees	64
70.	*Helicosporium* sp.	15,42,87,88
71.	*Heliscella stellatacula* Marvanová	73
72.	*Heliscina campanulata* Marvanová	64
73.	*Heliscus lugdunensis* Sacc. & Therry	73,75,76,78,79
74.	*Heliscus submersus* Hudson	48
75.	*Henicospora coronata* Kirk & Sutton	93
76.	*Hydrometrospora symmetrica* J Gönczöl & Révay	65,66
77.	*Ingoldiella hamata* Shaw	42,64,87,88,93
78.	*Isthmotricladia gombakiensis* Nawawi	26,42,64-66
79.	*Isthmotricladia laeensis* Matsushima	15,42,64
80.	*Jaculispora submersa* Hudson & Ingold	64
81.	*Laridospora appendiculata* (Anastasiou) Nawawi	64,65,93
82.	*Lateriramulosa uni-inflata* Matsushima	26,65,66
83.	*Lemonniera alabamensis* RC Sinclair & Morgan-Jones	80
84.	*Lemonniera aquatica* de Wild	48,64
85.	*Lemonniera cornuta* Ranzoni	73,75,76,80
86.	*Lemonniera pseudofloscula* Dyko	73,75,76,80
87.	*Lemonniera terrestris* Tubaki	73,75,76,80
88.	*Lunulospora curvula* Ingold	14,15,26,42,47,64-66,71-73,75,76, 78,87,88, 93
89.	*Lunulospora cymbiformis* Miura	15,42,64,73,79,87,88,93
90.	*Magdalaenaea monogramma* G Arnaud	80
91.	*Margaritispora aquatica* Ingold	72
92.	*Mycocentrospora acerina* (Hartig) Deighton	64
93.	*Mycoleptodiscus lateralis* Alcorn & B Sutton	66

94.	*Nawawia filiformis* (Nawawi) Marvnová	64,93
95.	*Obstipispora* sp.	87,88
96.	*Pestalotia* sp.	65,66
97.	*Pestalotiopsis submersus* Sati & N Tiwari	73,75,76,78
98.	*Phalangispora constricta* Nawawi & Webster	15,42,64-66,93
99.	*Pleurophragmium sonam* Sati & N Tiwari	73
100.	*Pyramidospora casuarinae* Nilsson	64
101.	*Pyramidospora constricta* Singh	26,64
102.	*Robillarda sessilis* Sacc.	47
103.	*Robillarda phragmitis* Cummel	64
104.	*Robillarda* sp.	93
105.	*Scolecobasidium anellii* Graniti	64
106.	*Setosynnema isthomosporum* Shaw & Sutton	73
.07.	*Speiropsis hyalospora* Subram. & Lodha	15,64-66
.08.	*Speiropsis irregularis* Peterson	26
09.	*Speiropsis pedatospora* Tubaki	15,64-66,87,88
10.	*Speiropsis scopiformis* Kuthub. & Nawawi	73
11.	*Speiropsis* sp.	93
12.	*Subulispora procurvata* Tubaki	64
13.	*Tetrachaetum elegans* Ingold	26,64,73,75,76,78
14.	*Tetracladium apiense* Sinclair & Eicker	73
15.	*Tetracladium marchalianum* de Wild.	64,73,75,76,78
16.	*Tetracladium nainitalensis* Sati & P Arya	74
17.	*Tetracladium setigerum* (Grove) Ingold	64-66,73,75,76,78,93
18.	*Tetraploa aristata* Berk. & Br.	15,42,47,64,65,66, 87,88,93
19.	*Tetraploa* sp.	15
20.	*Tricellula aquatica* J Webster	48

121.	*Tricladiopsis flagelliformis* Descals	80
122.	*Tricladium angulatum* Ingold	42,65,66
123.	*Tricladium brunneum* Nawawi & Kuthubutheen	15,64
124.	*Tricladium chaetocladium* Ingold	73
125.	*Tricladium splendens* Ingold	42,65,66,72,93
126.	*Tricladium terrestre* Park	94
127.	*Tricladium* sp.	93
128.	*Trinacrium incurvum* Matsush.	80
129.	*Trinacrium robustum* Tzean & Chen	42
130.	*Trinacrium subtile* Riess	42
131.	*Tripospermum cameleopardus* Ingold, Dunn & PJ McDougall	42,65,66,93
132.	*Tripospermum myrti* (Lind.) Hughes	64,71,73,87,88,93
133.	*Triscelophorus acuminatus* Nawawi	15,42,64-66,73,79,87, 88, 93
134.	*Triscelophorus konajensis* Sridhar & Kaveriappa	42,64-66
135.	*Triscelophorus magnificus* Peterson	72
136.	*Triscelophorus monosporus* Ingold	14,15,26,42,47,64-66,7 73,79,87,88, 93
137.	*Triscelophorus* sp.	15
138.	*Trisulcosporium acerinum* HJ Huds. & B Sutton	71
139.	*Varicosporium elodeae* W Kegel	42,71,72
140.	*Varicosporium* sp.	93
141.	*Vermispora cauveriana* Rajash., Bhat & Kaver.	64
142.	*Vermispora* sp.	15
143.	*Wisneriomyces javanicus* Koorders	26,65,66
144.	*Wisneriomyces laurinus* (Tassi) Kirk	15,64

145.	*Ypsilina graminea* (Ingold, PJ McDougall & Dann) Descals, J Webster & Marvanová	42

Foam occurring on water surfaces traps/accumulates scoleco-form and stauroform conidia and several species were initially described and isolated in pure culture from foam. Scoleco- and stauroform conidial shapes are also relatively common in terrestrial hyphomycetes occurring in the permanently or intermittently humid layer of forest plant litter (83). The presence of branches is believed to facilitate movement on thin water films on wet litter surfaces (4).

Aquatic conidial fungi - taxonomy

Conidial fungi are basically Fungi Imperfecti which are present in asexual phase or mycelial state. A number of these fungi carry out sexual recombination through para-sexuality. Conidia are asexual, non motile and deciduous propagules, formed either from undifferentiated hyphae or well defined conidiophores with or without any enlargement of the recognizable conidium initial. In several fungi, the conidiophores occur singly, while in a few they are grouped together in parallel fascicles called synnema. Shorter aggregations of conidiophores are identified as sporodochia. Conidia are also formed in fructifications like pycnidia or acervuli. Generally the conidia occurring in water are hyaline. Conidial ontogeny or conidiogenesis and other parameters form the basis for classification. Numerous researchers have worked on this aspect (10,16,48,52,57,92,96). Descals (20) has suggested that karyological, cytological and structural components have to be taken into consideration as their growth and form are genetically controlled. He has suggested that the anamorph should be studied as a whole i.e. mycelium, fertile organs and propagules. Further, he also recommended that the mature structures need to be described along with form, size, wall and or protoplasmic characters, insertion, direction, arrangement and the number along with developmental process. The growth, elongation, branching, transformation, cell division, cell lysis, secession etc of elements and organs also have to be considered as flexible criteria. According to the article 2 of ICBN the whole fungus has been termed as holomorph. Many conidial fungi exist in asexual phase (anamorph) while their perfect stages are either of Ascomycotina or Basidiomycotina (teleomorph). Our attempts of characterization should now be directed towards describing the whole fungus instead of proliferating new genera and species based on anamorph. Manoharachary and Rama Rao (48) while emphasizing the prevalence and powers of fungi in aquatic ecosystem have listed conidial fungi which they have isolated, cultured, identified and classified. However they have emphasized the importance of characterization of such fungi based on molecular techniques.

Freshwater aeroaquatic mitosporic fungi

Aeroaquatic fungi live in submerged plant litter in a diversity of shallow stagnant to slow-flowing freshwater bodies. Usual habitats with a high variety of aeroaquatic fungi comprise shallow woodland ponds and ditches (83). As an adjustment to this habitat, many aeroaquatic fungi can withstand prolonged periods of oxygen depletion and are reasonably tolerant against sulfide (23,24). Nevertheless, aeroaquatic fungi cannot complete their life cycle on submerged substrates, since their spores/dispersal units are produced aerially only above the water level. As a result of this they are dependent on periodical drying of their habitats. These spores/dispersal units are the most prominent characteristics for this group. They consist of intricate, usually multicellular, structures of an amazing morphological diversity which entrap air and hence are extremely buoyant (83). Studies on this group of fungi from Indian sub-continent are meager.

Miscellaneous freshwater mitosporic ascomycetes

Miscellaneous freshwater mitosporic ascomycetes comprise hyphomycetes and coelomycetes excluding aeroaquatic and aquatic hyphomycetes. They commonly comprise of hyaline and dematiaceous hyphomycetes whose conidia are not typically adapted for the aquatic environment. Four hundred and five species of miscellaneous mitosporic fungi have been reported from freshwater thus far [refer to Freshwater Ascomycetes database (http://www.life.uiuc.edu/fungi/) for a species list] (83).

Miscellaneous mitosporic ascomycetes in freshwater are found chiefly on decaying herbaceous plant material and woody debris. Submerged woody debris sustains a higher species diversity of miscellaneous mitosporic ascomycetes though this may be a sign of a lack of study of herbaceous debris employing methods that would help in the discovery of species in this group (83).

Park (59) classified miscellaneous mitosporic ascomycetes into two major groups as indwellers and immigrants. Species in several genera of freshwater mitosporic ascomycetes like *Aquaphila* (31), *Canalisporium* (56), *Candelosynemma* (39), *Clohesyomyces* (38), *Conioscyphopsis* (30), *Enridescalsia* (70), *Paracryptophiale* (46), and *Yinmingella* (32) etc can be classified as indwellers since they have been reported only from freshwater habitats. Whereas, species in genera such as *Acrodictys, Acrogenospora, Arthrobotrys, Bactrodesmium, Berkleasmium, Brachysporiella, Cordana, Dictyosporium, Ellisembia, Spadicoides, Sporidesmiella, Sporochisma,* and *Trichocladium* can be classified as immigrants because they are reported frequently from terrestrial as well as freshwater habitats (83).

Tree canopies and tree holes

In spite of wide-ranging studies on floristic and ecological functions of water-borne fungi in woodland streams, our understanding of their importance, vertical distribution and ecological functions in riparian tree canopies is insufficient (86). While these fungi are in danger of extermination due to unidirectional flow of water in streams, they overcome such situation by colonizing stationary substrates in the stream column (e.g. roots, wood) and adjacent to stream banks (e.g. wood, leaf litter) (42). On the other hand, tree canopy surfaces (e.g. leaves, bark) and habitats (e.g. tree holes, epiphytes) also serve as likely refuge for these fungi. Incidence of typical aquatic and water-borne conidial fungi have been reported from a variety of canopy habitats such as throughfall, stemflow, intact leaves, trapped leaf litter and canopy snow (17, 34, 86). Studies on their occurrence in tree canopies in India are meager. Sridhar et al (86) were the first in India to study fungi associated with the fern *Drynaria quercifolia* epiphytic on riparian trees of a coastal stream of southwest India. They recorded 16 water-borne hyphomycetes (Table 2). About 75% of the taxa in the stream were found to be associated with the fern.

Ando (1) based on his studies of dews and rain drops on intact tree leaves, surmised that some fungi with micronematous conidiophores bearing staurosporous conidia evolved on canopies rather than streams and termed them as 'terrestrial aquatic hyphomycetes' (e.g. *Arborispora, Ceratosporium, Dicranidion,*

Dwayaangam, Microstella, Tetraploa, Titaea, Tricladiella, Tricranium, Tripospermum, Trisulcosporium). Ando and Tubaki (2) are of the opinion that these hyphomycetes may live on intact leaf surfaces and are capable of using several forms of water (rain, fog, dew etc for sporulation) for their survival. These fungi have been adapted to hold water around the conidium, which increases the chances of conidial germination. Canopy water drain, such as throughfall and stemflow, furnishes a number of nutrients and minerals (e.g. nitrogen, phosphorus, calcium, potassium, magnesium) (81), thus meeting the mineral necessity of canopy fungi by intermittent water flows through the canopy.

Tree holes with a substantial amount of detritus (e.g. leaves, twigs, wood) and continuous flow of nutrient rich water in the rainy season makes them an ideal habitat for water-borne fungi (42). Karamchand and Sridhar (42) for the first time from India listed 34 water-borne conidial fungi from tree holes of west coast and Western Ghats of India (Table 2). They observed that banyan (*Ficus bengalensis*) tree holes are highly diverse during the summer and monsoon seasons in Konje, but only during summer in Sampaje. The majorities of conidia recovered from tree canopies were multiradiate and probably resist total removal from the leaf or bark surface in contrast to streams (87).

Endophytes

Microbes living inside tissues of healthy plants lacking disease symptoms are endophytes. They occupy a unique ecological niche and have major influence on distribution, ecology, physiology and biochemistry of plants (90). Fisher and Petrini (25) reported aquatic hyphomycetes like *Tricladium splendens* and *Campylospora parvula* as endophytes in *Alnus gluitinosa* roots. Later endophytic aquatic hyphomycetes were reported by many from abroad (50,51,84). In India aquatic hyphomycetes as root endophytes have been reported from a variety of host roots such as *Equisetum, Botrychium, Alnus glutinosa, Betula papyrifera, Picea glauca, Lyonia ovalifolia, Murraya koenjii,* strawberry, fern bryophytes etc from Himalayas (Table 2) (58,74-77). Endophytic fungi are also recognized to be a good resource of antibiotics secondary metabolites and it has been postulated that endophytic fungi may be advantageous to their host plant by antagonizing pathogens and / or by inducing plant defense responses (76).

Overestimation and underestimation

According to Shearer et al (83) the species biodiversity within and among habitats may be overestimated. Within aquatic habitats, meiosporic/mitosporic connections are not adequately known hence the same fungus may be counted twice, once as its meiosporic morph and a second time as its mitosporic morph. Synonymies amongst species also occur, perhaps in many a times due to the lack of published keys and complications of species descriptions for this ecological group; terrestrial species occasionally washed into aquatic habitats and reported as aquatic. Amongst habitats, various taxonomic specialists may work at different ends of an ecotone and the same fungal species may be counted twice in the overall estimation of fungal diversity (83).

The chances of underestimation of the number of aquatic fungal species probably far outweigh the chances of over estimation. The type of fungi one collects from an aquatic habitat is very technique dependant (82,83). Normally, a researcher uses one or two techniques for a single study, thereby ignoring all the species the techniques do not detect. This too is a source of under estimation (83). Water being a mobile medium several techniques need to be discovered to provide exact data on quantitative estimate or fungal biomass.

Siginficance

Aquatic hyphomycetous fungi play an important ecological role in energy flow in running waters with mainly decomposition of allochthonous organic matter (43). They contribute the most to the microbial biomass associated with the leaves in streams (35,36). They commonly occur in freshwaters and produce tetrara-diate, sigmoid or variously shaped conidia adapted to dispersal in flowing water and adhesion to decaying plant tissues (7). Aquatic hyphomycetes are responsible for the initial breakdown of leaves and for enhancing the palatability of leaf detritus, thus providing a more suitable food source for shredders which further contribute to the conversion of leaf litter into fine particles and other decomposition products (8,11,33,91).

Aquatic hyphomycetes are a major component of complex microbial communities implied to be key contributors to the conversion of organic compounds in aquatic ecosystems due to their ubiquity and ability to degrade natural organic matter wit the help of a number of extracellular hydrolytic enzymes (61,62 Aquatic fungi, thus, can help to lessen organic contaminants in th course of stimulation of degradative microbial populations.

Aquatic hyphomycetes dominate leaf decomposition i streams and their biomass is an important component in the die of leaf-eating invertebrates (5,85,100). Studies involving exposin the leaves in streams, demonstrated that fungal biomas (estimated by ergosterol levels) quickly increases to a peak of up t 17% of total detritus mass (27,28,85) and it may remain at th level for a while before slowly declining.

Aquatic hyphomycetes appear to have a certain degradativ potential for the breakdown of organic pollutants, for example th conversion of polycyclic aromatic hydrocarbon metabolites int hydroxylated and methylated compounds and certain phase I products. Further, intracellular enzymes, extracellular oxidoredu tases such as laccases found to be produced by diverse aquat hyphomycetes strains may also be implicated in the degradation organic pollutants acting as endocrine disruptors, as indicated b studies employing pure fungal cultures (63).

Fungi occurring in aquatic ecosystems play a significant ro in degradation of water contaminants with endocrine activit Technical nonylphenol (t-NP), an endocrine-disrupting compoun mixture was degraded by *Clavariopsis aquatica* and anoth mitosporic fungus from river water (41). They have potential f use in waste water treatment.

Some species can tolerate extreme conditions such as high p (8.6 and above), low oxygen concentration as % to saturation (4. mg 1^{-7} and below), high ion content (10.91–11.28 mg 1^{-7}), har water (170-204 mg 1^{-7}) indicating that some species of aquat hyphomycetes can tolerate adverse environmental conditio (6,53) and help in the degradation of detritus. According Rajashekhar and Kaveriappa (64) for some of the less commo species tropical water temperature or some factor may b unfavourable.

Pollution

Numerous studies have demonstrated that chemical alteration aquatic ecosystems can have deleterious effects on fung ommunities. Streams contaminated with heavy metals (12,22) nd organic compounds (3,69) have recorded a decline in aquatic yphomycetes diversity. Likewise in rivers polluted with mineral nd organic pollutants (29) leaf breakdown rates are affected and rastically reduced in acidified ecosystems (11,19,55). Such studies eed more authentic data through laboratory experiments and eld evaluation.

Quite a few studies have listed the impact of organic ollutants such as pesticides and herbicides on fungal decompo- tion of plant litter. Frequently a decline in species richness in olluted habitats was recorded by Kraus et al (45), as compared to npolluted habitats, indicating sensitivity towards such ntamination. On the basis of their studies on highly heavy metal olluted surface and groundwater of a district in central Germany ith mining history Kraus et al (45) suggest that some strains ay represent ecotypes, where distinct generic and physiological laptations to differentially contaminated habitats have been volved. Since culturing of these fungi poses problems, hence con- usive documentation could not be envisaged by several workers. owever, it is worthwhile to investigate such possibilities.

Moreirinha et al (54) studied the effects of Cadmium and nenanthrene mixtures on the activity and diversity of fungi sociated with decomposing leaf litter in streams and observed at they pose risk to aquatic biodiversity and stream ecosystem nctions.

ibstrate specificity

ibstrate specificity or preference among the hyphomycetes is tablished (62,95) to some extent. Though, involvement of more an one factor in limiting the occurrence of certain species cannot ruled out (85). Aquatic hyphomycetes are predominantly protrophs, but several taxa are capable of biotrophy (e.g. *ycocentrospora acerina, Gyoerffyella entomobryoides*) or myco- rasitism (*Crucells subtilis, Naiadella fluitans*) (83). Randomized oorts indicate several substrates and substrates like bird oppings, organic detritus and others are only accidental and nporary.

Identification and characterization

Classical identification of fungi requires cultivation of th organism concerned. It is not possible to culture all the organism. The use of molecular biological technique in the study of fung species is a relatively new science. Traditional methods fo identifying fungi involve studies of their morphology, and some ar grown in culture. These methods are generally slow and can ofte be inconclusive or even misleading. Molecular techniques, base on either immunology or DNA analysis, offer the possibility efficient and more reliable means of diagnosis. There ar opportunities to determine the true taxonomic relationships different fungi. Taxonomy or the classification and naming populations of organisms have developed to satisfy two needs. Th first is an inventory of biodiversity which later also seeks determine the evolutionary or phylogenetic relationships. Th second is the practical need of being able to identify groups organisms with shared characteristics, which are of ecological an or economic importance. One of the chief causes of taxonom uncertainty has been inadequate sampling of variation with populations. Comparative analysis of the base sequence ribosomal DNA (r DNA) is a well-established method for dete mining relationship with a group of organisms. These molecul contain a mosaic of regions, which have evolved at different rate It is thus possible by careful selection of the region of the molecu to be investigated to establish relationships between organism The organisation of the regions of DNA encoding the riboson genes is known. The 5s gene is too small. 16s gene and 23s ge contain a number of conserved regions. If similarities occur in t variable region for example in non-transcribed spacer regions, th indicates that the isolates must be closely related. Molecu methods applied to aquatic hyphomycetes have been reviewed Bärlocher (9) emphasizing their growing importance.

Several aquatic fungi have been employed in the analysis isozyme, amylase, protein pattern, genome organization a growth requirement. Based on the data phylogenetic interp tations have been envisaged. Individual conidia may be stud morphologically and analysed by DGGE (denaturing gradient electrophoresis or T-RFLP (terminal restriction fragment len polymorphism) for valuable check on the validity of vis identifications (68).The molecular techniques have given us ability to distinguish individual strains from one another. This is useful for patented organisms and for ecological studies in which one needs to know the distribution of a given strain. It is possible that certain uncertainties may be removed in fungal systematics using the molecular methods.

Conventional studies, based on microscopical examination and identification (18,98) can be employed to revise sequence databases and to complete original descriptions with molecular definitions (101). New molecular tools such as metagenomics and transcriptomics can provide us a peek into the by and large genetic diversity that is present and or active (13). Additional insight into diversity can also be obtained by ribosomal genes or rRNA by ribotag sequencing (97). Further, meta-sequence data involve detailed bioinformatic post experimental work and additional personnel, which greatly reduces its practicality (101). In the near future, several basic questions in aquatic mycology may be addressed with a combination of traditional and well-established molecular methods.

Acknowledgements

The authors are grateful to Ministry of Environment and Forests and UGC for financial assistance.

REFERENCES

1. Ando K. 1992. A study of terrestrial hyphomycetes. *Trans Br mycol Soc* 33: 415-425.
2. Ando K and Tubaki K. 1984. Some undescribed hyphomycetes in the rain drops from intact leaf surface. *Trans mycol Soc Japan* 25: 21-37.
3. Au DWT, Vrijmoed LLP and Hodgkiss IJ. 1992. Fungi and cellulo-lytic activity associated with decomposition of *Bauhinia purpurea* leaf litter in a polluted and unpolluted Hong Kong waterway. *Can J Bot* 70: 1071-1079.
4. Bandoni RJ. 1974. Mycological observations on the aqueous films covering decaying leaves and other litter. *Trans Japn mycol Soc* 15: 309-315.
5. Bärlocher F. 1985. The role of fungi in the nutrition of stream invertebrates. *Bot J Linn Soc* 91: 83-94.
6. Bärlocher F. 1987. Aquatic hyphomycete spora in 10 streams of New Brunswick

and Nova Scotia. *Can J Bot* 65: 76-79.

7. Bärlocher F. 1992. Research on aquatic hyphomycetes: historical background and overview. In: *The ecology of aquatic hyphomycetes* F Bärlocher (Ed.). Springer, Berlin, pp. 1-15.

8. Bärlocher F. 2005. Freshwater fungal communities. In: *The fungal community. Its organization and role in the ecosystem* J Dighton, JF White, P Oudemans (Eds.). 3rd edn. CRC, Boca Raton, FL, pp. 39-59.

9. Bärlocher F. 2007. Molecular approaches applied to aquatic hyphomycetes. *Fungal Biol Rev* 21: 19-24.

10. Barron GL. 1968. *The Genera of Hyphomycetes from Soil.* The Williams and Wilkins Co., Baltimore, 364 pp.

11. Baudoin JM, Guerold F, Felten V, Chauvet E, Wagner P and Rousselle P. 2008. Elevated Aluminium concentration in acidified headwater streams lowers aquatic hyphomycete diversity and impairs leaf-litter breakdown. *Microb Ecol* 56: 260-269.

12. Bermingham S, Maltby L and Cooke R. 1996. Effects of a coal mine effluent on aquatic hyphomycetes. I. Field Study. *J Appl Ecol* 33: 1311-1321.

13. Bhadauria V, Popescu L, Zhao W and Peng Y. 2007. Fungal transcriptomics. *Microbiol Res* 162: 285-298.

14. Bhairavanath D and Manoharachary C. 1985. Conidial fungi from fresh-water foam. *Indian J Bot* 8: 67-68.

15. Chandrashekar KR, Sridhar KR, Rajashekar M. and Kaveriappa KM. 2000. Thermal springs of Dakshina Kannada. *Proc Lake* 2000, 27th to 29th November 2000, Indian Institute of Science, Bangalore. Sec 2, paper 5.

16. Cole GT and Samson RA. 1979. *Patterns of Development in Conidial Fungi.* Pitman, San Francisco.

17. Czeczuga B and Orlowska M. 1999. Hyphomycetes in rain water, melting snow and ice. *Acta Mycologica* 34: 181-200.

18. Czeczuga B, Mazalska B, Godlewska A and Muszynska E. 2005. Aquatic fungi growing on dead fragments of submerged plants. *Limnologica* 35: 283-297.

19. Dangles O, Gessner MO, Guérold F and Chauvet E. 2004. Impact of stream acidification on litter breakdown: implication for assessing ecosystem functioning. *J Appl Ecol* 41: 365-378.

20. Descals E. 1985. Conidia as modified hyphae. *Ind Acad Sci* 94: 209-227.

21. de Wildeman E. 1893. Notes mycologiques Fasc II. *Ann Soc Belge Microsc* 17: 35-68.

22. Duarte S, Pascoal C and Cássio F. 2004. Effects of zinc on leaf decomposition by fungi in streams: studies in microcosms. *Microb Ecol* 48: 366-374.

23. Field JI and Webster J. 1983. Anaerobic survival of aquatic fungi. *Trans Br mycol Soc* 81: 365-369.

24. Field JI and Webster J. 1985. Effects of sulfide on survival of aero-aquatic and aquatic hyphomycetes from stagnant water. *Trans Br mycol Soc* 85:1 93-199.

25. Fisher PJ and Petrini O. 1989. Two aquatic hyphomycetes as endo-phytes in *Alnus glutinosa* roots. *Mycol Res* 92: 367-368.

26. Galiah K and Manoharachary C. 1987. Studies on conidial fungi of a stream from Andhra Pradesh. *Indian Phytopath* 40: 466-473.

27. Gessner MO. 1997. Fungal biomass, production and Sporulation associated with particulate organic matter in streams. *Limnetica* 13: 33-44.

28. Gessner MO and Chauvet E. 1997. Growth and production of aquatic hyphomycetes in decomposing leaf litter. *Ecology* 75: 1807-1817.

29. Gessner MO and Chauvet E. 2002. A case for using litter breakdown to assess functional stream integrity. *Ecol Appl* 12: 498-510.

30. Goh TK and Hyde KD. 1998. A new hyphomycete genus *Conioscyphopsis* from wood submerged in a freshwater stream and a review of *Conioscypha*. *Mycol Res* 102: 308-312.

31. Goh TK, Hyde KD and Ho WH. 1998. *Aquaphila albicans* gen. et. sp. nov., a hyphomycete from submerged wood in the tropics. *Mycol Res* 102: 587-592.

32. Goh TK, Tsui KM and Hyde KD. 1997. *Yinmingella mitriformis* gen. et. sp. nov., a new sporodochial hyphomycete from submerged wood in Hong Kong. *Can J Bot* 76: 1693-1697.

33. Golladay SW, Webster JR and Benfield EF. 1983. Factors affecting food utilization by a leaf shredding aquatic insect: leaf species and conditioning time. *Holarct Ecol* 6: 157-162.

34. Gönczöl J and Révay Á. 2006. Species diversity of rainborne hyphomycete conidia from living trees. *Fungal Div* 22: 37-54.

35. Gulis V and Suberkropp K. 2003. Effect of inorganic nutrients on the relative importance of fungi and bacteria in carbon flow from submerged decomposing leaf litter. *Microb Ecol* 45: 11-19.

36. Gulis V, Marvanová L and Descals E. 2005. An illustrated key to the common temperate species of aquatic hyphomycetes. In: *Methods for studying litter decomposition* MAS Graça, F Bärlocher, MO Gessner (Eds.). Kluwer Academic, Dordrecht. pp. 130-153.

37. Hawksworth DL. 1991. The fungal dimension of biodiversity: magnitude, significance, and conservation. *Mycol Res* 95: 641-655.

38. Hyde KD. 1993. Tropical Australian freshwater fungi. VI. *Tiarosporella paludosa* and *Clohesyomyces aquaticus* gen. et. sp. nov. (Coelomycetes). *Aust Syst Bot* 6: 169-173.

39. Hyde KD and Seifert KA. 1992. Tropical Australian freshwater fungi. III. *Candelosynnema ranunculosporum*, a new genus and species of synnematous hyphomycetes. *Aust Syst Bot* 5: 401-405.

40. Ingold CT. 1942. Aquatic hyphomycetes of decaying alder leaves. *Trans Br mycol Soc.* 25: 339-417.

41. Junghanns C, Moeder M, Krauss G, Martin C and Schlosser D. 2005. Degradation of the xenoestrogen nonylphenol by aquatic fungi and their laccases. *Microbiology* 151: 45-57.

42. Karamchand KS and Sridhar KR. 2008. Water-borne conidial fungi inhabiting tree holes of the west coast and Western Ghats of India. *Czech Mycol* 60: 63-74.

43. Kaushik NK and Hynes HBN. 1971. The fate of dead leaves that fall into streams. *Arch Hydrobiol* 68: 465-515.

44. Kirk PM, Cannon PF, Minter DW and Stalpers JA. 2008. *Dictionary of the Fungi*. 10th edn. CABI Publishing, UK, 771 pp.

45. Krauss G, Schlosser D and Krauss GJ. 2005. Aquatic fungi in heavy metal and organically polluted habitats. In: *Biodiversity of Fungi: their role in human life.* SK Deshmukh and MK Rai (Eds.), Science Publ Inc. 460 pp.

46. Kuthubutheen AJ and Nawawi A. 1994. *Paracryptophiale kamaruddinii* gen et. sp. nov. from submerged litter in Malaysia. *Mycol Res* 98: 125-126.

47. Manoharachary C and Murthy ABS. 1981. Foam

spora from Andhra Pradesh, India. *Curr Sci* 50: 378-379.

48. Manoharachary C and Rama Rao P. 1991. Prevalence and powers of fungi in aquatic system. In: *Proc ICA Symposium* GP Agarwal and SK Hasıja (Eds). Narendra Publ House, Delhi. pp. 77-88.

49. Manoharachary C, Sridhar K, Singh R, Adholeya A, Suryanarayanan TS, Rawat S and Johri BN. 2005. Fungal biodiversity: Distribution, conservation and prospecting of fungi from India. *Curr Sci.* 89: 58-71.

50. Marvanová L and Fisher PJ. 1991. A new endophytic hyphomycetes from alder roots. *Nova Hedwigia.* 52: 33-37.

51. Marvanová L, Fisher PJ, Aimer R and Segedin BC. 1992. A new *Filosporella* from alder roots and foam water. *Nova Hedwigia* 54: 151-158.

52. Mason EW. 1937. Annotated account of fungi received at the Imperial Mycological Institute List II (page 2) *Mycol Pap* 4: 31-67.

53. Medeiros AO, Pascoal C and Graca AS. 2009. Diversity and activity of aquatic fungi under low oxygen conditions. *Freshwater Biol* 54: 142-149.

54. Moreirinha C, Duarte S, Pascoal C and Ca´ssio F. 2010. Effects of Cadmium and Phenanthrene mixtures on aquatic fungi and microbially mediated leaf litter decomposition. *Arch Environ Contam Toxicol.* DOI 10.1007/s00244-010-9610-6

55. Mulholland PJ, Driscoll CT, Elwood JW, Osgood MP, Palumbo AV, Rosemond AD, Smith ME and Schofield CL. 1992. Relationships between stream acidity and bacteria, macro invertebrates, and fish: a comparison of north temperate mountain streams, USA. *Hydrobio-logia* 239: 7-24.

56. Nawawi A and Kuthubutheen AJ. 1989. *Canalisporium*, a new genus of lignicolous hyphomycetes from Malaysia. *Mycotaxon* 34: 475-487.

57. Nilsson S. 1964. Freshwater Hyphomycetes: taxonomy, morphology and ecology. *Symp Bot Upsal* 18: 130 pp.

58. Pargain N, Sati SC and Belwal M. 2006. Endophytic aquatic hyphomycetes of roots from forested plants of Nainital, Kumaun Himalaya. In: *Plant Science Research in India: Challenges and Prospect.* S Kumar (Ed.). pp. 213-220.

59. Park D. 1972. On the ecology of heterotrophic microorganisms in fresh water. *Trans Br mycol Soc* 58: 291-299.

60. Pascoal C and Cássio F. 2004. Contribution of fungi and bacteria to leaf litter decomposition in a polluted river. *Appl Environ Microbiol* 70: 5266-5373.

61. Pascoal C, Cássio F and Marvanová L. 2005. Anthropogenic stress may affect aquatic hyphomycete diversity more than leaf decomposition in low-order stream. *Archiv für Hydrobiologie* 162: 481-496.

62. Premdas PD and Kendrick D. 1991. Colonization of autumn –shed leaves by four aero-aquatic fungi. *Mycologia* 83: 317-321.

63. Prenafeta-Boldu F, Summerbell R and de Hoog G. 2006. Fungi growing on aromatic hydrocarbons: biotechnology's unexpected encounter with biohazard? *FEMS Microbiol Rev* 30:109-130

64. Rajashekhar M and Kaveriappa KM. 2003. Diversity of aquatic hyphomycetes in the aquatic ecosystems of the Western Ghats of India. *Hydrobiologia* 501: 167-177.

65. Ramesh C and Vijaykumar S. 2006. Obsverations on waterborne fungi of Uttara Kannada region. In: *Recent Mycological Researches.* SC Sati (Ed.), IK International Publishing House, Pvt. Ltd., New Delhi, pp. 61-76.

66. Ramesh Ch. and Kumar SV. 2000. Seasonal occurrence of water borne fungi in Kali stream, Karnataka. Proc Internat Conf. *Integrated Plant Disease Management for Sustainable Agriculture*, 11-15 Nov, 1997, Indian Phtopathological Soc, Golden Jubilee. pp. 147-149.

67. Rao MM and Manoharachary C. 1980. Two interesting fungi from aquatic habitat. *Curr Sci* 49: 490-491.

68. Raviraja NS, Nikolcheva LG and Bärlocher F. 2005. Diversity of conidia of aquatic hyphomycetes assessed by microscopy and by DGGE. *Microbial Ecol* 49: 301-307.

69. Raviraja NS, Sridhar KR and Bärlocher F. 1998. Breakdown of ficus and eucalyptus leaves in an organically polluted river in India: fungal diversity and ecological functions. *Freshw Biol* 39: 537-545.

70. Ruiz RFC, Guarro J and Mayayo E. 1998. *Enridescalsia*, a new genus of conidial fungi from submerged leaves in Cuba. *Mycol Res* 102: 42-44.

71. Sarma KN and Manoharachary. 1987. Conidial fungi from a stream of Araku valley, A.P., India. *Proc Nat Symp Fungal Ecol* 33-34.

72. Sarma TK and Manoharachary C. 1988. Water borne

conidial fungi from a stream of Andhra Pradesh, India. *Proc Nat Acad Sci,*India, 58(B) II: pp. 337-339.

73. Sati SC and Arya P. 2009. Occurrence of water-borne conidial fungi in relation to some physico-chemical parameters in a fresh water stream. *Nature Sci* 7(4): 20-28.

74. Sati S, Arya P and Belwal M. 2009. *Tetracladium nainitalense* sp. nov., a root endophyte from Kumaun Himalaya, India. *Mycologia* 101(5): 692-695.

75. Sati SC and Belwal M. 2005. Aquatic hyphomycetes as endophytes of riparian plant roots. *Mycologia* 97(1): 45-49.

76. Sati SC, Belwal M and Pargaein N. 2008. Diversity of water borne conidial fungi as root endophytes in temperate forest plants of Western Himalayas. *Nature Sci* 6 (3): 59-62.

77. Sati SC, Pargaein N and Belwal M. 2006. Three species of aquatic hyphomycetes as new root endophytes of temperate forest plants. *Nat Acad Sci Lett* 29: 9-10.

78. Sati SC, Pargaein N and Belwal M. 2009. Diversity of aquatic hyphomycetes as root endophytes on pteridophytic plants in Kumaun Himalaya. *J Am Sci* 5(4): 179-182.

79. Sati SC and Tiwari N. 1992. Colonization, species composition, and conidial production of aquatic hyphomycetes on chirpine needle litter in a frreshwater Kumaun Himalayan stream. *Internat Review Hydrobiol* 77(3): 445-453.

80. Sati SC, Tiwari N and Belwal M. 2002. Conidial aquatic fungi of Nainital, Kumaun Himalaya, India. *Mycotaxon* 81: 445-455.

81. Schroth G, Elias MEA, Uguen K, Seixas R and Zech W. 2001. Nutrient fluxes in rainfall, throughfall and stemflow in tree-based land use systems and spontaneous tree vegetation of central Amazonia. *Agric Ecosys Environ* 87: 37-49.

82. Shearer CA, Langsam DM and Longcore JE. 2004. Fungi in fresh-water habitats. In: *Biodiversity of fungi: inventory and monitoring methods.* GM Mueller, GF Bills, MS Foster (Eds.) Elsevier Academic Press, San Diego, CA, pp 513-532.

83. Shearer CA, Descals E, Kohlmeyer B, Kohlmeyer J, Marvanová L, Padgett D, Porter D, Raja HA, Schmit JP, Thorton HA and Voglymayr H. 2007. Fungal biodiversity in aquatic habitats. *Biodivers Conserv* 16: 49-67.

84. Sridhar KR and Bärlocher F. 1992. Endophytic aquatic hypho-mycetes of roots of spruce, birch and maple. *Mycol Res* 96: 305-308.

85. Sridhar KR and Bärlocher F. 2000. Initial colonization, nutrient supply, and fungal activity on leaves decaying in streams. *Appl Environmental Microbiol*, 66(3): 1114-1119.

86. Sridhar KR, Karamchand KS and Bhat R. 2006. Arboreal water-borne hyphomycetes on oak-leaf basket fern *Drynaria quercifolia*. *Sydowia* 58 (2): 309-320.

87. Sridhar KR and Kaveriappa KM. 1982. Aquatic fungi of the Western ghat forests in Karnataka (India). *Indian Phytopathol* 35: 293-296.

88. Sridhar KR and Kaveriappa KM. 1984. Aquatic hyphomycetes of the Western ghat forests in Karnataka (India). *Indian Phytopathol*, 37: 546-548.

89. Sridhar KR and Kaveriappa KM. 1987. Occurrence and survival of aquatic hyphomycetes in terrestrial conditions. *Trans Br mycol Soc* 9: 606-609.

90. Sridhar KR and Raviraja NS. 1995. Endophytes - A crucial issue. *Curr Sci.* 69: 570-571.

91. Suberkropp K. 1998. Microorganisms and organic matter processing. In: *River ecology and management: lessons from the Pacific Coastal Ecoregion* RJ Naiman, RE Bilby (Eds.). Springer, Berlin, pp 120-143.

92. Subramanian CV. 1971. *Hyphomycetes.* ICAR, New Delhi, 930 pp.

93. Subramanian CV and Bhat DJ. 1981. Conidia from fresh water foam samples from the Western Ghats, Southern India. *Kavaka* 9: 45-62.

94. Tiwari N and Sati SC. 1992. Two notable additions to the Indian aquatic hyphomycetes. *Nat Sci Acad Lett* 15(2): 31-32.

95. Thomas K, Chilvers GA and Norris RH. 1992. Aquatic hyphomycetes from different substrates: substrate preference and seasonal occurrence. *Aus J Mar Freshwat Res* 43: 491-509.

96. Tubaki K. 1958. Studies on the Japanese hyphomycetes V. Leaf and stem group, with a discussion of classification of hyphomycetes and their perfect states. *J Hattori Bot Lab* 20: 142-244.

97. Urich T, Lanzén A, Qi J, Huson DH, Schleper C and Schuster SC. 2008. Simultaneous assessment of soil microbial community structure and function through analysis of the metatranscriptome.

PLoS ONE 3:e2527, doi: 10.1371/journal.pone.0002527

98. van Ryckegem G, Gessner M and Verbeken A. 2007. Fungi on leaf blades of *Phragmites australis* in a brackish tidal marsh: diversity, succession, and leaf decomposition. *Microb Ecol* 53: 600-611.

99. Webster J. 1959. Experiments with spores of aquatic hyphomycetes. I. Sedimentation and impaction on smooth surfaces. *Ann Bot NS* 23: 595-611.

100. Webster J. 1992. Anamorph-teleomorph relationships. In: *The ecology of aquatic hyphomycetes* F Bärlocher (Ed.), pp. 99-117. Springer-Verlag, Berlin, Germany.

101. Wurzbacher CM, Bärlocher Felix and Grossart Hans-Peter. 2010. Fungi in lake ecosystems. *Aquat Microb Ecol* 59: 125-149. doi: 10.3354/ame01385

FRESH WATER MOULDS: OCCURRENCE, DISTRIBUTION, SEASONALITY AND ZOOSPORE RELEASE MECHANISM

R.V. Gandhe[1] *and Kanchanganga Gandhe*[2]

[1]Emeritus Scientist, Agharkar Research Institute,
Department of Mycology, Pune 411 004
[2] Ex--Reader, Post-graduate Research Centre, Department of Botany,
Modern College, Pune 411 005
E-mail: rvgandhe123@gmail.com

KEYWORDS: Aquatic fungi, lentic ecosystem, zoosporic fungi, spore discharge, classification.

ABSTRACT

Review of fresh water moulds from closed water bodies has been undertaken with reference to their occurrence, distribution and seasonality. They were studied with different physicochemical parameters to find out their interrelationships. They were subjected to pure cultures in sterile distilled water and on different solid and liquid media for working out their life cycles. Zoosporic fungi with their dominant asexual phase with zoosporangia were studied in detail for their development and zoospore discharge mechanism in different genera from Chytridiales such as *Rozella, Allomyces, Gonapodya, Blastocladia*, and from Oomycetes, *Achlya, Saprolegnia, Dictyuchus, Pythium, Aphanomyces*, and *Olpidiopis*. The mechanism of zoospore release in different genera is characteristic and is a possible criterion for generic classification. The species are categorized into: Summer season species – *Blastocladia arborata, Gonapodya polymorpha, G. prolifera.* Rainy season species - *Achlya flagellata, A. orion, A. klebsiana, A. oblongata, A. recurva, Allomyces arbuscula, A. javanicus, A. moniliformis, Brevilegnia longicaulis, Gonapodya prolifera, Olpidiopsis achlyae, Rozella allomycis.* All season species - *Achlya bisexualis, A. orion, A. diffusa, A. proliferoides, Aphanomyces laevis, Dictyuchus carpophorus, D. sterile, Gonapodya polymorpha, G. prolifera, Pythium catenulatum, P. torulosum, P. proliferum* and *Saprolegnia anomalies.* The occurrence, distribution and seasonality of water moulds from closed water bodies were compared to those from open water systems or riverine fungi so far studied. They were also studied for their origin such as Allians, Active allians, Immigrants and Migrants. The review of fresh water moulds mainly restricted to these aspects is presented here.

INTRODUCTION

'Zoosporic fungi' is a group of those coenocytic fungi which complete their all life events either in open or/and closed water bodies. Thus, they are commonly called 'water moulds'. Several research workers apply the term 'water moulds' especially to the members of Saprolegniaceae. Many mycologists, however, referred 'Zoosporic fungi' as 'Aquatic Phycomycetes' mostly to the members of 'Chytridiomycetes' and mainly 'Oomycetes' as 'water moulds'. They are mostly freshwater fungi but also occur in marine water. In case of freshwater fungi, depending upon the niches, they occur in lentic or closed water systems such as lakes, ponds, pools, temporary puddles, ditches etc., or in lotic ecosystem or open water systems such as streams, rivers, their tributaries etc. However, the traditional view that 'Zoosporic fungi' occur only in an aquatic system is now outdated as these members are found to occur in damp soil also where, either they colonize with various substrates or they invade many living aquatic hosts particularly plants and animals, showing saprophytic as well as parasitic mode of nutrition (77). They include a wide diversity of fungal forms, some of them show definite relationship to the flagellates, some are true moulds and some show close resemblance to colourless algae. Water moulds show their diverse habitats. They play an important role in decomposition and as saprophytes they decompose various types of organic matter including cellulose chitin, keratin, etc. (71) thereby, forming a nutrient pool and make it available to the producers so as to make water ecosystem a self-regulating system (4). They are also one of the important fungal biodegraders as they utilize variety of substrates, namely, algae, other aquatic Phycomycetes, decaying plant debris and insect exuviae. Chytrids are mostly aquatic, growing as saprobes on fragments of plant and animal remains in water or as parasites in the algal cells and small aquatic animals.

In general, the Blastocladiales and Monoblepharidales appear to function as decomposers of detritus in the aquatic and soil ecosystems. They often infect the algal members of Zygnematales, Oedogoniales and Cladophorales (77,43). Some water moulds being plant pathogens, root pathogens, fish parasites etc. are economically important causing tremendous economic loss throughout the world. *Phytophthora* causes very serious diseases such as dieback, potato blight, sudden oak death and rhododendron root rot. *Pythium* damping off is a very common problem in greenhouses where the pathogen kills newly emerged seedlings. *Pythium insidiosum* is known to infect mammals. *Saprolegnia* is a causative organism of the disease Saprolegniosis in many fishes. *Aphanomyces* species are reported as crab pathogens. Species of *Aphanomyces* are either saprophytic or parasitic and specifically pathogenic on vascular plants and on freshwater animals (42). *A. astaci* is a serious pest on the edible Crayfish (86). *A. cochlioides* is a pathogen on roots of beets (66) and *A. euteiches* on roots of peas (33). *A. laevis* has larvicidal activity in mosquito (69) and a probable solution to control mosquito population. Depending on the origin of potential inoculum, Dick (24) distinguished water moulds mainly in three types. Fungal propagules having potential inoculum entering lake or lentic environment from terrestrial reservoirs or lotic environment via streams and overland drainage are called 'Aliens'. The aliens when grow actively on substrates are called 'Active Aliens' and fungal propagules which are derived from or forming a part of a potential inoculum situated and produced in the lentic environment are called 'Inhabitants'. Park (67) categorized water moulds in relation to habitat. The water moulds, which spend all life cycle events in the same habitat, are called 'Indwellers or Indigenous'. The organisms whose main habitat is extra–aquatic (damp soil, mud near water reservoir) are 'Immigrants' in relation to water; while those that alternate periodically between aquatic and extra aquatic habitats are regarded as 'Migrants'; water-moulds which show irregular or haphazard movement between aquatic and extra aquatic habitats are called 'Transients' or 'Versatiles'. Manoharachary (55), Manoharachary and Ramarao (59), distinguished water moulds as 'Indwellers', 'Transients' and 'Aliens'.

Seasonal periodicity depends on the fluctuations of temperature, light, water and nutrients which are due to the annual sequence of months. Phenology of water moulds depends on the seasonal periodicity. Climatic conditions are correlated with the occurrence of the water moulds or the nature of sexual reproductive structure such as oospore type. The occurrence of water moulds is not only dependent on the change in temperature with seasons but also on physic-chemical variations in the particular habitat (42,44).

Coker (10), Lund (50), Waterhouse (89), (Ziegler 91), Perrott (70), Hughes (38), and many others from temperate countries worked on the seasonal periodicity of watermoulds and concluded that temperature is the major environmental factor which determines the seasonal occurrence of water moulds especially members of Saprolegniaceae. The occurrence of centric, subcentric and eccentric oospore forms is also dependent on seasonal periodicity. The forms from Blastocladiaceae, Pythiaceae and Leptomitaceae (Chytridiomycetes/Oomycetes) also have a seasonal periodicity. *Achlya* spp. thrived best in cold waters while *Saprolegnia* spp. did not seem so dependent on temperature of water. For temperate climate, Saprolegniaceous fungi (from soil and water) with oospores are dominant in warm weather and those which are of frequent occurrence in cooler weather. Saprolegniaceae can be divided into two groups depending on the seasonal periodicity, occurrence and variation in abundance. One group comprises species characterized by centric and subcentric oospores, which exhibit marked seasonal periodicity, whereas, the second group consists of species forming eccentric oospores, showing no clear periodicity pattern but temperature is a major environmental factor determining the periodicity pattern in Saprolegniaceae (38). Klich and Tiffany (49) stated that centric or eccentric type of oospores occurs at similar climatic conditions but temperature also plays an important factor responsible for the occurrence of water moulds. According to Roberts (74) and Dick and Newby (27) seasonal fluctuations affect the occurrence of watermould especially Saprolegniaceae. During warm seasons the frequency is low, however, the frequency gets increased in autumn and maximum in spring. According to Hunter (39) many species of Saprolegniaceae were more common in winter than the early spring but there was a gradual increase in frequency from the low in spring to a peak in summer. From the tropical countries also many research workers observed the seasonal periodicity of water moulds. Alabi (1) from Nigeria reported that centric and subcentric species are rainy season species while eccentric species are dry season species. El–Hissy and El–Nagdy (28) from Egypt reported that highest population of aquatic Phycomycetes were found throughout the year except May to July and September. They also found that different Oomycetous forms occur at different seasons.

Suzuki (81,82) from Japan worked on the occurrence of different genera in different seasons and stated some genera are mainly restricted to particular seasons, for example, *Saprolegnia*, *Dictyuchus* and *Isoachlya eccentrica* occurred principally in winter and early spring while *Achlya* appeared in summer months. Members of *Aphanomyces* were recovered throughout the year.

India is one of the subtropical countries of the world. From India, many research workers studied the occurrence of water moulds with respect to climatic conditions or water conditions. Dayal and Tandon (14, 15) reported that the watermoulds showed seasonal periodicity but the growth is mainly restricted to winter season. Maturation of oospores and the temperature was the important factor in the germination of the oospore. They also stated that the maximum water moulds were from August to March and then their number declines. The growth is favoured by a fall in temperature, accompanied by an increase in nutrients and dissolved oxygen. Srivastava (80) stated that periodicity pattern of Oomycetous forms is similar to that occurs in temperate countries but eccentric oosporic species do not show any seasonal variations. Khulbe and Bhargava (48) worked on the occurrence of water moulds from temperate and subtropical situations from the same region and claimed that periodicity is governed by the temperature and not by the seasons and grouped the watermoulds in 'constant species', 'low temperature species', 'moderate temperature species' and 'high temperature species'. Chowdhary and Agarwal (8, 9) also confirmed the seasonal periodicity especially in

Saprolegniaceae which were frequent during winter followed by rainy and summer season, whereas, centric and subcentric oospores were common during winter, and the eccentric oosporic species were recorded round the year and did not show seasonal variations. Mer et al (60) observed that maximum numbers of water moulds were obtained in the rainy season. Manoharachary and Ramarao (59) also have the same opinion regarding the occurrence of watermoulds. Based on the occurrence and distribution in seasons, they divided the aquatic fungi into monsoon, winter, summer and constant species. Mishra (62) stated that the maximum zoosporic forms occur during winter. Fungal colonization and multiplication is enhanced at lower temperatures during winter season (57). The occurrence and distribution of aquatic fungi is not only dependent on seasons but also equally on the chemical composition of water bodies (32, 35).

In Maharashtra, there are three main seasons, rainy season (generally from June to September), winter (generally from October to January) and summer (generally from February to May). Temperature of water bodies and atmospheric change according to the seasons, which may directly or indirectly affect the occurrence and distribution of aquatic fungi as well as the quality of water. In the present review, main focus is given on the same aspects from the closed water bodies and the data is compared with open water bodies from the same area.

Development of zoosporangium, zoospores and zoosporangial discharge play an important role in the taxonomy of water moulds but inadequate attention to these events by research workers has led to confusion. Padgett and Johnson (65) mentioned that zoosporic fungi were studied extensively throughout the world by traditional methods. Johnson ct al (42) has made world survey of water-moulds, documented and discussed these observations. Today due to new and advanced techniques it is possible to record and study all the events of the life cycle of zoosporic fungi. It is observed and mentioned in the present review that zoospore mechanism is different in different genera. This would be another probable criterion for new system of classification.

Materials and Methods

For the present review, a general survey of closed water bodies in and around Pune (Maharashtra) as well as several water bodies scattered in Maharashtra state was carried out. Of these, some water bodies were selected on the basis of nearby localities and by observing surrounding environmental conditions. Water samples were collected from the established sites after every fifteen days with some organic debris and soil in sterilized glass bottles of 120 ml capacity and baited with sterilized opium seeds, hemp seeds, grass leaves, insect limbs and mustard seeds. Water moulds were isolated in the laboratory from these water samples by applying baiting technique (3, 6) using sterilized opium seeds as baits. The mixed cultures were purified by hyphal tip technique and maintained in separate sterilized Petri dishes for the progressive growth in culture. Pure cultures were maintained by transferring hyphal tips to the fresh sterilized water.

For the growth of members of Saprolegniaceae, Dick's agar medium was used (21). In addition, nutrient rich media like Potato Dextrose Agar (PDA), Malt Extract Agar (MEA) and broth etc. were used. Isolated species were studied and identified with the help of selected monographs and relevant literature (10,11,18,41, 42,45,47,52,61,64,75,76,77,78,79).

To study the mechanism of zoospore release from the zoosporangia in live cultures, the overall observations of the colonies were made under the low power of microscope and mature zoosporangial colonies were isolated and placed in cavity slide containing water. Video recording of all the stages of the protoplasmic activity, zoospore formation, zoospore release and behaviour of zoospores before and after the release from zoosporangia was done by using Olympus Microscope CH 20 I with CCD camera attachment. The 'VCD Plus' software was used to record the live events of zoospore release mechanism.

There are no natural lakes and ponds in Pune. But most of them are formed and established by trickling of rain water and from streams in ditches or in quarries. Some are manmade reservoirs with artificial source of water. Water reservoirs with changing periphery were considered as lakes and artificially constructed reservoirs or water in quarries were considered as ponds, the water in these may be polluted and unpolluted. Based on these criteria, the data of zoosporic fungi from three polluted lakes(PL) and four nonpolluted ponds (NPP) were analysed for the present review.

Observations and Results

The occurrence, distribution and frequency of water moulds, mainly, Oomycetes was more significant as compared to Chytridiomycetes in closed water bodies. The members of Oomycetes were present in polluted as well in non-polluted water bodies. Thus, Oomycetes did not show any specificity for their distribution, whereas, Chytridiomycetes were mainly restricted to polluted water bodies.

Table 1. Fungal species collected from different closed water bodies during the study period showing percent occurrence of each order

No.	Order	Name of the species	Percent occurrence
1	Chytridiales	*Rozella allomycis*	3.50
2	Blastocladiales	*Allomyces arbuscula*	17.80
		A. javanicus	
		A. moniliformis	
		Blastocladia arborata	
		B. incrassata	
3	Monoblepharidles	*Gonapodya polymorpha*	7.10
		G. prolifera	
4	Saprolegniales	*Achlya oblongata*	57.14
		A. recurva	
		A. bisexualis	
		A. diffusa	
		A. proliferoides	
		A. sp.	
		A. flagellata	
		A. klebsiana	
		A. orion	
		A. laevis	
		Brevilegnia longicaulis	
		Dictyuchus sterile	
		D. carpophorus	
		Saprolegnia torulosum	

		S. ferax	
		S. anomalies	
5	Lagenidiales	*Olpidiopsis achlyae*	3.50
6	Peronosporales	*Pythium catenulatum*	10.71
		P. proliferum	
		P. torulosum	

In the present review, the occurrence, distribution and frequency of 10 genera with 28 species (Table 1) isolated from established closed water bodies has been studied and discussed. Oomycete members were isolated with three orders, namely, Saprolegniales, Lagenidiales and Peronosporales with 20 species and Chytridiomycete members with three orders having 8 species. Thus, in general, it seems that Oomycetes rule all over other aquatic fungi as they may or may not necessarily require organic substratum for their growth whereas most of the members of Chytridiomycetes especially *Gonapodya* and *Blastocladia* as in the present case require specific organic substratum like decaying, submerged *Acacia nilotica* pods for their growth. Table 2 shows genera like *Achlya, Allomyces, Aphanomyces* and *Pythium* were common in temporary (rain water) ditches and in the established sites, whereas, *Dictyuchus, Gonapodya, Blastocladia* and *Saprolegnia* were restricted only to the established sites. The species, *Achlya recurva, A. klebsiana* and *A. flagellata, Allomyces javanicus, A. moniliformis, Blastocladia incrassata, Brevilegnia longicaulis, Olpidiopsis achlyae, Rozella allomycis* however, occurred only in rainwater temporary ditches.

From the closed water bodies (Table 1) the Saprolegniales showed complete dominance (57.14%) with respect to the total fungal population followed by Blastocladiales (17.8%) with three species of *Allomyces* and two of *Blastocladia*. Peronosporales with three species of *Pythium* were next (10.71%); Lagenidiales and Chytridiales (3.5%) were least represented by one species each.

The genus *Achlya* with 9 species was dominant with 31% frequency occurrence (Pie Chart 1) followed by *Saprolegnia, Pythium* and *Allomyces* 11%. The % frequency of occurrence of *Achlya bisexualis* (8.22%), *A. proliferoides* (7.79%), *Dictyuchus carpophorus* (5.62%), *D. sterile* (8.6%), *Pythium catenulatum* (5.19%), *P. torulosum* (11.25%) and *Aphanomyces laevis* (24.6%), was more as compared to the remaining species. As *Aphanomyces laevis* showed maximum % frequency in closed water systems therefore, was considered as universal in occurrence (Table 3 and Fig. 1).

Table 2. Distribution and occurrence of fungal species in ponds, lakes and rainwater ditches

Ponds	Lakes	Rainwater ditches
Achlya bisexualis	-	-
A. diffusa	*A. diffusa*	-
-	*A. oblongata*	*A. oblongata*
-	*A. orion*	*A. orion*
A. proliferoides	*A. proliferoides*	-
Achlya sp.	-	-
-	-	*A. flagellata*
-	-	*A. klebsiana*
-	-	*A. recurva*
A. arbuscula	*A. arbuscula*	*A. arbuscula*
-	-	*A. javanicus*
-	-	*A. moniliformis*
A. laevis	*A. laevis*	*A. laevis*
Blastocladia arborata	-	-
B. incrassata	-	-
-	-	*Brevilegnia longicaulis*
Dictyuchus carpophorus	-	-
D. sterile	*D. sterile*	-
Gonapodya polymorpha	-	-
G. prolifera	-	-
-	-	*Olpidiopsis achlyae*
Pythium catenulatum	*P. catenulatum*	*P. catenulatum*
P. proliferum	*P. proliferum*	*P. proliferum*
P. torulosum	*P. torulosum*	*P. torulosum*
-	-	*Rozella allomycis*
Saprolegnia torulosa	-	-
-	*S. ferax*	-
-	*S. anomalies*	-

On the other hand, occurrence and distribution of certain watermoulds was restricted to particular water bodies (Table 4). *Gonapodya prolifera, G. polymorpha, Blastocladia arborata, B. incrassata, Saprolegnia anomalies* were restricted only to most polluted water. *Achlya bisexualis, Achlya* sp. *Blastocladia arborata, B. incrassata, Dictyuchus carpophorus, Gonapodya polymorpha, G. prolifera* and *Saprolegnia torulosa* occurred only in ponds whereas species *Achlya oblongata, A. orion, Saprolegnia anomalies* and *S. ferax* were restricted to lakes.

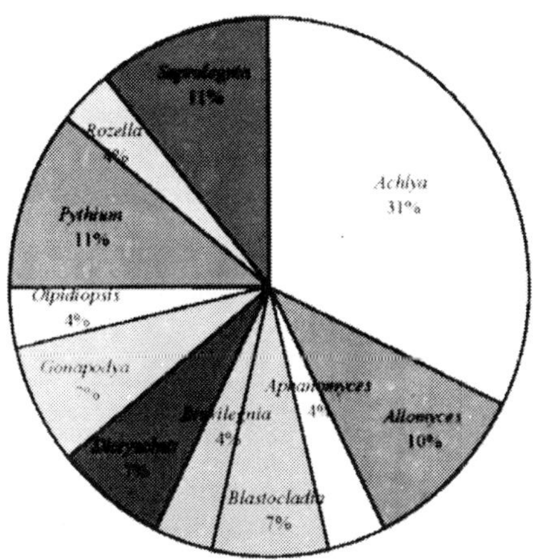

Pie Chart 1. Percentage of genera isolated from closed water bodies

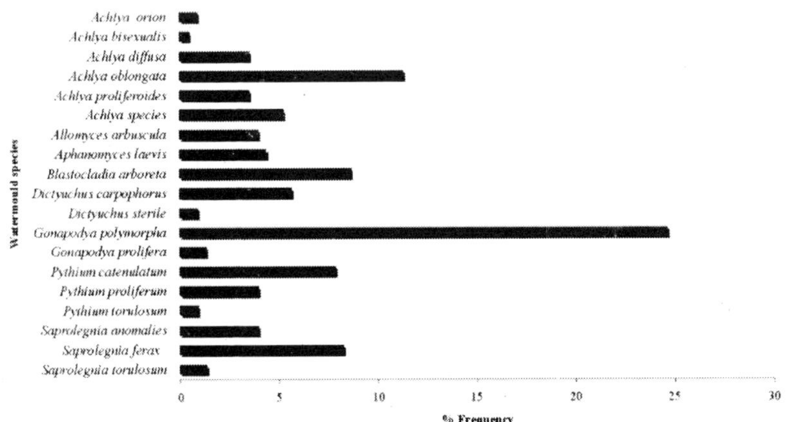

Fig 1: Percentage of occurrence of individual species from closed water bodies.

Table 3. Percentage of occurrence of fungal species in 3 Polluted Lakes (PL) and 4 Non-Polluted Ponds (NPP)

Name of sp.	a	b	c	d	e	f	g	h	i
Achlya orion	-	-	-	-	3	-	-	3	1.29
A. bisexualis	-	11	-	-	-	-	8	19	8.22
A. diffusa	-	4	-	-	-	2	-	9	3.89
A. sp.	-	2	-	-	-	-	-	2	0.86
A. oblongata	5	-	1	-	3	-	-	9	3.89
A. proliferoides	-	1	-	3	3	7	4	18	7.79
A. arbuscula	1	-	1	-	-	1	-	3	1.28
A. laevis	6	6	8	11	4	11	11	57	24.6
Blastocladia arborata	-	-	-	2	-	-	-	2	0.86
Dictyuchus carpophorus	-	4	-	-	-	-	9	13	5.62
D. sterile	-	6	2	4	5	3	-	20	8.6
Gonapodya polymorpha	-	-	-	10	-	-	-	10	4.32
G. prolifera	-	-	-	9	-	-	-	9	3.89
Pythium catenulatum	5	1	-	2	1	3	-	12	5.19
P. proliferum	-	-	4	-	-	-	4	8	3.46
P. torulosum	7	2	5	4	3	1	4	26	11.25
Saprolegnia anomalies	-	-	8	-	-	-	-	8	3.46
S. ferax	-	-	-	-	1	-	-	1	0.43
S. torulosa	-	-	-	-	-	2	-	2	0.86

a, c, e- Polluted Lakes (PL); b,d, f, g - Non Polluted Ponds (NPP); h - Total number of occurrence; i - Percentage of occurrence

Table 4. Distribution of water moulds in unpolluted and polluted waterbodies

Unpolluted water bodies	Polluted waterbodies
Class	
Oomycetes - 5	Oomycetes - 5
Chytridiomycetes - 1	Chytridiomycetes - 3
Orders	
Saprolegniales - 4	Saprolegniales - 4
Peronosporales - 1	Peronosporales - 1
Blastocladiales - 1	Blastocladiales - 2
-	Monoblepharidales - 1
Genera	
Achlya - 5 Species	*Achlya* - 4 Species
Allomyces - 1 Species	*Allomyces* - 1 Species
Aphanomyces- 1 Species	*Aphanomyces* - 1 Species
-	*Blastocladia* - 2 Species
Dictyuchus- 2 Species	*Dictyuchus* - 1 Species
-	*Gonapodya* - 2 Species
Pythium - 3 Species	*Pythium*- 3 Species
Saprolegnia -1 Species	*Saprolegnia* - 2 Species
Oospore types	
Eccentric - 5	Eccentric - 3
Subcentric -1	Subcentric -2
Centric - 1	Centric - 2
Species	
A. bisexualis	-
A. diffusa	*A. diffusa*
-	*A. oblongata*
-	*A. orion*
A. proliferoides	*A. proliferoides*
Achlya sp.	-
A. arbuscula	*A. arbuscula*
A. laevis	*A. laevis*
-	*B. arborata*

-	*B. incrassata*
D. carpophorus	-
D. sterile	*D. sterile*
-	*G. polymorpha*
-	*G. prolifera*
P. catenulatum	*P. catenulatum*
P. proliferum	*P. proliferum*
P. torulosum	*P. torulosum*
-	*S. anomalies*
-	*S. ferax*
S. torulosa	-

Out of these 28 species, only 10 species occurred in all seasons but most of them were sporadic. Seasonal periodicity at generic level was observed only in *Gonapodya* and *Saprolegnia*. Periodicity at particular season and at particular site was observed in *Achlya bisexualis, Dictyuchus carpophorus, G. polymorpha* and *G. prolifera* and *Saprolegnia anomalies*. During winter 14 species were isolated, of these, only three species, namely, *Achlya* sp., *Saprolegnia ferax* and *S. torulosa* were restricted only to winter. The species which were isolated in summer season did not have specific seasonal periodicity as they occur either in rainy or in winter. Maximum species were isolated, when the rain was more in the season, of these, 11 species were restricted only to rainy season.

In closed water systems, maximum number of zoosporic forms were isolated in the rainy season (25 species) followed by summer (13 species) and winter (12 species) season. Table 5 and 6 give the number of species isolated in each month of the study period (2002-2003 and 2003–2004) and the number of fungal species isolated each month from individual site.

In the closed water bodies under study, generally lakes have changing boundaries whereas ponds mainly have limited and permanent boundaries. 'Aliens' were more common in lakes than in ponds. *Allomyces arbuscula* was common in both lakes and ponds, *Achyla oblongata, A. orion, A. diffusa, Saprolegnia ferax* were 'aliens' in lakes and *Achyla proliferoides* and *Saprolegnia torulosa* were in polluted ponds and nonpolluted did not show any 'alien' species. *Dictyuchus sterile* was the only 'indweller' seen in slightly polluted lake, whereas, *Achyla bisexualis, A. diffusa, A. proliferoides, Dictyuchus carpophorus, D. sterile* were common especially in non-polluted garden ponds. *Gonapodya polymorpha, G. prolifera, Blastocladia arborata, B. incrassata* were common as 'indwellers' especially in polluted pond. 'Inhabitants' like *Aphanomyces laevis,* and *Pythium torulosum* were observed in slightly polluted lakes but not in ponds. *A. proliferoides* and *Pythium torulosum* were 'immigrants' observed in non-polluted garden pond.

Table 5. Fungal species isolated from all the sites during the study period

April 2002 - March 2003	April 2003 - March 2004
Achlya orion	*A. orion*
A. bisexualis	*A. bisexualis*
A. diffusa	*A. diffusa*
-	*A. flagellata*
A. sp.	*A.* sp.
-	*A. klebsiana*
A. oblongata	*A. oblongata*
A. proliferoides	*A. proliferoides*
-	*A. recurva*
A. arbuscula	*A. arbuscula*
-	*A. javanicus*
-	*A. moniliformis*
Aphanomyces laevis	*A. laevis*
Dictyuchus carpophorus	*D. carpophorus*
D. sterile	*D. sterile*
Gonapodya polymorpha	*G. polymorpha*
G. prolifera	*G. prolifera*
-	*Blastocladia arborata*
-	*B. incrassata*
-	*Brevilegnia longicaulis*
-	*Olpidiopsis achlyae*
Pythium catenulatum	*P. catenulatum*
P. proliferum	*P. proliferum*

P. torulosum	*P. torulosum*
-	*Rozella allomycis*
Saprolegnia anomalies	*S. anomalies*
-	*S. ferax*
-	*S. torulosum*

In the present studies, it is characteristically noted that summer season is favourable for the occurrence of fungal forms in all ponds, whereas, in all the lakes under study, the winter season proved to be best for the fungal growth and the fungal forms were minimum or zero in summer and maximum in winter season. It is confirmed that pond water harboured more fungal numbers than the lake water as they are less polluted; only Chytridiomycete members tolerate such polluted conditions.

Apart from 'All season species', some water moulds were isolated in a particular season and thus grouped them as 'winter season', 'summer season' and 'rainy season species'. The seasonal periodicity was maintained by many water moulds species.

Table 6. Occurrence of species in each month from each site during 2002–2004 from various sites.

		Apr	May	Jun	Jul	Aug	Sep	Oct	Nov	Dec	Jan	Feb	Mar	Total
Bhugaon lake	*A.oblongata*	-	-	1	-	2	1	-	-	1	-	-	-	5
	A. arbuscula	-	-	1	-	-	-	-	-	-	-	-	-	1
	A. laevis	-	-	-	-	-	1	1	1	-	2		1	6
	P.catenulatum	-	-	2	-	1	-	-	-	1	-	1	-	5
	P.torulosum	-	1	2	1	1	1	-	1	-	-	-	-	7
	Total	0	1	6	1	4	3	1	2	2	2	1	1	24
Empress Garden	*A.bisexualis*	2	1	-	2	-	-	-	1	1	2	1	1	11
	A. diffusa	-	-	-	-	1	1	-	-	1	-	1	-	4
	A. sp	-	-	-	-	-	-	-	1	-	1		-	2
	A. proliferoides	-	-	1	-	-	-	-	-	-	-	-	-	1
	A. laevis	1	1	1	1	-	1	-	-	-	-	-	1	6
	D.carpophorus	-	1	-	-	-	1	1	1	-	-	-	-	4
	D. sterile	2	-	-	1	1	1	-	-	1	-	-	-	6
	P.catenulatum	-	-	-	-	-	-	-	-	-	-	-	1	1
	P.torulosum	-	2	-	-	-	-	-	-	-	-	-	-	2
	Total	5	5	2	4	2	4	1	3	3	3	2	3	37

Ka-traj lake	*A. oblongata*	-	-	-	-	1	-	-	-	-	-	-	-	1
	A. arbuscula	-	-	-	1	-	-	-	-	-	-	-	-	1
	A. laevis	2	1	-	1	-	-	2	2	-	-	-	-	8
	D. sterile	-	-	-	-	1	-	1		-	-	-	-	2
	P.proliferum	-	-	-	-	1	1		1	-	-	1	-	4
	P.torulosum	1	1	-	-	-	-	2	1	-	-	-	-	5
	S. anomalies	-	-	-	1	-	-	-	2	1	1	2	1	8
	Total	3	2	0	3	3	1	5	6	1	1	3	1	29
Lak-aki lake	*A.proliferoides*	-	-	-	-	1	1	-	-	-	1	-	-	3
	A. laevis	1	-	2	1	1	2	1	1	-	2	-	-	11
	B. arboreta.	-	1	1	-	-	-	-	-	-	-	-	-	2
	D. sterile	2	-	-	-	-	1	-	-	-	1			4
	G. polymorpha	2	2	1	-	-	-	-	1	-	-	2	2	10
	G.prolifera	2	2	1	-	-	-	-	-	-	-	2	2	9
	P.catenulatum	1	-	-	-	1	-	-	1	-	-	-	-	3
	P.torulosum	1	-	1	-			1	-	-	-	-	-	3
	Total	9	5	6	1	3	4	2	3	-	4	4	4	45
Pas-han lake	*A. orion*	-	-	1	-	-	-	1	-	-	-	-	1	3
	A. diffusa	-	-	-	-	1	-	1	-	-	1	-	-	3
	A. oblongata	-	-	1	1	1	-	-	-	-	-	-	-	3
	A. proliferoides	-	-	-	-	-	-	-	2	-	1	-	-	3
	A. laevis	1	-	-	1	-	-	-	2	-	-	-	-	4
	D. sterile	-	-	1	-	1	-	-	2	-	1	-	-	5
	P.catenulatum	-	-	-	-	-	-	-	1	-	-	-	-	1
	P.torulosum	-	1	-	-	-	1	-	1	-	-	-	-	3
	S. ferax	-	-	-	-	-	-	-	1	-	-	-	-	1
	Total	1	1	3	2	3	1	2	9	-	3	-	1	26
Saras baug Pond	*A.diffusa*	-	-	1	-	-	1	-	-	-	-	-	-	2
	A. proliferoides	1	-	1	1	-	-	1	-	1	1	-	-	6
	A. arbuscula	-	-	1	-	-	-	-	-	-	-	-	-	1
	A. laevis	2	2	1	2	1	1	-	1	-	-	1	1	12
	D. sterile	-	-	-	-	1	1	-	-	-	1	-	-	3
	P.catenulatum	-	1	-	-	-	-	-	-	-	1	1	-	3
	P. torulosum	-	-	-	-	-	-	-	-	-	1	-	-	1
	S. torulosa	-	-	-	-	-	-	-	-	1	-	-	-	1
	Total	3	3	4	3	2	3	1	1	2	4	2	1	29
Univ-ersity gar-	*A. bisexualis*	2	1	1	-	1	1	-	-	-	-	2	-	8
	A.oblongata	-	-	-	-	-	-	-	-	-	-	-	-	0
	A.proliferoides	-	-	2	-	-	-	1	-	-	-	-	1	4

den	*A. laevis*	1	1	-	-	2	2	1	1	-	1	1	1	11
pond	*D. carpophorus*	1	2	1	2	-	-	-	1	-	1	1	-	9
	P. proliferum	1	1	1	-	1	-	-	-	-	-	-	-	4
	P. torulosum	1	-	1	-	-	-	-	-	-	1	1	-	4
	Total	6	5	6	2	4	3	2	2	0	3	5	2	40

Winter season species - *Achlya* spp., *Achlya oblongata, Blasto-cladia arborata, B. incrassata, Saprolegnia ferax* and *S. torulosum*

Summer season species - *Blastocladia arborata, Gonapodya polymorpha, G. prolifera.* The latter two species were recorded only once in other seasons.

Rainy season species - *Achlya flagellata, A. orion, A. klebsiana, A. oblongata, A. recurva, Allomyces arbuscula, A. javanicus, A. moniliformis, Brevilegnia longicaulis, Gonapodya prolifera, Olpidiopsis achlyae, Rozella allomycis.*

In the rainy season group most of the species were recorded only from rain water temporary ditches.

All season species - *Achlya bisexualis, A. orion, A. diffusa, A. proliferoides, Aphanomyces laevis, Dictyuchus carpophorus, Dictyuchus sterile, Gonapodya polymorpha, G. prolifera, Pythium catenulatum, P. torulosum, P. proliferum* and *Saprolegnia anomalies.*

Table 7. Percentage frequency of occurrence of eccentric, subcentric and centric oospore type

Oospore type	Months (2002 -2004)											
	Apr.	May	Jun.	Jul.	Aug.	Sep.	Oct.	Nov.	Dec.	Jan.	Feb.	Mar.
Subcentric species	10.6	7.57	9.09	10.6	12.12	12.12	7.57	12.12	1.51	7.57	3.03	6.06
Eccentric species	9.23	7.69	12.3	7.69	6.15	9.23	7.69	9.23	4.61	13.84	7.69	4.61
Centric species	0	0	0	9.09	0	0	0	27.27	18.18	18.18	18.18	9.09

It was noted that the type of oospores also has the periodicity of occurrence (Table 7). The eccentric oospore type with 66.66%

species was dominant. The centric oospore type with 20% and only 13.3% belonged to subcentric type. The species with eccentric oospore type did not show any seasonal periodicity as these were recorded throughout the year from each and every site; species with subcentric oospores were isolated in rainy as well as winter seasons. Centric oospores were strictly isolated in winter and early summer season.

Mechanism of Zoospore Release

In the present review, the video shooting of zoospore forma-tion and the mechanism of their discharge was recorded at 450x magnification in four genera of Chytridiomycetes and six genera of Oomycetes from the growing pure cultures. Dichotomous key is prepared mainly on the basis of mode of zoospore release mechanism and is as follows:

Allomyces Butler (Plate-I; Fig. 1)

Allomyces is a member of Blastocladiales of the class Chytridiomycetes. In *Allomyces* two cases of non-operculate type of zoospore discharge was observed. Zoosporangia of *Allomyces* are clearly distinct from the rest of the hypha.

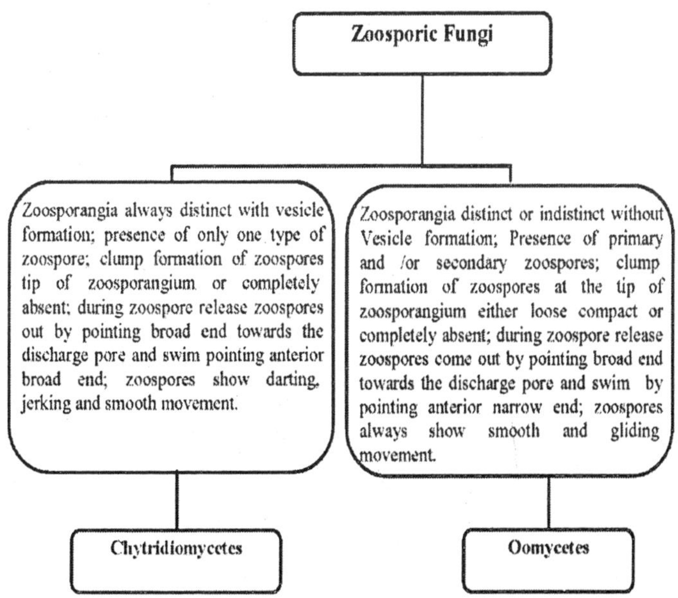

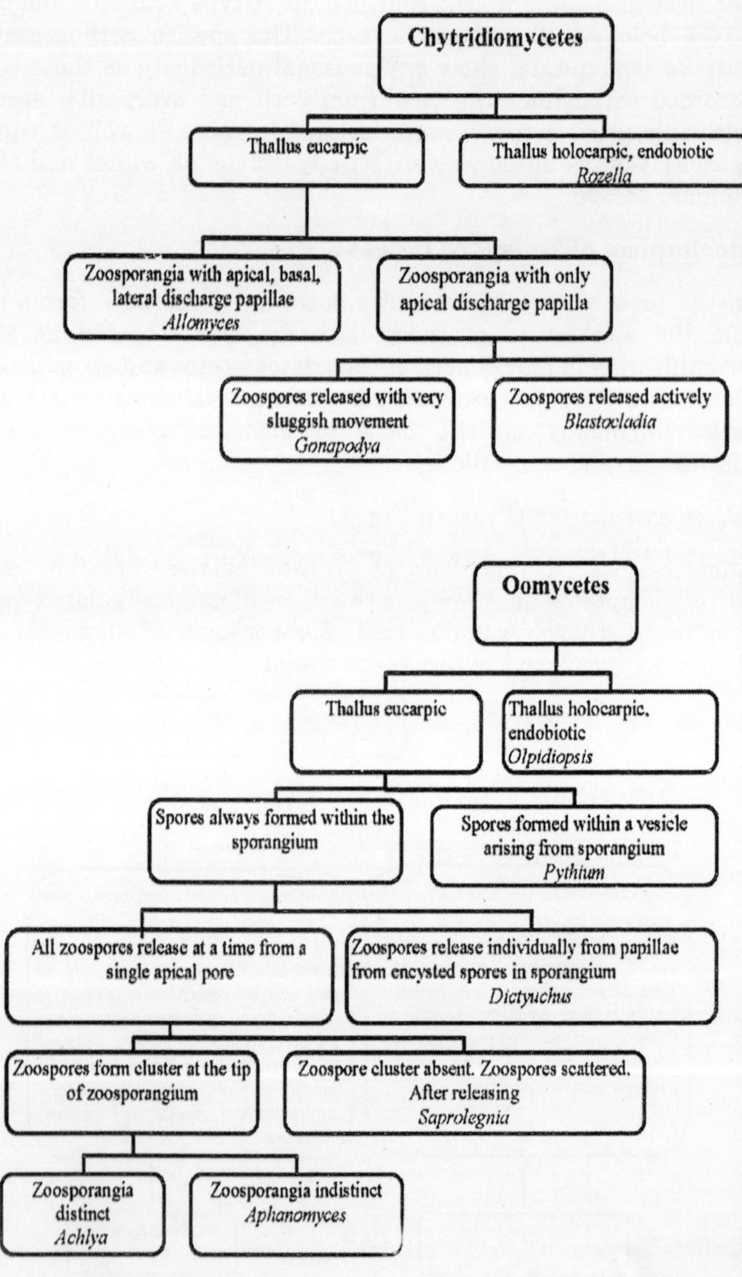
Chytridiomycetes
Thallus eucarpic
Thallus holocarpic, endobiotic
Rozella
Zoosporangia with apical, basal, lateral discharge papillae
Allomyces
Zoosporangia with only apical discharge papilla
Zoospores released with very sluggish movement
Gonapodya
Zoospores released actively
Blastocladia
Oomycetes
Thallus eucarpic
Thallus holocarpic, endobiotic
Olpidiopsis
Spores always formed within the sporangium
Spores formed within a vesicle arising from sporangium
Pythium
All zoospores release at a time from a single apical pore
Zoospores release individually from papillae from encysted spores in sporangium
Dictyuchus
Zoospores form cluster at the tip of zoosporangium
Zoospore cluster absent. Zoospores scattered. After releasing
Saprolegnia
Zoosporangia distinct
Achlya
Zoosporangia indistinct
Aphanomyces

In first case, the apical portion of the papilla ruptured and zoospores get released from the ovoid zoosporangium. At maturity the protoplasm activates and transformed into zoospores which at first showed slow movement within the sporangium and at a certain point the apical papilla dehisced and a few zoospores were released. These released zoospores formed a loose clump at the tip of the sporangium for a while and then quickly dispersed in the surrounding water. The remaining zoospores in the sporangium got released one by one and dispersed without forming a clump. In all 72 zoospores were released. Time required for the complete release of zoospores was 6 minutes, 21 seconds.

In the second case, two adjoining zoosporangia were with ateral papillae at the apical and basal part. These papillae uptured at different intervals. The upper zoosporangium was oval vith a truncated base and four discharge papillae, two on each side of the apical part and two on each side of the basal part. At naturity the protoplasm inside the sporangium got transformed nto zoospores which slowly moved inside the sporangium and at he critical point the basal lateral papilla dehisced to release oospores which formed a loose clump at the tip for a while as bserved in the first case and then quickly dispersed in the surrounding water with a darting movement. The remaining zoospores n the sporangium then emerged one at a time. The other papillae uptured at different intervals when the zoospore clump broke way after 2 minutes, 18 seconds. In all 67 zoospores were released. Time taken for the complete release was 10 minutes, 5 econds.

The lower zoosporangium had two apical lateral dehiscence apillae. At maturity the zoospores were formed inside the sporagium and then got released by rupturing of the papillae. Formtion of loose clump at the tip of the papilla was not observed. oospores got released one by one, uninterrupted till the sporanium emptied itself completely. In all 42 zoospores were released. ime required for the complete release was 6 minutes, 27 seconds.

The movement of zoospores within the sporangium was slow nd trembling at first, but as the zoospores from the sporangium ere released the movement of the remaining zoospores was rapid nd they moved all over the sporangium and on their release, they ispersed away in all directions. This mechanism was recorded in *llomyces moniliformis* and *A. arbuscula.*

Rozella Cornu (Plate I; Fig. 2)

Rozella is a member of Chytridiales of the class Chytridio-mycetes. It is an endoparasite on *Allomyces* with a holocarpic thallus. Inoperculate type of zoospore discharge was observed.

Zoosporangia of *Rozella* are formed at the distal ends as well as at the basal part of the host *Allomyces* by the segmentation and distinct swelling of the hyphae or sporangia of the host. The hyphal wall of *Allomyces* doubles as a wall for both the host and the parasite. At maturity zoospores get discharged from the sporangium with three different mechanisms: pore, pouch and plug formation, of these, video recording of the latter two are innovative

In type 1: The sporangium of the host considerably bulges at some points prior to zoospore release of endoparasite. The protoplasm activates and undergoes cleavages within its sporangium to form zoospores. At the maturity of the zoospores, pressure is exerted on a thin wall at the bulging point where the walls of the host and endoparasite get separated. Finally, due to the pressure of zoospores the thin area forms a discharge pore by rupturing and then a slight tilting movement of zoospores within the sporangium starts, accelerated and zoospores spill out till the sporangium is completely empty.

In type 2: The protoplasm of *Rozella* activates, undergoing cleavages within its sporangium to form zoospores. As soon as zoospores obtain maturity, the sporangial wall forms a pouch where the host wall is thin. The pouch is at first oriented inwardly and accommodated between the zoospores and the wall of the host. Zoospores start their movement and gradually accelerate the movement so as to exert a pressure on the pouch and the wall of the host. As a result, the pouch is flattened against the wall of the host and comes out forcefully by rupturing the thin wall of the host papilla, which is in line with that of *Rozella.* Motile, spherical or oval zoospores then start releasing continuously in a stream till the zoosporangium becomes empty. Generally, 3-4 zoospores are released at a time. Time required for the complete release of the zoospores from the sporangium was 2 minutes and 12 seconds. This type is a new record.

In type 3: The sporangia of *Rozella* develop partitions and these are closely appressed with the walls of the host. At maturity a plug like structure is formed on the sporangium of the parasite. It was observed that one of the swarming zoospores or a protozoan in the surrounding water stimulates the mechanism from outside the plug and later enters through the cavity developed by the removal of the plug. This triggered off the mechanism of release of zoospores, which after their release either dispersed in the surrounding water or remained in the form of a loose and elongated mass. The time required for the complete release of zoospores from the sporangium is 1 minute and 21 seconds. This type of mechanism is a new record, recorded in *Rozella allomycis*.

Blastocladia Reinsch (Plate- I; Fig. 3)

Blastocladia is a member of Blastocladiales of the class Chytridiomycetes. In *Blastocladia* non operculate type of zoospore discharge was observed.

Zoosporangia of *Blastocladia* are distinguished by broad, convcx apex that gradually tapers towards the base. At maturity, the protoplasm got transformed into spherical zoospores. These zoospores began to push one another and moved all over inside the sporangium before their release. At a critical point, they got released from the apical discharge pore and swam away in the surrounding water. This mechanism was recorded in *Blastocladia* species.

Gonapodya Fischer (Plate- I; Fig. 4)

Gonapodya is a member of Monoblepharidales of the class Chytridiomycetes. In *Gonapodya* non operculate type of zoospore discharge was observed. Zoosporangia of *Gonapodya* are distinct, pod shaped with an apical discharge pore. Protoplasm in the sporangium transformed into spherical zoospores. Zoospores were released from the apical discharge pore one by one with a sluggish movement for a while and then swam away vigorously in the surrounding water. This mechanism was recorded in *Gonapodya polymorpha.*

Achlya Nees (Plate- I; Fig. 5)

Achlya is a member of Saprolegniales of the class Oomycetes.

Primary zoospore release mechanism

Zoosporangia in *Achlya* are distinct from the rest of the hyphae. They are terminal, elongated and gradually tapering. The protoplasm inside the sporangium gets transformed into zoospores. At maturity zoospores get separated from the wall of the sporangium and get aggregated vertically in the centre of the sporangium, which is an indication of zoospore release. At the time of zoospore release the tapering portion of the sporangium forms a thin walled papilla like small elevation. At this time, zoospores show slow and steady movement due to which an internal pressure is developed, as a result most of the zoospores expelled out with force and get accumulated at the orifice in the form of a compact cluster. The lagged behind zoospores propel themselves and merge in cluster of zoospores at the orifice. Zoospores are pear shaped inside the sporangium but assume spherical shape immediately after release. Zoosporangium emptied completely within 15 to 40 seconds. The time required for the complete zoospore release may be dependent on the size of the sporangium and the environmental conditions.

The empty sporangium does not collapse immediately after zoospore release but remains intact with the apical cluster of zoospores, which, in the present study, is persistent.

This type of zoospore release is called Achlyoid, in which, zoospores after maturation are released with some force from the zoosporangium and form a cluster at the orifice. The formation of the cluster plays a very significant role in distinguishing *Achlya* from other genera of Saprolegniaceae. The zoospores released in this way are primary zoospores.

Secondary zoosporc release mechanism

The compact cluster of the primary zoospores remains at the orifice of the sporangium. After the resting period, the primary zoospores metamorphosed into secondary zoospores having a reniform shape and were released from the cyst. They then swim away in surrounding water with a smooth and gliding movement. The spores had to exert themselves a good deal to effect their release from the cyst. Some spores again rested on the cyst for a while after their release and then quickly moved away. The compactly arranged empty cysts lingered behind even after the complete release of secondary zoospores from the clump. This mechanism was recorded in *Achlya flagellata*.

Saprolegnia Nees (Plate- I; Fig. 6)

Saprolegnia is a member of Saprolegniales of the class Oomycetes.

Primary zoospore formation and zoospore release mechanism

Zoosporangia in *Saprolegnia* are distinct from the rest of the hypha. They are terminal, elongated and cylindrical with a blunt tip rimmed with a lid like structure through which zoospores get released.

Zoospores are formed from the homogeneous protoplasm of sporangium. The protoplasm gets cleaved to form penta- or hexagonal masses. These masses turn into primary zoospores, which in due course become mature and are evenly distributed. The whole process was completed in approximately 8-9 minutes. The mature zoospores show gradually increasing movement for 2-3 minutes. At the optimum point zoospores rupture the apical rim and numerous pear shaped spores get released. The spores swim away in the surrounding water with a smooth and gliding manner after their release from the sporangium.

It seems that most of the zoospores get released due to the great pressure inside the sporangium. The lagged behind zoospores propel themselves to outside. Time taken for the zoosporangium to empty itself completely was in one case two minutes thirty six seconds and three minutes twenty eight seconds in the other case. The difference in the two readings may be due to the size of the sporangium and environmental conditions. The empty sporangium did not collapse immediately after the zoospore release but remained intact.

This type of zoospore release is called Saprolegnoid type. In this, zoospores are released after maturation with some force from the zoosporangium without forming a cluster at the apical pore. Absence of the cluster formation at the exit orifice of the sporangium distinguishes *Saprolegnia* from other genera of Saprolegniaceae.

This mechanism was recorded in *Saprolegnia ferax*.

Aphanomyces de Bary (Plate-I; Fig.7)

Aphanomyces is a saprophytic, parasitic and pathogenic member of Saprolegniales of the class Oomycetes.

Primary zoospore release mechanism

Zoosporangia of *Aphanomyces* are indistinct from rest of the hypha. At maturity the protoplasm inside the hyphae get segmented isodiametrically and get separated from walls of the hyphae to form a single linear chain of approximately 60 to 70 cylindrical, rod shaped zoospores with one or two small protoplasmic thread like connections at one or both the ends. At maturity, zoospores come out, continuously, one by one with some force through the hyphal tube and form a compact clump. They become sickle shaped for a very short duration due to atmospheric pressure and immediately assume a spherical shape. These individual zoospores get encysted. The release of the first few zoospores (approximately 10-15 zoospores) was at first with a speed of about 25 seconds, however, all the zoospores get released with a high rate of velocity and the whole process completes in 50 seconds.

Secondary zoospore release mechanism

Encysted primary zoospores are, for a short while, at rest but soon begin to vibrate and tremble within the clump. The individual spherical zoospore comes out of the cyst through a pore in 19 seconds and then after about 1minute and 9 seconds they become reniform and again vibrate very rapidly for some time and then dispersed with smooth and gliding movement as a secondary zoospore leaving behind the cyst. The spores have to exert themselves a good deal to effect their release from the cyst. The whole clump emptied after about 10 minutes. This mechanism was recorded in *Aphanomyces laevis*.

Dictyuchus Leitgeb (Plate- I; Fig. 8)

Dictyuchus is a member of Saprolegniales of the Class Oomycetes.

The sporangia of *Dictyuchus* are distinct from the rest of the hypha. They are elongated and cylindrical. Primary zoospores are formed from the protoplasm and encysted within the sporangium Walls of these cysts form a net like structure in the sporangium and form a narrow, small tube like papilla through which secondary zoospores get released. At maturity the contents of the primary zoospores slowly pass through the papilla and form a small protrusion. At the same time, small vacuoles in the primary zoospores merged to form a big vacuole, which blocked the flow of contents in the protrusion for few seconds. The remaining contents pushed the vacuole in the protrusion, and the entire content entered the protrusion. As a result, a spherical zoospore is formed which droop down and get squeezed at the papilla, detached from it and remain there for a while with a trembling movement. After some time contraction of big vacuole took place which results in the decrease in size, assumed reniform shape and then swim in surrounding water as secondary zoospore. The whole process was completed within 1 to 1 minute and thirty seconds.

This type of zoospore release is called Dictyucoid type, and is characterised by encysted primary zoospores inside the zoosporangium and the release of secondary zoospores through the papilla. This mechanism was recorded in *Dictyuchus sterile*.

Olpidiopsis Cornu (Plate- I; Fig. 9)

Olpidiopsis is a member of Lagenidiales of the class Oomycetes. It is an endoparasite on *Achlya* species with holocarpic thallus.

Zoosporangia of the parasite *Olpidiopsis* are embedded in the swollen hypha of the host *Achlya* and are ellipsoidal or oval. Protoplasm inside the zoosporangium gets transformed into zoospores, which are very compact. At the same time, a discharge tube from the zoosporangial wall of the endoparasite is formed which extends right up to the surface or even beyond the surface of the host hypha. It is also filled with zoospores. At maturity zoospores begin to move slowly and steadily inside the sporangium. Due to the slow movement of the zoospores, it takes more time to create adequate pressure inside the zoosporangium (approx. 20 minutes). When the pressure reaches the peak point zoospores get released from the discharge tube with a spurt and a continuous stream and get accumulated at the tip of the extended discharge tube in the form of a loose clump. These zoospores are reniform. Initial zoospore release is due to the pressure inside the sporangium but the zoospores, which lag behind, propel themselves to the release point. Time required for the complete release of zoospores from the zoosporangium is 3 minutes and 10 seconds. The released zoospores within the clump start with a slow and jostling shaky movement, which gradually reach a rapid quivering movement. This state is maintained for 6-7 minutes after which all the spores rapidly disperse from the clump and swim away in the surrounding water. This mechanism was recorded in *Olpidiopsis achlyae.*

Pythium Pringsheim (Plate-I; Fig. 10)

Pythium is a member of Peronosporales of the class Oomycetes.

Mechanism of zoospore release: -

The zoosporangia in *Pythium* are distinct from the rest of the hypha. They are terminal, spherical with a small, nipple like apical tip. At the time of zoospore formation the tip of the sporangium distended into a thin vesicle into which simultaneously the protoplasm from the sporangium moves. The size of the vesicle is bigger than that of the sporangium. At the time of maturity the protoplasm in the vesicle cleaved gradually into zoospores. After some time, the spores began to move inside the sporangium due to which the vesicle as a whole showed a slow trembling movement. This movement gradually increased and the spores began to move rapidly in all directions within the vesicle pushing one another and against the wall of the vesicle. At a certain point, when the pressure inside the vesicle was maximum it bursts open releasing reniform zoospores which immediately dispersed in the surrounding water with a smooth and gliding movement. In all 15 reniform zoospores were released from the vesicle within 5 seconds. The formation of the zoospores in the vesicle and not in the zoosporangium is a characteristic feature of the genus *Pythium.* This mechanism was recorded in *Pythium proliferum.*

Discussion

The comparative account of closed water system (present study - 11 genera and 28 species) with open water system (19) (11 genera and 34 species) from same city showed that the latter has more diversity than that of the former.

Endoparasites *Rozella allomycis* and *Olpidiopsis achlyae* showed strict host specificity. *R. allomycis* was inoculated to the cultures of *Allomyces javanicus, A. moniliformis* and *A. arbuscula*, and it parasitised only on *A. arbuscula*. *O. achlyae* was inoculated in the cultures of *Achlya orion, A. oblongata, A. bisexualis, A. flagellate* and the latter seemed to be a specific host.

Powell and Blackwell (72) worked on the phenetic analysis of genera of the Saprolegniaceae and concluded that nonsexual characters dominantly provide intergeneric taxonomic separation and delineation of Saprolegniaceae, whereas, sexual (oogonial and antheridial) characters are generally more variable and perhaps of great utility when applied intragenerically. The video recording of zoospore (nonsexual structure) release from the zoosporangia of the different genera in the present studies give support to the statement made by Powell and Blackwell (72). The extensive studies of live sequential events like protoplasm differentiation leading to formation of zoospores, their movement within zoosporangium, actual mechanism of release of zoospores and their fate after release were recorded by using CCD camera with detailed specifications in different genera of zoosporic fungi. All these details and variations strongly support the above mentioned phenetic analysis put forth by Powell and Blackwell (72). Most striking feature of zoospore release was recorded in *Rozella allomycis* where zoospores get released from zoosporangia by three different methods: plug formation, pouch formation and pore formation, of these, video recording of the former two methods are innovative. The classificatory systems so far established are totally based on morphology and reproductive structures in the thallus but nobody has realized the importance of mechanism of zoospore release, which is probably an important criterion to give a proper status to the genus in classification. According to Hohnk (37) small vacuoles inside the zoospores get disappeared to create a strong pressure within the sporangium causing the passive escape of the zoospores. However, Dick (22,25) stated that the pressure of hydration and differential viscosity causes the zoospore release, however he did not give convincing proof.

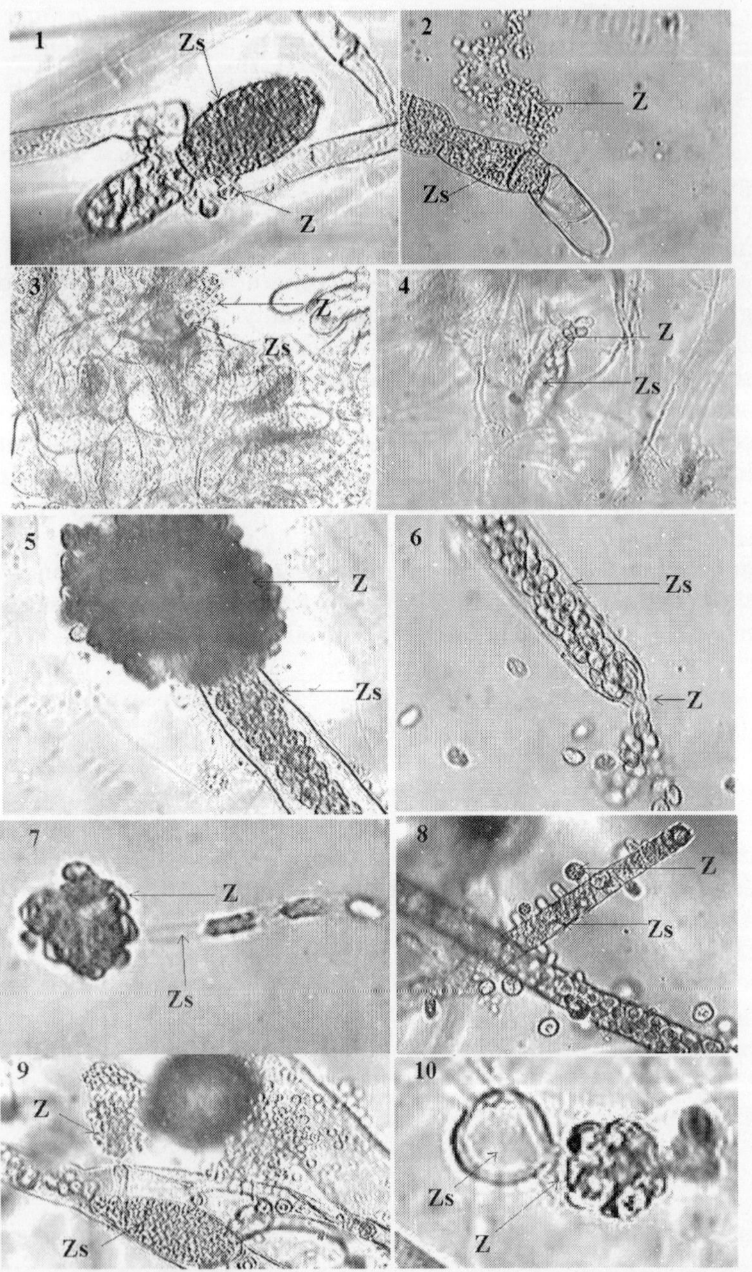

Photopage caption next page

Diversity in zoosporangia and Zoospore release mechanism:

1. Zoospore release in *Allomyces* from basal discharge papilla of oval zoosporangium. × 450; **2.** Zoospore release in *Rozella*: loose clump formation at the tip of discharge papilla: septate zoosporangia. × 450; **3.** Zoospore release in *Blastocladia* from the apical discharge pore of the zoosporangium. × 450; **4.** Zoospore release in *Gonapodya* from apical discharge pore of pod shaped zoosporangium. × 450; **5.** *Achlya*: achlyoid zoospore release with compact clump formation at the tip of sporangium. × 450; **6.** *Saprolegnia*: Saprolegnoid zoospore release in which zoospores dispersed. × 450; **7.** *Aphanomyces* : indistinct zoosporangium with cylindrical zoospores which form compact clump at the tip of sporangium. × 450; **8.** *Dictyuchus*: Dictyuchoid zoospore release with encysted primary zoospores and secondary zoospores get discharged from the primary by papilla formation. × 450; **9.** *Olpidiopsis* : discharge tube from sporangium extends beyond the host hypha and form loose clump at the tip. × 450; **10.** *Pythium*: zoospores form inside the vesicle at the tip of zoosporangium and get released by rupturing the wall of the vesicle. × 450

The present review was based on the video recording of zoospore release in live cultures of watermoulds, the presence of gradually increasing movement of all zoospores within their boundary inside the zoosporangium at the time of release create a pressure on the thin discharge papilla which is a convincing evidence to explain zoospore release which could be a realistic proof to put forth the hypothesis of zoospore release. This may be significant and new criterion for the classification of the genera in watermoulds.

The genera in class Chytridiomycetes and Oomycetes showed variations in the formation of zoospores, mechanism of release and behaviour of zoospores before and after their release from zoosprangia. In Chytridiomycetes, the zoosporangia are always distinct from rest of the hyphae and are mostly oval to ellipsoidal, except in *Gonapodya,* where, they are characteristically pod shaped. In the other members of the Chytridiomycetes the zoospores are formed within the sporangium by the transformation of protoplasm, whereas in Oomycetes, zoosporangia are distinct as well as indistinct from rest of the hyphae and are of different shapes. Zoospores are formed either in the sporangium or outside the sporangium in the vesicle formed at the tip of the sporangium. Before release, the zoospores show movement within the sporangium in both the classes. In Chytridiomycetes they show slow, slightly circular, rapid or sluggish movement. In most of the members of Oomycetes also though the movement is slow but it also shows trembling or very rapid movement, except in *Aphanomyces* where the movement is not seen clearly. Sporangial wall develops a special device for the release of zoospores. A short discharge tube or small, thin walled, nipple like papillae are developed on apical or basal part of sporangial wall in all Chytridiomycetes except in *Rozella allomycis.* In Oomycetes, the devices are different, having papilla like elevation, blunt rim with a lid, a tube or sometimes tip distend into thin walled vesicle near sporangium. But, in case of *Dictyuchus* (Oomycetes) spores get encysted within sporangium and instead of sporangial papillae cysts develop small tube like papilla from their wall to release individual zoospore. In Chytridiomycetes, zoospores are of only one type and are spherical or oval in shape and after release they either get scattered immediately in the surrounding water or form a loose mass around the discharge papilla whereas in Oomycetes zoospores are of two types. Pear shaped primary zoospores are either encysted within the sporangium or at the tip of the sporangium in the form of a compact clump except in *Saprolegnia*, where they immediately swim away in the surrounding water. The reniform secondary zoospores get released after the period of incubation, which varies from genus to genus. In *Pythium* and *Olpidiopsis* only secondary zoospores are formed. Zoospores after release from the sporangium show fast, jerking and darting movement in all members of Chytridiomycetes, except *Gonapodya* where it is smooth. In all the members of Oomycetes the movement is smooth and gliding.

Held (36) extensively studied the complete life cycle of *Rozella allomycis*, but did not express clear views regarding the release of zoospores. However, three types of mechanisms were recorded from the same genus from closed water temporary ditches, of which two are the new records in *Rozella.* This endoparasite develops its own devices for the release of its zoospores. Plug formation at the tip of discharge pore, formation of the pore and inwardly directed papillae from the sporangium of *Rozella* are commonly observed.

Sparrow (77) and Waterhouse (88,89) reported hopping darting and jerking as well as smooth movement in Chytridiomycetes. Same types of movements were recorded in the genera of Chytridiomycetes in the closed water system. Waterhouse (89) further discussed the zoospore discharge and various steps in the stages of development of zoospores right from initiation to maturation of zoospores with tissue disjunction and cytodieresis as intermediate steps in both Chytridiomycetes and Oomycetes. Similar steps were recorded during zoospore release in the forms which were isolated.

Thuret (85) described the events of the zoospore discharge for the first time in *Saprolegnia*, a member of Oomycetes. Hohnk (37) discussed planetism and dimorphism in relation to behaviour of zoospore and described 5 types based on the swarming of the spores. Type I and II (spores swimming immediately on discharge) occur in water. Type III (spores cluster at the exit orifice) occurs in swampy areas. Type IV (sporangium dissolve to release encysted spores) and Type V (spores are not motile) in swampy areas; where the Type IV and V are the transitional stages towards terrestrial habitat. Johnson et al (42) reviewed and discussed these events studied by many research workers.

In the present review, for the data analysis, ponds and lakes (closed water system) were treated as separate entities. The lakes under study have changing periphery as per the amount of rainfall during that year. However, the different water sources such as nearby rivers, streams and brooks also add some water in lakes. On the other hand, the ponds have stagnant water with limited periphery. Therefore, 'aliens', 'immigrants' and 'transients' were common in lakes whereas, 'indwellers' or 'indigenous' species were active in ponds. In polluted lakes one 'indweller', two 'inhabitants' and five 'aliens' were recorded. In unpolluted ponds two 'immigrants', six 'indwellers' and two 'aliens' were noted. In polluted pond one 'alien' and due to availability of substratum five 'indwellers' were recorded.

Manoharachary and Ramarao (59), Madhusudan Rao and Manoharachary (51) and Bhairavanath and Manoharachary (5) reported that *Achlya, Aphanomyces, Blastocladia, Olpidiopsis* and *Pythium* are confined to aquatic habitat and complete their life cycle in water and regarded as 'indwellers' or 'active forms' and 'inhabitants' of aquatic ecosystems as stated by Park (68) and Dick (26). They claimed that 'indwellers' form the major flora of aquatic environment. *Allomyces* might be 'transient' between aquatic and terrestrial ecosystem as stated by Madhusudan Rao and Manoharachary (51). In the present studies, species of the same genus may have different origin of potential inoculum. Based on observations different species in lakes and ponds are categorized differently. The 'indwellers' were restricted to ponds because of the submerged substratum and less wave action in water. The frequency of 'aliens' is more in lakes because of the addition of free flowing water from various sources. In lakes *A. arbuscula, A. oblongata, A. orion, A. diffusa* and *S. ferax* are considered as 'aliens'; *Aphanomyces laevis* and *P. torulosum* are 'inhabitants' and *D. sterile* is the only 'indweller' species. In ponds *A. proliferoides, S. torulosa* and *A. arbuscula* are the 'aliens'; *A. bisexualis, A. diffusa, D. carpophorus, D. sterile, A. laevis, G. prolifera, G. polymorpha, B. arborata* and *B. incrassata* are 'indwellers'; *A. proliferoides* and *P. torulosum* are 'immigrants'.

In the present investigations, the species diversity was more remarkable in rainy season followed by winter and least in summer. The maximum number of watermoulds was observed in most polluted pond and minimum number was observed in less polluted lakes. In all, 13 species were found to be 'all season species', 12 as 'rainy season', 6 as 'winter season' and 3 as 'summer season' species. *Allomyces laevis* was most common and found almost throughout the year. From the river water Desale (19) reported *P. carolinianum, P. proliferum, R. hyaline, B. simplex, D. sterile, A.* spp., *A. proliferoides, A. flagellata* are the constant species as occurring throughout the year. He isolated *P. carolinianum* most frequently and so considered it as universal in occurrence. *D. sterile, A. proliferoides* were the only two species, which were common in closed and open water bodies and occur throughout the year. Desale (19) grouped low temperature species designated as monsoon species and winter season species. In both open and closed waters *A. orion* and *A. arbuscula*, were found to be monsoon or rainy season species, whereas *S. ferax* was found commonly as winter season species. The species, namely, *A. klebsiana* and *O. achlyae* were isolated as winter season species from open water bodies but these were rainy season species from closed water bodies. *P. catenulatum* was the winter season species group from river water but it was all season species from the closed water bodies.

The occurrence, distribution and frequency of water moulds, mainly, Oomycetes was more as compared to Chytridiomycetes in closed water bodies. The members of Oomycetes were present in polluted as well as in non-polluted waterbodies, thus, Oomycetes did not show any specificity for their distribution, whereas, Chytridiomycetes were mainly restricted to polluted water bodies.

Rai and Mishra (73), Bhairavanath and Manoharachary (5), Mishra (62) and Mishra and Dwivedi (63) worked on alkaline ponds and lakes from different localities in India except Maharashtra and reported that in general Oomycetes ruled over all other aquatic fungi. But Mishra and Dwivedi (63) observed the dominance of Chytridiomycetes over Oomycetes. However, the present study showed the dominance of Oomycetes in closed water bodies from Maharashtra. Most of the members of Chytridiomycetes especially *Gonapodya* and *Blastocladia* require prominent organic substratum like angiosperm fruits (84,40,70, 56,13,87) for their growth. The decaying submerged *Acacia nilotica* pods were observed to support the growth of these fungi which is a new substratum for these water moulds.

The occurrence and distribution of watermoulds was restricted to particular water bodies. *G. prolifera, G. polymorpha, B. arborata, B. incrassata, S. anomalies* were restricted only to most polluted water. *A. bisexualis, Achlya* sp. *B. arborata, B. incrassata, D. carpophorus, G. polymorpha, G. prolifera* and *S. torulosa* occurred only in ponds whereas species *A. oblongata, A. orion, S. anomalies* and *S. ferax* were restricted to lakes. Seasonal periodicity at generic level was observed only in *Gonapodya* and *Saprolegnia*. Periodicity at particular season and at particular site was observed in *A. bisexualis, D. carpophorus, G. polymorpha, G. prolifera* and *S. anomalies*.

The genera *Achlya, Allomyces, Aphanomyces* and *Pythium* were common in temporary ditches and in ponds and lakes, whereas, *Dictyuchus, Gonapodya, Blastocladia* and *Saprolegnia* were restricted only to the ponds and lakes. The species, *A. recurva, A. klebsiana* and *A. flagellata, A. javanicus, A. moniliformis, B. incrassata, B. longicaulis, O. achlyae, R. allomycis* occurred only in rainwater ditches.

The species which were isolated in summer season did not have specific seasonal periodicity as they occur either in rainy season or in winter. In rainy season maximum species were isolated in the second year of study, of these 11 species were restricted only to this season. During winter 14 species were isolated, of these, only three species, namely, *Achlya* spp., *S. ferax* and *S. torulosa* were restricted only to winter. Out of the 28 species, only 10 species occurred in all seasons but most of them were sporadic.

Periodicity is the occurrence of species at particular environmental conditions. Some species are isolated frequently from any particular site within a particular niche and some are restricted to particular season, thus, indirectly related to climatic conditions.

Thus, survey of the seasonal occurrence, distribution and periodicity pattern of the aquatic fungi indicates that exhaustive work has been done on this aspect from all over the world mainly in temperate countries. However, Indian workers paid very meager attention towards this aspect while studying aquatic fungi. From Maharashtra, Desale (19) did significant and important contributory work over the seasonal occurrence, distribution and periodicity of aquatic fungi from riverine water. From river water *Achlya* was recorded to be dominant with 14 species. Saprolegniales were recorded with 4 genera and 16 species and the percentage occurrence was more than 60%, Blastocladiales with three species of *Allomyces* and two of *Blastocladia* were next.

From closed water bodies the Saprolegniales showed 57.14% occurrence with respect to the total fungal population followed by Blastocladiales with 17.8% and with three species of *Allomyces* and two of *Blastocladia*. Peronosporales with three species of *Pythium* showed 10.71% and Lagenidiales and Chytridiales were 3.5%. In case of open water bodies 19, Saprolegniales were dominant with 76.47% occurrence with respect to total fungal population followed by Peronosporales with 9.92% occurrence. Lagenidiales and Blastocladiales were with 5.92% occurrence. Minimum level of occurrence was observed in the case of Chytridiales with 2.91%. Thus, the Saprolegniales have more percentage occurrence in open water system than that in closed water system. However, Blastocladiales was more in closed than that in open water body.

The genus *Achlya* with 9 species was dominant with 31% frequency occurrence (Pie Chart No. 1) followed by *Saprolegnia*, *Pythium* and *Allomyces*. The percentage of frequency occurrence of *A. bisexualis* (8.22%), *A. proliferoides* (7.79%), *D. carpophorum* (5.62%), *D. sterile* (8.6%), *P. atenulatum* (5.19%), *P. torulosum*

(11.25%) and *A. laevis* (24.6%), was more as compared to the remaining species. As *A. laevis* showed maximum % frequency therefore is considered as universal in occurrence.

Dominance of *Achlya* in both closed and open water system from the same region of subtropical country like India confirms the conclusion of Manoharachay (56) that *Achlya* forms the most dominant genus of Saprolegniaceae while other dominant member of Saprolegniaceae, namely, *Saprolegnia* is absent from freshwaters of subtropical countries. Previously, Alabi (1) in tropical country like Nigeria observed the same. Willoughby (90) reported that in temperate countries, *Saprolegnia* forms the predominant genus of Saprolegniaceae followed by *Achlya.* Dick (20,26) confirmed the same from Europe. He stated that *Saprolegnia* forms the most dominant member of Saprolegniaceae as compared to *Achlya* and it may be largely absent from warm temperate and tropical climate.

Saprolegniales was recorded with 4 genera and 16 species from different closed water bodies, whereas from river water 6 genera with 26 species were recorded. In closed and open water bodies *Achlya* was dominant with 9 species and 14 species respectively. *Saprolegnia, Pythium* and *Allomyces* had 11% occurrence; *Gonapodya, Blastocladia* and *Dictyuchus* with 7% and *Olpidiopsis*, *Aphanomyces, Rozella* and *Brevilegnia* with 4% of occurrence in closed water system.

Desale (19) recorded somewhat similar observations in open system. From riverine water he collected 26 species of Saproleg-niaceae of which eccentric species were dominant with 58.33% occurrence followed by centric species with 36.36% and the minimum occurrence was of subcentric species with 4.66%. He also noted that the eccentric oospore species did not show seasonal variations while species with centric oospores were isolated in winter and some part of rainy season.

The seasonal periodicity of different oospore types was recorded by many research workers from temperate and tropical–subtropical countries. From temperate countries, Ziegler (91) reported that eccentric species of Saprolegniaceae were most abundant in warm weather, whereas those with subcentric and centric oospore were more frequently observed in cooler weather. However, Maestres and Nolan (53) and Klich and Tiffany (49) have observed that species with centric and subcentric oospores had a distinct seasonal periodicity with maxima during summer and decreasing in the spring and fall. Hughes (38), Dick and Newby (27) and Fox and Wolf (30) reported that eccentric species had no seasonality. In tropical and subtropical countries, Dayal and Thakur Ji (16,17), Srivastava 80), Alabi (1,2) and Dick (26), Chowdhary and Agarwal (8,9) noted that species with centric and subcentric oospores are dominant in rainy and winter seasons and those with eccentric oospores were dominant in summer. In the closed water system similar type of seasonal periodicity for the species with different oospore types was noted.

In temperate countries maximum abundance of water moulds is in winter or in autumn and spring and decreases in summer (29, 70,74,89,). But in tropical countries like India, maximum numbers of water moulds was recorded in rainy season followed by winter and least in summer in closed system (5,8,9,14,15,32,54,55,58, 59,60, 80).

Similar types of results were observed during the study of watermoulds from closed water bodies. Whereas, Dayal and Tandon (14,15) stated that winter is the ideal season for Phycomycetes. Alabi (1,2), in tropical countries like Nigeria, noted that most of the fungi from Saprolegniaceae were recorded in dry season when temperatures were relatively high. Das Gupta (12) and Manoharachary (56) noted that period immediately after monsoon rains and winter were best for the collection of watermoulds in India. Our observations are more or less similar.

Eccentric forms were dominant in ponds and mostly in unpolluted ones whereas subcentric and centric forms were more in lakes and in polluted water. Eccentric species did not show seasonal periodicity, whereas, species with subcentric oospore had it in rainy and winter seasons; those with centric oospores had it only in winter. Hughes (38), Alabi (1,2), Manoharachary and Ramarao (59) claimed that in tropical conditions the species with eccentric oospores are more frequent than the species with centric or subcentric oospore. Similar results were observed in the closed water bodies in the study. In closed water bodies, distinct periodicity is not observed in eccentric species but marked periodicity is observed in centric and subcentric species by Chaudhuri (7), Zeigler (91), Dick and Newby (27), Hughes (38), Srivastava (80), Alabi (1,2), Fox and Wolf (30), Maestres and

Nolan (53) and Klich and Tiffany (49). Similar type of periodicity was also seen in the closed water bodies under study. However, Bhairavanath and Manoharachary (5) and Manoharachary (57) have reported that species with eccentric oospore are common in winter.

Desale (19), from river waters, recorded maximum number of fungal forms in the winter and rainy seasons and the number was decreased remarkably in summer. Hence, occurrence of fungal forms in lake waters and river water were similar as far as seasons are concerned. The occurrence of species may vary from water body to water body depending upon the niche and the influence of the environmental conditions.

Achlya, Allomyces, Aphanomyces, Blastocladia, Dictyuchus and *Pythium* are the common watermoulds in closed waterbodies from India. In tropical countries the genus *Achlya* is dominant with maximum species, whereas, in temperate countries *Saprolegnia* is dominant mostly in cool water. Harvey (34) stated that certain species of *Saprolegnia* did not survive in heavily polluted waters. On the contrary, in the present studies, *S. ferax* and *S. anomalies were* isolated from the polluted water. These seemed to be pollution indicator species. Khulbe (46) also found the former species in the polluted water. These species are thus considered as indicators of water pollution.

Aphanomyces is recovered from temperate as well as tropical countries throughout the temperature range in all seasons over the year. However, Thakur Ji and Dayal (83) observed it more frequently in winter. *A. laevis* is the only species, which occurred in all concentrations of physicochemical parameters. Its occurrence is in polluted and non-polluted sites. Dick (23) observed nonsexual reproductive structures on insect exuviae, whereas in present studies, abundant oogonia were seen on cockroach cadaver. *A. laevis* is present in closed as well as open water bodies and soil, so considered as universal in occurrence. The larvicidal activity of *A. laevis* on mosquito larvae in several ditches and the laboratory was tested.

Waterhouse (89) reported Blastocladiaceae, Pythiaceae and Leptomitaceae from the Hogsmill River and commented that, most of the watermoulds have a seasonal periodicity. Hunter (39) found maximum number of Saprolegniales followed by Peronosporales.

Only one member of Chytridiomycetes was observed in her studies. Desale (19) and Gandhe and Gandhe (31) reported similar observations. In the present studies, though Saprolegniales was dominant, Chytridiomycetes occurred in good number. Thus the lentic system proved to be favourable for the growth of Chytridiomycetes. *Aphanomyces laevis* occurs in polluted as well as unpolluted river water as well as in closed water bodies. *Achlya* was the dominant genus in both the water systems. However, generic and species diversity was more in riverine system with remarkable fluctuations in their occurrence, distribution and development of thallus. Strong water currents or wave actions often influence the open water systems and zoospores are migrated from one site to other. However, the closed water systems are with limited diversity, which remains unchanged for a longer period till fluctuations in physicochemical parameters.

The occurrence of water moulds species from the closed (collected data) water bodies was compared with open water bodies (generated data) and noted the maximum fungal forms were present in open water systems with the maxima in occurrence of members of Saprolegniaceae. The species commonly observed in both the closed and open water systems were, namely, *A. orion, A. flagellata, A. klebsiana, A. oblongata, A. proliferoides, A. arbuscula, A. laevis, D. carpophorus, D. sterile, P. proliferum* and *S. ferax.*

The species which are isolated only in open water systems are *A. americana, A. caroliniana, A. conspicua, A. dayali* sp. nov., *A. debaryana, A. dubia, A. imperfecta, A. megasperma, A. punensis* sp. nov., *A. scaber, B. simplex, D. monosporus, B. diclina, O. saprolegniae* var. *levis, P. paradoxa,. P. carolinianum, R. hyaline, S. delica, S. irregularis, S. lapponica, S. parasitica, T. clavata* (19).

Some species are restricted or commonly found in the closed water bodies, these were, *A. bisexualis, A. diffusa, A. recurva, A. javanicus, A. moniliformis, Achlya* spp., *B. arborata, B. incrassata, B. diclina, G. polymorpha, G. prolifera, R. allomycis, S. anomalies* sp. nov. and *S. torulosum* are mostly from temporary rainwater ditches. The Chytridiomycetes members were dominant in closed water bodies than the open water bodies may be due to the absence of wave action and the availability of the submerged substratum.

Thus, open water system harboured maximum and diversified fungal flora than the closed water systems.

Acknowledgement

Senior author (RVG) is thankful to Prof. V.S. Rao, Director Agharkar Research Institute, Pune for providing necessary facilities. The authors are thankful to Prof. C. Manoharachary, Department of Botany, Osmania University, Hyderabad for valuable suggestions and are grateful to Prof. M.J. Powell, University of Albama, Tuscaloosa, AL, USA, for valuable suggestions in the discussion of zoospore discharge mechanism. The authors also thank the students for providing important data of aquatic fungi of closed and open water systems.

REFERENCES

1. Alabi RO. 1971. Factors affecting seasonal occurrence of Saprolegniaceae in Nigeria. *Trans Br mycol Soc* 56 (2), 289–299.
2. Alabi RO. 1971. Seasonal periodicity of Saprolegniaceae at Ibadan, Nigeria, *Trans Br mycol Soc* 56 (3): 337-341.
3. APHA, AWWA and WEE 1992. Standard methods for the examination of water and waste water. *Amer Health Assoc Publ New York.* 18th Edn.
4. Bhairavanath D and Manoharachary C. 1983. Myco ecological studies of two water-bodies from Andhra Pradesh, India. *Ind Phytopathol* 36 (1): 69-78.
5. Bhairavanath D and Manoharachary C. 1985. Mycoecological studies of two water-bodies from Andhra Pradesh, India. *Ind J Bot.* 8 (1): 69–78.
6. Butler EJ. 1907. An account of the genus *Pythium* and some Chytridiaceae. *Mem Dept Agric India, Bot* Ser No. 51: 1-160.
7. Chaudhuri H. 1942. A new genus of Saprolegniaceae. *Proc 29th Indian Sci Congr* Pt. IV: 19.
8. Chowdhary PN and Agarwal GP. 1980. Studies on distribution of some aquatic fungi in India. *Ind Phytopathol* 33(1): 107–109.
9. Chowdhary PN and Agarwal GP. 1980. Studies in seasonal variation on aquatic fungi from Delhi. *Ind Phytopathol* 33(4): 614–615.
10. Coker WC. 1923. *The Saprolegniaceae with notes on other watermoulds.* The Univ North

Carolina Press, Chapel Hill, North Carolina, 201pp.

11. Couch JN. 1941. The structure and the action of the cilia in some aquatic phycomycetes. *Amer J Bot* 28: 704-713.

12. Das Gupta SN. 1982. Discourse on aquatic Phycomycetes of India. *Ind Phytopathol* 35: 193-216.

13. Dasgupta SN and John. 1988. A contribution to our knowledge of the genus *Blastocladia*. *Ind Phytopathol* 41: 521–547.

14. Dayal R. and Tandon RN. 1962. Ecological studies of some Phycomycetes. *Hydrobiologia* 20(2): 121-127.

15. Dayal R and Tandon RN. 1963. Ecological studies of some Phycomycetes. II. Fungi in relation to chemical factors of the water. *Hydrobiologia* 22: 324-329.

16. Dayal R and Thakur Ji. 1968. Studies in aquatic fungi of Varanasi. III. Observations on some parasitic aquatic Phycomycetes. *Sydowia* 22: 278-283.

17. Dayal R. and Thakur Ji. 1968. Studies in aquatic fungi of Varanasi V - A taxonomic study. *Proc Nat Acad Sci India* 38 (B) : 32-38.

18. Dayal R and Usha Kiran. 1988. *Zoosporic fungi of India*. Inter India Publ, New Delhi, 297 pp.

19. Desale MJ. 1996. *Study of the aquatic fungi of the rivers Mula and Mutha, Pune, India*. Ph.D. Thesis, Pune Univ, Pune 297 pp.

20. Dick MW. 1963. The occurrence and distribution of Saprolegniaceae in certain soils of south east England III. Distribution in relation to pH and water content. *J Ecol* 51: 75-81.

21. Dick MW. 1965. The maintenance of stock cultures of Saprolegniaceae. *Mycologia* 57: 828-831.

22. Dick MW 1966. The Saprolegniaceae of the Environs of Blelham Tarn: Sampling Techniques and the estimation of propagule numbers. *J Gen Microbiol* 42: 257–282.

23. Dick MW. 1970. Saprolegniaceae on insect exuviae. *Trans Brit mycol Soc* 55: 449-458.

24. Dick MW. 1971. The ecology of Saprolegniaceae in Lentic and Littoral muds with a general theory of fungi in the lake ecosystem. *J Gen Microbiol* 65: 325-337.

25. Dick MW. 1973. Saprolegniales. In: *The Fungi. An Advanced Treatise.* Ainsworth, G.C., F.K. Sparrow and A.S. Sussman (Eds.). Vol. IV B.

Academic Press, New York and London. pp. 113–144.

26. Dick MW. 1976. The ecology of aquatic Phycomycetes. In: *Recent Advances in Aquatic Mycology*. EBG Jones (Ed.). Elek Science, London. pp. 513-542.

27. Dick MW and Newby HV. 1961. The occurrence and distribution of Saprolegniaceae in certain soils of South-East England. I. Occurrence. *J Ecol* 49: 403-419.

28. EL–Hissy F.T. and EL–Nagdy MA. 1983. Aquatic Phycomycetes on the mud of the River Nile (Assiut, Egypt). *Sydowia* Ser. II. 36: 118–124.

29. Forbes EJ. 1935. Observation on some British watermoulds (Saprolegniales and Blastocladiales). *Trans Brit mycol. Soc.* 19: 221-239.

30. Fox NC and Wolf FT. 1977. Aquatic Phycomycetes of Radnor Lake, Nashville, Tennessee. *J Tennessee Acad Sci.* 52: 100-104.

31. Gandhe RV and Gandhe KR. 2003. Occurrence, distribution and seasonality of riverine fungi. *Geophytology* 31 (1 & 2): 53-57.

32. Gupta S and Dubey T. 1987. Effect of chemical contents of water on seasonal occurrence of aquatic fungi. In: *Perspectives in Mycological Research*–I. Prof. GP Agarwal Festschrift volume: 63-68.

33. Hanseler CM. 1928. Reduction in yield of peas due to root rot caused by *Aphanomyces euteiches. New Jersey Agric Coll Exp Sta Ann Rep* No. 49: 273-275.

34. Harvey JV. 1952. Relationship of aquatic fungi to water pollution, *J water pollution control* 24: 1159-1164.

35. Hasija SK and Khan MA. 1987. Seasonal occurrence and distribution of aquatic fungi in relation to the water quality of Jabalpur lakes. In: *Perspectives in Mycological Research*–I. Prof. GP Agarwal Festschrift volume. pp. 79-87.

36. Held AA. 1980. Development of *Rozella* in *Allomyces*: a single zoospore produces numerous zoosporangia and resistant sporangia. *Can J Bot* 58: 959-979.

37. Hohnk W. 1933. Polyplanetism and zoospore germination in the Saprolegniaceae and *Pythium*. *Amer J Bot* 20: 45-62.

38. Hughes GC. 1962. Seasonal periodicity of the Saprolegniaceae in the south eastern united states. *Trans Br mycol Soc* 45 (4): 519–231.

39. Hunter RE. 1975. Water moulds of the river Great Ouse and its tributaries.

Trans Br mycol Soc 65(1): 101-108.

40. John R. 1958. Studies in the Indian aquatic fungi. III. Sexual reproduction in *Gonapodya polymorpha* Thaxter. *Proc Ind Acad Sci* B 50, 38: 253-259.

41. Johnson TW. 1956. *The genus Achlya morphology and taxon-omy*. Univ Michigan Press, Ann Arbor, Michigan. 180 pp.

42. Johnson TW Jr, Seymour RL, Padgett DE. 2002. *Biology and systematics of the Saprolegniaceae.* Published at http:// www.uncw.edu/ people/padgett.

43. Kadlubowska, JZ. 2001. Trends in plant sciences. Remarks on European Zygnematales. *Phytomorphology* golden jubilee issue. 457-466.

44. Kanouse BB. 1925. On the distribution of watermolds, with notes on the occurrence in Michigan of members of the Leptomitaceae and Blastocladiaceae. *Papers, Michigan Acad Sci Arts, Ltrs.* 5: 105-114. (1926).

45. Khulbe RD. 1980. Occurrence of parasitic watermoulds in some lakes of Nainital, India *Hydrobiologia* 70: 119-121.

46. Khulbe RD. 1983. Aquatic fungal flora in rhizosphere and non rhizosphere soils in some cultivated fields of Nainital. *Sci Culture* 49: 105–106.

47. Khulbe RD. 2001. *A manual of aquatic fungi* (Chytridiomycetes & Oomycetes). Daya Publishing House, Delhi, 204 pp.

48. Khulbe RD and Bhargava KS. 1977. Distribution and periodicity of watermoulds in some lakes in Nainital, India. *Hydrobiologia,* 54: 67–72.

49. Klich MA and Tiffany LH. 1985. Distribution and seasonal occurrence of aquatic Saprolegniaceae in Northwest Iowa. *Mycologia* 77(3): 373-380.

50. Lund A. 1934. Studies on Danish freshwater Phycomycetes and notes on their occurrence particularly relative to the hydrogen ion concentration of the water. *Mem Acad Roy Sci, Danemark,* Sect. Sci. 9: 1-97.

51. Madhusudan Rao M and Manoharachary C. 1983. Ecology of freshwater phycomycetes from Andhra Pradesh, India. *Ind Phytopathol* 36 (3): 431-437.

52. Maestres LE. 1977. *A quantitative ecological study of aquatic fungi from Broadcove River with emphasis on representatives of the genus Saprolegnia.* Mas-ter's thesis,

Memorial Univ St John's, Newfoundland. 155 pp.

53. Maestres LE and Nolan RA. 1978. A quantitative study of aquatic fungi from the Broadcove river (New found land) with emphasis on representatives of the order Saprolegniales. *Occas Pap Biol, Memorial Univ. Newfoundland,* No. 1. 22 pp.

54. Manoharachary C. 1979. Ecological studies on some species of *Allomyces* from Andhra Pradesh, India, *Nova Hedwigia* 63: 179–182.

55. Manoharachary C. 1979. Taxo–ecological studies on some freshwater pond mud fungi of Hyderabad, India, *Nova Hedwigia* 63: 197-207.

56. Manoharachary C. 1981. The taxonomy and ecology of fresh water Phycomycetes from India. *Ind Rev Life Sci.* 1: 3-21.

57. Manoharachary C. 1985. Ecolo-gical account of some oomycetous fungi. *Ind bot Reptr* 4(1): 50-51.

58. Manoharachary C, Bhairavanath D and Madhusudan Rao M. 1983. Taxoecological studies on some aquatic fungi from India. *Bibliotheca Mycologica* 91: 457-462.

59. Manoharachary C and Ramarao P. 1981. Seasonal variation and distribution of fungi in two freshwater ponds of Andhra Pradesh, India, *Proc Ind. Acad Sci* (Plant Sci.), 90(3): 237-243.

60. Mer GS, Sati SC and Khulbe RD. 1980. Occurrence, distribution and seasonal periodicity of some aquatic fungi of Sattal (Nainital), India, *Hydrobiologia* 76: 201-205.

61. Miller CE. 1961. Some aquatic Phycomycetes from lake Texoma. *J Elisha Mitchell Sci Soc* 77: 293-298.

62. Mishra JK. 1983. Occurrence, distribution and seasonal periodicity of aquatic fungi as affected by water temperature in certain alkaline ponds of India. *Indian J Plant Pathol* 1(2): 133-140.

63. Mishra KC and Dwivedi RS. 1987. Aquatic moulds from Gujar Lake, Jaunpur. *Proc Nat Acad Sci* 57 (1): 92-94.

64. Mitra A. 1935. Studies on Indian aquatic fungi. *Proc. 22nd Indian Science Congress*. p. 252.

65. Padgett DE and Johnson TW. 2004. Zoosporangial discharge in a *Protoachlya hypogyna* (Saprolegniaceae) isolate from southeastern North Carolina. *Mycologia* 96 (2): 205–207.

66. Papavizas GC and Ayers WA. 1974. *Aphanomyces* species and their root diseases in pea

and sugar beet. USDA Techn Bull No. 1485. 158 pp.

67. Park D. 1972. Methods of detecting fungi in organic detritus in water. *Trans Brit mycol Soc* 58: 281-290.

68. Park D. 1972. On the ecology of heterotrophic micro organisms in fresh water. *Trans Br Mycol Soc* 58 (2): 291-299.

69. Patwardhan AA. 2004. *Taxo-ecology, cell wall analysis and biological control of Aquatic fungi.* Ph.D. Thesis. University of Pune, Pune. 193 pp.

70. Perrot PE. 1960. The ecology of some aquatic phycomycetes. *Trans Brit mycol Soc* 43: 19-30.

71. Powell MJ. 1993. Looking at Mycology with a Janus Face: A glimpse at Chytridiomycetes active in the environment. *Mycologia* 85 (1): 1-20.

72. Powell MJ and Blackwell WH. 1998. Phenetic analysis of genera of the Saprolegniaceae (Oomycetes). *Mycotaxon* 68: 505-516.

73. Rai JN and Misra JK. 1977. Aquatic fungi from alkaline ponds and soils. *Kavaka,* 5: 73-78.

74. Roberts RE. 1963. A study of the distribution of certain members of the Saprolegniales. *Trans Brit mycol. Soc* 46: 213-224.

75. Seymour RL. 1970. The genus *Saprolegnia. Nova Hedwigia* 19: 1-124.

76. Sparrow, FK Jr. 1935. Recent contributions to our knowledge of the aquatic Phycomycetes. *Biol Rev* 10: 152–186.

77. Sparrow FK Jr. 1960. *Aquatic Phycomycetes.* 2nd edn. Univ. Michigan Press, Ann Arbor. 1187 pp.

78. Sparrow FK Jr. 1968. Ecology of Freshwater fungi, In: *The Fungi. An Advanced Treatise.* Ainsworth GC, Sparrow FK and Sussman AS (Eds.). Vol. II Academic Press New York and London. pp 95-105.

79. Sparrow FK Jr. 1973. Mastigomycotina (Zoosporic Fungi) In: *The Fungi. An Advanced Treatise.* Ainsworth GC, Sparrow FK and Sussman AS (Eds.). Vol. IVB. Academic Press, New York and London. pp. 61-73.

80. Srivastava GC. 1967. Ecological Studies on Some Aquatic Fungi of Gorakhpur, India. *Hydrobiologia* 30: 385-404.

81. Suzuki S. 1960. Seasonal variation in the amount of zoospores of aquatic Phycomycetes in Lake Shinseiko. *Botan Mag* (Tokyo) 73: 483-486.

82. Suzuki S. 1960. The seasonal variation of aquatic fungi in

Shensun-ike Pond. *Jap J Limnol* 21: 271-278.

83. Thakur Ji and Dayal R. 1966. The occurrence and distribution of reproductive spores of Saprolegniales in certain ponds of Varanasi. *Nova Hedwigia* 12: 509-517.

84. Thaxter R. 1895. New or peculiar aquatic fungi. II. *Gonapodya* Fischer and *Myrioblepharis*, gen. nov. *Bot Gaz* 20: 477-485.

85. Thuret G. 1850. Researches sur les zoospores des algues et les antheridies des cryptogams. *Ann Sci Nat Bot* (3e ser.) 14: 214-260.

86. Unestam T. 1965. Studies on the Crayfish plague fungus *Aphanomyces astaci*. I. Some factors affecting growth *in vitro*. *Physiol Pl* (Copenhagen) 18: 483-505.

87. Usha Kiran and Dayal R. 1992. A new host record of *Blastocladia* and *Gonapodya. Proc Nat Acad Sci,* India 62 (B) III: 503-506.

88. Waterhouse GM. 1942. Some water molds of the Hogsmill River collected from 1937 to 1939. *Trans Brit mycol Soc* 25: 315–325.

89. Waterhouse GM. 1962. The zoospore. *Trans Brit mycol Soc* 45 (1): 1-20.

90. Willoughby LG. 1962. The ecology of some lower fungi in the English lake district. *Trans Br mycol Soc* 45 (1): 121-136.

91. Ziegler AW. 1958. The Saprolegniaceae of Florida. *Mycologia* 50: 693-696.

BIODIVERSITY OF DERMATOPHYTES: AN OVERVIEW

Kamal Rai Aneja, Radhika Joshi, Chetan Sharma, Parveen Surain and Ashish Aneja***

*Department of Microbiology, Kurukshetra University, Kurukshetra-136119, Haryana, India. Email: anejakr@yahoo.ca
**Medical Officer, University Health Center, KUK

KEY WORDS: Dermatophyte, dermatophytoses, tinea, *Trichophyton*, *Microsporum*, *Epidermophyton*, *Arthroderma*, keratinophilic

ABSTRACT

Dermatophytes (Ringworm fungi) are anamorphic hyphomycetous, keratinophilic fungi that parasitize keratinized tissue (hair, skin, nails) of humans and animals and cause dermatophytoses (dermatomycoses). These fungi occur widely in soil and other keratin containing substrata such as bird's nests and horns of some mammals. These belong to three anamorphic genera; *Trichophyton*, *Microsporum* and *Epidermophyton* consisting of 39 species with over 1000 names. Some species of *Microsporum* and *Trichophyton* produce *Arthroderma* as their teleomorph. The diseases caused by these organisms are generally named after the part of the body infected and are known as tineas. Eight types of tineas are known to occur in humans, such as tinea barbae, tinea capitis, tinea corporis, tinea cruris, tinea faciei, tinea pedis, tinea unguium and tinea manuum. Dermatophytes are grouped into three categories; zoophilic, anthropophilic and geophilic, on the basis of host preferences and natural habitat. The majority of dermatophytes are identified phenotypically based on two characteristics; colony characteristics on Sabouraud glucose agar (SGA) and microscopic morphology that includes the examination of the types of conidia; microconidia (presence or absence) and characteristics of macroconidia, such as the surface of the wall (rough verses smooth), number of septa and size of a conidium. Species of dermatophytes are identified on the basis of: *In vitro* hair perforation test, special nutritional requirements, urea hydrolysis, growth on polished rice grains and BCP-milk-solid glucose medium (dermatophytes milk agar medium) and temperature tolerance. Molecular characteristics are also being currently used in identification of a species.

1. INTRODUCTION

Dermatophytes (Greek for skin plants) or "Ringworm fungi" are anamorphic hyphomycetous, keratinophilic fungi that parasitize keratinized tissue (hair, skin, nails) of humans and animals cause dermatophytoses. These fungi occur widely in soil and other keratin containing substrata such as bird's nests and horns of some mammals. Dermatophytes are among the very few fungal species that cause contagious, directly host to host transmissible diseases of humans and animals (9, 36, 20).

The dermatophytes belong to three anamorphic genera *Trichophyton*, *Microsporum* and *Epidermophyton*. There are ove 1000 names for the 39 species now recognized as valid in the thre genera. Some dermatophytes reproduce sexually by producin ascomata (ascocarps) with asci and ascospores. The perfect stat (called teleomorph) produced by some species of *Microsporum* an *Trichophyton* belongs to *Arthroderma*, a member of the famil *Arthrodermataceae* (38, 20).

2. Dermatophytoses

Dermatophytoses (or dermatomycoses), commonly referred t as "*ringworm*" or "*tinea*" are the fungal infections of the kerat nized tissues of living animals. These are the most common caus of fungal infections worldwide, resulting in treatment costs of clos to half a billion dollars annually in the USA. The World Healt Organization estimates global prevalence of dermatomycoses to b approaching 20% (15, 3). The diseases caused by these organism are generally named after the part of the body that is infecte rather than the infecting organism. For example, tinea ped refers to athlete's foot and tinea unguium refers to a nail infection. Eight types of tineas are known to occur in humans, such as tinea barbae, tinea capitis, tinea corporis, tinea cruris, tinea faciei, tinea pedis, tinea unguium and tinea manuum (Fig. 1). Dermatophyte infections are generally superficial, but immunocompromised patients can experience severe, disseminated disease (27). Although dermatophyte infections are treatable, there is a high rate of recurrence following treatment with an antifungal, it remains to be determined whether this is due to relapse (the fungus not being completely eradicated during treatment) or a new infection (14).

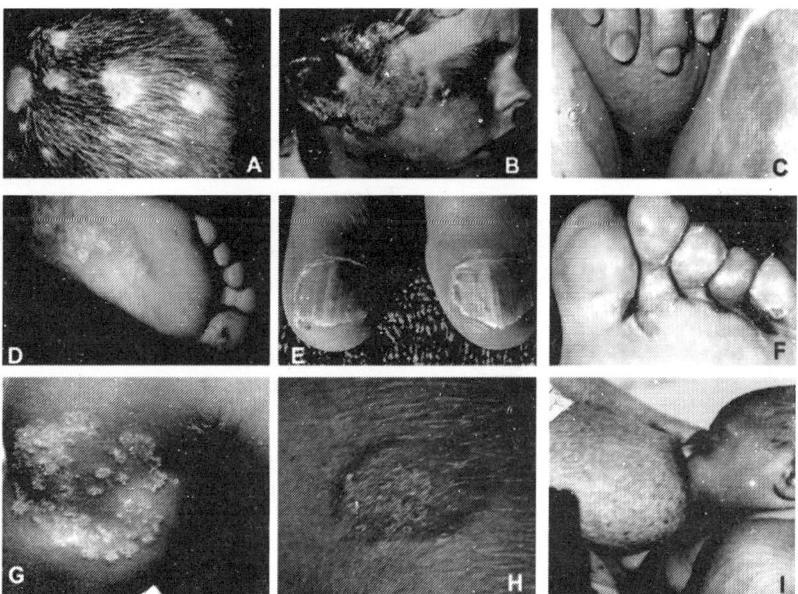

Figure 1. Clinical types of dermatophytoses based on the part of the body infected: fungal infections of hair (tinea capitis) (A); face (tinea capitis / ringworm) (B); groin (tinea cruris) (C); foot (tinea pedis / moccasin type) (D); nails (onycomycosis) (E); toe webbing (tinea pedis) (F); arm (tinea corporis) (G); close-up of ringworm (H); torso with concentric rings (tinea imbricata /tinea corporis) (I).

3. Prevalence and distribution

Fungal infections of skin and nails are a common global problem. The majority of fungal infections seen in both temperate and tropical countries are superficial infections of the skin. The prevalence of superficial mycotic infections shows that 20-25% of the world's population has skin mycoses, making one of the most frequent forms of infection. A majority of the superficial fungal infections of skin are caused by 5 or 6 species of dermatophytes, of which *Trichophyton rubrum* is the most common (16). The dominant species of dermatophytes vary according to their clinical localization as described in the "Clinical manifestations and global patterns of superficial fungal infections" section and listed in table 1 (11).

Table 1: Dominant species of dermatophytes according to clinical localization.

Clinical types	Localization of the body	Species
Tinea capitis	scalp	*Trichophyton violaceum* *T. tonsurans* *T. soudanense* *Microsporum canis* *M. audouinii*
Tinea pedis	feet	*T. rubrum* *T. mentagrophytes* (var. *interdigitale*) *Epidermophyton floccosum*
Tinea cruris	groin	*T. rubrum* *T. mentagrophytes* (var. *interdigit et granulosum)*
Tinea corporis	arms, legs and trunk	*T. rubrum* *M. canis* *T. tonsurans* *T. verrucosum*
Tinea unguium	finger-nails and toe-nails	*T. rubrum* *T. mentagrophytes* (var. *interdigitale)* *E. floccosum* *T. violaceum* *M. gypseum* *T. tonsurans* *T. soudanense*

Although dermatophytes are found throughout the world, the most prevalent strains and the most common sites of infection vary by region. Hot, humid climates and overcrowding predispose populations to skin diseases, including tinea infections. Developing countries have high rates of tinea capitis, while developed countries have high rates of tinea pedis and onychomycosis. Low socioeconomic conditions are strongly linked to higher prevalence rates for skin infections, including tinea infections. High prevalence rates of tinea pedis and onychomycosis have been linked to increased urbanization, community showers, sports, and the use of occlusive footwear (35, 16, 39). Although tropical and subtropical climates have a higher overall prevalence of skin mycoses, tinea pedis and onychomycosis are rare in India and rural Africa (4).

In northern India, anthropophilic dermatophytes are the predominant pathogens causing tinea capitis. *Trichophyton violaceum* (38%) is the most commonly isolated pathogen following *M. audouinii* (34%), *T. schoenleinii* (10%) and *T. tonsurans* (10%) (30). Tinea capitis caused by *T. violaceum* is endemic in South India (17). Unlike in Europe, tinea pedis is comparatively rare in India, especially in the poorer areas. As in Europe, the most important causative agents are *T. rubrum, T. interdigitalis* and *E. floccosum* (31).

4. Ecological groupings and transmission

Dermatophytes which have reservoirs in soil are classified into three ecological groupings on the basis of their host preferences and natural habitat: geophilic, zoophilic and anthropophilic. *Geophilic dermatophytes* are those that usually are saprobic in soil, where they are associated with decomposing hair, features, hooves and other keratin sources. They infect both humans and animals. *Zoophilic dermatophytes* primarily infect animals other than humans. They often are found on a single host or very limited host range, outside which they are found only in exceptional circumstances. *Anthropophilic dermatophytes* which are usually restricted to humans, are mainly associated with community life. The spread of anthropophiles is more common in communities like schools, barracks, prisons and the family (25, 10, 26, 23). Examples of common dermatophytes classified in three groups are given in Table 2.

Table 2: Examples of dermatophytes classified in three types: anthropophilic, zoophilic and geophilic[a].

Anthropophilic	Zoophilic	Geophilic
Epidermophyton floccosum	*Microsporum* spp.	*Microsporum* spp.
	M.canis	*M.gypseum complex*
	M.gallinae	
		M.praecox
Microsporum spp.	*M.nanum*	*M.racemosum*
M.audouinii	*M.persicolor*	*M.vanbreuseghemii*
M.ferrugineum		
	Trichophyton spp.	*Tricophyton vanbreuseghemii*
Trichophyton spp.	*T.equinum*	
T.concentricum	*T.erinacei*	
T.megninii	*T.mentagrophytes* complex (granular isolates)	
T.mentagrophytes complex (velvety and cottony isolates)	*T.simii*	
	T.verrucosum	
T.rubrum		
T.schoenleinii		
T.soudanense		
T.tonsurans		
T.violaceum		

[a] Normally nonpathogenic, soil-associated dermatophytes such as *T.terrestre* and *M.cookei* are not included in this table.

The transmissions of dermatophyses or tinea occur by direct contact with infected animals and humans or by indirect contact with contaminated fomites. The dermatophytes are usually carried out by arthroconidia that have formed in or on infected host tissues. These conidia may spread by direct skin-to-skin contact or via fomites containing free arthroconidia, shed skin scales or hairs. Typical fomites include such divergent materials as hats shoes, shower, room floors, bedding, clothing of nursing staff in chronic care institutes, and farm fenceposts used by animals for scratching. Tissue invasion is normally cutaneous; dermatophytes tend to grow in an annular fashion on most affected skin regions, producing a ringworm infection form with a more or less raised and erythematous active area at the periphery and a relatively scurfy inactive zone at the centre of established lesions. Skin in addition to acting as an effective physical barrier also produces secretions, including sweat, sebum, transferring and antimicrobial peptides called defensins (e.g. human-beta-defensin) which has the ability to kill many bacterial and fungal species (23).

5. Taxonomy

Sabouraud in his classic monograph on dermatophytes in 1910 adopted four fungal genera for the dermatophytes: *Achorion*, *Trichophyton*, *Microsporum* and *Epidermophyton*, on the basis of three features: clinical aspect of the disease, cultural characteristics and morphological characteristics of the fungus (28). Langeron and Milochevitch (21) did not accept the classification of Sabouraud and transferred the species of *Achorion* to the genus *Trichophyton*, the classification later accepted by Emmons (9). George (12) recognised 16 species in the three genera based on physiological characteristics, morphology and nutritional requirements. Ajello (1), who carried out extensive work on the dermatophytes recognised 39 species in the three genera. Some species of *Microsporum* and *Trichophyton* produced ascomycetous perfect states which belonged to *Nannizzia* (anamorph *Microsporum*) and *Arthroderma* (anamorph *Trichophyton*) classified in the family *Arthrodermataceae*. The number of species in the three anamorphic genera ranges between 35 and 45 (9, 29). Kirk *et al.* (20) have accepted 39 species of dermatophytes maximum belonging to *Trichophyton* (22 spp.) followed by *Microsporum* (15 spp.) and minimum to *Epidermophyton* (2 spp.). The perfect state (teleomorph) of some of species has been found which belongs to a single genus *Arthroderma*. The macroconidia produced by the dermatophytes are phragmoseptate aleuroconidia (34).

7. Teleomorphic state

Dermatophytes are known to be heterothallic, either a species with two mating types or producing heterokaryotic ascospores. The compatibility is determined by a single pair of allels **A** and **a** or (+) and (-) (7,33, 38, 18, 19). Matruchot and Dassonville (22) were the first scientists to suggest the possible relationship between the dermatophytes and *Gymnoascaceae.* It was Nannizzi who in 1927 described cleithothecia in *Microsporum gypseum* and named the perfect state as *Gymnoascus gypseus* (24). It was followed by discoveries of Stockdale (32) and Dawson and Gentles (7) who reported the perfect states for three keratinophilic fungi viz. *Arthroderma uncinatum* for *Trichophyton (= Keratinomyces) ajelloi, A. quadrifidum* for *Trichophyton terrestre* and *Nannizzia obtuse* for *Microsporum nanum.* In both the teleomorphic genera *Arthroderma* and *Nannizzia*, asci are produced in ascocarps referred to as *gymnothecia* where wall is composed of loose interwoven hyphae. Unicellular, smooth, oblate ascospores are produced in ovoid, deliquescent asci, characteristic features of the family *Arthrodermataceae* (9, 6). Previously, the teleomorph with *Microsporum* anamorphs were placed in the genus *Nannizzia* and teleomorph of *Trichophyton* in the genus *Arthroderma.* Currently, the teleomorph of *Microsporum, Trichophyton* and possibly *Epidermophyton* belongs to a single genus *Arthroderma* Curr. with 25 species (6, 2, 5, 20). It is characterized by the cleistothecial wall composed of hyphae that bear ossiform cells (the cells shaped like small bones because each cell is swollen at the ends), and coiled hyphal appendages extending from the cleistothecial peridium (38). According to the 10th edition of the *Dictionary of the Fungi* (20) and Circa (5) - *Epidermophyton, Trichophyton* and *Microsporum* are considered as anamorphic *Arthrodermataceae* (20). The classification of the dermatophytes is given in the Table 3.

Table 3. Classification of *Epidermophyton, Microsporum* and *Trichophyton* (anamorphs) and *Arthroderma* (teleomorph) (20, 5).

Taxono-mic rank	Kirk *et al.*, (2008)	Circa (2009)	
	Teleomorph & anamorphs	Teleomorph	Anamorphs
	Arthroderma, Epidermophyton, Microsporum and *Trichophyton*	*Arthroderma*	*Epidermophyton, Microsporum* and *Trichophyton*
Kingdom	Fungi	Fungi	Fungi
Sub-kingdom	-	*Eumycotera / Dikariomycotera*	*Eumycotera / Dikariomycotera*

Phylum	*Ascomycota*	*Ascomycota*	*Fungi Imperfecti / Deuteromycota*
Subphylum	*Pezizomycotina (Euascomycotina / Ascomycotina*	*Euascomycotina / Pezizomycotina*	—
Class	*Eurotiomycetes*	*Plectomycetes / Eurotiomycetes*	*Hyphomycetes*
Subclass	*Eurotiomycetidae*	*Eurotiomycetidae*	—
Order	*Onygenales*	*Onygenales*	*Moniliales*
Family	*Arthrodermataceae*	*Gymnoascaceae / Arthrodermataceae*	*Moniliaceae*

6. Identification and description of dermatophytes

The great majority of dermatophytes are identified phenotypically based on two characteristics: 1) Colony characteristic on Sabouraud glucose agar (SGA); 2) Microscopic morphology that includes the examination of two types of conidia; microconidia (presence or absence); macroconidia (surface of the wall, rough verses smooth; number of septa and size of the conidium). In addition to these, other features used in identification include; *in vitro* hair perforation test, special nutritional requirements, urea hydrolysis, growth on polished rice grains and BCP-milk- solid glucose medium (commercially available as dermatophytes milk agar medium), and temperature tolerance. In recent years molecular identification techniques such as Restriction fragment length polymorphism (RFLP) generated from PCR products, amplification of the DNA topoisomerase II gene followed by nested PCR and restriction fragment digestion and Repetitive sequence based PCR performed via the Patented Diversi Lab system are being tried for the identification of the dermatophytes (36).

Trichophyton Malmsten

The genus *Trichophyton* is the largest of the three dermatophyte genera with 22 species. These are found on humans and animals, and in soil and river sediments. The genus produces both macroconidia and microconidia (Fig 2). The macroconidia are phragmoseptate aleurioconidia, clavate (cigar-shaped), 8-50 μm × 4-8 μm and smooth walled The microconidia are spherical (2.5 to 4 μm) or clavate (2-3×3-4 μm) (Fig. 2). Some species rarely produce macroconidia. The teleomorph belongs to *Arthroderma*.

Microsporum Gruby

The genus *Microsporum* contains 15 species parasitic on humans and other mammals and produces both micro- and macro-conidia (Fig 2). The macroconidia are fusiform or spindle shaped, thick- walled, phragmoseptate aleurioconidia. The outer surface of the wall is pitted, a sperulate or spiny. The macroconidia are 35-125 µm (sometimes upto 160 µm)× 7-20 µm with 4 to 15 septa. However, the essential distinguishing feature of this genus is the echinulations on the macroconidial wall. The microconidia are clavate, 4-7µm × 2.5 µm. The teleomorph belongs to *Arthroderma.*

Epidermophyton Sabouraud

The genus contains two species which are parasitic on humans. No microconidia are produced (Fig 2). Macroconidia are club shaped with broadly rounded apex with fewer than five septa, 8-15 µm × 6-10 µm, borne singly or in clusters on the conidiophore. The teleomorph is not known.

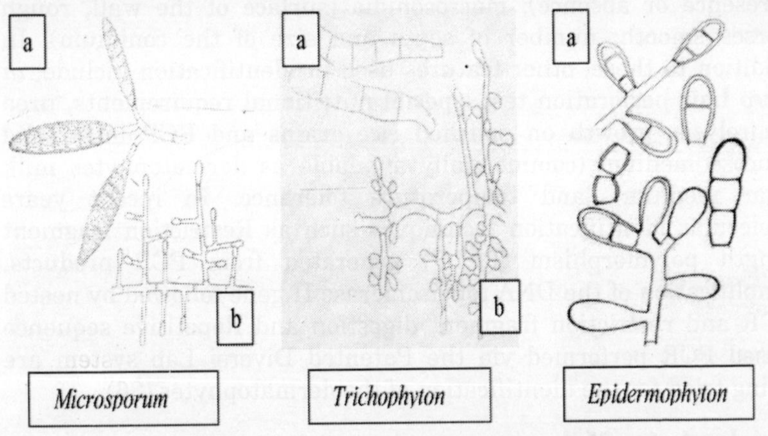

Fig. 2: Conidial types in the three dermatophytes.
(a) Macroconidia; (b) Microconidia.

The three anamorphic genera of dermatophytes *Trichophyton, Microsporum* and *Epidermophyton* can be distinguished on the basis of shape, size, septation and wall of macroconidia in addition to the presence or absence of microconidia as given in the Table 4.

Table 4: Differentiation of the dermatophytes genera, disease caused and mode of transmission.

	Features	*Trichophyton*	*Microsporum*	*Epidermophyton*
Macro-conidia	Shape	Clavate or cigar shaped	Fusiform or spindle shaped	Club shaped
	Size	(8-50μm×4-8μm)	(7-20×35-125 μm, upto 160 μm long)	(6-10×8-15μm)
	Nature of wall	Smooth walled	Thick echinulate walled	Thin walled
	Septation	0-4 septate	5-15 septate	Less than 5 septate
Micro-conidia	Shape	Spherical or clavate	Clavate	Absent
	Size	(2.5-4μm) (2-3×3-4μm)	(2.5-3.5×4-7μm)	–
Part of the body infected		Ringworm of scalp, body, beard and nails, Athlete's foot	Ringworm of scalp and skin	Ringworm of groin and nails
Principal targets		Hair, skin, nails	Scalp hair, skin	Skin, nails
Transmission		Human to human, animal to human	Animal to human, soil to human, human to human	Strictly human to human

The characteristic feature of dominant species of dermatophytes based on the microscopic appearance and types of conidia, colonial morphology on SGA and physiological tests are presented in Table 5.

Table 5: Colonial, conidial and growth characteristics of dominant clinically isolated species of dermatophytes

Fungus		Growth	Colonial characteristics	Conidial characteristics
Anamorph	Teleomorph			
Trichophyton ajelloi	*Arthroderma uncinatum*	Moderately rapid	Powdery, rich tan to medium orange brown in colour; reverse pale, brownish or with purple black pigment	Macroconidia Smooth, thick walled, cylindrical, often more than 7 cells long; with cellular compartments rather than true cross walls
T.rubrum	-	Moderately slow	Cottony to velvety, white to reddish surface, typically wine red reverse	Macroconidia seldom seen, pencil shaped; Microconidia drop shaped, abundant
T.mentagrophytes	*A.benhamiae, A.vanbreuseghemii*	Rapid	Powdery to cottony, yellow cream to white surface, pale to red brown reverse	Macroconidia Uncommon, club shaped, smooth; Microconidia drop shaped, abundant, produced mainly in dense tufts, spiral appendages present but rare in very cottony isolates
T. tonsurans	-	Moderately slow	Powdery to velvety, white to yellowish or red brown surface, reverse chest nut	Macroconidia uncommon, Pencil shaped; Microconidia abundant
T.soudanense	-	Moderately slow	Flat, bright yellow to wine red colony with radial striations	Macroconidia not seen, Microconidia drop shaped, scarce or absent

Microsporum audouinii	-	Moderately rapid	Flat to velvety, thin, pale salmon to pale brownish reverse	Rare, deformed macroconidia, often with beak, constricted mid region Microconidia drop shaped and aerial arthroconidia may be present; apiculate terminal chlamydospores often seen
M.canis	*A.otae*	Rapid	Flat to velvety, thin, pale to yellow, with yellow reverse	Macroconidia thick walled, roughened and beaked; Microconidia drop shaped
Epidermophyton floccosum	-	Moderately rapid	Flat, slightly granular at first, soon developing white puffs sandy to olive brown; reverse pale to yellow	Macroconidia abundant, club shaped with broadly rounded apex, usually with fewer than 6 cells; no microconidia formed; many chlamydospores in primary isolates

Conclusions

Dermatophytoses are the most frequent fungal infections worldwide, affecting individuals in various age groups. *Trichophyton rubrum* is the main causative agent of tinea pedis, tinea unguium, tinea corporis, and tinea cruris, and given the growing incidence of tinea pedis and tinea unguium, it is likely to remain the dominant dermatophyte worldwide. Knowledge of epidemiologic trends of dermatophytoses and understanding their causation can lead to the implementation of public health measures to halt the increase in these infections. *T. rubrum, T. interdigitale (=T. mentagrophytes* var. *interdigitale), M. canis, M. audouinii, T. tonsurans* and *T. verrucosum* account for most dermatophytoses globally, but the attack rates and incidence of specific mycoses can vary widely. The highest incidence of *T. rubrum* has been reported in Europe, whereas *T. mentagrophytes* has more commonly been reported in Asia. Increasingly frequent tinea pedis caused by *T. rubrum* and *T. interdigitale* (*mentagrophytes* var. *interdigitale*) infection seems to be a characteristic of highly developed countries with booming sport and fitness facilities. Dermatophytes research is posed to take off. The sequencing of seven dermatophyte genomes has recently been completed, and the sequence information is now publically available. During the past decades, an extensive research has been conducted on dermatophytes and dermatophytosis which has largely improved our knowledge on the epidemiology of these infections. The need of the hour is to lay emphasis on research for fast and proper identification of dermatophytes and control of dermatophytes by plant based novel drugs.

REFERENCES

1. Ajello L. 1968. A taxonomic review of the dermatophytes and related species. *Sabouraudia* 6: 147-159.
2. Alexopoulos CJ, Mims CW and Blackwell M. 1966. *Introductory Mycology*. 4th ed. John Wiley and Sons, New York.
3. Achterman RR and White TC. 2012. Dermatophyte Virulence Factors: Identifying and Analyzing Genes that May Contribute to Chronic or Acute Skin Infections. *Inter J Microbiol*. 1-8.
4. Ameen M .2010. Epidemiology of superficial fungal infections. *Clin Dermatol*. 28: 197–201.
5. Circa,2009.(See,http://www.sbs.utexas.edu/mycology/bio329/pdf_files/SP2009/RefGuidFungi%20Word%20SPR09.pdf).
6. Currah RS. 1985. Taxonomy of the Onygenales: Arthrodermataceae, Gymnoascaceae, Myxotrichaceae and Onygenaceae. *Mycotaxon* 24: 1-216.
7. Dawson CO and Gentles JC. 1961. The perfect states of *Keratinomyces ajelloi* Vanbreuseghem, *Trichophyton terrestre* Durie and Frey and *Microsporum nanum* Fuentes. *Sabouraudia* 1: 49-57.
8. Emmons CW. 1934. Dermatophytes; natural grouping based on the form of the spores and accessory organs. *Arch Dermatol Syphil* 30: 337-362.
9. Emmons CW, Binford CH, Utz JP and Kwon- Chung KJ. 1977. *Medical Mycology*. 3rd ed. Lea & Febiger, Philadelphia.
10. English MP. 1980. Ecological aspects of dermatophytes

regarded essentially as anthropophilic. In: Preusser JH, Ed *Medical Mycology*, Zbl Bkt Suppl 8. Gustav Fisher Verlag, New York, 53-59.

11. Freedberg IM, Eisen AZ, Wolff K, Austen KF, Goldsmith LA, Katz S. 2003. In *Dermatology in General Medicine*, 6th edn. Vol 2 (Ed. T.B. Fitzpatick). McGraw-Hill, New York.
12. Georg, L. 1957. *Dermatophytes. New methods in classification.* Atlanta. U.S. Public Health Service.
13. Georg LK, Kaplan W, Ajello L, William WM and Tilden EB. 1959. The parasitic nature of the soil fungus *Keratinomyces ajelloi. J Investigative Dermatol* 32: 539-544.
14. Gupta AK and Cooper EA. 2008. "Update in antifungal therapy of dermatophytosis," *Mycopathologia*. 166(5-6):353–367.
15. Graser Y, Scott J, Summerbell R. 2008. The new species concept in dermatophytes- a polyphasic approach. *Mycopathologia* 166: 239-256.
16. Havlickova B, Czaika VA and Markus Friedrich M. 2008. Epidemiological trends in skin mycoses worldwide. *Blackwell Publishing Ltd. Mycoses*. 51(4):2–15.
17. Kamalam A, Thambiah AS. 1979. Tinea capitis in a Moslem school. *Indian J Med Res*. 70: 40–3.
18. Kwon-Chung KJ. 1967a. Studies on the sexuality of *Nannizzia*, 1. Heterothallism vs. Fertile isolates. *Sabouraudia* 6:5-13.
19. Kwon-Chung KJ. 1967b. Segregation of two mating type components in colonies of fertile isolates of heterothallic dermatophytes. *Sabouraudia* 6: 37-41.
20. Kirk PM, Cannon PF, David JC and Stalpers JA. 2008. *Dictionary of the Fungi*. 10th ed. Wallingford: CABI.
21. Langeron, M and Milochevitch S. 1930. Morphologie des dermatophtes. *Ann Parasitol Hum Comp* 8: 465-508.
22. Matruchot L and Dassonville C. 1899. Sur le champignon de l'herpes (*Trichophyton*) et les forms voisines, et sur la classification des Ascomycetes. *Bulletin Social Mycol France* 15: 240-253.
23. Murray PR, Baron EJ, Jorgensen JH, Landry ML, Pfaller MA. 2007. *Manual of Clinical Microbiology*. 9th ed (2). ASM Press, Washington, D.C.
24. Nannizzi A. 1927. Riecerche sull'origine saprofitica dei funghi delle tigne. *Atti Acad Fisiocr Sienna* 10: 89-97.
25. Philpot CM. 1977. The use of nutritional tests for the identification of dermatophytes. *Sabouraudia* 15:141-150.
26. Rippon JW. 1982. *Medical Mycology: The Pathogenic Fungi and the Pathogenic Actinomycetes*. London, WB Saunders, 154-248.

27. Rodwell GEJ, Bayles CL, Towersey L and Aly R. 2008. The prevalence of dermatophyte infection in patients infected with human immunodeficiency virus. *Inter. J. Dermatol.* 47(4): 339–343.
28. Sabouraud R. 1910. Les Teignes. Paris, Masson et Cie.
29. Simpanya MF. 2000. Dermatophytes: their taxonomy, ecology and pathogenicity. *Revista Iberoamericana de Micologia*. 1-12.
30. Singal A, Rawat S, Bhattacharya SN et al. 2001. Clinicomyocological profile of tinea capitis in North India and response to griseofulvin. *J Dermatol*. 28: 22–6.
31. Singh AK, Srivastava KS et al.1994. A clinico-mycological study on tinea pedis at Ranchi. *Indian J Dermatol Venerol Leprol*. 60: 68–71.
32. Stockdale PM. 1961. *Nannizzia incurvata* gen. nov., sp. nov., a perfect state of *Microsporum* (Bodin) Guiart et Grigorakis. *Sabouraudia* 1: 41–48.
33. Stockdale PM. 1963. The *Microsporum gypseum* complex. *Sabouraudia* 3: 114-126.
34. Seifert KA and Gams W. 2001. *The taxonomy of anamorphic fungi*. Eds: McLaughlin, Mc Laughlin, Lemke. *The Mycota* VII part A. *Systematics and Evolution*.
35. Seebacher C, Bouchara JP, Mignon B .2008. Updates on the epidemiology of dermatophyte infections. *Mycopathologia* 166: 335–352.
36. Summerbell RC, Weitzman I, Padhye AA. 2007. *Trichophyton, Microsporum, Epidermophyton* and agents of superficial mycoses. Eds: Murray PR, Baron EJ, Jorgensen JH, Landry ML, Pfaller MA. *Manual of Clinical Microbiology*. 9th ed (2). ASM Press, Washington, D.C.
37. Weitzmans I. 1964. Variation in *Microsporum gypseum* I, a genetic study of pleomorphism. *Sabouraudia* 3:195-204.
38. Weitzman I, McGinnis MR, Padhye AA and Ajello L. 1986. The genus *Arthroderma* and its later synonym *Nannizzia*. *Mycotaxon* 25: 505-518.
39. World Health Organization. 2005. Epidemiology and management of common skin diseases in children in developing countries. Department of Child and Adolescent Health and Development. WHO reference number WHO/FCH/ CAH/05.12. 54 p.

EXPLORING DIFFERENT AVENUES OF *TRICHODERMA* AS A POTENT BIO-FUNGICIDAL AND PLANT GROWTH PROMOTING CANDIDATE-AN OVERVIEW

H.B. Singh, Brahma N. Singh, S.P. Singh and B.K. Sarma*

Department of Mycology and Plant Pathology, Institute of Agricultural Sciences, Banaras Hindu University, Varanasi-221 005
*Corresponding author: hbs1@rediffmail.com

KEYWORDS: *Trichoderma* spp.; lytic enzymes; Secondary metabolites; Induced systemic resistance; Plant growth promotion; Commercilization

ABSTRACT

Trichoderma species are free-living, ubiquitous hyphomycete fungi commonly found in all types of soils and root ecosystems especially those rich in organic matter. Due to their ability to protect plants and contain pathogen populations under different soil conditions, these fungi have been widely studied and commercially marketed as biopesticides, biofertilizers and soil amendments. Different species of *Trichoderma* also produces numerous biologically active compounds, including cell wall degrading enzymes, and secondary metabolites. Moreover, faster metabolic rates, anti-microbial metabolites, and physiological conformation is key factors which chiefly contribute to antagonism of these phytopathgenic fungi. Mycoparasitism, spatial and nutrient competition, antibiosis by enzymes and secondary metabolites, and induction of plant defence system are typical biocontrol actions of these fungi. Information on the classification of the genus, *Trichoderma*, mechanisms of antagonism and role in plant growth promotion has been well documented. A number of successful products based on different species of *Trichoderma* have been commercialized in India and elsewhere. Whatever limitations these biocontrol products may have, it can be addressed by enhancing biocontrol through manipulation of the environment, using mixtures of beneficial organisms, physiological and genetic enhancement of the biocontrol mechanisms, manipulation of formulations, and integration of biocontrol with other alternative methods that alone do not provided adequate protection but in combination with biocontrol provide additive or synergistic effects. However, fast paced current research in this field should be carefully updated for the full-proof commercialization of the fungi. The aim of this review is to sum up the biocontrol activities of these fungi and to shed light on commercial production processes. This review thus focuses on different aspectes of *Trichoderma* spp. including mechanisms of plant disease control and, growth promotion, their delivery systems and commercilization. The current review is thus an attempt to summarize the latest information on *Trichoderma* research and commerilization.

1. INTRODUCTION

Trichoderma species are free-living, ubiquitous hyphomycete fungi commonly found in all types of soils and root ecosystems especially those rich in organic matter (1). They are mostly known for their ability to produce enzymes responsible for cell wall degradation of target pathogenic fungi, the phenomenon mostly utilized for biological control in agriculture (2). They are basically used as antagonists against some soil-borne phytopathogenic fungi but can have a large impact on overall growth and development of plants where they are inoculated through various mechanisms revealed recently. Apart from suppressing pathogen's growth by antagonism, they also induce resistance in the hosts by mediating in development of signals in the host for induction of local as well as systemic resistance against some foliar pathogens (3).

Trichoderma also competes well with other microflora in the rhizosphere for nutrient and space by deploying various strategies that helps them to emerge out as winner in the competition. They inhibit growth of other microorganisms especially the pathogenic ones by mobilization and uptake of nutrients present in very limited quantity in the soil and thereby depriving the other micro-organisms for those nutrients. The versatility in their metabolic ability along with efficient nutrient utilization facilitates rapid colonization resulting in occupation of more space and thereby restricts the development of other microorganisms (4).

Some *Trichoderma* strains has the special ability to colonize the root surfaces of the host plants in a permanent manner and protects the plants from invading pathogens by creating a physical barrier as well as production of antimicrobial metabolites in a concentrated way in the rhizoplane. Similarly, root colonization also enhances performance of *Trichoderma* in triggering Induced Systemic Resistance (ISR) as the interaction of *Trichoderma* with the host is more stable upon their colonization (3,4). *Trichoderma* species are also known for cleaning the environment from harmful chemicals. Due to their ability to protect plants and contain pathogen populations under different soil conditions, these fungi have been widely studied and commercially marketed as biopesticides, biofertilizers and soil amendments. This review presents a compilation of the most recent advances in understanding the mechanisms involved in the interaction of *Trichoderma* spp. with phytopathogenic fungi and plants. We are specially emphasizing the biological, biochemical and molecular aspects of the mechanisms, with particular attention paid to the molecular factors involved in the natural crosstalk occurring in soil and root environment. Description of the current status of research and application of *Trichoderma* would enhance the practical application techniques of these beneficial microorganisms for plant disease control in different agro ecosystems.

2. *TRICHODERMA* AS A BIOCONTROL AGENT

In general, the current literature indicates that *Trichoderma* spp. have been mostly used as biocontrol agents globally (Table 1). Weindling (5), for the first time implicated the role of *T. lignorum* in biological control of citrus the seedling disease caused by *Rhizoctonia solani*. Since this pioneering work, several reports on successful biocontrol by *Trichoderma* spp. have accumulated. *T. harzianum, T. viride* and *T. virens* are the most widely used /cited for biological control (1). They are reported effective in controlling root rots/wilt complexes and foliar diseases in several crops and are reported to inhibit a number of soil borne fungi like *Rhizoctonia, Pythium, Sclerotinia, Sclerotium, Fusarium, Macrophomina*, etc. and recently the root knot nematode, *Meloidogyne* spp. as well (6,7). However, the first report on its mycoparasitism was made by Coley-Smith et al. (8) who by means of microtome sections had shown that medulla of infected sclerotia of *Sclerotium delphinii* were completely replaced by hyphae and chlamydospores of *Trichoderma hamatum* on agar plates. Likewise, Henis et al. (9) reported mycoparasitism (penetration and infection) of *Trichoderma* spp. against *Sclerotium rolfsii*, where chlamydospores were abundantly produced in contrast to conidia within the infected fungal sclerotia. One of most interesting aspects of studies on *Trichoderma* is the varied mechanisms employed by *Trichoderma* species to arrest disease development. A different member of *Trichoderma* deploys an array of diverse biocontrol mechanisms. In addition to being parasite of other fungi, recent studies also shows that they are opportunistic plant symbionts. They produce or release a variety of compounds that induce localized or systemic resistance responses (7). Biocontrol activity of *Trichoderma* is generally achived through its ability to serve as antagonist, plant growth promoter, plant defense inducer, rhizosphere colonizer and neutralizer of pathogen's activity favouring infection.

So far, *Trichoderma* spp. are among the most studied fungal bio-control agents (BCAs) and commercially marketed as biopesticides, biofertilizers and soil amendments (10,11,12). Depending upon the strain, the use of *Trichoderma* in agriculture can provide numerous advantages: (i) colonization of the rhizosphere by the BCA ("rhizosphere competence") allowing rapid establishment within the stable microbial communities in the rhizosphere; (ii) control of pathogenic and competitive/deleterious microflora by using a variety of mechanisms; (iii) improvement of the plant health and (iv) stimulation of root growth (11).

3. PRESENT STATUS OF *TRICHODERMA*

Although, *Trichoderma* spp. have been known for a long time since 1865 (13), the taxonomy and species identification were vague until around 1969 (Rifai, 1969). In fact, Druzhinina and Kubicek (14) have extensively reviewed species concepts and biodiversity in *Trichoderma* fungi. Although some of the *Trichoderma* species fungi are difficult to distinct morphologically, the phylogenetic classification has increased over time and rapidly reached more than 100 (15), and number is expected to increase consistently. In this context, the advancements as well as limitations of modern methods like genealogical concordance phylogenetic species recognition (GCPSR) and DNA-barcode system for safe identification of *Trichoderma* spp. warrant future investigations. The GCPSR requires the analysis of trees of several unlinked genes, whereas, DNA-barcode system is based on the defined nucleotide sequence differences of different *Trichoderma* spp. Nevertheless, application of *Trichoderma* spp. as BCAs in the environment as well as the reported epidemics of commercially grown mushroom (*Agaricus bisporus*) (16) and harmful effects on immunocompromised mammals (17) are reasons which necessitates efficient and reliable species identification for *Trichoderma* fungi.

The widespread application of *Trichoderma* spp. as BCAs has been exploited and reported only lately against several soil-borne phytopathogenic fungi (18). Akin to most fungal BCAs, *Trichoderma* spp. can be efficiently used as spores (especially, conidia), which are more tolerant to adverse environmental conditions during product formulation and field use, in contrast to their mycelial and chlamydospore forms as microbial propagules (19). Nevertheless, the presence of a mycelial mass is also a key component for the production of antagonistic metabolites (20). Conidia and mycelia can be produced in either solid-state or liquid fermentation (21). In general, liquid fermentation is more suitable method over solid-state fermentation for large scale production and there is plenty of, still special techniques for abundant conidia production.

Table 1. Suppression/management of plant pathogens/diseases by *Trichoderma*

Crop	Disease	Pathogen	Possible Biological control agents	Reference
Cereals				
Barley	Foot and Root rot	*Sclerotium rolfsii*, *Fusarium*, *Curvularia*, *Pythium*,	*T. viride* *T. pseudokoningii*	(22)

		Aspergillus, Penicillium		
Rice	Blast	*Pyricularia oryzae*	*Trichoderma* spp.	(23)
	Bunt	*Neovossia indica*	*T. harzianum, T. viride, T. virens, T. deliquescens*	(22)
	Kernel smut	*Tilletia barclayana*	*T. harzianum, T. viride, T. virens, T. deliquescens,*	(22)
	Sheath blight	*Rhizoctonia solani*	*T. harzianum, T. viride, T. virens*	(24,22,25)
	Brown spot	*Drechslera oryzae*	*T. viride*	(26)
Wheat	Karnal bunt	*Neovossia indica*	*T. viride, T. harzianum, T. pseudokoningii, T. koningii*	(27,22)
	Loose smut	*Ustilago segetum*	*T. viride, T. harzianum, T. koningii, T. lignorum*	(27,22)
	Root rot	*S. rolfsii, F. oxysporum*	*T. harzianum*	(22)
	Spot blotch	*Drechslera sorokiniana*	*T. reesei, T. pseudokoningii, T. viride*	(22)
	Take-all	*Gaeumannomyces graminis var. tritici*	*T. harzianum, T. koningii*	(22)
Maize	Charcol rot, Banded blight	*Macrophomina phaseolina, R. solani*	*T. viride* *Trichoderma* spp.	(28,22)
Pulses				
Pigeon pea	Wilt	*Fusarium udum*	*T. viride, T. hamatum, T. harzianum, T. koningii*	(29,22,30,31)
	Seedbore disease	*Xanthomonas campestris pv. vinae radiatae*	*T. viride, T. harzianum*	(22)

Chickpea	Wilt	*F. oxysporum* f.sp. *ciceri*	*T. viride*, *T. harzianum*, *T. virens*	(32,22,25)
	Root rot	*Rhizoctonia solani/ M. phaseolina*	*T. viride*, *T. harzianum*	(33,22)
	Collar rot	*Sclerotium rolfsii*	*T. viride*, *T. harzianum*	(22,33,34)
	Grey mould	*B. cinerea*	*Trichoderma* spp.	(22)
	Stem rot	*S. sclerotiorum*	*T. harzianum*	(22)
Blackgram	Dry rot	*M. phaseolina*	*T. harzianum*	(22)
	Damping off	*S. rolfsii*	*T. viride*	(22)
Cowpea	Wilt	*F. oxysporum* f. sp. *ciceris*	*T. viride*	(22)
	Charcol rot and wilt	*M. phaseolina, F. oxysporum* f. sp. *tracheiphilum*	*T. viride*, *T. harzianum*, *T. koningii*, *T. pseudokoningii*	(22)
Soybean	Dry root rot	*M. phaseolina*	*T. viride*, *T. harzianum*	(35)
Horse gram	Damping off	*M. phaseolina*	*T. viride*	(22)
Lentil	Wilt complex; collar rot	*R. solani*, *F. oxysporum*, *S. rolfsii*	*T. viride*	(22)
Moth bean	Blight	*M. phaseolina*	*T. harzianum*, *T. viride*	(22)
Mung-bean	Root rot	*M. phaseolina*	*T. harzianum*, *T. viride*	(22,36)
Flower crops				
Gladiolus	Yellows and Corm rot	*F. oxysporum gladioli*	*T. viride*, *T. harzianum*	(22,21)
Jasmine	Root rot	*M. phaseolina*	*T. harzianum*, *T. viride*	(22)
Rose	Grey mould	*Botrytis cinerea*	*T. harzianum*, *T. viride*	(22)

Carnation	Wilt	*F. oxysporum* f. sp. *dianthi*	*Trichoderma* spp.	(37)
Oil seed crops				
Groundnut	Crown rot, Stem rot, Pod rot	*A. flavus, S. rolfsii, A. niger*	*T. viride, T. harzianum*	(38,39,40, 41,42)
	Aflatoxin	*Aspergillus flavus*	*Trichoderma* spp.	(43)
	Late leaf spot	*Phaeoisariopsis personata*	*T. harzianum*	(44,45)
	Root and collar rot	*R. solani*	*T. virens, T. longibrachiatum*	(22)
	Rust	*Puccinia arachidis*	*T. harzianum*	(44)
	Wilt complex seed rot	*S. rolfsii, F. solani, F. oxysporum*	*T. harzianum, T. virens, T. viride*	(22)
Castor	Wilt	*Fusarium oxysporum* f.sp. *ricini*	*T. viride*	(25)
	Grey rat	*Botrytis cinerea*	*T. viride*	(46)
Mustard	Damping off	*Pythium aphanidermatum*	*T. harzianum, T. viride*	(38)
Safflower	Root rot	*M. phaseolina*	*T. viride*	(47)
Seasamum	Blight	*Phytophthora* spp.	*T. harzianum, T. viride*	(48) (49)
	Root rot	*M. phaseolina*	*Trichoderma* spp., *Gliocladium* spp.	(50,51)
Sunflower	Blight	*Alternaria helianthii*	*T. virens*	(52)
	Root/ collar rot	*S. rolfsii, R. solani, S. sclerotiorum*	*T. harzianum, T. hamatum*	(22)
Vegetables				
Bean	Seedling rot	*Pythium* spp., *S. sclerotiorum, B. cinerea, R. solani*	*T. koningii, G. catenulatum*	(53)

Brinjal	Damping-off	*Phytophthora* or *Pythium* spp. *Fusarum*	*T. viride, T. harzianum, T. koningii*	(54)
	Collar rot	*Sclerotinia sclerotiorum*	*T. viride, T. virens*	(55)
Cabbage	Damping off	*R. solani*	*T. viride, T. harzianum, T. koningii*	(56,57)
	Alternaria blight	*Alternaria brassicola*	*T. virens, T. longibrachiatum*	(58)
Cauli-flower	Damping off	*R. solani* *P. aphanidermatum*	*T. harzianum*	(59,25,60)
Chilli	Root rot	*S. rolfsii, Phytop-hthora capsici; Phytophthora erythroseptica*	*T. harzianum*	(61,62)
	Fruit root and die back	*Clletotrichum capsici*	*T. viride, T. harzianum, T. koningii, T. pseudokoningii, T. hamatum, T. longibrachiatum, T. pileatus*	(63,62)
Cucumber	Seedling diseases	*Phytophthora* or *Pythium* spp. *Fusarium oxysporum* f. sp. *cucumerinum*	*T. harzianum*	(20)
Egg plant	Wilt, Damping off	*F. solani, P. aphanid-ermatum*	*T. viride, T. harzianum, T. konningii*	(54)
	Collar rot	*S. sclerotiorum*	*T. viride,T. virens*	(55)
Fenu-greek	Root rot	*R. solani*	*Trichoderma* spp., *T. viride*	(64,65)
French bean	Root rot	*R. solani*	*T. viride,T. harzianum, T. hamatum, T. viride*	(66)

Pea	Seed and Collar rot	*Pythium* sp., *R. solani*	*T. harzianum*, *T. hamatum*	(67)
Potato	Black-scurf	*R. solani*	*T. viride*, *T. harzianum*, *T. koningii*, *T. viride*	(68,69)
Radish	Seedling-rot/Damping-off	*Pythium* spp., *R. solani*	*T. harzianum*, *T. hamatum*	(70)
Tomato	Damping off and wilt	*P. indicum* *F. oxysporum* f. sp. *lycopersici*	*T. viride*, *T. harzianum*	(71,72)
	Grey mold	*B. cinerea*	*T. harzianum*	(71)
	Root knot	*Meloidogyne incognita*, *M. javanica*	*T. harzianum*	(73,70)
Plantation crops				
Arecanut palm	Fruit rot	*Phytophthora* spp.	*Trichoderma* spp.	(22)
	Foot rot/ anabe	*Ganoderma lucidum*	*T. harzianum*	(74)
Black papper	Foot rot and root rot	*Phytophthora capsici*	*T. harzianum*, *T. virens*	(75)
Card-amom	Capsule rot	*Phytophthora meadii*, *P. incotianae* var. *nicotianae*	*T. harzianum*	(76)
Coconut	Stem bleeding	*Thielaviopsis paradoxa*	*T. virens*	(77)
	Basal stem rot	*Ganoderma lucidum*/*G. applanatum*	*T. harzianum*	(78)
	Bud rot	*P. palmivora*	*Trichoderma* spp.	(79)
Coffee	Collar rot	*R. solani*	*T. harzianum*	(80)
Ginger	Rhizome rot	*Pythium aphanidermatum*	*T. harzianum*, *T. virens*	(81)

Mulberry	Leaf spot	*Cercospora moricola*	*T. harzianum, T. viride*	(82)
	Stem canker & die back	*Botryodiploidia* spp.	*T. harzianum, T. virens, T. pseudokoningii*	(83)
	Cutting rot	*F. solani*	*T. harzianum, T. virens, T. pseudokoningii*	(83)
	Collar rot	*Phoma sorghina*	*T. harzianum, T. virens, T. pseudokoningii*	(83)
Rubber	Brown root	*Phelinus noxius*	*T. harzianum, T. virens, T. hamatum*	(84)
Fruit crops				
Apple	White root rot	*Dematophora necatrix*	*T. viride, T. harzianum, T. virens*	(85)
	Blue and gray mold/ fruit rot	*Penicillium expansum*	*T. harzianum*	(85)
Banana	Panama disease	*Fusarium oxysporum* f. sp. *cubense*	*T. virede*	(86)
Citrus (Mand-arin)	Root rot	*Phytopthora nicotianae* pv. *parasitica, P. colocasiae*	*T. harzianum,T. viride, T. virens*	(87)
Guava	Wilt	*Gliocladium roseum*/ *Fusarium solani*	*T. harzianum*	(88) (89)
	Fruit rot	*Lasiodiplodia theobromae, C. gloeosporioides, Pestalotiopsis versicolor, Phomopsis psidi, Rhizoctonia arrhizus*	*Trichoderma* sp.	(90)
Mango	Fruit rot	*L. theobromae, R. arrhinus*	*Trichoderma* sp., *T. harzianum*	(91)

Melon	Wilt	*F. oxysporum*	*T. viride*	(92)
Mulberry	Stem cancer and Die back	*Botryodiplodia theobroae*	*T. pseudokoningii*	(93,66)
	Root rot	*F. solani, F. oxysporum*	*T. harzianum*	(93,85,94)
	Cutting rot	*F. solani*	*T. pseudokoningii*	(93,94)
	Collar rot	*Phoma mororum / P. sorghina*	*T. pseudokoningii*	(93)
	Root knot	*M. incognita*	*V. chlamydo-sporium*	(93)
Musk-melon	Wilt	*F. oxysporum* *F. solani, R. solani*	*T. harzianum*	(92)
Orange	Blue mould	*Penicillium italicum*	*T. harzianum*	(95)
Straw berry	Grey mould	*B. cinerea*	*T. harzianum*	(96)
Water melon	Wilt	*F. oxysporum* f. sp. *solani*	*T. viride*	(97)
Plantation crops				
Cottan	Bacterial blight	*Xanthomonas campestris* pv. *malvacearum* *Verticillium dahliae*	*T. harzianum*	(98)
	Root rot	*Rhizoctonia* sp./*M. phaseolina*	*T. viride, T. harzianum, T. virens*	(99,100, 22)
	Wilt	*F. oxysporum* f. sp. *vasinfectum, S. rolfsii, R. solani*	*T. harzianum, T. virens V. dahliae*	(22)
Sugarcane	Red rot	*Colletotrichum falcatum*	*T. harzianum, T. viride*	(22)
	Root rot	*Pythium graminicola*	*T. viride*	(22)

	Sett rot	*Ceratocystis paradoxa*	*Trichodarma* spp.	(101)
	Seedling rot	*Pythium* sp.	*T. viride, T. harzianum*	(102)
	Wilt	*F. moniliformae*	*T. viride, T. harzianum, T. longibrachiatum*	(22)
Sugarbeet	Damping off	*P. aphanidermatum*	*T. harzianum*	(22,103)
	Root rot/wilt	*S. rolfsii*	*T. harzianum*	(22,104)
Condiments, Spices, Nacrotics, Stimulants crops				
Black pepper	Foot rot	*Phytophthora capsici*	*T. harzianum, T. virens*	(65,105)
	Slow decline	*Radopholus similis, Meloidogyne incognita, P. capsici*	*T. harzianum, T. virens*	(65)
	Collar rot	*Phytophthora capsici*	*Trichoderma* spp.	(22,75)
	Root knot	*M. incognita*	*Trichoderma* spp.	(106)
Carda-mom	Capsule rot (Azhukal)	*Phytophthora mead:i,P. nicotianae*	*T. harzianum, T. viride, T. hamatum*	(22,65,107)
	Damping off	*F. moniliforme, P. aphanidermatum*	*T. harzianum, T. viride,*	(22,75)
	Clump rot	*Pythium vexans, R. solani, M. incognita*	*T. harzianum*	(65)
	Rhizome rot	*Pythium vexans, R. solani, F. oxysporum*	*T. harzianum, T. viride*	(107)
Coriander	Wilt	*F.oxysporium* f. sp. *corianderii*	*T. viride, T. harzianum*	(22,65)
Cumin	Wilt	*F. oxysporum* f. sp. *cumin*	*Trichoderma / Glioc -ladium*	(65,75)

Ginger	Rhizome rot	*F. oxysporum* f. sp. *zingiberi* *Pythium pleroticum* *P. aphanidermatum,* *P. myriotylum*	*T. harzianum, G. virens, T. viride*	(22,65)
	Yellowing	*F. oxysporum* f. sp. *zingiberi*	*T. harzianum, T. virens, T. hamatum*	(22,75)
	Dry root rot	*Pratylenchus coffeae, Fusarium complex, M. pharcolesia.*	*T. harzianum*	(65)
Turmeric	Rhizome rot	*P. graminicolum* *Fusarium* sp. *P. aphanidermatum,* *M. incognita, R. similis*	*T. harzianum, T. viride*	(65)
Poppy	Downy mildew	*Peronospora arborescens*	*Trichoderma* spp.	(22)
	Sclerotinia rot	*S. sclerotiorum*	*T. harzianum, T. viride*	(26)
Tobacco	Damping off	*Pythium aphani-dermatum*	*T. harzianum*	(22)
Vanilla	Damping off	*P. meadii, F. oxys-porum* f. sp. *vanille*	*T. harzianum*	(65)
Other crops				
Betelvine	Foot and root rot	*Phytophthora parasitica* var. *piperina*	*T. viride, T. harzianum*	(26,108)
	Collar rot	*S. rolfsii*	*T. harzianum*	(109)
Chir pine	Pre and post dam-ping-off	*F. solani*	*T. viride*	(110)
	Collar rot	*S. rolfsii*	*T. harzianum, T. virens*	(108)
	Wilt	*S. rolfsii*	*T. harzianum*	(111)

Moth bean	Blight	*M. phaseolina*	*T. harzianum, T. viride*	(112)
Passion fruit	Collar rot	*R. solani*	*T. harzianum, Trichoderma* spp.	(113)
Pearl-millet	Ergot	*Claviceps fusiformis*	*T. viride, T. harzianum, T. virens*	(22)
Cumin	Wilt	*F. oxysporum* f. sp. *cumini*	*T. harzianum*	(114)

4. TAXONOMY

Trichoderma for the most part, classified as imperfect fungi, as they produce only asexual spores. The sexual stage, when found, is within the Ascomycetes in the genus *Hypocrea*. Tradi-tional taxonomy was based upon differences in morphology, prim-arily of the asexual sporulation apparatus. Rifai (115) outlined the speciation concept within the genus *Trichoderma* and described nine species aggregates: *T. piluliferum* Webster & Rifai, *T. polysporum* (Link) Rifai, *T. virens* Gedden & Foster, and *T. hamatum* (Bon.) Bain. *T. koningii* Oudem. Apud Oudem. Et Koning, *T. aureoviride* Rifai, *T. harzianum* Rifai, *T. longibrachiatum* Rifai, *T. pseudokoningii* Rifai and *T. viride* Pers: Fr. However, with the use of molecular approaches particularly sequencing of internal transcribed spacer-1 and 2 (ITS1 and ITS2) the taxa recently have gone from nine to at least 35 species (Table 2).

Trichoderma cultures are best grown on 2% malt extract agar in day light or under UV. In the dark they quickly loose the capacity to sporulate. For the examination of phialide structure, very young cultures usually five days old must be employed, while conidial roughing is best judged in two weeks old culture.

Table 2: Sections/ groups within the genus *Trichoderma*

Section (group)	Characteristics	Species aggregate
Trichoderma (*T. atroviride-T. koningii* complex or *T. atroviride-T. viride* complex)	The species in this section have narrow & flexuous conidiophores with branches and phialides uncrowded, frequently	*T. viride, T. koningii, T. aureoviride, T. atroviride, T. asperellum*

	paired and seldom in verticils of more than three.	
Longibrachiatum (*T. longibrachiatum* group)	In this section the species have conidiophores which are sparingly and irregularly branched with the phialides also irregularly disposed and not usually in whorls or vertical. Species may also produce distinctive yellow greenish pigments in the reverse of cultures.	*T. longibrachiatum, T. pseudokoningii, Trichoderma* anamorphs of *H. schweinitzil*
Saturnisporum	The species have a *T. viride* like branching pattern with branches and phialides uncrowded and frequently paired but with inflated phialides and compact conidiogenous pustules as in section Pachybasium. They are further differentiated by the bullate or wing like conidial ornamentation.	*T. saturnisporum, T. ghanensi*
Pachybasium (*T. harzianum*- *T. hamatum* complex)	It includes species with highly ramified, broad conidiophores and phialides broad or inflated, relatively short and disposed in crowded verticiles. Some species are further characterized by the production of sterile conidiophore extensions and many isolates produce compact conidiogenous pustules with adjacent conidiophore anastomosing.	*T. hamatum, T. inhamatum, T. polysporum, T. piluferum, T. harzianum, T. filanitosa. Fasciculatum, T. flavofuscum, T. crassum, T. virens, T. croceum, T. semorbis, T. minutisporum, T. tomentosum, T. fertile, T. spirale, T. longipilis, T. oblongisporum, T. strigosum, T. strictipilis, T. pubescens, T. aggressivum**, *T. aggressivum f. europaeum**

Hypocreanum	It accommodated *Hypocrea* anamorphs, mostly from section Homalocrea and occasional isolates from soil or wood that are characterized by effuse, usually sparse conidiation, sparingly branched conidiophore and cylindrical to subulate phialides frequently borne in *Verticillium*- like divergent vertical.	*Hypocrea anum*

5. MECHANISMS OF ACTION

Against fungal pathogens, *Trichoderma* species rely on three major mechanisms, *viz.*, mycoparasitism/ hyperparasitism, antibiosis and competition.

(i) Mycoparasitism/hyperparasitism

Trichoderma species are able to derive nutrition from other fungi by parasitizing and degrading their cell wall. Sequential events leading to host recognition, attack, penetration, and killing of the host takes place during the process. *Trichoderma* spp. attach to the host hyphae via coiling, hooks and appressorium like bodies, and penetrate the host cell wall by secreting lytic enzymes. The interaction is specific and is not merely a contact response. *Trichoderma* recognizes signals from the host fungus, triggering coiling and host penetration (Fig. 1). A biomimetic system consisting of lectin-coated nylon fibers was used to study the role of lectins in mycoparasitism. Using this system identification of specific coiling-inducing molecules was done. *Trichoderma* recognizes their host and grows chemotactically towards them (116). The recognition is partially achieved through detection of specific oligomers of the pathogen cell wall released by constitutive cell wall degrading enzymes of *Trichoderma* (117).

Weindling (5) for the first time ascribed the biocontrol of *R. solani* (causing citrus seedling disease) by *Trichoderma lignorum* to mycoparasitism. Rather than coiling, hyphae of *Trichoderma* may also grow attached with hyphae of *R. solani*, form haustoria, which may penetrate host fungal cell to draw nutrients. Same isolates of *Trichoderma harzianum,* against *R. solani*, may show both coiling and haustoria formation, however, one or other mechanism may dominate depending upon isolate of the antagonist (108). *Trichoderma* spp. are the most widely studied mycoparasitic fungi. However, their mycoparasitism is difficult to demonstrate *in situ* until very recently due to technical difficulties in making *in situ* microscopic observations (*e.g.*, fluorescence imaging and differential staining), such as at the soil–root interface. Moreover, techniques involving antibodies, such as combined baiting-ELISA (enzyme linked immunosorbant assay) techniques to detect *Trichoderma* spp. in composts would certainly increase our understanding of the mycoparasitic interaction of these fungi (118).

Studies on the molecular and cellular aspects of the process of mycoparasitism indicate that it is extremely complex process involving several steps and numerous separate genes and gene products. *Trichoderma* can detect its host from a distance and on detection it starts branching in an atypical way towards the fungus. This process is probably induced by nutrient gradients arising from the host. Dennis and Webster (119,120) conducted experiments using plastic threads similar in diameter to *Pythium ultimum* hyphae and concluded that the coiling of *Trichoderma* is not merely a contact stimulus. *Trichoderma* hyphae were never observed to coil around plastic threads. Later studies done by Barak *et al.* (121) indicated the role of lectins in the process of host recognition by *Trichoderma*. Elad (122) isolated a lectin in the process of host recognition by *Trichoderma.* Elad (122) isolated a lectin from *R. solani* hyphae and culture filtrate which they concluded, binds to the galactose residues in cell walls of *Trichoderma*. Extracts from *S. rolfsii* also exhibited agglutinin activity but its properties were different from that of *R. solani*. Inbar and Chet (1994) and Khan et al. (123,124) provided direct evidence for the role of lectins in mycoparasitism. They observed that *T. harzianum* coiled around nylon fibres, which had been treated with concanavalin A, a lectin, purified from *S. rolsfii*. Attachment of *Trichoderma* to host hyphae is followed by a series of degenerative events and the host cell wall structure is disrupted which promotes osmotic imbalances triggering cell disruption.

Moreover, the mycoparasitism shown by *Trichoderma* sp. was host specific (*e.g.*, *Pythium oligandrum*) (125). Recently, the role of extracellular enzymes has been well documented by several researchers (*e.g.*, proteolytic enzymes (126,127). β-1,3-glucanolytic system (128) and chitinase (129). The complex group of extracellular enzymes has been reported to be a key factor in pathogen cell wall lysis during mycoparasitism.

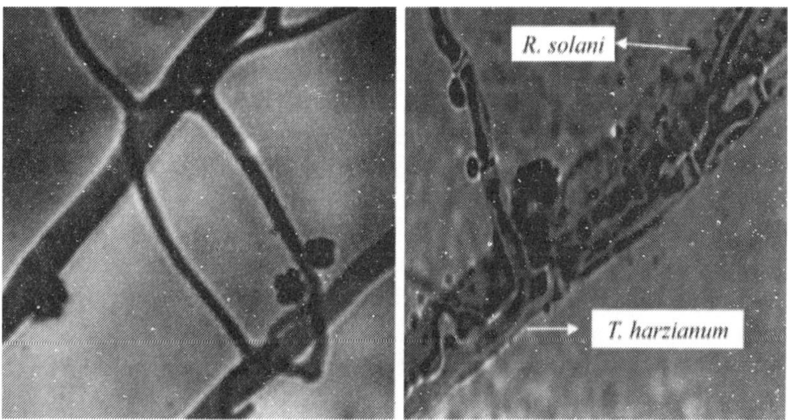

Fig. 1. Mycoparasitism by *T. harzianum* against soil borne pathogen *R. solani*

(ii) Hydrolytic enzymes

Most of the pathogenic fungi contain chitin and b-glucans in their cell walls. Dissolution or damage of these structural polymers has adverse effects on the growth of these fungi. Lysis of the host cell wall of the plant pathogenic fungi has been demonstrated to be an important step in the mycoparasitic attack (116,130). Its biocontrol mechamism involves a complementary action of antibiosis, nutrient competition and cell wall degrading enzymes such as chitinases, β-1,3-glucanases and proteases. Since chitin is the major component of most fungal cell walls, a primary role has been attributed to chitinases in the biocontrol activity of *Trichoderma* (10). Studies on the molecular structure and characteristics of genes encoding enzymes of the chitinase complex in *Trichoderma* will contribute to the better understanding of the relationships of the different enzymes involved in the biocontrol mechanisms.

Recent research work has implicated a major role of enzymes in biological control by *Trichoderma* species and the secretion of enzymes is reported to be an integral step of the mycoparasitic process of *Trichoderma. Trichoderma* species secrete a number of hydrolytic enzymes, which includes chitinases, proteases, cellulases, glucanases and xylanases (108). Lorito et al. (131) listed 10 separate chitinolytic enzymes alone. Similar levels of diversity are reported to exist for β-1,3 glucanases (132). Elad et al. (133) tested the secretion of chitinases and β-1,3 glucanases by *T. harzianum* and observed that the enzymes degraded hyphae of *S. rolfsii*. Harman (10) studied the involvement of "chitinase and β -1, 3-glucanase" in *Trichoderma* medicated biological control. Geremia *et al.* (134) purified and biochemically characterized a serine protease enzyme. Elad and Kapat (135) suggested the role of proteases in biocontrol of *B. cinerea* by *T. harzianum.* For mycoparasitism of Pythiaceous funge, β-1,4-glucanases may play an important role. Recently, Ait-Lahsan *et al.* (136) isolated and characterized an exo-α-1,3-glucanase enzyme from *T. harzianum* that degrade a-glycosidic linkage of polysaccharides of cell wall of fungi.

The various enzymes secreted by *Trichoderma* are reported to act synergistically *in vitro*. These enzymes not only act synergistically amongst themselves but also with chitinolytic enzymes from *T. virens* as well as with antibiotics. Lorito *et al.* (137) purified two enzymes; endochitinase and cellobiosidase, and observed their effect on spore germination and hyphal elongation of *Botrytis cinerea, Fusarium solani, Uncinula necator* and *Ustilago avenae*. He observed that combining the two enzymes resulted in a synergistic increase of antifungal activity. Di Pietro *et al.* (138) reported the synergism between enzymes and the antibiotic gliotoxin.

Baek *et al.* (139) disrupted or over-expressed the gene coding for chitinase (*cht42*) in *T. virens* (Gv29-8). Transformations with reduced enzyme activity or over-expression of the enzyme gave significantly decreased or enhanced biocontrol activity, respectively, against *R. solani* incited cotton seedling disease. However, the differences were not great and probably indicate that other factors were also at work. In a similar study, Woo et al. (140) also disrupted chitinase (*ech42*) activity in *T. harzianum* (P1) and showed reduced biocontrol activity against *Botrytis cinerea* on bean leaves. However, the transformant was as effective as the wild type against *P. ultimum* and exhibited enhanced activity against *R. solani* when compared with the control. The authors concluded that reactions between *T. harzianum* strains and various fungal hosts were based on different mechanisms. This, again, indicates that factors other than chitinase activity are important to the biocontrol process (108). The possible role of chitinolytic enzymes in biocontrol is further supported by the work of Lorito *et al.* (141), who transferred the gene encoding endochitinase from *T. harzianum* (P1) into tobacco and potato and demonstrated a high level and broad spectrum of resistance against a number of plant pathogens. Bolar *et al.* (142) demonstrated enhanced resistance to apple scab incited by *Venturia inaequalis* in transgenic apple plants that had been *T. atroviride* (P1). Kenerley (quoted in 34) transformed cotton cultivar cokar 312 with a gene from *T. virens* (Gv29-8) encoding a 42-k Da endochitinase under control of the CaMV 35 S promoter. Transgenic lines were screened for fungal chitianse activity, and several high producers were identified. Homozygous lines were generated that retained high enzyme activity and expression of his gene in the roots, shoots, and leaves. Seedlings of these lines were screened against the seedling pathogens *R. solani* and *Thielaviopsis basicola*, and detached leaves were screened against *Alternaria alternata*. Selected transgenic lines demonstrated significant levels of resistance against all three pathogens when compared with the parental control and a commercial cotton cultivar.

(iii) Antibiosis and secondary metabolites

This is the second major mechanism implicated in the biocontrol of pathogens by *Trichoderma*. *Trichoderma* is a wonderful fungal biocontrol agent for controlling variety of soilborne pathogens by producing toxic metabolites (Fig. 2 & Table 3). *Trichoderma* produces a plethora of secondary metabolites with biological activity (143,144). The term “secondary metabolite” includes a heterogeneous group of chemically different natural compounds possibly related to survival functions for the producing organism, such as competition against other micro- and macroorganisms, symbiosis, metal transport, differentiation, etc. (145). Included in this group are antibiotics, which are natural products able to inhibit microbial growth. Antibiotic production is often well correlated with biocontrol ability, and the application of purified antibiotics (Fig. 2) was found to show effects on the host pathogen similar to those obtained by using the corresponding living microbe. Ghisalberti et al. (146) demonstrated that the biocontrol efficacy of *T. harzianum* isolates against *Gaeuma-nnomyces graminis* var. *tritici* is related to the production of pyrone-like antibiotics.

The production of secondary metabolites by *Trichoderma* spp. is strain dependent and includes antifungal substances belonging to a variety of classes of chemical compounds. They were classified by Ghisalberti and Sivasithamparam (143) into three categories: (i) volatile antibiotics, i.e. 6-pentyl-a-pyrone (6PP) and most of the isocyanide derivates; (ii) water-soluble compounds, i.e. heptelidic acid or koningic acid;and (iii) peptaibols, which are linear oligopeptides of 12–22 amino acids rich in a-aminoisobutyric acid, N-acetylated at the N-terminus and containing an amino alcohol (Pheol or Trpol) at the C-terminus (147). Recently, Vinale et al. (148) isolated and characterized the main secondary metabolites obtained from culture filtrates of two commercial *T. harzianum* strains (T22 and T39), and their production during the antagonistic interaction with the pathogen *R. solani* was also investigated. According to the secondary metabolite produced, Howell *et al.* (149) divided strains of *T. virens* into two groups: the "Q" strains able to produce the antibiotic gliotoxin and the "P" strains that produce a related compound, gliovirin, instead of gliotoxin (150). Gliotoxin has a broad spectrum of antibiotic activity, while gliovirin is a specific potent inhibitor of Oomycetes and its production was positively correlated with biocontrol efficacy of "P" group strains to control *Pythium* damping-off of cotton (151,152). On substrates with high C/N ratios, both "P" and "Q" strains of *T. virens* produce a phytotoxin similar to viridin, that is called viridiol. The viridiol-producing strains may be applied to surface soil as bio-herbicide for weeds, where they do not affect the crop plant that is planted in the treated soil (117). Other observations indicated that the biological control of pre-emergence damping-off by *T. virens* could also be related to its ability to degrade seed-emitted compounds that stimulate pathogen propagule germination (153). On the other hand, the induction of plant defence responses by some strains of *T. virens* plays a pivotal role in successful disease control of *R. solani* on cotton (154). In a recent study, it has been indicated that "P" strains unable to induce the production of phytoalexins in cotton were ineffective as BCAs and pathogenic to susceptible cultivars. Conversely, "Q" strains inducing high levels of phytoalexin synthesis showed improved biocontrol efficacy and were not pathogenic to cotton roots. Phytoalexin synthesis in cotton is elicited by a protein produced by *T. virens* (155), but the exact biochemical processes involved are not yet understood.

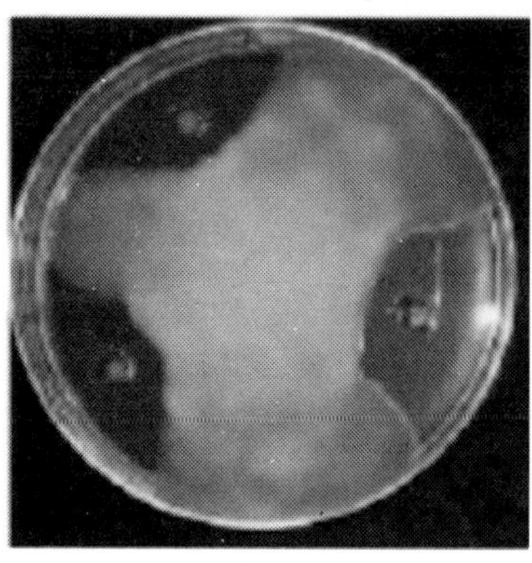

Fig. 2. Growth inhibition of *P. ultimum* by the antibiotic of *T. harzianum*.

The chemical structures of *Trichoderma* antibiotics may suggest two different mechanisms of action. The production of low molecular weight, non-polar, volatile compounds (i.e. 6PP) results in a high concentration of antibiotics in the soil environment that have a relatively long distance range of influence on the microbial community. On the contrary, a short distance effect may be due to the polar antibiotics and peptaibols acting in close proximity to the producing hyphae. Lorito et al. (156) demonstrated that peptaibols inhibited β-glucan synthase activity in the host fungus, while acting synergistically with *T. harzianum* β-glucanases. The inhibition of glucan synthase prevented the reconstruction of the pathogen cell wall, thus facilitating the disruptive action of β-glucanases. The synergism existing between enzymes and polar antibiotics is strictly related to their mechanism of action (157). Although the role and the effects of peptaibols are clear, the mode of action of other *Trichoderma* secondary metabolites (i.e. pyrones), and their possible synergisms with other compounds have not yet been elucidated (152).

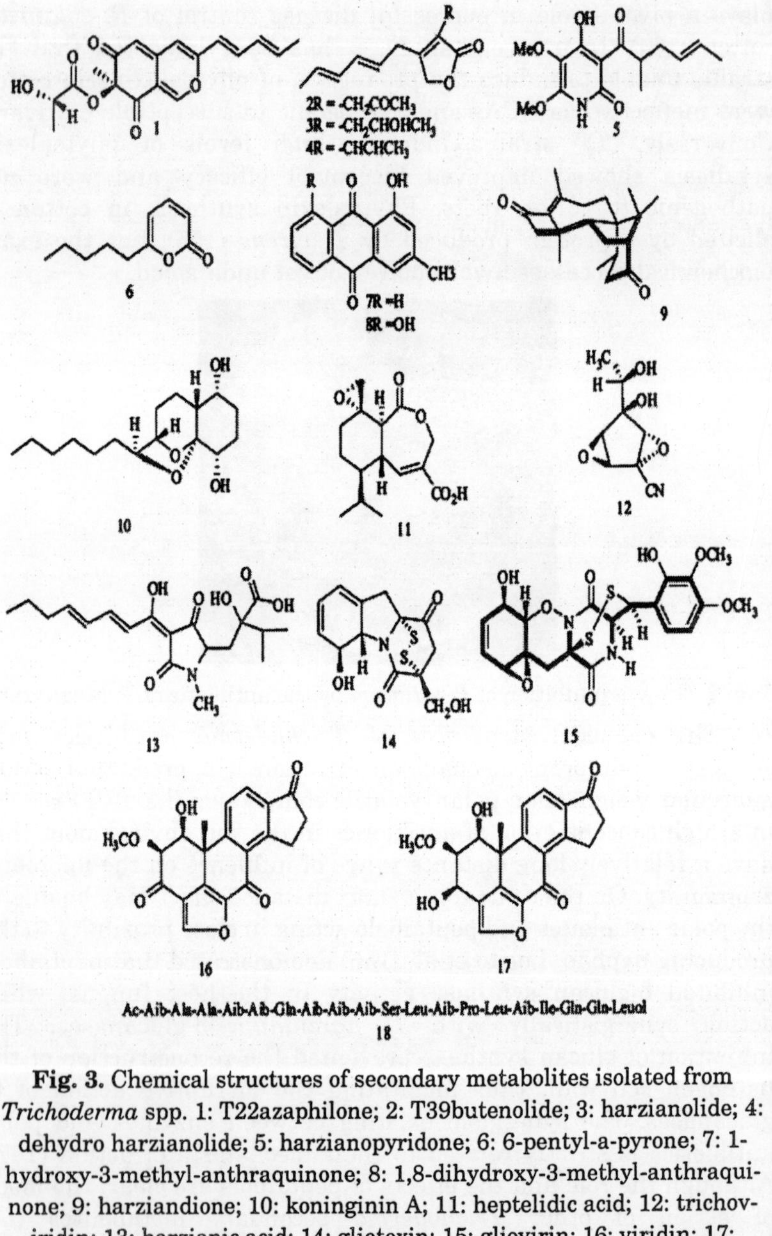
1
2R = -CH2COCH3
3R = -CH2CHOHCH3
4R = -CHCHCH3
5
6
7R = H
8R = OH
9
10
11
12
13
14
15
16
17
Ac-Aib-Ala-Ala-Aib-Aib-Gln-Aib-Aib-Aib-Ser-Leu-Aib-Pro-Leu-Aib-Ile-Gln-Gln-Leuol
18

Fig. 3. Chemical structures of secondary metabolites isolated from *Trichoderma* spp. 1: T22azaphilone; 2: T39butenolide; 3: harzianolide; 4: dehydro harzianolide; 5: harzianopyridone; 6: 6-pentyl-a-pyrone; 7: 1-hydroxy-3-methyl-anthraquinone; 8: 1,8-dihydroxy-3-methyl-anthraquinone; 9: harziandione; 10: koninginin A; 11: heptelidic acid; 12: trichoviridin; 13: harzianic acid; 14: gliotoxin; 15: gliovirin; 16: viridin; 17: viridiol.

Table 3. List of secondary metabolites produced by *Trichoderma* spp.

Isolated from	Secondary metabolites	Biological activities	Other activities
T. hamatum	Mannitol		Antimutagenic
T. pseudokoningii			
T. pseudokoningii	2-hydroxymalonic acid		
T. koningii	Methyl benzoate p-hydroxy benzyl alchohol		
G. virens	Ferulic acid	Antiviral, bactericide, fungicide	
T. pseudokoningii	2,5-dimethoxybenzoquinone		Cytotoxic
G. roseum ACC 650	Dihydrocoenzyme Q10 Coenzyme Q10		Cardiotonic
T. pseudokoningii	Succinic acid Itaconic acid		
T. album	Pencolide		
Trichoderma sp.	Carolic acid		
T. viride ATCC74084	Viridiofungin A Viridiofungin B Viridiofungin C	Antifungal Antifungal Antifungal	Squalene synthase inhibitor
T. pseudokoningii	Methyl-2,4,6- octatriene caarvizykate Trichodermene A		
G. zaleskii	2,4,6,8-nonatetrone-2,8-bis-ethyleneketal	Antibiotic	Phytotoxin
G. roseum	2,3-dihydroxy-5,6-dimethyl benzoquinone	Antibiotic	
G. roseum	2,3-dimethoxy-5,6-dimethyl benzoquinone	Antibiotic	
G. roseum	2-methoxy-3-hydroxy-5,6-dimethyl benzoquinone	Antibiotic	

G. roseum	2,3-dimethoxy-5,6-dimethyl quinhydrone	Antibiotic	
G. roseum	2,3-dimethyl-5,6-dimethoxy-2,3-dihydro benzoquinone	Antibiotic	
G. roseum CMI 93065 G. roseum	3,5-dihydroxy toluene	Antibiotic	
	1,2-dimethyl-3,4-dihydroxy benzene	Antibiotic	
	2,3-dimethyl-4,6-dihydroxy benzoic acid		
T. viride PRL 2233 *T. harzianum* IMI 311089	1-hydroxy-3-methyl anthraquionone	Bactericide	
T. viride PRL 2233 *T. harzianum* IMI 311089	1,8-dihydroxy-3-methyl anthraquinone	Anticeptic, viricide	Cytotoxic
T. viride PRL 2233	1,6,8-trihydroxy-3-methyl anthraquinone		
T. viride	1,3,6,8-tetrahydroxy anthraquinone 1,3,6,8-tetrahydroxy-4-aacetyl anthraquinone	Antibacteria	
Trichoderma spp.	Trichodermaol		Phosphodiesterase inhibitor
Trichoderma sp. SC2051	Dimeric xanthone		
T. longibrachiatum. ATCC2449	Sorbicillin Bisvertinolone Trichodimerol Trichodermolide Sorbiquinol		Inhibitor tumor necrosis factor
G. vermoesenii IMI40231	Nectriapyrone		
G. vermoesenii IMI40231	Vermopyrone	Antifungal	Plant growth regulator

T. harzianum IMI298371	Harzianopyridone	Antifungal	
T. harzianum IMI 311092	Harzianolide	Antifungal	
T. harzianum IMI 298371			
T. harzianum			
T. harzianum IMI 311092	Dehydro harzianolide	Antifungal	
T. harzianum SY-307	Harzianic acid		Hyperchole-sterimic
T. harzianum	Trichoharzin	Antifungal, Antimic-robial	Plant growth regulator
T. longibrachiatum	Compactin	Antifungal	
T. pseudokoningii			
T. harzianum IMI275950 *T. harzianum* IMI284726 *T. harzianum* ATCC20672 *T. koningii* IMI308475 *Trichoderma* spp *Trichoderma* spp. *T. viride* *T. viride* 0101	6-pentyl-a-pyrone	Antifungal	
T. harzianum IMI275950 *T. harzianum* IMI284726 *T. viride*	6-pent-1-enyl-a-pyrone	Antifungal	
Trichoderma spp.	Massoilactone	Antifungal	Plant growth regulator
Trichoderma spp.	d- decenolactone	Antifungal	
T. harzianum IMI 311090	Koninginin E	Antifungal	Plant growth regulator

T. koningii ATCC46314			
T. harzianum IMI 311090	Koninginin D	Antifungal	
T. harzianum IMI 311090	Koninginin B	Antifungal	Plant growth regulator
T. koningii ATCC 46314			
T. harzianum IMI 311090	Hydroxy koninginin B		Plant growth regulator
T. harzianum IMI 311090 *T. koningii* ATCC46314	Koninginin A		
T. koningii ATCC46314	Koninginin C		
T. harzianum IMI 311090	Seco-koninginin		
T. harzianum IMI 311090 *T. koningii* ATCC 46314 *T. koningii*	Cyclonerodiol		
T. polysporum	Cyclonerodiol oxide	Antibiotic	Antitumor
T. polysporum	Epicyclonerodiol oxide		
G. virens	Gliocladic acid	Antibacterial & antibiotic	
G. virens	Cadlene hydroxy acid		
G. virens *T. koningii* M3947 *T. viride*	Heptelidic acid	Antifungal	
T. koningii	Tricho-acorenol, coccinol		K channel agonist
G. virens IFO9166	3,4-dihydroxycarotane		Mycotoxin
T. virens ATCC74180			

T. virens ATCC74180	3,4,14-trihydroxycarotane-14-oleate	Antifungal	Antitrichomal, Mycotoin
T. polysporum CMI40624	Trichodermol	Antifungal	
T. sporulosum CMI104643			
T. virens			
T. polysporum CMI40624	Trichodermin	Antifungal	Mycotoxin
T. sporulosum CMI104643			
T. reesei P-12			
T. viride			
T. virens			
T. harzianum ATCC 90237	Harzianum A		
T. lingorum	Mycotoxin T2		
T. harzianum IMI 311090	Harziandione		
T. pseudokoningii	Lanosta-3,21-diol		
G. roseum			
T. hamatum	Ergosterol	Antibiotic	
T. polysporum			
T. pseudokoningii			
T. sporulosum			
T. pseudokoningii	Pyrocalciferol	Antifungal	
Gliocladium spp.	Helvolic acid	Antifungal	
T. koningii	Ergokonin A	Antibiotic	Inhibitor fungal spore, germination
T. koningii	Ergokonin B		
T. viride IFO 31137			
G. fimbriatum CMI101525	Viridin		Phytotoxin

T. viride			
G. flavofuscum IMI 100714			
G. virens ACC 213			
G. virens GL-21			
T. koningii			
T. viride			
G. flavofuscum IMI 100714	2-epiviridin		
G. deliquesc. CMI101523 *G. fimbriatum CMI101525* *T. viride* NRRL 1828 *G. virens* GL-21 *G. virens* ACC 213	Viridiol		
G. virens	Virone	Antibiotic	
T. pseudokoningii	Epifridelenol	Antibiotic	
T. hamatum HLX 1379	Isonitrinic acid F	Antibiotic	Immunosuppr-essive
T. hamatum HLX 1360	Dermadin	Antibiotic	
T. koningii TK-1			
T. viride UC 4875			
T. polysporum	Dermadin methyl ester	Antibiotic	
T. hamatum HLX 1379	Epoxy diol	Antibiotic	
T. hamatum HLX 1379	Spirolactone	Antibiotic	
T. hamatum HLX 1379	Diol isocyanide	Antibiotic	
T. hamatum HLX 1379	Epidiol isocyanide	Antibiotic	
T. hamatum IMI 3208	Isonitrin A	Antibiotic	

T. hamatum *T. harzianum IMI3198*			
T. koningii TK-163 *Trichoderma* Leo AK 5139 *Trichoderma* sp. *T. viride* IFO 8951	Isonitrin B, deoxytrichoviridin	Antibiotic	
T. hamatum HLX 1379	Hydroxy sporolactone	Antibiotic	
T. hamatum IMRL 3200 *T. koningii* IMRL 3201	Isonitrin C, trichoviridin	Antibiotic	Inhibit melanin synthesis
T. hamatum HLX 1379	Tetrahydroxy isocyanide	Antibiotic	
T. harzianum	MR304A	Antibiotic	Induction ofoospores
T. harzianum	Isonitrin D	Antibiotic	Induction ofoospores
T. koningii	Homothallin I	Antibiotic	
T. harzianum	Homothallin II	Antibiotic & antifungal	
T. koningii	Amine from Homothallin II	Antibiotic	
T. koningii	Formamide from Homothallin II	Antibiotic & antiviral	Acetolactate synthase inhibitor, Imm-unomodulator
T. koningii	N,N-dimethylamine from homothallin II	Antibiotic	
T. album	3-methoxy-5hydroxy-5-allyl-cyclopentenone		PAF inhibitor
G. fimbriatum *G. virens* IMI 101525	Gliotoxin	Antibiotic & antiviral	

G. virens GL-21 *G. lingorum* *G. hamatum*			
G. virens IMI 101525 *G. virens* GL-21	Gliotoxin E		PAF inhibitor
G. virens IMI 101525	bisdethiobis(methylthio)glio toxin		
G. virens IMI 101525	Didehydrogliotoxin		
G. virens IMI 101525	Bisdethiobis (methylthio) didehydrogliotoxin		
G. virens IMI 101525 *G. deliqescens*	Bis-N-norgliovictin		
G. virens IMI 101525	Phenol		
G. virens IMI 101525 *G. deliqescens*	Cyclo-(glycyl-O-3-methylbut-2-enyl-L-tyrosyl	Antibiotic	
G. virens IMI 101525 *G. deliqescens*	3-methylbut-2-enyl ether		
G. virens IMI 101525 *G. deliqescens*	3-hydroxymethylbut-2-enyl ether		Antitumor activity, Immunosuppre-ssive activity
G. virens	Gliovirin		Antitumor activity, Immunosuppre-ssive activity
T. koningii	Cyclo-(L-pro-L-Leu)		Antitumor activity, Immunosuppre ssive activity

Gliocladium sp. SCF-1168	Verticillin A		
Gliocladium sp. SCF-1168	Homoverticillin A	Antibiotic	
Gliocladium sp. SCF-1168	Hydroxyhomoverticillin A	Antibiotic	
G. deliqescens			
T. polysporum	Trichopolyn II		
T. polysporum	Trichopolyn I		
G. deliqescens	3-hydroxy-3,4-dimethylpentanoic acid		Melanin biosynthesis inhibitor
T. harzianum	Uracil		
T. harzianum	Melanoxadin		
T. harzianum	Ceramide		Melanin biosynthesis inhibitor
T. polysporum	Valinotricin		HIV inhibitor
Trichoderma sp. *ATF*	Melanoxazal		HIV inhibitor

(iv) Competition for nutrients and rhizosphere competence

Competition is considered as a 'classical' mechanism of biological control. It involves competition between antagonist and plant pathogen for space and nutrients (1). Competition for carbon, nitrogen and other growth factors, together with competition for space or specific infection sites, may be also used by the BCA to control plant pathogens. Celar (158) conducted a study on the forms of nutrients commonly available to phytopathogenic and antagonistic fungi. Earlier study reconfirmed the findings of Blakeman (159) that shortage of easily accessible nutrients for microorganisms, especially of those living in soil and on plant surfaces, could result in explicit nutrient competition among microorganisms (160). The presence of *Trichoderma* in agricultural and natural soils throughout the world proves that it must be an excellent competitor. In addition, *Trichoderma* spp. could compete and sequester ions of iron (the ions are essential for the plant pathogen, *Serpula lacrymans* as part of a non-enzymatic complex) by releasing compounds known as siderophores (161). Thus, the cited examples confirmed that significance of competition for nutrients between *Trichoderma* and pathogenic fungi. Several authors have highlighted the significance of lytic enzymes in BCA activity and studied isolates of *Trichoderma* spp. with cellulose and chitin degradation characteristics (162). Hutchinson (163) and Hanson and Howe (164) have reported the significance of secondary metabolites (antibiotic activity) in antagonistic action of *Trichoderma* spp. against pathogenic fungi *P. ultimum* and *R. solani*. However, there seems to be a general consent on the combined synergistic effect of the two factors (enzymes and antibiotic compounds) (165). Competition for nutrients is the major mechanism used by *T. harzianum* to control *F. oxysporum* f. sp. *melonis* was also studied. Moreover, *Trichoderma* has a strong capacity to mobilize and take up soil nutrients, thus making it more efficient and competitive than many other soil microbes. Ultraviolet light irradiation was used to produce mutants of *T. virens*, deficient for both mycoparasitism and antibiotic production. However, the mutants still retained biocontrol efficacy equal to that of the parent strain against both *P. ultimum* and *R. solani* causing cotton seedling disease. This indicated that neither mycoparisitism nor antibiosis is the principal mechanisms involved in the biocontrol of seedling disease in cotton.

Moreover, *T. harzianum* is able to control *B. cinerea* on grapes by colonizing blossom tissue and excluding the pathogen from its infection site (166). Sivan and Chet (167) demonstrated that competition for nutrients is the major mechanism used by *T. harzianum* to control *F. oxysporum* f. sp. *melonis*. Moreover *Trichoderma* has a strong capacity to mobilize and take up soil nutrients, thus making it more efficient and competitive than many other soil microbes (130). The biotic components of the soil environment have relevant effects on the biocontrol activity of *Trichoderma* against plant pathogens. Bae and Knudsen (168), by using a GFP-tagged mutant, showed that higher levels of microbial soil biomass induced a shift from hyphal growth to sporulation in *T. harzianum*, thus reducing its biocontrol efficacy. This effect may be associated with a phenomenon known as "soil fungistasis", which is largely dependent on the soil microbial community composition (169). In particular, the production of antibiotic compounds and the presence of bacteria belonging to the genus *Pseudomonas* seem to be essential for the development of this phenomenon. In this context a detailed study of the metabolites produced by microorganisms present in the soil environment should be performed in order to avoid the suppression of BCAs.

(v) Siderophore production

Siderophores (Gr. "iron-bearers") are defined as 'low molecular weight virtually ferric specific legands, the biosynthesis of which is carefully regulated by iron and the function of which is to supply iron to the cell' (170). The structural diversity among the different siderophores is quite considerable and depends on the producing microorganisms. However, a common feature of all siderophores is that they form six coordinate octahedral complexes with ferric ion. Iron is generally present in the microbial environment as the ferric ion (Fe (III)), which is virtually insoluble in the presence of O_2 and therefore, is not available for microbial growth. Siderophore chelate Fe (III) and microbial membrane receptor proteins specifically recognize and take-up the siderophore-Fe-complex (171,170). This results in making iron unavailable to rhizosphere microorganisms, including plant pathogens, which produce less or different siderophores with lower binding coefficients. The result is less pathogen infection and biological control. Siderophores also help in improving antagonistic activities, rhizosphere competence and plant growth. *Trichoderma virens* is reported to produce three types of hydroxymate siderophores: a monohydroxamate (*cis*-and *trans*-fusarinines), a dipeptide of *trans*-fusarinine (dimerum acid), and a trimer disdepsipeptide (copragen) (137,172).

(vi) Signal transduction

The ability of *Trichoderma* to sense and respond to different environmental conditions, including the presence of a potential host, is essential for successful colonization of soil, organic material, and developing plant roots. Sensing of such environmental conditions may occur through a variety of transduction pathways, which determine the adequate cellular response. Mitogen-activated protein kinase (MAPK) cascades and G-protein a subunits transducer a large variety of signals, including those associated with pathogenesis.

Molecular mechanism: Signal transduction through the G-protein/cAMP and MAP kinase pathways have long been known to be involved in the parasitism of plants by pathogenic fungi. Since eukaryotic signaling mechanisms are well conserved, it is interesting to examine the role of these signaling elements in mycoparasitism, and hence biocontrol. The first direct evidence on the role of a G-protein came from Rocha-Ramirez et al. (173). Antisense-mediated gene silencing of Tgal attenuated mycoparasitism of *T. atroviride* against *R. solani*. On the other hand, transgenic strains carrying multicopies of the gene overgrew *R. solani* colonies at a faster rete. Using gene knockout, Reithner et al. (174) demonstrated that Tgal modulates chitinase formulation and secondary metabolism in *T. atroviride*. The deletion of another G protein, Tga 3 resulted in loss of mycoparasitism in *T. atroviride*. Mukherjee et al. (170) studied the role of the G-proteins TgaA and TgaB in *T. virens*. Deletion of these genes individually had no effect on hyphal coiling of *R. solani*, but TgaA was involved in the parasitism of sclerotia of S. *rolfsii*. Deletion of the MAPK TmkA in *T. virens* resulted in attenuation of sclerotial parasitism of *S. rolfsii* and *R. solani*, while the hyphal parasitism was unaltere. The TmkA mutants also had reduced ability to induce resistance in cucumber seedlings, even though there was no effect on root colonization. The mutants also had reduced biocontrol of *S. rolfsii* in greenhouse tests. In contrast, however, Mendoza-Mendoza et al. (175) reported improvement in biocontrol potential of *T. virens* through inactivation of the MAP kinase Tvk 1. Whether this apparent contradiction is due to strain differences needs to be examined carefully. Recently, a *T. harzianum* stress-response MAPK ThHOG1 has been identified to be involved in osmotic and oxidative stress response. Recently, adenylate cyclase-encoding gene tac 1 of *T. virens* is cloned and obtained knockout mutants through homologous recombination that has been cloned. The mutants grew extremely slowly, failed to germinate in water, were impaired in mycoparasitism and produced lower amounts of secondary metabolites. This study proved that the cAMP signaling is involved in growth, germination and biocontrol properties in *T. virens*. Using suppression subtractive hybridization, the genes regulated by signaling genes like the *TmkA/Tvk 1* in *T. virens* have been identified. These target genes include some novel genes like the *mrsp 1* (MAPK Repressed Secreted Protein) with an expansin-like domain that could be involved in *Trichoderma*-plant root interaction. Recently Reithner et al. (177) have examined the function of the *tmk 1* gene encoding a MAPK during fungal growth, mycoparasitic interaction, and biocontrol conidiation, and displayed de-regulated infection structure formation in the absence of a host-derived signal. In confrontation assays, tmk 1 deletion caused reduced comparable to the parental strain. Under chitinase-inducing conditions, *nag 1* and *ech42* transcript levels and extracellular chitinase activities were elevated in a *Dtmk 1* mutant, whereas upon direct confrontation with *R. solani* or *B. cinerea* a host-specific regulation of *ech42* transcription was found and *nag 1* gene transcription was no more inducible over an elevated basal level. *Dtmk 1* mutants exhibited higher antifungal activity caused by low molecular weight substances, which was reflected by an over-production of 6-pentyl-a-pyrone and peptaibol antibiotics. In biocontrol assays, a *Dtmk 1* mutant displayed a higher ability to protect bean plants against *R. solani* (177). These findings strongly suggest the presence of further, still unknown, mycoparasitism related factors which are missing in our *Dtmk 1* mutant and which are therefore affected by a signaling pathway involving *Tmk 1*.

6. *TRICHODERMA*–PLANT INTERACTION AND ROOT COLONIZATION

In addition to the beneficial effects that occur in direct interactions with plant disease agents, some *Trichoderma* species are also able to colonize root surfaces and cause substantial changes in plant metabolism (11). It is well documented that some strains promote plant growth, increase nutrient availability, improve crop production and enhance disease resistance (11). The physical interaction between *Trichoderma* and the plant was observed by electron microscopy to be limited to the first few cell layers of plant epidermis and root outer cortex. The hyphae of the BCA penetrate the root cortex but the colonization by *Trichoderma* is stopped, probably by the deposition of callose barriers by the surrounding plant tissues (178). It appears that this interaction evolves into a symbiotic rather than a parasitic relationship between the fungus and the plant, whereby the fungus occupies a nutritional niche and the plant is protected from disease. A very active, direct molecular cross-talk occurs between the fungus and the plant. Elicitors from *Trichoderma* activate the expression of genes involved in the plant defence response system, and promote the growth of the plant, root system and nutrient availability. This effect in turn augments the zone for colonization and the nutrients available for the biocontrol fungus, subsequently increasing the overall antagonism to plant pathogens (155,11).

7. *TRICHODERMA* AS A BIOFERTILIZERS & PLANT GROWTH PROMOTER

Many BCAs, such as fungi, bacteria and viruses, are not only able to control the pathogens that cause plant disease, but are also able to promote plant growth and development. Apart from the direct inhibition of plant pathogens, *Trichoderma* spp. are reported to improve crop health by promotion of plant growth (both root and shoot). It is reported to enhance growth in a number of plant species like rice, wheat, sorghum, tomato, brinjal, soy-bean, chickpea, pea, rajma, chilli, etc. (Tables 4,5,6,7). However, this growth promotion effect was not only dependent on isolate of *Trichoderma* but also on plant species cultivar involved. When applied as seed treatment maximum shoot growth promotion in tomato, brinjal, chilli, and pea was caused by *T. harzianum* isolates (Fig. 4; H.B. Singh, unpublished information). Similarly when *T. harzianum* isolate was applied as seed treatment it resulted in different degree of growth promotion in different cultivars of rice.

In greenhouse and field trials, the ability of *T. harzianum* T22 and *T. atroviride* P1 to improve the growth of lettuce, tomato and pepper plants under field conditions was investigated (179). Crop productivity was increased up to 300%, as determined by comparing the treated plots with the untreated controls and measuring fresh/dry root and above ground biomass weights, height of plants, number of leaves and fruits. This study also demonstrated the compatibility of *T. harzianum* T22 and *T. atroviride* P1 with pesticides conventionally used in organic farming by monitoring the effect on mycelia growth in both liquid

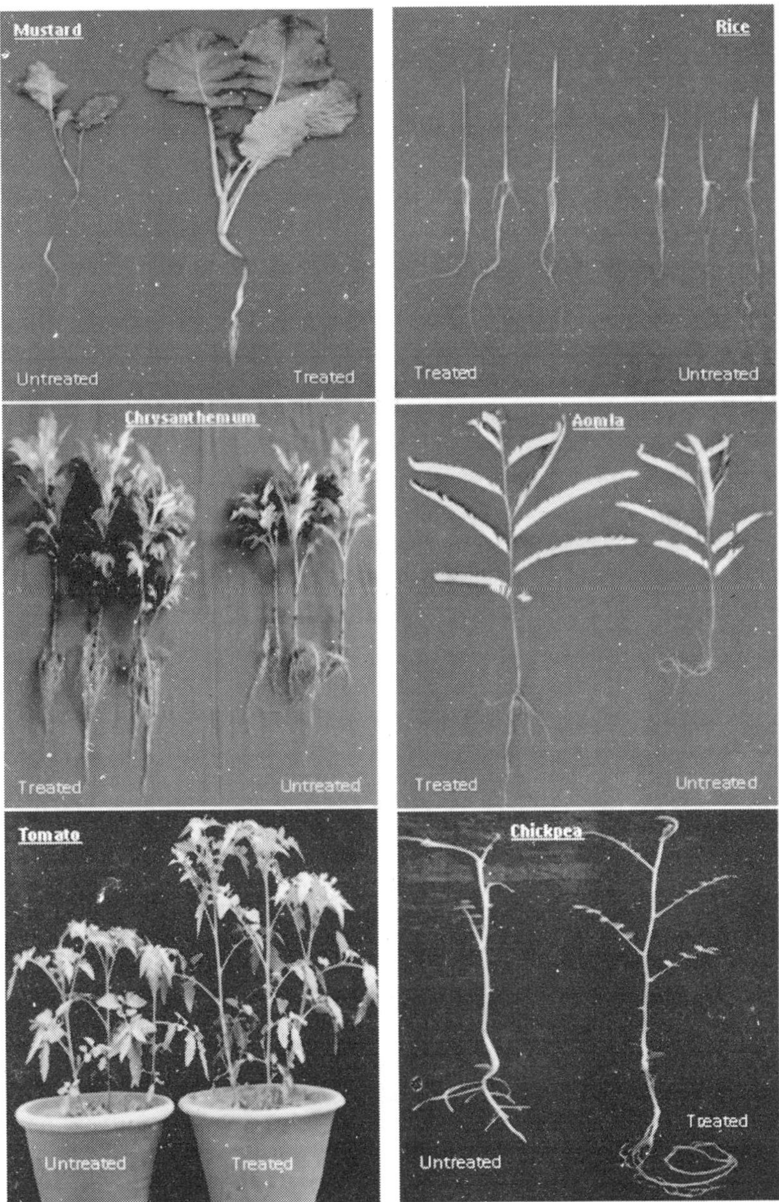

Fig. 4: Effect of *Trichoderma* on palnt growth and root development of different crops

and solid media. Results indicated a high level of tolerance by *Trichoderma* strains to concentrations of copper oxychloride varying from 0.1 up to 5mM (179). These positive effects of *Trichoderma* may be obtained with different plant species, thus the genetic base of such interactions seems not to be predominant. Conversely, at least in maize the plant growth promotion effect is genotype specific and some inbreds respond negatively to different strains. A yield increase was also observed when plant seeds were exposed to *Trichoderma* conidia that were separated from them by cellophane, suggesting that *Trichoderma* metabolites can influence the plant growth (130). On the other hand, only a few reports deal with the ability of antagonistic fungal strains to produce compounds acting as growth promoting factors. Cutler et al. (180) reported the isolation, identification and biological activity of secondary metabolites produced by *Tricoderma koningii* (koninginin A) and *T. harzianum* (6-pentyl-apyrone), that acted as plant growth regulators. Both metabolites significantly inhibited the growth of etiolated wheat coleoptiles at a relatively high concentration (10^{-3} M), but no effect was registered at lower doses (range from 10^{-4} to 10^{-3} M).

It is hypothesized that such *Trichoderma* secondary metabolites may act as auxin-like compounds, which typically have an optimum activity between at 10^{-5} and 10^{-6} M while having an inhibitory effect at higher concentrations (181), and/or are involved in the production of auxin inducers. The dose–effect response of such compounds on plant growth and development requires further investigation. *Trichoderma* spp. also produce organic acids, such as gluconic, citric or fumaric acids, that decrease soil pH and permit the solubilization of phosphates, micronutrients and mineral cations like iron, manganese and magnesium, useful for plant metabolism (130,11).

Table 4: Plant growth promotion in sunflower by *Trichoderma*

Treatment	Plant height (cm)	No. of grains/ head	Dry wt. of 1000 grains	Yield (q/ha)	% increase in yield over control
Inoculated	126.3	994.00	51.00	13.14	32.59
Control	108.6	718.40	45.00	9.91	-

Table 5: Effect of different fungal antagonists on corm rot diseases and plant growth of Gladiolus

Starins	% Germin-ation	% Mortality	Palnt Height (cm)	Spike Length (cm)	No. of Cornels/ Plant
T. harzianum (MTCC-792)	100 ± 0.00	2.33 ± 0.33	72.33 ± 0.68	85.33 ± 0.52	26.33 ± 0.58
T. virens (MTCC-794)	96 ± 1.93	3.33 ± 0.66	70.87 ± 1.10	82.10 ± 1.66	21.67 ± 0.37
T. harzianum (MTCC-795)	97 ± 0.56	3.50 ± 0.25	70.67 ± 1.15	82.56 ± 2.11	22.33 ± 0.33
T. koningii (MTCC-796)	98 ± 0.33	2.64 ± 0.20	70.67 ± 1.33	80.50 ± 1.50	21.00 ± 1.15
T. viride (MTCC-800)	96 ± 0.57	6.66 ± 0.47	67.50 ± 1.15	79.53 ± 0.88	15.25 ± 0.15
T. harzianum (MTCC-801)	100 ± 0.00	6.67 ± 0.67	66.00 ± 0.00	75.30 ± 0.68	16.69 ± 0.98
T. harzianum (up)	97 ± 1.67	8.77 ± 0.57	65.00 ± 1.10	75.66 ± 1.86	16.00 ± 1.15
Control	79 ± 2.31	28.67 ± 0.67	50.80 ± 0.46	45.80 ± 0.56	12.67 ± 0.88

Table 6: Effect of different fungal antagonists on sclerotinia rot disease and plant growth of opium poppy

Strains	% Plants infected (weeks after)						Plant height (cm)		Yield (mg/plant)	
	1st year			2nd year						
	4	8	16	4	8	16	1st year	2nd year	1st year	2nd year
TH-792	1.07	4.54	8.63	1.58	4.98	9.88	97±0.94	98 ±0.45	159±4.60	162 ± 1.37
TH-795	1.99	6.02	9.01	2.02	5.16	9.96	95±0.54	98 ±0.95	169±1.05	172 ± 2.57
TK-796	1.10	4.01	8.23	0.00	3.02	8.72	96±1.38	98 ±0.99	115±7.06	123 ± 1.29
TH-801	2.03	5.96	7.98	3.10	6.48	8.01	97±0.09	98 ± 0.81	114±9.41	120 ± 0.80
NBRI-TH-1	1.59	3.86	8.99	1.98	3.22	7.52	97±1.15	97 ± 0.89	191±2.37	200 ± 1.69

NBRI-TH-111	3.08	5.86	7.93	2.68	5.11	9.01	99±0.92	99 ± 1.25	135±1.79	125 ± 1.69
NBRI-TH-X	2.88	4.96	8.02	3.54	6.01	9.99	96±0.92	100±0.27	173±1.94	128 ± 0.70
NBRI-TH-XIV	3.09	5.82	9.99	2.86	5.33	10.01	95±0.52	97±0.07	128±1.86	122 ± 0.78
NBRI-TH-XV	3.42	4.02	7.86	2.30	2.88	6.89	94±1.37	99 ± 1.37	143±2.26	184 ± 1.91
NBRI-TH-XVII	0.00	1.94	5.63	0.00	2.06	6.54	96±0.76	96 ± 0.76	181±5.41	133 ± 1.40
Control	19.68	29.09	35.63	21.68	32.40	39.96	85±0.41	85±2.54	94±2.68	92.40±1.55

Table 7: Plant growth promotion by *Trichoderma* spp. in different crops (182).

Crop	Biocontrol Agents	Special Features
Bean	*T. harzianum*	Increase in size and weight of plants
Brinjal	*T. harzianum*	Promotion of shoot and root length
Cauliflower	*T. harzianum*	Increase in root and shoot growth
Chickpea	*T. harzianum*	Increase in root and shoot growth
Cucumber	*T. harzianum*	Increase in plant growth
Chrysanthemum	*T. harzianum*	Increase in plant height and no. of flowers
Lentil	*T. harzianum*, *T. virens*	Increase in plant growth
Pigeon pea	*T. harzianum*	Increase in root and shoot growth
Radish	*T. harzianum*	Increased vigour and emergence of seedlings
Rice	*T. virens*, *T. longibrachiatum*, *T. harzianum*	Increase in plant growth
Tomato	*T. harzianum*	Promotion of shoot and root length

8. *TRICHODERMA* AS AN ANTAGONIST OF NEMATODES

Trichoderma species have been attributed with the ability to control diseases caused by nematodes (183). Root dipping in antagonist's suspension not only reduced root knot severity caused by *Meloidogyne* but also enhance seedling growth in tomato, brinjal and capsicum (184). Culture filtrate of *T. harzianum* and *T. virens* suppressed hatching and release of second stage juveniles of *Meloidogyne*. *Trichoderma harzianum* formed loops and trapped second stage juveniles of *M. incognita*. *Trichoderma* penetrated nematode body by forming haustoria like structures and colonized internally replacing all internal organs with fungal mycelia resulting in death of the nematode (1). Egg masses are also penetrated and colonized by *T. harzianum* (185). Hyphae of *T. harzianum* were attracted towards nematode body in *Anguina tritici*. This chemotactic response was not recorded against second stage juveniles of *Meloidogyne*. This may be because of rapid motility of juveniles in suspension or on agar medium.

Kalra *et al*. (183) conducted green house experiments on the potential of *T. harzianum* to control root knot nematode, *M. javanica*. They reported that root galling was reduced and top fresh weight increases in *T. harzianum* pretreated soils. The mycelium of *T. harzianum* coiled around the second stage juveniles of root knot nematode *Meloidogyne javanica* and penetrated them by forming haustoria like structures. Protease production by *T. harzianum* has been associated with the reduction in root galling. However, field experiments are still required to prove the potential of *Trichoderma* as an effective antagonist against nematodes.

9. INDUCTION OF PLANT DEFENCE RESPONSE

The induction of plant defence responses mediated by the antagonistic fungus has been well documented (155). Various plants, both mono- and dicotyledonous species, showed increased resistance to pathogen attack when pre-treated with *Trichoderma* (11). Plant colonization by *Trichoderma* spp. reduced disease caused by one or more different pathogens, at the site of inoculation (induced localized acquired resistance, LAR), as well as when the biocontrol fungus was inoculated at different times or sites than that of the pathogen (induced systemic resistance or

ISR). *Trichoderma* strains capable of establishing such interaction also induce metabolic changes in plants that increase resistance to a wide range of plant-pathogenic microorganisms and viruses (Fig. 5). This response is believed to be beneficial for several plants, indicating that there is little or no plant specificity. Several studies have shown that root colonization by *Trichoderma* strains lead to massive changes in the plant genome and metabolism and these changes in plant metabolism lead to the accumulation of antimicrobial compounds. In addition to the ability of *Trichoderma* spp. to attack or inhibit the growth of plant pathogens directly, recent discoveries indicate that they can also induce systemic and localized resistance to a variety of plant pathogens (Table 7). Magnitude and period of induction is good enough to provide protection to the plants based on this mechanism alone. These new findings are dramatically changing our knowledge of the mechanisms of action and uses of these fungi.

Fig. 5: Diseases resistance effect of *Trichoderma* strains on potato crop against late *Phytophthora*.

The induction of plant resistance by colonization with some *Trichoderma* species is similar to that elicited by rhizobacteria, which enhance the defence system but do not involve the production of pathogenesis-related proteins (PR proteins) (Harman et al., 2004). In a recent work Alfano and co-workers (186) investigated at a molecular level the plant genes involved in *T. hamatum* 382 resistance inductions by using a high-density oligonucleotide microarray approach. Interestingly, *Trichoderma*-induced genes were associated with biotic or abiotic stresses, as well as RNA, DNA, and protein metabolism. In particular, genes that codes for extensin and extensin-like proteins were found to be induced by the BCA, but not those codes for proteins belonging to the PR-5 family (thaumatin-like proteins), which are considered the main molecular markers of SAR.

Studies on biochemical elicitors of disease resistance show the presence of three classes of compounds that are produced by *Trichoderma* strains and induce resistance in plants. At least three classes of substances that elicit plant defense responses have been identified which include peptides, proteins and low-molecular weight compounds (187). The systemic response in plants occurs through the JA/ethylene signaling pathway in a similar manner to the rhizobacteria-induced systemic resistance (34). These are proteins with enzymatic or other functions, homologues of proteins encoded by the avirulence (Avr) genes, and oligosaccharides and other low-molecular-weight compounds that are released from fungal or plant cell walls by the activity of *Trichoderma* enzymes (188).

It is now well documented that treatment of plants with various agents (e.g., virulent or avirulent pathogens, non-pathogens, cell wall fragments, plant extracts, and synthetic chemicals) can lead to the induction of resistance to subsequent pathogen attack, both locally and systemically (189). The ability of *Trichoderma* spp. to induce local and systemic resistance has been shown with *T. harzianum* in agricultural crops such as bean, cotton, tobacco, lettuce, tomato, and maize (190,155,11), with *T. asperellum* in cucumber (20), and with *H. virens* (*T. virens*) in cotton (154). According to Harman et al. (11) contrary to previously held opinions on biocontrol mechanisms, direct effects on plant pathogens are only one mechanism of biocontrol, and are perhaps less important than induced resistance. They also concluded that, it is not yet known how long the induced resistance response lasts in the absence of the inducing *Trichoderma* strain. However, with rhizosphere-competent strains that grow continuously with the plant, long-term systemic resistance can occur. In our field trials with sunflower and mustard using *T. harzianum* NBRI-IV, disease caused by natural infection of *Alternaria solani* was substantially reduced on foliage by root application of *T. harzianum*. So, the data indicate that induced resistance (localized or systemic) is an important component of plant disease control by *Trichoderma* spp. However, different mechanisms might be responsible for biocontrol caused by different strains, in different plants, and with different pathogens. The molecular basis of this resistance – which makes this fungus an active colonizer of toxic environments and a strong competitor-has been partially elucidated with the recent discovery that *Trichoderma* strains produce a set of ATP-binding cassette (ABC) transporters (191). In *Trichoderma* spp., ABC transporters have recently been shown to be important in many processes. These include resistance to environmental toxicants that are produced by soil microflora or introduced by human activity and secretion of factors (antibiotics and cell-wall-degrading enzymes) that are necessary for establishment of a compatible interaction with a host fungus, or for the creation of a favorable microenvironment.

Several studies revealed that some biocontrol agents including *Trichoderma* spp. are also able to reduce disease through a plant-mediated mechanism that is phenotypically similar to SAR, since the resistance is systemically activated and extends to aboveground plant parts. This type of induced resistance, which is activated by biocontrol agents, is often referred to as induced systemic resistance (ISR) (192). In one of the first comprehensive studies on induction of resistance by *Trichoderma* spp. and the accompanying changes in the host plant, Yedidia *et al.* (184) demonstrated that inoculating roots of 7 days old cucumber seedlings in an aseptic hydroponic system with *T. harzianum* T-203 spores initiated plant defense responses in both the roots and leaves of treated plants. They observed that *T. harzianum* penetrated the epidermis and outer cortex of the cucumber roots and the treated plants were more developed compared to the untreated plants throughout the experiment. The plant response was marked by an increase in the peroxidase and chitinase activity and by the deposition of callobiose and cellulose enriched wall appositions on the inner surface of cell wall even in areas beyond the site of fungal penetration.

The induction of defense response in plants by *Trichoderma* spp. is often associated with accumulation of various antimicrobial compounds like phytoalexins, PR proteins along with the strengthening of cell walls and other barriers in the plant cells. Yedidia et al. (184) reported the accumulation of mRNA of two defense genes (phenylpropanoid pathway gene and lipoxygenase pathway gene) in cucumber plants on treatment with *T.*

asperellum. They further reported the accumulation of phenolic secondary metabolites, which may also play a role in the defense response of the plant to various pathogens. Similarly, Howell *et al.* (154) demonstrated that seed treatment of cotton with biocontrol preparations of *T. virens* or application of *T. virens* culture filtrate to cotton seedlings radicles induced synthesis of terpenoids deosxyhemigossypol, hemigossypol and gossypol in developing roots in very high concentration and also led to increased peroxidase activity as compared to that of control. These compounds were inhibitory to the cotton seedlings pathogen *R. solani* at quite low concentr-ations. Various groups of compounds secreted by *Trichoderma* spp. may act as elicitors for the induction of defense responses in plants. Xylanase from *Trichoderma* spp. were reported to induce systemic resistance in cotton, tobacco, grapevine, etc. (154). Elad (122) reported that cellulose produced by *T. harzianum* acts as elicitor for SAR by triggering peroxides and chitinase activity. *Fusarium oxysporum* f. sp. *melonis* and powdery mildew infections on green house melon plants were reduced in plants treated with cellulases.

Table 7. Evidence for, and effectiveness of, induced resistance in plants by *Trichoderma* spp.

Speciesand strain	Plant	Pathogens	Evidence or effects	Time after appli-cation	Efficiency
Trichoderma virens G-6, G6-5 and G-11	Cotton	*Rhizoctonia solani*	Protection of plants; induc-tion of fungi-toxic terpenoid phytoalexins	4 days	78% reduction in disease; ability to induce phytoale-xins for maximum biocontrol activity
Trichoderma harzianum T-39	Bean	*Colletotrichum lindemuthianum Botrytis cinerea*	Protection of leaves when T-39 was present only on roots	10 days	42% reduction in lesion area; 41 nu-mber of spreading lesions reduced
Trichoderma harzianum T-39	Tomato, pepper, tobacco, lettuce, bean	*B. cinerea*	Protection of leaves when T-39 was present only on roots	10 days	25-100% reduction Grey-mould symptoms

Trichoderma asperellum T-203	Cucumber	*Pseudomonas syringe.* pv. *lachrymans*	Protection of leaves when T-203 was presently only roots	7 days	Upto 80% reduction in disease on leaves; 100-fold reduction in level of pathogenic bacterial cells in leaves
Trichoderma harzianum T-22; *T. atroviride* P1	Bean	*B. cinerea Xanthomonas campestris* pv. *phaseoli*	Protection of leaves when T-22 or P1 was present only on roots; production of antifungal compounds in leaves	5 days	69% reduction in grey-mould (*B. cinerea*) symptomps with T22; lower level of control with P1. 54% reduction in bacterial disease symptoms.
Trichoderma harzianum T-1 and T-22; *T. virens* T3	Cucumber	*Green-mottle mosaic virus*	Protection of leaves when *Trichoderma* strains were present only on roots	7-10 days	Disease-induced reduction in growth eliminated
Trichoderma harzianum T-22	Tomato	*Alternaria solani*	Protection of leaves when T-22 was present only on roots	7 days	Upto 80% reduction in early blight symptoms from natural field infection
Trichoderma harzianum T-22	Maize	*Colletotrichum graminicola*	Protection of leaves when *Trichoderma* strains were present only on roots	3 months	44% reduction of lesion size on wounded leaves; no disease on non-wounded leaves
Trichoderma GT3-2	Cucumber	*C. orbiculare, P. syringae* pv. *lachrymans*	Protection of leaves when *Trichoderma* strains were present only on roots; induction of lignification and superoxide generation	14 days	59% and 52% protection from disease caused by *C. orbiculare* or *P. syringae*, respectively

Trichoderma harzianum	Pepper	*Phytcphthora capsici*	Protection of stems when *Trichoderma* strains were presents only on roots; enhanced production of the phytoalexin capsidiol.	1 day	34-50% reduction in disease
Trichoderma harzianum NF-9	Rice	*Magnaporthe grisea; Xanthomonas oryzae* pv. *oryzae*	Protection of leaves when NF-9 was present only on Roots	9 days	40-78% reduction in disease severity
	Rice	*Rhizoctonia solani*	Protection of steath when applied to roots	14 days	~70% reduction in disease and 80% increase in shoot growth
Trichoderma harzianum PBAT-7,8, 11,39 & 43**				3 days	>80% reduction in disease effect lasted for>45 days

Induced-resistance systems in plants are complex. There are three generally recognized pathways of induced resistance in plants. Two of these pathways involve the direct production of pathogenesis-related (PR) proteins; in one pathway, the production of PR proteins is generally the result of attack by pathogenic microorganisms, and in other pathway, PR proteins are generally produced as a result of wounding, or necrosis inducing plant pathogens- for example, herbivory by insects- although both pathways can be induced by other mechanisms. Typically, the pathogen-induced pathway relies on salicyclic acid produced by the plant as a significant molecule, whereas the herbivory-induced pathway relies on jasmonic acid as the signaling molecule (73). These compounds, and their analogues, induce similar responses when they are applied exogenously, and there is considerable crosstalk between the pathways (11). The jasmonate-induced pathway is designated as induced systemic resistance. The jasmonate- and salicylate-induced pathways are characterized by the production of a cascade of PR proteins. These include antifungal chitinases, glucanases and thaumatins, and oxidative enzymes, such as peroxidases, polyphenol oxidases and lipoxygenases. Low-molecular-weight compounds with anti-microbial properties (phytoalexins) can also accumulate. The triggering molecules in the *Trichoderma* responses are unknown. The third type of induced resistance has been best-described as being induced by non-pathogenic, root-associated bacteria, and is termed as rhizobacteria-induced systemic resistance (RISR) (11). It is phenotypically similar to the jasmonate- and salicylate-induced systems, as it results in systemic resistance to plant diseases. However, it is functionally very different, as the PR proteins and phytoalexins are not induced by root colonization by the rhizobacteria in the absence of attack occurs, the magnitude of the plant response to attack is increased and disease is reduced. Thus, RISR results in a potentiation of plant defense responses in the absence of the cascade of proteins that is typical of the jasmonate- or salicylate-induced systems. Similar observations were recorded by Shoresh *et al.* (3) who studied the pathway involved in the induction of resistance by *Trichoderma* spp. in cucumber. They reported that treatment with an inhibitor of ethylene action strongly inhibited the protective effect of *Trichoderma* on plants thus indicating that ethylene signal is required for ISR. Moreover, application of jasmonic acid production inhibitor completely abolished the protective effect of *Trichoderma* on plants. These experiments confirmed that like in case of rhizospheric bacteria, induction of resistance by *Trichoderma* also occurs through the jasmonic acid/ ethylene signaling pathway.

The role of a mitogen activated protein kinase TmkA in inducing systemic resistance in cucumber against a bacterial pathogen *Pseudomonas syrinagae* pv. *lacrymans* was investigated by Viterbo *et al.* (187) using *tmkA* loss-of function mutants of *Trichoderma virens*. They observed that the mutants were able to colonize the plant roots as effectively as the wild type strain, but failed to induce a full systemic resistance against the leaf pathogen. Interactions with the plant roots enhanced the level of *tmkA* transcript in *T. virens* and its homologue in *T. asperellum*. At the protein level activation of two forms reacting to the phosphor-p44/42 MAPK antibody were detected. They further demonstrated that the *tmkA* mutants retained their biocontrol potential in soil against *Rhizoctonia solani,* but were not effective in reducing disease incidence against *Sclerotium rolsfii*. They concluded that unlike in many plant-pathogen interactions, *Trichoderma* TmkA MAPK is not involved in limited root colonization. *Trichoderma*, however, needs MAPK signaling in order to induce full systemic resistance in the plant. Regardless of the exact mechanism responsible, this study demonstrates that a conserved fungal signal transduction pathway is involved in the three-way interaction between biocontrol fungus, plant pathogen, and plant and that different signals control mycoparasitic activity to induce plant systemic resistance. Therefore, it may be concluded that the interactions of *Trichoderma* with its host fungi are very complex often influenced by the particular strain of *Trichoderma* involved, the host fungus in question and maybe several other ecological factors.

10. GENETIC IMPROVEMENT OF *TRICHODERMA*

Biocontrol agents can be improved for desirable characteristics like better antagonistic ability, wider host range, tolerance to pesticides, and survival ability in the environment, rhizosphere-competence, and tolerance to adverse environmental conditions, vigorous growth and longer shelf life. Different methods used for strain improvement are as follows:

(i) Selection

A wide variability exists in natural population of different biocontrol agents with respect to various characteristics. Majority of the strains being used currently are selection from natural population. This biodiversity is a great asset and can be further exploited to select efficient strains for different diseases, crops and environmental conditions. Natural biodiversity can also be exploited to target important diseases. Therefore, there is a need to collect, characterize properly and maintain as many isolates of biocontrol agents as possible. Most of the commercial strains available today are selections from natural populations.

(ii) Mutation

N-methyl-*N*-nitro-*N*-nitrosoguanidine (*NTG*) has been the most widely used chemical for inducing mutation in biocontrol fungi. By exposing the conidia of *Trichoderma* spp. to NTG, and generated mutants that were highly rhizosphere competent and superior to wild types in respect of controlling *P. ultimum* in barley, radish, cucumber, pea and tomato and also in increasing fruit weight of cucumber. Liu (193) used several chemicals (diethyl sulphate, ethyl methane sulphonate, sodium nitrate) to induce benomyl-resistant mutants of *T. koningii* which could tolerate up to 1000 mg/ml of benomyl and also without affecting antagonistic ability including lysis and antibiosis against *R. solani* and *S. rolfsii*. Mukherjee *et al.* (194) developed benomyl-tolerant mutants of *T. viride* through chemical mutagenesis. The stable mutants produced more antifungal substances and were equally effective as the wild type strain in disease control potential. A vinclazolin-tolerant strain to *T. viride* was also developed showing better potential than the wild one.

Several genetic variants of fungal biocontrol agents have also been developed by using radiations. Troutman and Matejka (195) made the first attempt of this kind by inducing benomyl-tolerant strains of *T. viride* by exposing the conidia to gamma radiation. The application to ultraviolet (UV) radiations for this purpose has been most extensive (196). Several genetic variants of *T. harzianum* and *T. viride* have been developed. Mukherjee and Mukhopadhyay (197) developed seven stable mutants of *T. virens* by exposing the cultures to 125 k rad of gamma radiation. The mutants differed from the wild type strain in phenotype, growth rate, sporulation and antagonistic potential. Ultraviolet irradiation of *T. viride* induced mutants with increased biomass, conidia and chlamydospores at higher pH levels (198). Triazole (tebuconazole)- tolerant biotypes of *T. viride* developed by UV-irradiation showed a wide variation in enzyme activity especially β-1,3-glucanase, β-1,4-endoglucanase and chitinases (199).

(iii) Protoplast fusion

Strain improvement of biocontrol agents can also be achieved by protoplast fusion. The major advantage of protoplast fusion is higher possibility of obtaining recombinants, and testing their large numbers in a short time (200). This technology also allows the induction of the para sexual cycle at high frequency not only for the strains of same species but also for different species. One of the commercially most successful strains of *Trichoderma* i.e. *T. harzianum* T-22 was developed by protoplast fusion of two strains i.e. T-95 (a rhizospere competent strain from Columbian soil) and T-12 (from New York soil) (201).

(iv) Genetic engineering

Genetic modifications can introduce or improve desirable traits in fungal and bacterial biocontrol agents and this is, therefore, an attractive option for the improvement of these strains. The biosynthesis of cell wall degrading enzymes like chiti-nases, glucanases, proteases etc., which are involved in mycopa-rasitism, is controlled mainly at transcriptional level (202) and responsible genes are present as single copy genes (203). To over produce these enzymes, their gene copy number had been increased by transformation (204). Transformants over-producing these enzymes are more efficient as biocontrol agents (205). Biocontrol efficiency of *Trichoderma* has been improved by transformation with genes *prb1* (basic protease), *egl1* (ß-1, 4-glucanase) and *chit33* (chitinase) (206).

11. COMMERCIALISATION OF *TRICHODERMA* BASED BIOPESTICIDE

Our research in the field of agriculture immediately after the independence has been mainly basic. It was only after green revolution that the research in agricultural disciplines started to move towards applied areas and the scientists started to categorize the effective researches on the basis of their reach to the farmers, how easily it could be applied in the field and the cost benefit ratio etc. In our country and elsewhere N, P, K has been used since ancient times though the use of organic manure and disease management practices were in sync with the use of traditional simple techniques. But it was only after the chemical fungicides began to be produced on commercial scale did the production of crops burgeoned. Same is the case with the biological control agents. The scientists have screened a large number of antagonists, selected the effectiveness of biocontrol agents has been slow to take off and is still in infancy as compared to the production of chemicals.

After the identification of an effective biocontrol system it is essential that it can be and should be commercialized. As a biocontrol researcher, it has become clear to our group that solving the technological problems is perhaps the easiest part of developing biocontrol agents and systems. Other considerations like legal and commercial ones of a particular country must also be kept in mind while commercializing a product based on microbes. Over the past 10 years, we have filed a number of patents through the Council of Scientific and Industrial Research, New Delhi on the utilization of microbes as biopesticide and biofertilizers that covered the relevant technology. Protection of intellectual property is an essential component of the further development of biopesticides and if the technology cannot be protected it is doubtful whether any commercial venture will be interested in taking it up and develop it further.

By 2001, we had demonstrated that strain *T. harzianum* NBRI-1055 had several desirable features for its development as a successful biocontrol agent and also that a seed treatment using the process of solid matrix priming permitted this strain to be a highly effective seed protectant. During the early part of this century one of the Industrial houses of our country, Gujarat State Fertilizers and Chemicals Limited, Vadodara, engaged in agricultural products became interested and they took up the technology for mass production of *T. harzianum* at industrial scale in 2004. However, they could launch the product "Sardar Ecogreen Biofungicide" only in March, 2006 due to some legal formalities (CIB registration under Sections 9 (3B) and 9(3) which must be fulfilled before any biopesticides is given to the farmers).

(i) Selection of Potential Trichoderma Strain

The discovery of fungal antagonists has led to new challenges in research, development, and registration of biocontrol products in a market where chemical pesticides dominate. Bringing a biocontrol product to market is a multilayered process that includes discovery, efficacy trials, toxicological testing, mass production, formulation, registration, etc. (109,108). Screening of potential strains can be conducted in four ways: selection of effective strains in relation to plant pathogens; screening of isolates which have high biotechnological indices; analysis of pathogens' properties specific for plants, useful insects, animals and peoples; search for substrates which are economically viable, convenient for mass multiplication of fungus and maintain high colony forming units (cfu) for longer period (26,108). The first step in developing a fungal biocontrol agent is the discovery, through empirical or targeted screening. Isolation and screening of potential fungal biocontrol agents have been identified in numerous cropping systems or natural ecosystems (207). Following isolation of potential strain of *Trichoderma* spp. for a particular or broad range of diseases, screening of the fungal isolates for biocontrol activity is generally performed *in vitro*, along with tests to establish identity and relatedness of the newly found strains. This is most often accomplished using various standard techniques available in literature. The relative ease of finding the right antagonist strain of *Trichoderma* as well as other biocontrol microorganisms under laboratory and field condition is an added advantage but it does not ensure that the agent would work consistently and effectively in field. The development of T-22 strain of *T. harzianum* by Harman and his associate took more than a decade and it was only in last decades of last century that the sale of *Trichoderma* based strain of *Trichoderma*, we have isolated several strains of *Trichoderma* belonging to different species such as *T. harzianum*, *T. viride*, *T. atroviride*, *T. koningii*, etc. They have been found effective against a large number of fungal phytopathogens affecting several economically important crops (21). Kalra et al. (183) had developed a potential strain of *T. harzianum* (ATCC-PTA3701) which is useful as nematode inhibitor, biofungicide and plant growth promoter (US Patent No. 6,475,772). The application of this strain as a bio-control agent of soil-borne fungal pathogens have additional advantage of improvement of plant growth and economic yield of crop plants; contributing to the reduction of deleterious nematode population in the host tissue and rhizosphere and thereby reducing the severity of root knot disease and its use as soil amendment in reducing the application of hazardous chemical fungicides and nematicides which disturb the natural beneficial soil microflora and pollute the soil and soil water. The application of this strain in nursery has also been reported to reduce the input of chemical fungicides, which sometimes inhibit the rooting of cuttings. Similarly, Singh and Singh (2004) screened the two strains, viz. *T. harzianum* (IMI No. 359869) and *T. virens* (ITCC No. 1066.95) having the potential to control the collar rot disease of *Mentha* species caused by *S. rolfsii* and also increase the oil yield significantly which is otherwise drastically reduced in diseased plants. Further, control of collar rot of groundnut (208) and wilt disease of cumin caused by *Fusarium oxysporum* f. sp. *cumini*

(114) by the application of *T. harzianum* (NBRI 1055) and *T. harzianum* Rifai was successfully reported. It would be good to the industries associated with the development of biocontrol products, if such potential strain could be produced successfully at large scale and marketed for use by the farmers.

Once the identity of the fungus is established, it is often desirable to screen different isolates to assess diversity within the species. This will aid in selecting the most efficient and antagonistic strains for biocontrol ability in vivo. Overall, ecological fitness is a fundamental requirement for BCAs because of the relatively narrow window of parameters particularly relative humidity, type of soil, osmotic potential, and temperature. Adaptability to these factors is a major requisite for *Trichoderma* strains to grow efficiently in natural environment. However, isolates with the highest level of biocontrol in vitro may not perform as well in vivo since environmental conditions and competition with other microorganisms are much more restrictive. It is, therefore, essential to select fungal isolates under a range of conditions. Also, should the selection process be limited to a narrow genetic basis, it is foreseeable that isolates with unknown potential would be left out.

(ii) Mass Production, Formulation, and Quality Control

Before a BCA can be successfully introduced into the market, a series of studies must be carried out. Concomitantly, production and formulation of an inoculum with an acceptable shelf-life has to be achieved, which is frequently a major bottleneck. With the availability of large quantities of inoculum, field and glasshouse tests on relevant crops can be done to determine the reproducibility of control. Until recently, the commercialization of *Trichoderma* spp. as a BCA has been hampered by the lack of a cost-effective way of producing sufficient amounts of fungal material to be used in field and glasshouse trials. Therefore, current research activities have been aimed at the mass production of fungal spores of *Trichoderma* and the development of mass production facilities. Many researchers have reviewed various growth media viz., grain bran, wheat straw, wheat bran, mushroom spent, wheat bran-saw dust, sorghum grain, wheat bransaw dust modified medium molasses, brewer's yeast, lignite, pod pericarp of pulse crops and stillage for mass multiplication of *Trichoderma* spp. In a study, inoculated mycelial discs of *T. viride* and *T. harzianum* on dried, powdered and sterilized 18 agricultural by products and wastes with moisture content adjusted to 50% except for tapioca rind and refuse and rice wheat ran for which the moisture content was adjusted to 90% and incubated for 20 days to test their suitability as sustrates for mass multiplication. The colony forming unit (cfu) was assessed by dilution plate technique on 21st day (Table 5). Large number of agri-wastes had been identified for mass production of biocontrol fungi using solid state fermentation technology and the formulation has been tested on several soil-borne phytopathogens of ornamental and industrial crops (Table 6). The most effective formulations were multplied and distributed to farmers for diseas control caused by soil-borne plant pathogens (26).

Table 5: Different agricultural by-products and wastes used for multiplication of *Trichoderma* species

S. No.	Species	Substrates	(x 10^6) Cfu/g
1.	*T. viride*, *T. harzianum*	Tapioca rind	27.2, 26.4
2.	*T. viride*, *T. harzianum*	Tapioca refuse	26.4, 25.8
3.	*T. viride*, *T. harzianum*	Well composed farm-yard manure	25.9, 25.6
4.	*T. viride*, *T. harzianum*	Press mud	25.2, 24.8
5.	*T. viride*, *T. harzianum*	Cowdung gas slurry	25.0, 24.8
6.	*T. viride*, *T. harzianum*	Mushroom Spent straw	24.3, 23.5
7.	*T. viride*, *T. harzianum*	Paddy chaff	22.3, 23.1
8.	*T. viride*, *T. harzianum*	Wheat bran	21.9, 22.0
9.	*T. viride*, *T. harzianum*	Groundnut shell	20.0, 19.2
10.	*T. viride*, *T. harzianum*	Rice bran	18.1, 17.5
11.	*T. viride*, *T. harzianum*	Sugarcane bagaase	16.9, 15.1
12.	*T. viride*, *T. harzianum*	Wheat straw	15.7, 12.3
13.	*T. viride*, *T. harzianum*	Sheep manure	13.1, 12.5
14.	*T. viride*, *T. harzianum*	Poultry manure	0.7, 1.0
15.	*T. viride*, *T. harzianum*	Shelled maize cob	11.2, 10.7

16.	*T. viride*, *T. harzianum*	Paddy Straw	10.1, 11.0
17.	*T. viride*, *T. harzianum*	Chick pea husk	8.4, 5.2
18.	*T. viride*, *T. harzianum*	Peat soil	10.0, 11.0

Table 6: Mass production and formulation of antagonistic organism standardized/used

Strain	Substrate	Method and scale of produ-ction	Cfu/g	Formulation of antagonistic organism			
				Carriers usd	Addi-tives added	Cfu/g	Shelf life (Months)
T. harzianum (MTCC 3841)	Banana pseudo-stem	Solid stae fermen-tation, small scale	5.8x10^6	Banana pseudo-stem	None	5.8X10^6-2.6X10^4	5
	Compost	-do-	7.3X10^6	Compost	-do-	7.3X10^6-2.5X10^4	5
	Maize cob	-do-	5.4X10^6	Maize cob	-do-	5.4X10^6-2.4X10^4	6
	Maize meal	-do-	3.3X10^7	Maize meal	-do-	3.3X10^7-8.3X10^4	6
	Rice husk	-do-	6.4X10^7	Rice husk	-do-	6.4X10^7-5.2X10^4	6
	Saw dust	-do-	4.4X10^7	Saw dust	-do-	4.4X10^6-2.4X10^4	5
	Sorghum grain	-do-	2.1X10^7	Sorghum grain	-do-	2.1X10^7-2.1X10^4	4
	Used tea leaves	-do-	8.0X10^8	Used tea leaves	-do-	8.0X10^8-6.1X10^4	6
	Wheat bran	-do-	2,4X10^7	Wheat bran	-do-	2.4X10^7-5.7X10^4	6
	Wheat bran-saw dust	-do-	2.3X10^7	Wheat bran-saw dust	-do-	2.3X10^7-2.9X10^6	6

	Sorghum grain	-do-	$2.5X10^7$	Sorghum grain	$2.5X10^7$- $2.2X10^4$	6
T. harzianum (MTCC 3843)	Citrus fruit-pulp	-do-	$5.5X10^8$	Citrus fruit pulp	$5.5X10^8$- $1.7X10^3$	

Since *Trichoderma* sporulates relatively poorly in liquid media and sporulates well on various solid substrates, Solid Substrate Fermentation (SSF) processes were developed for its cultivation. A significant advantage of *Trichoderma* spp. is that they are culturable on a wide range of carbon and nitrogen sources and can be induced to produce fungal biomass, suitable for a variety of applications. *Trichoderma* produces three types of propagules: hyphae, chlamydospores and conidia. Since hyphae cannot survive after rapid drying, hyphal biomass is not useful (210). Chlamydospores have been suggested as promising propagules for biological control application since they are able to survive in soil for extended period of time (211). Conidia have been the most widely employed propagules in biological control applications. Conidial biomass can be obtained by either submerged (212) or solid-substrate (211) cultivation technique. For mass production of *Trichoderma* spp. solid media have been frequently used. Among the methods employed to produce the *Trichoderma* propagules *en*-mass the utilization of SSF technology has been found most important, successful, ideal and the best adapted making it easier for commercialization of *Trichoderma* which could be easily applied at field level (26,108).

There is another aspect regarding the use of fermentation technology for mass production of *Trichoderma* strains. Technologically liquid state fermentation technology of biocontrol agents is more superior in the sense that it is easier to control the process parameters and go to guarantee homogenous conditions for development of the bioagent. However, solid state fermentation has its own advantages. The advantages associated with solid substrates technologies are numerous; SSF seems to have theoretical advantages over liquid substrates fermentation (LSF). Nevertheless, SSF has several important limitations. Table 1 shows advantages and disadvantages of SSF compared to LSF.

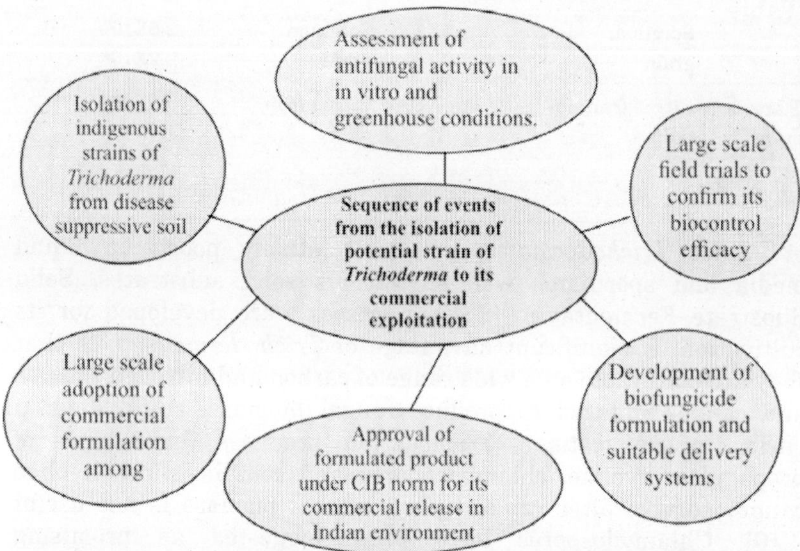

Fig. 6: Schematic diagram showing the sequences of events from the isolation of *Trichoderma* strains from disease suppressive soils, through their development and improvement, to their marketing as biopesticides to make it boom for sustainable agriculture in India.

For the successful development of formulation of biocontrol agents, it is important not only to provide a substrate that will promote the synthesis of the desired enzymes which helps in its biocontrol mechanism but also to provide sufficient substrates so as not to limit the synthesis of the enzymes at the time they are required. The mass multiplication of *Trichoderma* on solid substrate promotes the synthesis of the desired enzymes which helps in the biocontrol mechanism (208). A closer evaluation in recent years in several research centers throughout the world has revealed the enormous economical and practical advantages of SSF technology over submerged fermentation (SmF) technology (Table 7). These include utilization of large number of agrowastes as substrate for the production of *Trichoderma en*-casse, use of a wide variety of matrices, low capital cost, low energy expenditure, less expensive downstream processing, less water usage and lower wastewater output, potential higher volumetric productivity, higher concentrations of the products required, high reproducibility, lesser fermentation space and easier control of contamination, etc. (213).

Table 7. Comparison between liquid and solid substrate fermentation

Factors	Solid Substrate Fermentation	Liquid Substrate Fermentation
Substrates	Cheap substrates locally available	Soluble substrates (sugars)
Water	Limited Consumption of Water	High volumes of water consumed & effluents discarded
Aeration	No effluent Easy aeration and high surface exchange air/substrate	Limitation of soluble oxygen. High level of air required
Mechanical agitation	Static conditions preferred	Required good homogenization
Energetic consideration	Low energy consuming	High energy consuming.
Volume of Equipment	Low volumes & low costs of equipments	High volumes and high cost technology
Effluent & pollution	No effluents, less pollution	High volumes of polluting effluents
Concentration/ Products	Consisting of spores	Consisting of chlamydospores and blastospore

Numerous attempts have been made to control several soilborne pathogens by incorporating natural substrates colonized by antagonists of the pathogens into soil. The results of these attempts have varied accordingly to substrate. Lowered disease severity, increased yield, and a decreased pathogen population resulted from incorporation of different substrates colonized by *Trichoderma* spp. in nurseries and field experimental plots in fields. The importance of substrates in mass production of *Trichoderma* cannot be determined. The *Trichoderma* production process must also end up in biomass with excellent shelf life even under adverse storage conditions. In several respects, the requirements for production of products for agricultural use are more difficult to meet than those required for pharmaceutical products. If agricultural materials are to be successful, they must not be very expensive, able to produced in large quantities, and maintain good viability without specialized storage systems. With *Trichoderma* spp. we have been studying the physiology of desiccation tolerance, the mechanisms of shelf life, and processes for large-scale production of these fungi. This research can be completed on the laboratory scale at National Botanical Research Institute and now has been transferred to different industries for implementation and large scale production. In our studies on mass multiplication, we found variations in time of colonization, the type of enzymes produced and the shelf life of *Trichoderma* (unpublished data). The type of substrates used also showed variations in ability of antagonists to control different plant pathogens.

12. DELIVERY SYSTEM OF *TRICHODERMA*

Seed coating with biocontrol agents has emerged as a feasible way of delivering the antagonist for the management of plant diseases (Fig. 7). As this technique of disease management represents a living system, the viability of the biocontrol agent may pose a limitation to its commercialization. One approach for this might be supplying the coated seeds to the farmers directly by the seed companies/agencies. A considerable time gap between coating seeds by the seed supply agencies and sowing such seeds by farmers is bound to occur and it should be ensured that sufficient propagules remain viable on coated seeds at the time of sowing. Spore/cell suspension as well as dry powder has been used to coat the seeds with potential antagonists (67). Propagules of biocontrol agents germinate on the seed surface and colonize roots of germinated seedlings (214). For commercial purpose, dry powders of antagonists are used @ 3 to 10 g powder per kg seed, based on seed size and formulations of antagonist. *Trichoderma hamatum, T. harzianum, T. virens* and *T. viride* are effective seed protectants against *Pythium* spp. and *R. solani* (215).

Seed biopriming (treating of seeds with biocontrol agents and incubating under warm and moist conditions until just prior to radicle emergence) has potential advantages over simple coating of seeds as it results in rapid and uniform seedling emergence. *Trichoderma* conidia germinate on the seed surface and form a layer around bioprimed seeds. Such seeds tolerate adverse soil conditions better. Biopriming could also reduce the amount of biocontrol agents that is applied to the seed (Fig. 8). Population of

T. harzianum and *T. virens* on the surface of treated seeds of tomato, brinjal, soyabean and chickpea increases from @ 10^4 cfu per seed up to 10^6-10^7 cfu per seed after biopriming (216).

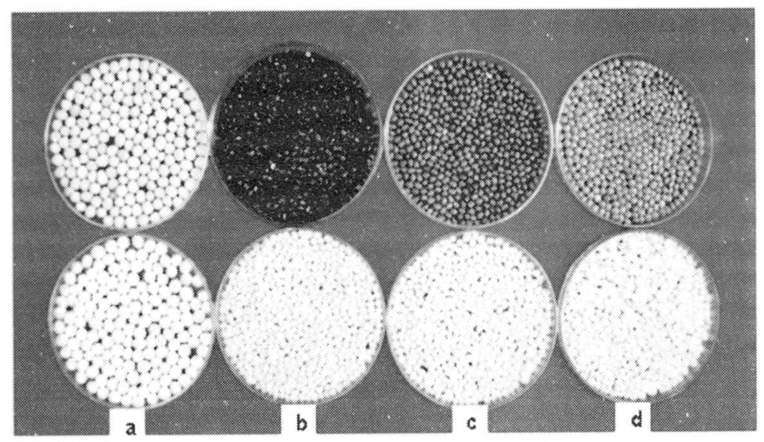

Fig. 7: Seed coating with *Trichoderma* on: (a) Pea; (b) Black gram; (c) Moong; (d) Lentil

Treatment of planting materials with beneficial microorganisms is becoming increasingly important. Seedling rots can be treated with spore or cell suspension of antagonists either by drenching the bioagent suspension before transplanting. This method is generally used for the vegetable crops, rice etc. where transplanting is practiced. There are also reports on the reduction of sheath blight disease severity in rice by treatment of seedlings in nursery before transplanting. Root dipping of tomato seedlings in suppression of antagonists reduces the severity of root knot caused by *Meloidogyne incognita*. Root dipping in antagonists suspension not only reduces disease severity but also enhances seedlings growth in rice, tomato, brinjal, chilli and capsicum (26).

There are several reports on the application of biocontrol agents to the soil and other growing media either before or at the time of planting for control of a wide range of soil-borne fungal pathogens (217). Such applications are ideally suited for green house and nursery but because of the bulk requirement, cost and problem of uniform distribution, feasibility of field application is less. Granular or pellets preparations have been used directly for soil application and they have provided effective control of diseases both under green house and field conditions (218).

Trichoderma is capable of colonizing farmyard manure (FYM). Therefore, for soil application, first it is added to moist FYM, incubated for 10 to 15 days to allow colonization and colonized FYM is distributed in field followed by irrigation. It is most effective method of application of *Trichoderma* particularly for the management of soil-borne diseases. *T. harzianum*, which was ineffective when applied through rhizome, effectively suppressed ginger rhizome rot when applied through colonized compost.

Existence of epiphytic micro-flora on plant surface including leaves and flowers is a natural phenomenon. However, *Trichoderma* is primarily a soil fungus. Nevertheless it has been used, less frequently, as a spray for the management of foliar diseases (219). There are several reports showing effectiveness of biocontrol agents applied as foliar spray against different plant pathogens. However, success of any antagonists on leaf/sheath surface depends largely on its ability to colonize these surfaces. Even environmetnal factors like humidity, temperature and sunlight affecting this colonization affect bioefficiency of antagonists. Because of these reasons, foliar application is preferred during evening hours and it is more successful in rainy season crops.

13. COMMERCIAL DEVELOPMENT OF *TRICHODERMA* BASED BIOPESTICIDE

Interest in biological control research reflects the desire of multiple constituencies to develop sustainable methods for controlling plant disease. Growing concerns about environmental health and safety have led to substantial regulatory changes in the past several years. Such concerns have led to increased restriction on a variety of chemical pesticides, including some of those used to suppress plant diseases. It also encourages the development of potential biopesticides (Fig. 8).

In India, the majority of growers continue to express interest in biologically-based pest management strategies of all types as central components of an IPM approach. Such market realities promote the development of biocontrol products. Still, the path to developing and applying effective biocontrol methods is a long one, fraught with a variety of difficulties. In addition to that scientific, regulatory, business management and marketing issues must all be handled effectively for a biocontrol product to be successful in the market place.

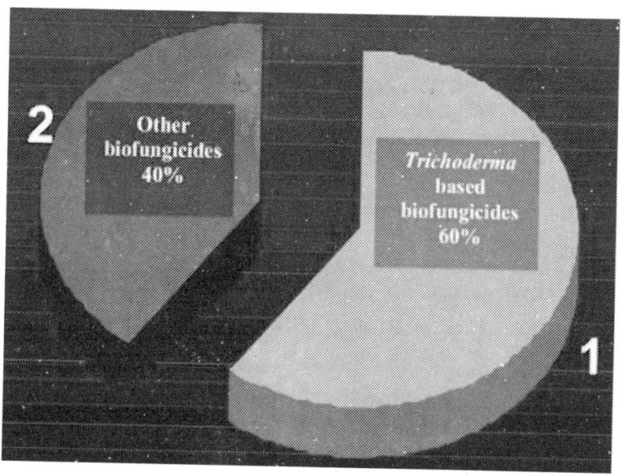

Fig. 8: *Trichoderma* spp. based biofungicide market statistics. Other biofungicides include bacteria, nematodes and virus.

Currently in India, a large number of biologically-based products are being sold for the control of plant diseases. A growing number of companies are also developing new products that are in the process of being registered with Central Insecticide Board (Pesticides regulatory authority in India). Many of these companies are small, privately owned firms with a limited product-line. In India *Trichoderma* based biopesticides biocontrol products are marketed as wettable powder (WP) based on single strain of *Trichoderma* spp. The utilization of *Trichoderma* at large scale may be possible if they are sold instead as plant strengtheners or growth promoters without any specific claims regarding disease control, this will help in improving the global market perception of *Trichoderma* as effective product. The use of *Trichoderma* based product is also suitable for organic production of food crops, which is a very important component of sustainable agriculture. More information on commercially available biocontrol products in India can be obtained from http:/ / www.cibrc.nic.in (Table 8). Globally the list of commercialized products based on microorganisms or compounds thereof may be accessed at http:/ / www.oardc.ohio-state.edu/apsbcc/productlist2003USA.htm

Other web sites that list information on commercially available biological control products and other biopesticides can be found at http://pestdata.ncsu.edu/ir-4/apps/news/BioCompaines.htm. (1)

14. COMMERCIAL REQUIREMENT FOR *TRICHODERMA* BIOPESTICIDE

For successful commercial production of *Trichoderma* based biopesticides following points are essential.

- Abundant and cost effective production of microbial propagules.
- Optimization of culture conditions for producing high yield and high quality conidian biomass.
- Development of lower cost production, storage, and distribution systems.
- Ability to survive downstream processing, particularly drying, which is required to avoid contamination.
- Stability and adequate shelf life of the final product upon storage, preferably without refrigeration.
- Ability to withstand environmental variations in temperature, desiccation, relative humidity in order to survive and establish active populations in soil.
- Consistent efficacy under varying agroclimatic conditions at commercial feasible states.
- Integration of biocontrol into current agronomic practices.

Table 8. List of bio-products based on *Trichoderma* spp. registered under Central Insecticide Board (CIB, www.cibrc.nic.in)

S.No.	Product description	Manufactured by	Section
1	*Trichoderma viride* 1.0% WP	M/s Advance Crop care P. Ltd.	9(3B)
2	*Trichoderma harzianum* 1% WP	M/s Super Pesticides & Agro Pvt. Ltd.	9(3B)
3	*Trichoderma viride* 1% WP	M/s Prathibha Biotech	9(3B)
4	*Trichoderma viride* 1% WP	M/s Krishna Industrial Corporation Ltd.	9(3B)
5	*Trichoderma viride* 1% WP	M/s Sri Biotech Laboratories India Pvt. Ltd., Hyd.	9(3B)
6	*Trichoderma viride* 1% WP	M/s Vishwa Mithra Bio Agro (P) Ltd., Guntur	9(3B)

7	*Trichoderma viride* 1% WP	M/s Shree Jee Biotech Agri.& Equipments, Wardha.	9(3B)
8	*Trichoderma viride* 1% WP	M/s Samiridhi Bioculture Pvt. Ltd., Indore	9(3B)
9	*Trichoderma viride* 1% WP	M/s Rajshree Sugars & Chemicals Ltd., Coimbatore	9(3B)
10	*Trichoderma viride* 1% WP	M/s. Shri Ram Solvent Extractions (P) Ltd.	9(3B)
11	*Trichoderma viride* 1% WP	M/s. Vaibhavalaxmi Bio-Control Lab, Wardha	9(3B)
12	*Trichoderma viride* 1% WP	M/s Sainath Agro Vet Ind. Ltd., Ahmednagar	9(3B)
13	*Trichoderma viride* 1% WP	M/s Modi Agro Products, Bhopal	9(3B)
14	*Trichoderma viride* 1% WP	M/s Shilabati Horticulture & Agriculture, Kolkata	9(3B)
15	*Trichoderma viride* 1% WP	M/s Jai Shree Rasayan Udyog Ltd., New Delhi	9(3B)
16	*Trichoderma virde* 1.0% WP	M/s Maa Bhagwati Biotech & Chemcials, Wardha	9(3B)
17	*Trichoderma viride* 1% WP	M/s Pramukh Agri Clinic, Gujarat	9(3B)
18	*Trichoderma viride* 1% WP	M/s Sujay Biotech Pvt. Ltd., Andhhra Pradesh	9(3B)
19	*Trichoderma viride* 1% WP	M/s Kundu Agro Chem (P) Ltd. Kolkata	9(3B)
20	*Trichoderma viride* 1% WP	M/s Indo Sikkim Bio Agro Industries, Gangtok	9(3B)
21	*Trichoderma viride* 1% WP	M/s Aastha Biotech Pvt. Ltd., Kolkata	9(3B)
22	*Trichoderma viride* 1% WP	M/s Chhatisgarh Agro Biotech Lab, Raipur	9(3B)
23	*Trichoderma viride* 1% WP	M/s Nirmal Organo Bio-Tech Pvt. Ltd., Mumbai	9(3B)
24	*Trichoderma viride* 1% WP	M/s Bio Agro Ferticons, Pune	9(3B)
25	*Trichoderma viride* 1% WP	M/s Shri Ram Solvent Extractions Pvt. Ltd., Jaspur	9(3B)

26	*Trichoderma viride* 1% WP	M/s Sun Agro Biosystem Pvt. Ltd., Chennai	9(3B)
27	*Trichoderma viride* 1% WP	M/s Liebigs Agro Chem Pvt. Ltd., Kolkata	9(3B)
28	*Trichoderma viride* 1% WP	M/s M.S. Industries, Amra vati	9(3B)
29	*Trichoderma viride* 1.00% WP	M/s Om Agro Organics, Yavatmal	9(3B)
30	*Trichoderma viride* 1% WP	M/s Sai Agrotech, Yavatmal	9(3B)
31	*Trichoderma viride* 1% WP	M/s Biotech International Ltd., New Delhi	9(3)
32	*Trichoderma harzianum* 1% WP	M/s Mac Hi-Tech, Kerala	9(3B)
33	*Trichoderma viride* 1% WP	M/s. Maa Bhagwati Biotech & Chemicals, Wardha	9(3B)
34	*Trichoderma viride* 1% WP	M/s. Sai National Rural Development and Research Institute, Allahabad	9(3B)
35	*Trichoderma viride* 1% WP	M/s. Kan Biosys Pvt. Ltd., Pune	9(3B)
36	*Trichoderma viride* 1% WP	M/s. Choudhari Agro-Tech(I), Nagpur	9(3B)
37	*Trichoderma viride* 1% WP	M/s. Jai Biotech & Research Centre, Jaipur	9(3B)
38	*Trichoderma viride* 1% WP	M/s. Cab-Tech Labs, Hyderabad	9(3B)
39	*Trichoderma virde* 0.5% WP	M/s Pest Control (India) Pvt. Ltd. Bangalore	9(3)
40	*Trichoderma viride* 1% WP	M/s Ishwar Agro, Dhule	9(3B)
41	*Trichoderma viride* 1% WP	M/s. Ecosense Labs (I) Pvt. Ltd., Mumbai	9(3B)
42	*Trichoderma viride* 1% WP	M/s. Sai Agrotech, Yavatmal	9(3B
43	*Trichoderma viride* 1% WP	M/s Durga Bio Tech, Nagpur	9(3B
44	*Trichoderma viride* 1.0% WP	Co M/s Rajshree Sugars & Chemicals Ltd., Coimbatore	9(3B
45	*Trichoderma viride* 1.0% WP	M/s Jai Kissan Agro. Indore	9(3B
46	*Trichoderma virde* 1.0% WP	M/s Krishi Rasayan Export Pvt. Ltd., New Delhi	9(3B

47	*Trichoderma harzianum* 1.0% WP	M/s Indian Institute of Horticultural Research, Bangalore	9(3B)
48	*Trichoderma viride* 1.15% WP	M/s Romvijay Bio Tech Private Ltd., Pondicherry	9(3B)
49	*Trichoderma viride* 1.15% WP	M/s K.N. Bio Sciences (India) Pvt. Ltd., Hyderabad	9(3B)
50	*Trichoderma viride* 1% WP	M/s Jai Biotech Industries, Nasik	9(3B)
51	*Trichoderma viride* 1% WP	Government of Andhra Pradesh, Department of Agriculture	9(3B)
52	*Trichoderma viride* 1.% WP	M/s K.N. Biotech Pvt. Ltd., Hyderabad	9(3B)
53	*Trichoderma viride* 1% WP	M/s Sneha Biotech,	9(3B)
54	*Trichoderma viride* 1.15% WP	M/s Prathista Inndustries Ltd., Secundrabad	9(3B)
55	*Trichoderma viride* 1% WP	M/s Govinda Agrotech Ltd., Nagpur	9(3B)
56	*Trichoderma viride* 1% WP	M/s Monarch Bio-Fertilizers and Research Centre, Chennai	9(3B)
57	*Trichoderma viride* 1% WP	M/s Basarass Biocon (India) Pvt. Ltd., Chennai	9(3B)
58	*Trichoderma viride* 1% WP	Institute of Natural Organic Agriculture, Pune	9(3B)
59	*Trichoderma viride* 1% WP	M/s Agri Life, Medak	9(3B)
60	*Trichoderma viride* 1% WP	M/s S & S Bio-tech, Nagpur	9(3B)
61	*Trichoderma viride* 1% WP	M/s Pravara Agro Bio-tech, Ahmednagar	9(3B)
62	*Trichoderma viride* 1% WP	M/s Kaveri Seed Company Limited, Secunderabad	9(3B)
63	*Trichoderma viride* 1.0% WP	M/s Varssha Bioscience & Technology, Hyderabad	9(3B)
64	*Trichoderma viride* 1% WP.	Modern Bio-tech, Jalgaon	9(3B)
65	*Trichoderma viride* 1% WP	Advance Biotech Industries & research Inputs, Indore	9(3B)

66	*Trichoderma viride* 1% WP	Esvin Advanced Technologies Ltd., Chennai	9(3B)
67	*Trichoderma viride* 1% WP	Bio-Pest Management Pvt. Ltd., Banglore	9(3B)
68	*Trichoderma viride* 1% WP	Bio-Tech Agri Science, Hyderabad	9(3B)
69	*Trichoderma virde* 1.0% WP	M/s Transgene Biotech Ltd	9(3B)
70	*Trichoderma viride* 1.0 % WP	M/s Green Plus Biotech	9(3B)
71	*Trichoderma virde* 1.0% WP	M/s Anshul Agro Chemicals Bangalore,	9(3B)
72	*Trichoderma harzianum* 1.0% WP	M/s Western Organics Indore,	9(3B)
73	*Trichoderma harzianum* 1.0% WP	M/s Bio tech International Ltd, New Delhi	9(3B)
74	*Trichoderma harzianum* 1.0% WP	M/s Crystal Phoshates Ltd., Delhi	9(3B)
75	*Trichoderma viride* 1.0 % WP	M/s Antecedent Pabulum Inc. Punjab	9(3B)
76	*Trichoderma harzianum* 0.5% WS	M/s Pest Control (India) (P) Ltd., Banglore	9(3)
77	*Trichoderma viride* 1.0% WP	M/s Surya Bio Products, Eluru, AP	9(3B)
78	*Trichoderma viride* 1.0% WP	M/s Abhinav Biotech. Akola,	9(3B)
79	*Trichoderma viride* 1.0 % WP	M/s Shree Shiva Bio-tech Pudukottai	9(3B)
80	*Trichoderma viride* 1.0% WP	M/s Lila Agrotech	9(3B)
81	*Trichoderma viride* 1.0% WP	M/s Biocontrols, Hyderabad	9(3B)
82	*Trichoderma viride* 1.% WP	M/s Yash Krishi Takniki Ewam Vigyan Kendra, Allahabad.	9(3B)
83	*Trichoderma viride* 1.15% WP	M/s Chaitra Agri Organics, Mysore,	9(3B)
84	*Trichoderma viride* 1.% WP	M/s Harit Bio-control Lab., Yavatmal.	9(3B)
85	*Trichoderma viride* 1.% WP	M/s Vidharbha Bio Tech Lab., Yavatmal.	9(3B)

86	*Trichoderma viride* 1.% WP	M/s Central Biotech , Nagpur.	9(3B)
87	*Trichoderma viride* 1.% WP	M/s Ocean Agro (I) Ltd., Baroda.	9(3B)
88	*Trichoderma viride* 1.% WP	M/s Jai Kissan Agro, Indore	9(3B)
89	*Trichoderma viride* 1.0% WP	M/s Microplex (India) Ltd., Wardha.	9(3B)
90	*Trichoderma viride* 1.0% WP	M/s Om Organics, Yavatmal.	9(3B)
91	*Trichoderma viride* 1.0% WP	M/s Indore Biotech Inputs & Research (P) Ltd., Indore	9(3B)
92	*Trichoderma viride* 1.0% WP	M/s Universal Agro Biotech, Nohar	9(3B)
93	*Trichoderma viride* 1.0%	M/s Camson Bio Technologies	9(3B)
94	*Trichoderma harzianum* 2.0% WP	M/s Deptt. of Agriculture Govt. of Uttar Pradesh	9(3B)
95	*Trichoderma viride* 1% WP	J.R. Biocontrol Lab	9(3B)
96	*Trichoderma viride* 1% WP	M/s M.S. Industries	9(3B)
97	*Trichoderma viride* 1% WP	Samridhi Bioculture Pvt. Ltd.	9(3B)
98	*Trichoderma viride* 1% WP	Sheer Agro Bio-Fertilizer	9(3B)
99	*Trichoderma viride* 1% WP	Kalpavruksha Biosystems	9(3B)
100	*Trichoderma viride* 1% WP	M/s Microplex (India)	9(3B)
101	*Trichoderma viride* 1.15% WP	M/s Krishna Biotech Fertilizers	9(3B)
102	*Trichoderma viride* 1.00% WP	M/s State Bio-fertilizer Quality Control Laboratory	9(3B)
103	*Trichoderma viride* 1.15% WP	M/s Grace Bio-care Pvt. Ltd.	9(3B)
104	*Trichoderma viride* 1.00% WP	M/s Ashwamedh Agritech & Farm Solutions	9(3B)
105	*Trichoderma viride* 1.00% WP	M/s Sri Laxmi Narayan Chemical & Fertilizers Pvt. Ltd., Hubli	9(3B)
106	*Trichoderma viride* 1.00% WP	M/s Agri Gold Organics Pvt. Ltd.	9(3B)
107	*Trichoderma viride* 1.00% WP	M/s Arvind Biotech, Buldhana	9(3B)
108	*Trichoderma viride* 1.00% WP	M/s Maharashtra Insecticides Ltd.	9(3B)
109	*Trichoderma viride* 1.00% WP	M/s Juna Life Sciences Pvt. Ltd	9(3B)

110	*Trichoderma viride* 1% WP	M/s Multiplex Agricare Pvt. Ltd., Bangalore	9(3B)
111	*Trichoderma viride* 1% WP	M/s Pandian Biosol, Mathura	9(3B)
112	*Trichoderma viride* 1% WP	M/s Tripti Biotech, Balaghat	9(3B)
113	*Trichoderma viride* 1% WP	M/Gujarat Life Science Pvt. Ltd	9(3B)
114	*Trichoderma viride* 1% WP	M/s Chirayu Biotech, Pune	9(3B)
115	*Trichoderma viride* 1% WP	M/s Sri Aurobindo Institute of Rural Development, Nalagonda	9(3B)
116	*Trichoderma viride* 1% WP	M/s National Bio-control Laboratories	9(3B)
117	*Trichoderma viride* 1% WP	M/s Crop Health Products Ltd.	9(3B)
118	*Trichoderma viride* 1% WP	Sri Bio-Tech, Hyderabad	9(3B)
119	*Trichoderma viride* 1% WP	M/s Indore Biotech Inputs & Research Pvt. Ltd.	9(3B)
120	*Trichoderma viride* 1% WP	Biotech International Ltd.,	9(3B)
121	*Trichoderma viride* 1% WP	Mitcon Biotechnology Centre, Pune	9(3B)
122	*Trichoderma viride* 1.00% WP	M/s Insecticides India Ltd	9(3B)
123	*Trichoderma viride* 1.00% WP	M/s Vasundhara Bio-Products, Latur	9(3B)
124	*Trichoderma viride* 1% WP	M/s Sun & Ocean Agro (India) Pvt. Ltd.	9(3B)
125	*Trichoderma harzianum* 1% WP	M/s Indian Institute of Horticulture Research	9(3B)
126	*Trichoderma viride* 1% WP	M/s Pushpanjali Agri Input Technologies	9(3B)
127	*Trichoderma viride* 1% WP	M/s Vaibhavlaxmi Bio Control lab	9(3B)
128	*Trichoderma viride* 1% WP	M/s Nafed Bio fertilizer, Bharatpur	9(3B)
129	*Trichoderma viride* 1% WP	M/s Sun Agro Bio-system Pvt. Ltd.	9(3B)

130	*Trichoderma viride* 1.00% WP	M/s International Panacea Ltd.	9(3)
131	*Trichoderma viride* 1.00% WP	M/s Shreebiotech & Research Inputs (India), Mandsaur (MP)	9(3B)
132	*Trichoderma viride* 1% WP	M/s Krishna Industrial Corp. Ltd., Nidadavole, West Godavari Distt., A.P.	9(3B)
133	*Trichoderma viride* 1% WP	M/s Ganesh Bio-Control System, Rajkot	9(3B)
134	*Trichoderma viride* 1% WP	M/s Tari Bio-Tech, Thanjavur	9(3B)
135	*Trichoderma viride* 1% WP	M/s Nath Krupa Bio-Control Lab, Nagpur	9(3B)
136	*Trichoderma viride* 1% WP	M/s Enpro Bio Sciences Pvt. Ltd., Nashik	9(3B)
137	*Trichoderma viride* 1% WP	M/s Siddhant Biotech Lab, Amravati	9(3B)
138	*Trichoderma viride* 1% WP	M/s Vidyas Biotech Laboratories, Nagpur	9(3B)
139	*Trichoderma viride* 1.00% WP	M/s Directorate of Oilseed Research (ICAR), Hyderabad	9(3B)
140	*Trichoderma viride* 1.00% WP	M/s Pruthvi Fertilizers Pvt. Ltd., Anand (Gujarat)	9(3B)
141	*Trichoderma viride* 1.00% WP	M/s Poabs Envirotech Pvt. Ltd., Kerala	9(3B)
142	*Trichoderma harzianum* 2.00% WP	M/s Balaji Cropcare Pvt. Ltd., Rangareddy (AP)	9(3B)
143	*Trichoderma viride* 1.00% WP	M/s IPM Biocontrol Labs, Secunderabad	9(3B)
144	*Trichoderma viride* 1.00% WP	M/s Prakash Seeds Agro Division, Osmanabad(M.S.)	9(3B)
145	*Trichoderma viride* 1.00% WP	M/s Arya Bio Technologies, Aurangabad	9(3B)
146	*Trichoderma viride* 1.00% WP	M/s Honey Dew Biotechnologies, Krishna (AP)	9(3B)
147	*Trichoderma viride* 1.00% WP	M/s Ajay Biotech India Ltd., Pune	9(3B)

148	*Trichoderma viride* 1.00% WP	M/s Ellora Biotech & Agro Services, Aurangabad	9(3B)
149	*Trichoderma viride* 1% WP	M/s Elbitec Innovations Ltd., Chennai	9(3B)
150	*Trichoderma viride* 1% WP	M/s Bio-Pest Management Pvt. Ltd., Bangalore	9(3B)
151	*Trichoderma harzianum* 0.5% WS	M/s Pest Control (I) Pvt. Ltd., Mumbai	9(3B)
152	*Trichoderma viride* 1% WP	M/s Agriland Biotech Ltd., Gujarat.	9(3B)
153	*Trichoderma viride* 1.00% WP	M/s.RPC Biotech Industries, Kolkata	9(3B)
154	*Trichoderma viride* 1.00% WP	M/s Soman Biofertilizers, Pune	9(3B)
155	*Trichoderma viride* 1.00% WP	M/s. ECI Agrochem Pvt. Ltd., Kolkata	9(3B)
156	*Trichoderma viride* 1.00% WP	M/s Khodke agro Oriduct Pvt Limited, Amravati(MS),	9(3B)
157	*Trichoderma viride* 1.00% WP	M/s Avishkar Biofarm, MS	9(3B)
158	*Trichoderma viride* 1.00% WP	M/s Krishi Vigyan Kendra, Baramati	9(3B)
159	*Trichoderma viride* 1% WP	M/s KCP Sugar and Industries Corpn. Ltd., Andhra Pradesh	9(3B)
160	*Trichoderma viride* 1% WP	M/s Prathibha Biotech, Hyderabad	9(3B)
161	*Trichoderma viride* 1% WP	M/s Super Pesticides & Agro (I) P. Ltd.	9(3B)
162	*Trichoderma viride* 1% WP	M/s Rajshree Sugars & Chemicals Ltd,. Coimbatore	9(3B)
163	*Trichoderma viride* 1% WP	M/s International Panacea Ltd., New Delhi	9(3B)
164	*Trichoderma viride* 1% WP	M/s Vishwa Mitra Bio Agro (P) Ltd.	9(3B)
165	*Trichoderma harzianum* 1% WP	M/s Vishwa Mitra Bio Agro (P) Ltd.	9(3B)

166	*Trichoderma viride* 1.00% WP	M/s Shri Ram Solvent Extractions Pvt. Ltd., Uttaranchal	9(3B)
167	*Trichoderma viride* 1% WP	M/s Plantrich Chemicals & Fertilizers	9(3B)
168	*Trichoderma viride* 1.00% WP	M/s Jai Biotech & Research Centre, Jaipur	9(3B)
169	*Trichoderma viride* 1.00% WP	M/s Liebigs Agro Chem Pvt. Ltd., Kolkata	9(3B)
170	*Trichoderma viride* 1.00% WP	M/s Bio Chaudhari Agro Tech (I)	9(3B)
171	*Trichoderma viride* 1.00% WP	M/s Margo Biocontrols Pvt. Ltd., Bangalore	9(3B)
172	*Trichoderma viride* 1% WP	M/s Sri Biotech, Hyderabad	9(3B)
173	*Trichoderma viride* 1.00% WP	M/s CAB Tech. Labs., Hyderabad	9(3B)
174	*Trichoderma viride* 1.00% WP	M/s Bio Agro Ferticons, Pune	9(3B)
175	*Trichoderma viride* 1.00% WP	M/s Shree Jee Biotech Agriculture & Equipment, Wardha	9(3B)
176	*Trichoderma viride* 1% WP	M/s Kan Biosys Pvt. Ltd., Pune	9(3B)
177	*Trichoderma viride* 1% WP	M/s Anjali Biotech, Amravati	9(3B)
178	*Trichoderma viride* 1.00% WP	M/s Deepa Farm Inputs (P) Ltd., Trivandrum	9(3B)
179	*Trichoderma viride* 1% WP	M/s Sai Agrotech, Yavatmal	9(3B)
180	*Trichoderma viride* 1% WP	M/s Nirmal Organics Biotech P. Ltd., Mumbai	9(3B)
181	*Trichoderma viride* 1% WP	M/s Kalpavruksha Biosystems, Bangalore	9(3B)
182	*Trichoderma viride* 1% WP	M/s Vasundhra Agrotech, Aurangabad (Maharashtra)	9(3B)
183	*Trichoderma viride 1% WP*	M/s Chhattisgarh Agro Biotech Lab., Raipur	9(3B)
184	*Trichoderma viride* 1% WP.	M/s Sai National Rural Development & Research Institute, Allahabad	9(3B)

185	*Trichoderma viride* 1% WP.	M/s. Institute of Natural Organic Agriculture, Pune	9(3B)
186	*Trichoderma viride* 1% WP,	M/s. Sri Biotech, Hyderabad	9(3B)
187	*Trichoderma viride* 1% WP	M/s. Micoplex Biotech & Agrochem Pvt. Ltd., Wardha	9(3B)
188	*Trichoderma viride* 1% WP	M/s Ruchi Biochemicals, Gondia, Maharashtra	9(3B)
189	*Trichoderma viride* 1% WP	M/s International Panaacea Ltd., New Delhi	9(3B)
190	*Trichoderma viride* 1% WP	M/s Sudarshan Chemical Industries Ltd., Pune	9(3)
191	*Trichoderma viride* 1% WP	M/s J.R. Biocontrol Laboratories, Yavatmal	9(3B)
192	*Trichoderma viride* 1% WP	M/s S&S Biotech, Nagpur	9(3B)
193	*Trichoderma viride* 1% WP	M/s Agri Life, Secunderabad	9(3B)
194	*Trichoderma viride* 1% WP	M/s Nomin Agri Bio Pvt.. Ltd	9(3B)
195	*Trichoderma viride* 1% WP	M/s R.B. Herbal Agro Satana, Distt. Nashik (Maharashtra)	9(3B)
196	*Trichoderma viride* 1% WP	M/s Parvara Agro Bio Tech, Sangamer (Maharashtra)	9(3B)
197	*Trichoderma viride* 1% WP	M/s Jai Bio Tech Ind.	9(3B)
198	*Trichoderma harzianum* 2% WP	M/s Gujarat Green Revolution Company Ltd.	9(3B)
198	*Trichoderma viride* 1% WP	M/s Esvin Advance Technologies Ltd., Chennai	9(3B)
200	*Trichoderma viride* 1% WP	M/s M.S. Industries, Amravati (MS)	9(3B)
201	*Trichoderma viride* 1% WP	M/s Jai Kissan Agro, Indore	9(3B)
202	*Trichoderma viride* 1% WP	M/s Sheer Agro Bio-Fertilizer Creative Organization, Purba Medinipur (WB)	9(3B)
203	*Trichoderma viride* 1% WP	M/s Central Bio Tech, Nagpur	9(3B)
204	*Trichoderma viride* 1% WP	M/s Indore Biotech Inputs & Research (P) Ltd, Indore	9(3B)

205	*Trichoderma viride* 1% WP	M/s Yash Krishi Takniki Ewam Vigyan Kendra, Allahabad	9(3B)
206	*Trichoderma viride* 1% WP	M/s K.N. Bio Tech Pvt Ltd., Hyderabad	9(3B)
207	*Trichoderma viride* 1% WP	M/s Modern Biotech, Jalgaon (MS)	9(3B)
208	*Trichoderma viride* 1% WP	M/s Amit Bio Tech. Kolkata	9(3B)
209	*Trichoderma viride* 1% WP	M/s Universal Agro Bio-Tech., Nohar (Rajasthan)	9(3B)
210	*Trichoderma viride* 1% WP	M/s Durva Biotech, Nagpur	9(3B)
211	*Trichoderma viride* WP	M/s Agriland Biotech Limited	9(3B)
212	*Trichoderma viride* 1.5% WP	M/s Romvijay Bio Tech Pvt. Ltd.	9(3B)
213	*Trichoderma viride* 1% WP	M/s Multiplex Agricare Pvt. Limitd	9(3B)
214	*Trichoderma viride* 1% WP	M/s Bio Tech. International Ltd.	9(3B)
215	*Trichoderma viride* 1% WP	M/s Margo Bio control Pvt Ltd.	9(3)
216	*Trichoderma viride* 1% WP	M/s K.N. Bio Science (India) Ltd.	9(3B)
217	*Trichoderma viride* 1% WP	M/s Pragathi Bio Fertilizers	9(3B)
218	*Trichoderma viride* 1% WP	M/s Sri Biotech, Hyderabad	9(3B)
219	*Trichoderma viride* 1% WP	M/s Vishva Mithra Bio Agro Pvt. Ltd., Guntur	9(3B)
220	*Trichoderma harzianum* 1% WP	M/s Vishwa Mithra Bio Agro Pvt. Ltd., Guntur	9(3B)
221	*Trichoderma harzianum* 1% WP	M/s Greentech Agro Services Pvt. Ltd., Coimbatore	9(3B)
222	*Trichoderma harzianum* 1% WP	M/s Sri Biotech, Hyderabad (formerly M/s Vermigreen Bio-fertilizer)	9(3B)
223	*Trichoderma harzianum* 1% WP	M/s Bio-pest Management Pvt. Ltd., Bangalore under	9(3B)
224	*Trichoderma viride* 1% WP	M/s Varsha Bio Science and Technology, Hyderabad	9(3B)

225	*Trichoderma viride* 1% WP	M/s Ecosense Labs Pvt. Ltd., Mumbai	9(3B)
226	*Trichoderma viride* 1% WP	Tamil Nadu Agricultural University- ICAR	9(3B)
227	*Trichoderma viride* 1.15% WP	M/s T.Stanes & Co.	9(3)
228	*Trichoderma viride* 1% WP	M/s Green Care Bio technologies & M/s Kaveri Agri Tech Bio Division,	9(3B)
229	*Trichoderma viride* 1.15% WP	M/s Sun Agro Industries Ltd.	9(3B)
230	*Trichoderma harzianum* 0.5% WP	M/s Pest Control India Pvt. Ltd.	9(3B)
231	*Trichoderma viride* 0.5% WP	M/s Pest Control India Pvt. Ltd.	9(3B)
232	*Trichoderma viride* 1% WP	M/s Crop Health Products Ltd.	9(3B)
233	*Trichoderma viride* 1% WP	M/s International Panacea Ltd., New Delhi	9(3B)
234	*Trichoderma viride* 1% WP	M/s Pushpanjali Agri Input Technologies Kurnool (AP)	9(3B)
235	*Trichoderma viride* 1%WP-	Maharashtra Manufacturers Association for Biocontrol Agents	9(3B)
236	*Trichoderma viride* 1%WP	Vidharbha Marathwada Agro Biotech Association	9(3B)
237	*Trichoderma viride* 1%WP	M/s Nafed Biofertilizer, Bharatpur (Rajasthan)	9(3B)
238	*Trichoderma viride* 1% WP	M/s Mitcon Consultancy Ltd., Pune	9(3B)
239	*Trichoderma viride* 1% WP	M/s Monarch Bio Fertilizers & Research Centre, Chennai	9(3B)
240	*Trichoderma viride* 1% WP	M./s Maa Bhagwati Biotech & Chemical	9(3B)
241	*Trichoderma viride* 1% WP	M/s K.N.S. Biotech, Degloor, Distt. Nanded (Maharashtra)	9(3B)
242	*Trichoderma viride* 1% WP	M/s Harit Bio-Control Lab., Yavatmal	9(3B)

243	*Trichoderma viride* 1% WP	M/s Sneha Biotech, Vijayawada	9(3B)
244	*Trichoderma viride* 1% WP	M/s Sigma Bio Laboratory, Nagpur	9(3B)
245	*Trichoderma viride* 1% WP	M/s Vaibavlaxmi Biocontrol Lab., Wardha (MS)	9(3B)
246	*Trichoderma viride* 1% WP	M/s Advance Bio-tech Industries and Research Input, Indore	9(3B)
247	*Trichoderma viride* 1% WP	M/s Jain Biotech, Nagpur	9(3B)
248	*Trichoderma viride* 1% WP	M/s Ocean Agro (India) Ltd., Nandesari, Baroda	9(3B)
249	*Trichoderma viride* 1% WP	M/s Samridhi Bioculture Pvt. Ltd., Indore	9(3B)
250	*Trichoderma viride* 1% WP	M/s Vidarbha Biotech Lab., Yauatmal	9(3B)
251	*Trichoderma viride* 1% WP	M/s Ecophila Biotech, Ahmednagar	9(3B)
252	*Trichoderma viride* 1% WP	M/s Om Agro Organic, Yavatmal	9(3B)

15. DEVELOPMENT OF MIXED FORMULATIONS OF BIOCONTROL AGENTS

A very important prerequisite for the possible use of antagonists in practical farming is that performance be consistent under varying and unpredictable weather conditions. Because of the different sensitivities of biocontrol agents vis-à-vis environment conditions, the use of antagonists mixtures could contribute to consistency of performance of antagonists preparations. There are four main approaches to achieve the goal-(1) select strains of biocontrol agent with wide host range, (2) modify the genetics of the biocontrol agent to add mechanisms of disease suppression that are operable against more than one pathogen, (3) alter the environment to favour the biocontrol agent and to disfavour competitive micro-flora, and (4) develop strain mixtures with superior biocontrol activity (220). Several strategies for developing mixtures of biocontrol agents could be envisioned including mixtures of antagonists with different mechanisms of disease suppression, mixtures of taxonomically different organisms, mixtures of organisms (antagonists) with different optimum temperature, pH, or moisture conditions for rhizosphere/ phyllosphere colonization. The mixture should be composed of organisms that could help each other to get a broader organism range and a higher level of protection against different phytopathogens. If one organism fails in any circumstances due to any unfavorable conditions, the other will be effective under these conditions. Moreover, these combinations might have a synergistic effect. The majority of strategies for biocontrol of soilborne plant pathogens rely on a single microbial biocontrol agent for pathogens. Unfortunately, biocontrol agents applied individually are not likely to perform consistently against all pathogens of the crop or under diverse rhizospheric and environmental conditions. An approach to overcoming this inconsistent performance is to include a combination of biocontrol agents in a single preparation. A combination of biocontrol agents is more likely to have a greater variety of traits responsible for suppression of one or more pathogens and also is likely to have these traits expressed over a wide range of environmental conditions. The long-term goal of our research is to develop individual conditions for management of important diseases of important crops and provide plant growth promotion.

Our result indicate that the consortium formulation of *Trichoderma* consisting of three potential strains of *T. harzianum* is capable of suppressing important diseases of sun flower, mustard, maize, chickpea, soyabean, caused by the fungal pathogens in diverse agro climatic zones more effectively (Unpublished data). We also studied compatibility among these isolates in disease suppression assays in rhizosphere coexistence assays. *T. harzianum* used alone or in combination also resulted in increase plant growth and overall yield parameters. Rhizosphere colonization pattern of *T. harzianum* strains used individually or in combination also differ significantly. Likewise, some bacterial strains have also found to be very effective in plant disease control and growth promotion (221,192). However, we have found in our studies that compatible fungal and bacterial antagonists prove more effective in management of plant disease and plant growth enhancement (222). This suggests that introduction of consortium formulation resulted in greater adaptability in soil in presence of soil microbial load or it can withstand biotic and abiotic stresses. The enhanced vigor in plants tested is also found associated with an increase in overall phonolic content, antioxidant activity, chlorophyll content, which may contribute to, improved lignifications and antioxidant response. Therefore, it is essential to investigate microbial interactions that enhance or detract from biocontrol to understand and predict the performance of mixtures of specific biocontrol agents.

It is likely that most cases of naturally occurring biological control results from mixtures of antagonists, rather that from high population of a single antagonist. For example, mixtures of antagonists are considered to account for protection in disease suppressive soils (223). Combinations of biocontrol agents for plant diseases include mixtures of fungi (224) and mixtures of fungi and bacteria (225). Most of these reports on mixtures of biocontrol agents showed that combining antagonists resulted in improved biocontrol. However, there also are reports of combinations of biocontrol agents that do not result in improved suppression of disease compared with the individual antagonists (226). Incompatibility of the co-inoculants can arise because biocontrol agents may also inhibit each other as well as the target pathogen or pathogens. Thus an important prerequisite for successful development of strain mixtures appears to be the compatibility of the co-inoculated microorganisms.

T. harzianum (TH) and *P. fluorescens* (PsF) exhibit better efficacy against soil-borne plant pathogens in acidic and neutral to alkaline soils, respectively. In a study, compatibility of 12 isolates of TH was tested against 41 strains of PsF under *in vitro* condition. In general, PsF suppressed the growth and sporulation of TH but 4 neutral combinations were identified. One of these combinations TH strain PBAT-43 and PsF strain PBAP-27 were used to develop mixed formulation (Plant Biocontrol Agent-3), which was equally or more effective than individual formulations both under green house and field conditions.

16. ANTAGONISTIC GENES FROM *TRICHODERMA* SPP. FOR DEVELOPING DISEASE RESISTANCE IN PLANTS

Trans-kingdom transfer of genes for biocontrol from *Trichoderma* to plants to enhance disease resistance was, for the first time, demonstrated by Lorito et al. (141). An endochitinase-encoding gene *ech42*, expressed in tobacco and potato provided near total protection against *Alternaria alternate, A. solani, B. cinerea* and *R. solani*. The high degree of broad-spectrum resis-tance was attributed to high fungitoxicity of *Trichoderma* chitinase, relative to the plant endogenous chitinases. This was followed by several reports on transfer of *Trichoderma* genes to plants. The genome of mycoparasites, which has evolved specifically to attack other pathogens but not plants, represents a potential source of powerful antifungal genes. Genes encoding a serine protease and endochitinase (*ThEn-42, ech42, Chit-42*) have been cloned and characterized (Hayes *et al.*, 1994 227). Recently, transgenic potato and tobacco plants expressing chitinase *Chit-42* from *Trichoderma* were shown to be highly tolerant or resistant to foliar pathogens (*Alternaria alternate*, *A. solani* and *B. cineria*) and to the soil-borne pathogen *R. solani* (141). On the basis of results obtained from this experiment they have pointed out that: (i) there was a very high level of chitinolytic activity in some transgenic lines; (ii) the endochitinase concentrated in the extracellular space had a strong antifungal activity; (iii) the *Trichoderma* enzyme is highly synergistics with antifungal pathogenesis-related proteins in plants, such as tobacco osmotin (228); and the *Trichoderma endochitinase* in plant tissues may release from the cell wall of invading fungal compounds that elicit the plant defense response (141).

Bolar et al. (229) produced transgenic apple resistant to *Venturia inaequalis* by expression of both an endochitinase and an exochitinase, and observed synergistic interaction. *T. harzinaum* endochitinase, transferred to broccoli produced transgenic plants resistant to *Alternaria* leaf spot (230). Liu and Yang (231) produced transgenic rice resistant to blast and sheath blight by expressing *ech42*, *nag70* and *gluc78*, in different combinations. A *T. virens* endochitinase gene *ech42* transferred to cotton enhanced resistance against *A. alternate* and *R. solani* (232). Noel et al. (233) introduced an endochitinase gene and hybrid poplar (*Populus nigra* XP, maximowiczii) by *Agrobacterium*-mediated transformation. Fifteen transgenic black spruce lines and six poplar lines were obtained. Northern hybridization analysis showed an increased accumulation of the transcript encoding the recombinant endochitinase gene in all the transgenic plants tested. Endochitinase activity 55-115 times the level of the control was detected in transformed poplar leaves. Embryogenic tissue of transgenic black spruce showed endochitinase activity two to eight times that of the non-transgenic line, despite stronger basal endogenous activity. *In vitro* assays using inoculated leaf disks demonstrated that the transgenic poplars had increased resistance to the leaf rust pathogen *Melampsora medusae.* Seedlings of transgenic spruce lined showed an increased resistance to the spruce root pathogen. These results suggest that constitutive expression of the *ech42* gene from *T. harzianum* could be exploited to enhance resistance to fungal pathogens in important forest tree species. Thus tree genetic engineering with endochitinase genes could provide an alternative to the use of fungicides and help reduce tree growth losses caused by phytopathogenic fungi. In a very recent report, expression of *T. harzianum* endochitinases to tobacco was shown to enhance resistance not only against pathogens, but also against abiotic stress, presumably through the release of some elicitors (de las Mercedes Dana et al. 2006). Contrary to these reports, expression of *T. atroviride ech 42* in transgenic alfalfa did not yield resistance against Phoma medicaginis var medicaginis, even though there was a 50-to-2650-fold greater chitinase activity in transgenic plants (234). The protection provided by expression of mycoparasitism-related genes in plants thus may not be universal.

Promoter analysis can be used to confirm molecular models of gene regulation deriving from studies carried out under various in vitro conditions. Published studies for the promoter regions of genes involved in biocontrol have focused on either promoter sequences or regulatory proteins. Some investigators have studied the promoter sequence of a gene in order to confirm the involvement of previously identified motifs in the regulation of its transcription under biocontrol conditions. Electromobility Shift Assays (EMSAs), *in vitro* footprinting, and /or promoter deletion analysis (235) are the techniques used. Regulatory proteins can influence gene transcription either directly (by binding to the promoter sequence or indirectly via signal transmission). The molecular tools are also used to inactivate genes coding for regulatory proteins. Peterbauer et al. (236) found that inactivation of the seb 1 gene does not modify transcription of the *nag 1*, *chit33*, and *ech42* genes. They also showed that other proteins can bind to the 5'-AGGGG-3' promoter motifs of *nag 1* and *ech42* in the disrupted strain. In *in vitro* studies, examined how inactivating two nitrogen activated protein kinases (MAPKs) affect the mycoparasitisic properties of *T. virens*. In many fungal species,

MAPK proteins participate in cascade signals involved, e.g. in plant parasitism. The role of two G-protein ά-subunits, *TgaA* and *TgaB*, in biocontrol by *T. virens* has been studied by Mukherjee et al. (170). G-proteins play an important role in intracellular signaling. They amplify receptor responses and influence the amplitude and duration of cellular signals. Using null-TgaA and null-TgaB strains, these authors showed that TgaA is involved in the biocontrol activity against *S. rolfsii*, but that neither TgaA nor TgaB is required for its activity against *R. solani*. The authors conclude that the involvement of G-proteins in biocontrol agent is in contact. Zeilinger et al. (237) have shown that the tga3 gene of *T. atroviride*, also coding for a G-proteins ά-subunit, is involved in this biocontrol agent's vegetative growth and mycoparasitic activity.

Zhou et al. (238) have recently used restriction enzymes mediated integration (REMI) techniques to construct mutants with improved cyanide-degradation ability from biocontrol fungus *T. koningii* strain T30. This successful insertional mutagenesis of the cyanide-biodegrading agent, *Trichoderma* spp., led to the creation of mutants with deficient and enhanced cyanide-degrading properties. Liu et al. (239) transformed three genes encoding for fungal cell wall degrading enzymes (CWDE), *ech42*, *nag70* and *glue78* from the biocontrol fungus *T. atroviride* into rice mediated by *Agrobacterium tumefaciens* singly and in all possible combinations. These results indicated that expression of several genes in one T-DNA region interfered with each other and expression of exogenous gene in recipient plant was a complex behaviour. It has been suggested that target gene must avoid being list in transgenic process so as to be sure of expressing in transgenic plants and on the other hand, gene breaking and segregation in transgenic process can be used to delete selective gene so as to enhance transgenic security. This approach and the biological materials thus obtained could find a variety of applications in the discovery and manipulation of genes and gene products from *Trichoderma*.

17. *TRICHODERMA* AS A MAJOR COMPONENT OF INTEGRATED DISEASE MANAGEMENT SYSTEM

An integrated pest management (IPM) system entails simultaneous or sequential use of several methods of control. Biological control is of particular interest as a component of, and can best be exploited within the framework of an IPM system. Biocontrol agents have distinct advantages in being compatible with most of the agricultural practices and hence, can be successfully utilized as a part of total crop management practices or broadly, as a component of the agro-ecosystem management. Different biocontrol agents have been integrated with cultural practices, soil solarization, fungicides and disease resistant varieties for managing different crop diseases. Combination of the seed/root application of *T. harzianum* or *P. fluorescens* with soil solarization was very effective in management of seed and seedling diseases of tomato, brinjal and capsicum in nursery at farmers' field. Wilt and root-rot complex of chickpea, lentil and pigeonpea were successfully managed by integration of *T. harzianum* or *T. virens* with carboxin (240). Integration of fertilizers or herbicides with biocontrol agents to control plant diseases has also been attempted. *Trichoderma* spp. are intensive to fungicides like carboxins, oxycarboxins, metalaxyl, tricyclazoles, carpropamid, host defense inducers (e.g. benzothiadiazole), etc. A necessary prerequisite for this type of combined treatment is that the antagonist should be resistant against fungicides (241,242). The use of *Trichoderma* together with fungicides may provide the basis for a system of integrated disease management in seedling production (Table 9). Currently, there is little knowledge on the combined effects of *Trichoderma* and fungicides. In general, the efficacy of the *Trichoderma* treatments has been found to be more variable and less effective than chemical fungicide treatments in some cases. However, we found that integration of *Trichoderma* with compatible fungicides reduces the corm rot disease of gladiolus significantly. The disease control was found to be more than when either the antagonists or fungicides were used alone (20). Apparently, biocontrol with *Trichoderma* has to be implemented with other disease control measures to reach a satisfactory level of disease control on food crops. Combining biocontrol agents with selective fungicides that are not inhibitory to the antagonist will help it to become established in the soil and achieve better disease control. We screened the compatibility of a number of commercially available pesticides with our strain, *T. harzianum* NBRI-1055 and found that fungicides such as Captaf, and Sulfex (243) and insecticides such as Hilmida and Confidor (244) were highly compatible and successful in integration trials.

Table 9: Seed treatment with *Trichoderma* and pesticides for major crops

Name of Crop	Pest / Disease	Seed Treatment
Sugarcane	Root rot, wilt	Carbendazim (0.1%); *Trichoderma* spp. @ 4.6 g/kg seed
Maize	Soil & seed borne diseases	*T. viride, T. harzianum* @ 4 g/kg seed
Groundnut	Stem rot, Seed rot, Seedling rot	Soil application of castor cake @ 1000kg/ha or Neem cake or Seed treatment with *T. viride* @ 4 g/kg seed
	White grubs	Chlorpyriphos/Quinalphos @ 2.5 to 1.2 ml/ kg seed treatment with *T. viride, T. harzianum* @ 4 g/kg seed
Cotton	Soil & Seed borne diseases	Acid delenting should be followed before sowing @ one liter commercial H_2SO_4 for 10 kg seed, *Trichoderma* spp. @ 4 gm/kg seed,
	Black gram	Captan 3g/kg seed, Carbendazim 2g/kg seed treatment with *T. harzianum* @ 4 g/kg seed
Rice	Root rot disease & other insects pests	*Trichoderma* 5-10g/kg seed (before transplanting) Chlorpyriphos 2.5 to 12 ml/kg seed or 3-10g/kg seed.
	Root knot nematode	Seed soaking in 0.2% solution of monocrotophos for 6 hours and treated with *T. harzianum* @ 4 g/kg seed
	White tip nematode	Seed soaking in 0.2% solution of monocrotophos plus *T. viride* @ 4 g/kg seed
Chillies	*Anthracnose* spp. *Pseudomonas* spp.	Seed treatment with *T. viride* 4g/kg, Carbendazin @ 1g/100 g seed
	Soil borne infection of fungal diseases	Captan 3g/kg seed, Carbendazim 2g/kg seed plus *T. viride, T. harzianum* @ 2g/kg seed.
Pigeon pea	Wilt Blight	Carbendazim 2g/kg seed plus *Trichoderma* spp. @ 4g/kg seed.
Pearl Millet (Bajra)	Soil borne disease	*T. harzianum T. viride* @ 4g/kg seed.
Maize	Soil borne disease	*T. harzianum T. viride* @ 4g/kg seed.

Sesame	Root rot disease, Seedling blight, Cercospora leaf spot, Dry root rot, *Alternaria* leaf spot, Bacterial blight, Bacterial leaf spot	*T. viride* @ 4g/kg seed; Seed treatment with Thiram 2.2-2.5g/kg; Agrimycin-100 (250ppm) or Streptocycline suspension 0.05%.
Sorghum	Soft borne	*T. harzianum T. viride* @ 4g/kg seed.
Pea	Root rot	Seed treatment with *Bacillus subtilis* or *Pseudomonas fluorescens* Soil application @2.5-5g/kg in 100 kg FYM
		Or
	White rot	Carbendazin or Captan 2g/kg seed, *T. harzianum* @4g/kg seed or Thiram + Carbendazium 2g/kg seed
Onion	Smut	*T. viride* @2g/100g seed. Benlate or Vitavax @0.01%.
Bhindi	Root knot nematode, Collar Rot	Carbosulfan (25 ST) @ 30% (w/w) *T. viride* @ 2g/100g seed
Tomato	Soil borne infection of fungal disease Early blight Damping off	*T. harzianum, T. viride* @ 2g/100g seed plus systemic fungicides
Coriander	Wilt	*T. harzianum, T. viride* @ 4g/100g seed plus fungicide
Brinjal	Soil borne infection of fungal disease	*T. viride, T. harzianum,* @ 2g/100g seed plus systemic fungicides
Cucurbits	Soil borne disease	*Trichoderma viride,* @ 2g/100g seed
Leguminous Vegetables	Soil borne infection	*T. viride,* @ 2g/100g seed
	Nematode	Carbofuran/Carbosulfan 3% (w/w)
Sunflower	Diseases	*T. harzianum, T. viride* @ 4g/100g seed
Soyabean	Seedling disease	*Rhizobium* spp. and phosphate solubilizing bacteria (PSB @ 5+5 g/kg seed)
Wheat	Termite	Treat the seed before sowing with any one of the following: Chlorpyriphos @4ml/kg seed or Endosulfan @7ml/kg seeds plus *T. harzianum* @ 4g/kg seed

Sorghum	Soil/seed borne diseases	Seed treatment with *T. harzianum T. viride* @ 4g/kg seed
Cruciferous	Soil/seed borne diseases	Seed treatment with *T. harzianum* , *T. viride* @ 2ml/100 kg seed
Gram	Root Knot nematode Lesion Nematode Wilt	Seed treatment with Combination of Carbendazim with Carbosulfan @ 0.2%, Carbendazim with Thiram 2.5g/kg
Potato	Soil and Tuber borne diseases	Seed treatment with boric acid 3% for 20 minutes before storage and *T. harzianum T. viride* @ 4g/kg seed
Citrus	Seed borne diseases	Use *T. viride, T. harzianum* with organic matters in the ratio of 1:40 @ 2 kg mixed culture/plant
Apple	Root rot & collar rot	Use *T. viride, T. harzianum* at later stage with fungicides

18. DEVELOPMENT OF RURAL TECHNOLOGY FOR PRODUCTION OF BIOPESTICIDE AT FARM SITES

The on farm / on sites production of *Trichoderma* inoculum, close to the site for application using cheap local resources can be the most appropriate method of biopesticides production and application at low cost (Singh et al., 108,109). Our group has popularized the mass production of *Trichoderma*-based bio-pesticides at village level. The farmers have been trained in the production of the fungal bioagents (*T. harzianum*) on locally available substraits like cow dung and different agriculture wastes. They have also been trained in the delivering method (seed treatment, seed biopriming and furrow treatment) of bioagents in field and a large chunk of farmers have been using the formulations developed by them for organic cultivation of crops.

19. CONCLUSION

Biological control involves the use of beneficial organisms, their genes, and/or products, such as metabolites, that reduce the negative effects of plant pathogens and promote positive responses by the plant. Disease suppression, as mediated by biocontrol agents, is the consequence of the interactions between the plant, pathogens, and the microbial community. *Trichoderma* spp. play major role as biocontrol agents, owing to their capabilities of ameliorating crop-yields by multiple role, such as biopesticide and plant growth promotion. Faster metabolic rates, anti-microbial metabolites, and physiological conformation are key factors which chiefly contribute to antagonism of these fungi. Mycoparasitism, spatial and nutrient competition, antibiosis by enzymes and secondary metabolites, and induction of plant defence system are typical biocontrol actions of these fungi. Information on the classification of the genus, *Trichoderma*, mechanisms of antagonism and role in plant growth promotion has been well documented. However, fast paced current research in this field should be carefully updated for the fool-proof commercialization of the fungi. In order to enhance marketability of these fungi as BCAs, feasible commercial production processes are of utmost importance. Pursuit for cheaper and alternative substrates and optimal operating parameters to increase conidia production is on, and several encouraging results are being reported by researchers worldwide. Thus, it is expected that in near future, exploitation of these interesting BCAs would be maximized. Area awaiting attention of scientists and policy makers include:

1. Biodiversity in *Trichoderma* strains.
2. Improvement of BCA *Trichoderma* by using molecular tools.
3. Improvement in mass multiplication, formulation, shelf-life and delivery system.
4. Integration of biocontrol with other management practices.
5. Bio-priming of seeds with *Trichoderma* spp. must be explored and exploited.
7. There is need to select *Trichoderma* strains, which are not only good antagonists but also inducers of good growth and plant defense.

Acknowledgemkents

The authors are greatful to Department of Agriculture, Government of U.P. and Department of Science and Technology (DST), New Delhi for providing financial assistance.

REFERENCES

1. Singh HB, Singh BN, Singh SP, Singh SR, Sarma BK. 2009. Biological control of plant diseases Current status and future prospects. In: *Recent Advances in Biopesticides (Biotechnological applications)*. (Ed. Johri JK.). New India Publishing Agency, New Delhi pp. 193-304.
2. Woo SL, Scala F, Ruocco M, Lorito M. 2006. The molecular biology of the interactions between *Trichoderma* spp., phytopathogenic fungi, and plants. *Phytopathology* 96:181–185.
3. Shores M, Yedidia I, Chet I. 2005. Involvement of jasmonic acid/ethylene signaling pathway in the systemic resistance induced in cucumber by *Trichoderma asperellum* T 203. *Phytopathology* 95:76-84.
4. Harman GE, Kubicek CP. 1998. *Trichoderma* and *Gliocladium*. Taylor & Francis, London, pp. 278.
5. Weindling R. 1932. *Trichoderma lignorum* as a parasite of other soil fungi. *Phytopathology* 22:834-845.
6. Henis Y, Adams PB, Papavizas GC, Lewis JA. 1982. Penetration of sclerotia of *Sclerotium rolfsii* by *Trichoderma* spp. *Phytophathology* 72:70–74.
7. Singh HB. 2006a. Achievements in biological control of diseases with antagonistic organisms at National Botanical Research Institute, Lucknow. In: Current Status of Biological Control of Plant Diseases using Antagonistic Organisms in India (eds, B. Ramanujan and R.J. Rabindra). Technical Document, Project Directorate of Biological Control, Bangalore, India, pp. 329-40.
8. Cole-Smith JR, Ghaffar A, Javed ZUR. 1971. The eeffect of dry conditions on subsequent leakage and rotting of fungal sclerotia. *Soil Biol. Biochem.* 6:307.
9. Henis Y, Gaffar A, Baker R.1978. Integratede control of *R. solani* damping off of radish and effect on successive plantings, PCNB and *Trichoderam* on pathogen and disease. *Phytopathology.* 68: 900-907.
10. Harman GE. 2000. Myths and dogmas of biocontrol: changes in perceptions derived from research on *Trichoderma harzianum* T-22. *Plant Dis.* 84:377–393.
11. Harman GE, Howell CR, Vitrevo A, Chet I, Lorito M. 2004. *Trichoderma* species- Opportunistic Avirulent Plant

symbionts. *Nat. Rev. Microbiol.* 2:43-56.

12. Lorito M, Woo SL, Scala F. 2004. Le biotecnologie utili alla difesa sostenibile delle piante: i funghi. Agroindustria 3:181–195.

13. Bisby GR. 1939. *Trichoderma viride* Pers. ex. Fries, and notes on *Hypocrea*. *Trans. Br. Mycol. Soc.* 33:149–168.

14. Druzhinina IS, Kubicek CP. 2005. Species concepts and biodiversity in *Trichoderma* and *Hypocrea*: from aggregate species to species clusters? *J. Zhejiang Univ. Sci.* 6 :100–112.

15. Druzhinina IS, Kopchinskiy AG, Kubicek CP. 2006. The first 100 *Trichoderma* species characterized by molecular data. *Mycoscience* 47: 55–64.

16. Samuels GJ, Dodd SL, Gams W, Castlebury LA, Petrini O. 2002. *Trichoderma* species associated with the green mold epidemic of commercially grown *Agaricus bisporus*. pp. 146–170.

17. Kredics L, Zsuzsanna A, Ilona D. Manczinger L, Kevei F, Elisabeth N. 2003. Clinical importance of the genus *Trichoderma*-a review. *Acta Microbiol. Immunol. Hung.* 50:105–117.

18. Baker KF, Cook RJ. 1974. Biological Control of Plant Pathogens, Freeman, San Francisco pp. 433.

19. Amsellem Z, Zidack NK, Quimby Jr, PC, Gressel J. 1999. Long-term dry preservation of viable mycelia of two mycoherbicidal organisms. *Crop Prot.* 18:643–649.

20. Yedidia I, Benhamou N, Kapulnik Y, Chet I. 2000. Induction and accumulation of PR proteins activity during early stages of root colonization by the mycoparasite *Trichoderma harzianum* strain T-203. *Plant Physiol. Biochem.* 38:863–873.

21. Singh HB, Rehman A, Srivastava S, Nautiyal CS. 2006. A synergistic composition useful as bionoculent. (US Patent Filed NF No. 0020NF2006/IN).

22. Singh US, Mishra DS, Varshney S, Zaidi NW, Khan T, Joshi D, Sharma R, Singh S, Singh N. 2003b. Potential and effectiveness of fungi and bacteria as biocontrol agents for plant disease management. In: Integrated Pest Management: Principles and application (In press).

23. Vidhyasekaran P, Rabindran R, Muthamilan M, Nayar K, Rajappan K, Subramanian N, Vasumathi K. 1997. Development of a powder formulation of *Pseudomonas fluorescens* for control of rice

blast. *Plant Pathol.* 46:291-297.

24. Thara KV. 1994. Biological control of rice sheath blight in India. Ph.D dissertation. University of Madras.

25. Sen B. 2003. *Aspergillus niger* – An effective antagonist for disease management. In: Proceeding of the group meeting on antagonistic organisms in plant disease management held at PDBC, Bangalore on 10-11th July 2003 pp. 109-124.

26. Singh HB, Kalra A, Patra NK, Kumar S, Pandey P, Khanuja SPS, Shasany AK. 2003. Process for the prepa-ration of novel growth media from distillation and other plant wastes for mass multiplication of bio-control fungi. (US Patent No. 6,511,821).

27. Aggarwal R, Singh DV, Srivastava KD. 2001. Wheat disease management through biological control agents. In: *Plant Pathology* (Eds. Pravin, Chandra, Trivedi), Pointer publishers, Jaipur, pp. 51-71.

28. Sanker P, Sharma RC. 1999. Effect of non-volatiole metabolites of *Trichoderma viride* isolates on *M. phaseolina* in maize. India Phytopathological Sc., 51st Annual Meeting and National Symposium on "Seed Health Care and Phytosanitation for Sustainable Agriculture", Feb. 17, 1999. IISR, Lucknow.

29. Sumitha R, Gaikward SJ. 1995. Checking *Fusarium* wilt of pigeonpea by biological means. *J. Soils and Crops* 5:163-165.

30. Upadhyay RS, Rai B. 1992. Wilt of pigeonpea. In: Plant Diseases of International Important. (Eds. Singh US, Mukhopadhyay AN, Kumar J, Chaube HS.). Vol. 1 Prentice Hall, New Jersey, pp 388-414.

31. Podile AR, Laxmi VDV. 1998. Seed bacterization with *Bacillus subtilis* AE 1 increases phenylalanine ammonia-lyase and reduced the incidence of *Fusarium* wilt in pigeonpea. *J.Phytopathol.* 146:255-259.

32. De RK, Chaudhry Naimuddin RG. 1996. Comparative efficacy of biocontrol agents and fungicides for controlling chickpea wilt caused by *F. Oxysporum* f.sp. *ciceri*. *Indian J. Agri. Sci.* 66 (6):370-373.

33. Prasad D, Rangeswaran R. 2001. Biological control of root and collar rot of chickpea caused by *S. rolfsii*. *Ann. Plant Prot. Sci.* 9:297-303.

34. Sarma BK, Singh BP, Mehta S, Singh HB, Singh UP. 2002. Plant growth promoting rhizobacteria-mediated alternation in phenolic profile of chickpea (*Cicer arietinum*)

infected by *Sclerotium rolfsii. J Phytopathol.* 150(4&5):277-282.

35. Vyas SC. 1994. Integrated biological and chemical control of dry root rot on soybean. *Indian J. Mycol. Pl. Path.* 24:132-134.
36. Raghuchander T, Samiappan R, Arjunan G. 1993. Biocontrol of Macrophomina root rot of mungbean. *Indian Phytopathol.* 46:379-382.
37. Sharma P. 2000. An integrated approach for the management of carnation wilt caused by *F. oxysporium* f. sp. *dianthi* (pril. And Del.) Snyd. And Hans. *New Botanist.* 27:143-150.
38. Ebenezar EG, Alice D, Sivaprakasam K. 1996. Biolgical control of damping-off disease in mustard caused by Pytium aphanidermatum. *J. Ecobiol.* 8(1):55-57.
39. Karthikeyan A. 1996. Effect of organic amendments, antagonist *Trichoderma viride* and fungicide on seed and collar rot of gdroundnut. *Plant Dis. Res.* 11(1):72-74.
40. Kishore GK, Pandey S, Podile AR, Rao JN. 2001. Biological control of rot of groundnut by *Trichoderma harzianum* and *T. viride. Int. Arachis Newsletter* 21:39-40.
41. Rajappan K, Raguchander T, Manickam K. 1996. Efficacy of UV induced mutants of *Trichoderma viride* against *Sclerotium rolfsii. Plant Dis. Res.* 11(1):97-99.
42. Muthamilan M, Jeyarajan R. 1996. Integrated management of Sclerotium root rot of groundnut involving *Tricho-derma harzianum, Rhizobium* and carbendazim. *Indian J. Mycol. Pl. Pathol.* 26(2):204-209.
43. Desai S, Thakur RP, Rao VP, Anjaiah V. 2000. Characteriz-ation of isolates of *Tricho-derma* for biocontrol potential against *Aspergillus flavus* infection in groundnut. Inter-national-Arachis-Newsletter, 20:57-59.
44. Ghewande MP, Desai S, Narayan P, Ingle AK. 1993. Integrated Management of foliar diseases of groundnut (*Arachis hypogaea* L.). *India. Inter. J. Pest Manag.* 39:375-378.
45. Pandey S, KIshore GK, Rao JN, Podile AR. 2003. Biological control of late spot of groundnut. In: Proceedings of Plant Pathology Held on 2-7th Febraury, 2003 at Christchurch, New Zealand, pp. 35.
46. Raoof MA, Mehtab Y, Rana K. 2003. Potential of biocontrol agents for the management of

castor grey mold, *Botrytis ricini* Godfrey. *Indian J. Plant Prot.* 31:124-126.

47. Parashanthi SK, Sreenivasa MN, Kulkarni S. 1997. Integrated management of root rot disease of safflower caused by *Rhizoctonia bataticola*. Environment and Ecology, 15(4): 800-802.

48. Gaikwad SJ, Kapgate DJ. 1990. Biological control of *Sclerotium* root rot of sesamum. *PKV Res.J.* 14(1):87-89.

49. Gupta IJ, Cheema HS. 1990. Effect of microsclerotia of Macrophomina phaseolina and seed dressers on gormination and vigoutr of fsesame seed. Seed Res. 18(2):169-172.

50. Sankar, P. Jeyarajan, R., 1996a. Seed treatment formulation of *Trichoderma* and *Gliocladium* for biological control of M. paseolina in sesamum. Indian Phytopathology, 49:148-151.

51. Sanker, P. and Jeyarajan, R. 1996b. Biological control of sesamum root rots by seed treatment with *Trichoderma* spp. and *Bacillus subtilis*. *Indian J. Mycol. Plant Pathol.* 26: 217-220.

52. Narayan A, Behira R. 2000. Antagonistic potential of Gliocladium virens and Streptoverticillium sp. on selected phytopathogenic fungi. In: Proceeding of Indian Phytopathological Society-Golden Jubilee, International conference on Integrated Plant Disease Management for Sustainable Agriculture, pp. 294-296.

53. Mukhopadhyay AN, Mukherjee PK. 1998. Biological control of plant diseases: Status in India. In:S.P.Singh and S.S. Husain (eds.), Biological Suppression of Phytopathogens, Nematodes and Weeds. Project Deirectorate of Biological Control, Bangalore.

54. Sheela J, Sivaprakashan K, Seetharaman K. 1995. Biological control of *Fusarium* wilt of egg plant. *Madras Agric. J.* 82:199-201.

55. Phookan AK, Chalia K. 2000. Biological control of Collar rot of brinjal caused by *Sclerotinia sclerotiorum*. In: Proceeding of Indian Phytopathological Society-Golden Jubilee, International conference on Integrated Plant Disease Management for Sustainable Agriculture, p. 411.

56. Das BC, Roy S, Bora LC. 2000. Biological seed treatment with *Trichoderma* spp. for management of damping off of cabbage caused by *Rhizoctonia solani*. In: Proceeding of Indian Phytopathological society-Golden Jubilee,

International Conference on "Integrated Plant Disease Management for Sustainable Agriculture. pp. 409-410.

57. Sharma P, Singh L, Adlakha D. 2001. Antagonistic potential of *Trichoderma* and *Aspergillus* spp. on *S. sclerotiorum* (Lib.) de Bary causing rots in cabbage and Cauliflower. Pesticides Information. 24:41-44.

58. Bhargava YR. 1993. Pleiotropic effect of seed pelleting in controlling seed borne fungal disease in *Brassica oleracea* var. *capitata. Adv. Plant Sci.* 6:62-65.

59. Kundu PL, Nandi B. 1993. Inhibition of Rhizoctonia solani by Streptomyces arenae and S. chibaensis in soil. *Uournal Mycopathol. Res.* 31(1):35-42.

60. Mukherjee PK, Upadhyay JP, Mukhopadhyay AN. 1989. Biological control of *Pythium* damping-off of cauliflower by *Trichoderma harzianum. J. Biol. Cont.* 3:119-124.

61. Mukherjee, PK, and Raghu, K. 1997. *Trichoderma* sp. as a microbial suppressive agent of i on vegetables. World Journal of Microbiology and Biotechnology, 13:497-499.

62. Sreenivasa MN. 1994. Biological deterrent activities of VA-mycorrhiza and *Trichoderma harzianum* on *Sclerotium rolfsii* at different P-levels in chilli. *Environ. Ecol.* 12(2):319-321.

63. Sharma P, Kulshrestha G, Gopal M, Chandra S, Mukherjee I, Singh SB, Pant S, Kadu LN, Singh V. 2003. Integrated management of chilli die back and anthracnose in Delhi region at presented in National Symposium IPM in Horticultural Crops for protection of Environment & Biodiversity (Feb 10-12, 2003) at Zonal Agricultural Research Station, Chandangaon, Chhindawara (M.P.) Won the best paper award.

64. Rajan FS, Vedamuthu PGB, Khader MDA, Jeyarajan R. 1991. Management of roo disease of fenugreek. *Southern Indian Hort.* 39(4):221-223.

65. Sarma YR. 2003. Recent trends in the use of antagonistic organisms for the disease management in spice crop. In: Proceeding of the group meeting on antagonistic organisms in plant disease management held at PDBC, Bangalore on 10-11th July 2003 pp. 49-73.

66. Gupta VP, 1998. 'Nursery-Guard': A bio-fungicide for nursery disease management. Indian Silk 36:8-9.

67. Harman GE, Chet I, Baker R. 1980. *Trichoderma hamatum* effects on seed and seedling disease in duced in radish and pea induced; by *Pythium* spp. and *Rhizoctonia solani*. Phytopatholoty, 70:1167-1172.

68. Arora RK. 1999. Evaluation of bioagents for control of soil and tuber diseases of potato. *Indian Phytopathology* 52:310.

69. Bari FMA, Mukhopadhyay AN. 1988. Biocontrol efficacy of *Trichoderma* spp. in controlling black scurf of potato. *Indian J. Mycol. Pl. Pathol. 19:92.*

70. Sharma P. 2002. Use of improved *Trichoderma harzianum* isolate against major soil, borne diseases of vegetable crops. In: 2nd International Agronomy Congress on Balancing Food and Environmental Security A Continuing Challenge, New Delhi, Nov. 26-30, 2002 Extended Summaries Vol. 2: pp 1014-1016.

71. Krishnamoorthy AS, Bhasakaran R. 1990. Biological control of damping-off disease of tomato caused by *Pythium indicum*. *J. Bio. Cont.* 4:52-54.

72. Varshney S, Chaube HS, Mishra DS, Sriwastava P. 1999. Evaluation of biocontrol agents against soilborne plat pathogens. In: National Symposium on "Challenges & Prospects of Plant Pathology in the Coming Millennium", held at Lucknow, Dec. 9-11, 1999.

73. Sharon E, Bar-Eyal M, Chet I, Herrera-Estrella A, Kleifeld O, Spiegel Y. 2001. Biological control of the root knot nematode *Meloidogyne javanica* by *Trichoderma harzianum*. *Phytopathology* 91:687-693.

74. Ramanujam B. 2000. Final Report of ICAR adhoc scheme on "Studies on *Ganoderma* affected palms and other perennial crops". CPCRI, Kasaragod.

75. Sarma YR, Anandaraj M, Vanugopalan MN. 1996. Biological control of diseases of spices. In: Biological control in Spices. Anandaraj M, Peter KV. (eds.) IISR, Calicut, Kerala pp. 1-19.

76. Bhai SR, Thomas J, Sarma YR, 1997. Biological of capsule rot of cardamom. Paper presented in International Conference on Integrated Plant Diseases Management for Sustainable Agriculture. pp 268. 10-15 Nov. 1997, New Delhi.

77. Ramanujam B, Nambiar KKN, Lyer R. 1997. Management of stem bleeding disease of coconut with fungicides and biocontrol agents. *J. Plant. Crops* 25(2):175-179.

78. Bhaskaran R. 1993. Integrated Management of basal stem rot disease of coconut. Indian Coconut Journal; 24:5-8.

79. Moosa H. 2002. Biocontrol of *Pythopthora palmivora* (Bitl.), the causal organism of bud rot of coconut palm (*Cocus nucifera*, L.). Ph.D. Thesis submitted to Mangalore University, Manglore, Karnataka, pp. 268.

80. Venkatasubbaiah P, Saffeulla KM. 1984. *Aspergillus niger* for biocontrol of *Rhizoctonia solani* of coffee seedlings. Trop. Pest Management, 30(4): 401-406.

81. Usman NM, Balakrishnan P, Sharma YR. 1996. Biocontrol of rhizome rot of ginger. *J. Plant. Crops* 24:184-191.

82. Shivapratap HR, Philip T, Sharma DD. 1996. In vitro antagonism of *Trichoderma* sp. against mulberry leaf spot pathogen, *Cercospora moricola*. *Indian J. Seri.* 35(2):107-110.

83. Gupta VP, Sarkar A. 1997. Biological control of nursery diseases of mulberry (*Morus alba* L.). International Conference on Integrated Plant Disease Management for Sustainable Agriculture, 10-15 Nov. 1997, New Delhi, India, Abst. P5A-032, pp. 207.

84. Jacob CK, Joseph A, Jayarathnam K, Joseph A. 1991. Effect of fungal antagonists on F phellinus noxius causing brown root disease of *Hevea*. *Indian J.Nat. Rubber Res.*4(2):142-145.

85. Gupta VP. 2000. Biological control of mulberry diseases-The current status, pp. 145-148. In: Dasgupta MK. (ed.), Diseases of Plantation Crops, Spices, Betelvine and Mulberry. Visa-Bharati Publications, Sriniketan, India.

86. Raguchander T, Shanmugam V, Samiyappan R. 2001. Biological control of Panama wilt disease of banana. *Madras Agric. J.* 87:320-321.

87. Sawant I. Sawant SS, Nayana KA. 1995. Biological control of *Phytopthora* root rot of coorg mandrians (*Citrus reticulata*) by *Trichoderma* species grown on coffee waste. *Indian J. Agric. Sci.* 65: 342-346.

88. Prakash O, Misra AK, Shukla SK. 2002. *Penicillium citrinum*, a potent pathogen against wilt disease of guava. Asian congress of Mycology and Plant Pathology, Dept. of Applied Botany, University of Mysore, October 104:118.

89. Shukla SK, Prasad B, Prasad D, Misra AK. 2003. A cheap substrate for the mass multiplication of bioagents of guava wilt. Zonal Conference

(East Zone) I.S.M.P.P. and Seminar on "Plant Diseases of National Importance with Special Reference to Guava Wilt and Mango Malformation", CISH, Lucknow, April 4-5, 2003: 54-55.

90. Majumdar VL, Pathak VN. 1995. Biological control of post harvwest diseases of guava fruits by Trichoderma sppp. *Acta Botanica Indica* 23:263-267.

91. Patil RK. 1992. Studies on Rhizopus rot and Botryodiploidia rot of Mango fruits. Ph.D. Thesis, Rajsthan Agril. University, Bikaner.

92. Chattopadhyay C, Sen B. 1996. Integrated management of *Fusarium* wilt of muskmelon caused by *F. oxysporum*. *Indian J. Mycol. Plant Pathol.* 26:162-170.

93. Gupta VP, Dandin SB. 2003. Achievements in biological control of diseases of mulberry with antagonistic organisms at Central Sericulture Research and Training Institute, Mysore. In: Proceeding of the group meeting on antagonistic organisms in plant disease management held at PDBC, Bangalore on 10-11 July 2003 pp. 319-326.

94. Gupta VP, 2001. Diseases of mulberry and their management pp. 130-164. In: Trivedi PC. (Ed.), Plant Pathology. Pointer Publishers, Jaipur, India.

95. Godara SL, Pathak VN. 2000. Integrated disease management of post harvest rots of sweet orange fruits. Increase in root and shoot growth Gopinathan, S. 1995. Biological control of *Rhizoctonia* sp. root rot of *Casuarina equisetifolia* seedlings by *Frankia* sp. strains. Biology and Fertility of soils, 20(4):221-225.

96. Waraitch, K.S. 1987. Efficacy of different fungicides and Trichoderma harzianum, Rifai applied as soil drench for the control of root rot of sugar beet (Beta vulgaris L.). Pesticides, 21(9): 37-38.

97. Furgo PA, Patil VH, Mandokhot FAM. 2000. Management of *Fusarium* wilt of watermelon. In: Proceeding of Indian Phytopathological Society-Golden Jubilee, International conference on Integrated Plant Disease Management for Sustainable Agriculture, pp. 713-714.

98. Mondal KK, Singh RP, Verma JP. 1999. Beneficial effects of indigenous cotton rhizobacteria on seed germinability, growth promotion and suppression of blight disease. *indian Phtyopath.* 52:228-235.

99. Laha GS, Verma JP. 1998. Role of fluorescent pseudom-

onads in the suppression of root-rot and damping off of cotton. *Indian Phytopathol.* 51:275-278.

100. Mongia D, Sheoraj 2000. Integrated management of root rot of cotton. In: proceeding of Indian Phytopathological Society-Golden Jubilee, International Conference on Integrated Plant Disease Management for Sustainable Agriculture. Indian Phytopathological Society, IARI, New Delhi, pp. 910-911.

101. Thennarasu R. 1997. Biological control of sett rot (*Ceratocystis paradoxa*) of sugarcane. M.Sc. (Ag.) Thesis, TNAU, Coimbatore, p 93.

102. Nallathambi P, Mophanraj D, Prasad NR. 1997. Biological control of seedling rot of sugarcane using *Trichoderma* and *Pseudomonas*. International Symposium on Integrated Management for Sustainable Agriculture, Nov. 10-15, New Delhi.

103. Sawant IS, Mukhopadhyay AN. 1991. Integration of metalaxyl with *Trichoderma harzianum* for the control of *Pythium* damping-off in sugarbeet. *Indian Phytopathol.* 43: 535-541.

104. Upadhyay JP, Mukhopadhyay AN. 1986. Biological control of *S. rolfsii* by *Trichoderma harzianum* in sugarbeet. Trop. *Pest Management* 32:215-220.

105. Saju KA. 2004. Factors affecting the biological control of *Phytopthora capsici* infections in black pepper (*Piper nigrum* L.). Ph.D. Thesis, University of Calicut, Calicut, pp. 232.

106. Beena B, Eapen SJ, Ramana KV. 2001. Antagonistic rhizobacteria of root knot nematodes infesting black pepper (*Piper nigrum* L.). National Congress on "Centenary of Nematology in India. December 5-7, 2001. IARI, New Delhi.

107. Thomas J, Vijayan AK. 2003. Achievements in biological control of diseases of spice crops with antagonistic organisms at Indian Cardamom Research Institute, Myladumpara. In: Proceeding of the group meeting on antagonistic organisms in plant disease management held at PDBC, Bangalore on 10-11th July 2003 pp. 249-253.

108. Singh HB, Singh A, Singh SP, Nautiyal CS. 2004. Commercialization of biocontrol agents: The necessity and its impact on agriculture. In: Eco-Agriculture with Bioaugmentation: An emerging concept [Eds. S.P. Singh and H.B. Singh]. Rohitashwa Printers, Lucknow, India.pp. 1-20.

109. Singh HB, Singh A, Nautiyal CS. 2002. Commercialization of biocontrol agents: problems and prospects. In: Frontiers of Fungal Diversity in Indian Subcontinent (Prof. Kamal Fetchrift). [Eds. Rao GP. et al.]. International Book Distributing Company, Lucknow, India, pp.847-861.

110. Kumar RN. 1993. Biological control of damping off disease in chir pine. *Indian J. Mycol. Pl. Pathol.* 23(3):237-240.

111. Singh HB, Kalra A, Patra NK. 2001. Potential for biological control of wilt and rot in mints by *Trichoderma harzianum* and *Gliocladium virens*. *Kavaka* 28&29:45-48.

112. Majumdar VL, Jat JR, Gour HN. 1996. Effect of biocontrol agents on the growth of *Macrophomina phaseolina*, the incitant of blight of moth bean. *Indian J. Mycol. Pl. Pathol.* 26(2):202-203

113. Sawant IS, Sawant SD, Ganapathy MM. 1995a. Collar rot of passion fruit caused by *Rhizoctonia solani* and its control. *Indian Phytopathol.* 48(2):202-205.

114. Singh HB, Srivastava S, Singh A, Katiyar RS. 2007b. Field efficacy of *Trichoderma harzianum* Rifai application on wilt disease of cumin by *Fusarium oxysporum* f. sp. *cumini*. *J. Biol. Contl.* 21:317-319.

115. Rifai MA. 1969. A revision of the genus *Trichoderma*. *Mycol. Pap.* 116 1–56.

116. Chet I, Benhamou N, Haran S. 1998. Mycoparasitism and lytic enzymes. In: *Trichoderma* and *Gliocladium*. Enzymes, Biological Control and Commercial Applications (Harman GE & Kubicek CP, eds.), pp. 153-172. Taylor & Francis Ltd, London.

117. Howell CR. 2006. Understanding the mechanisms employed by *Trichoderma virens* to effect biological control of cotton diseases. *Phytopathology* 96:178–180.

118. Thornton CR, Pitt D, Wakley GE, Tablot NJ. 2002 Production of a monoclonal antibody specific to the genus *Trichoderma* and closely related fungi, and its use to detect *Trichoderma* spp. in naturally infested composts. *Microbiology* 148:1263–1279.

119. Dennis C, Webster J. 1971a Antagonistic properties of species-groups of *Trichoderma* I. Production of non-volatile antibiotics. *Trans. Brit Mycol.Soc.* 57:25.

120. Dennis C, Webster J. 1971b Antagonistic properties of species-groups of *Trichoderma* III. Hyphal interactions

Trans. Brit. Mycol.Soc. 57:363.

121. Barak R, Elad Y, Merelman B, Chet I. 1985. A possible basis for specific recognition in *Trichoderma Sclerotium rolfsii* interaction. *Phytopathology* 75:458-463.

122. Elad Y. 1995. Mycoparasitism. In: *Pathogenesis and Host Specifiecity in plant diseases: Histopathologicsal, Biochemical, Genetic and Molecular Basis*. Eudaruotes, Vol II. Pergamon, Elsevier, Oxford, UK, pp. 289-307.

123. Inbar J, Chet I. 1994. *Microbiology*. 140:651-67.

124. Khan J, Ooka JJ, Miller SA, Madden LV, Hotinik HAJ. 2004. Systemic resistance induced by *Trichoderma hamatum* 382 in cucumber against *Phytophthora* crown rot and leaf blight. *Plant Disease* 88:280-286.

125. Benhamou N, Rey P, Picard K, Tirilly Y. 1999. Ultrastructural and cytochemical aspects of the interaction between the mycoparasite *Pythium oligandrum*, and soilborne plant pathogens. *Phytopathology* 89:506–517.

126. Kredics L, Zsuzsanna A, Szekeres A, Hatvani L, Manczinger L, Cs V. Erzs´ebet N. 2005. Extracellular proteases of *Trichoderma* species—a review. *Acta Microbiol. Immunol. Hung.* V52 169–184.

127. Pozo MJ, Baek JM, Garcia JM, Kenerley CM. 2004. Functional analysis of tvsp1, a serine protease-encoding gene in the biocontrol agent *Trichoderma virens. Fungal Genet. Biol.* 41 336–348.

128. Kubicek CP, Mach RL, Peterbauer CK, Lorito M. 2001. *Trichoderma*: from genes to biocontrol. *J. Plant Pathol.* 83:11–23.

129. Hoell IA, Klemsdal SS, Vaaje-Kolstad G, Horn SJ, Eijsink VGH. 2005. Overexpression and characterization of a novel chitinase from *Trichoderma atroviride* strain P1. *Biochim. Biophys. Acta* 1748:180–190.

130. Benıtez T, Rinco´ n, AM, Limo´ n, MC, Codo´ n, AC. 2004. Biocontrol mechanisms of *Trichoderma* strains. *Int. Microbiol.* 7:249–260.

131. Lorito M, Harman GE, Hayes CK, Broadway RM, Tronsmo A, Woo SL, Di Pietro A. 1993b. Chitinolytic enzymes produced by *Trichoderma harzianum*: Antifungal activity of purified endochitinase and chitobiosidase. *Phytopathology* 83:302-307.

132. Benfez LJ, Delgado-Jarana J, Rey M. 1998. Glucanoluytic and other enzymes and their

control. *In Trichoderma and Gliocladium Vol. 2*, ed. GE Harman, C.P.Kubicek, pp. 185-204, London: Taylor and Francis.

133. Elad Y, Chet I, Henis Y. 1982. Degradation of plant pathogenic fungi by *Trichoderma harzianum*. *Can. J. Microbial*. 28:719-725.

134. Geremia RA, Goldman GH, Jacobs D, Ardiles W, Wila SB, Van Montagu M, Herrera-Estrella A. 1993. Molecular characterization of the proteinase encoding gene *prb 1*, related to mycoparasitism by *Trichoderma harzianum*. *Mol. Microbial*. 8:603-613.

135. Elad Y, Kapat A. 1999. Role of *Trichoderma harzianum* protease in the biocontrol of *Botrytis cineria*. *Eur. J. Plant Pathol*. 105:177-189.

136. Ait-Lahsen H, Soler A, Rey M, de La Cruz, Monte E, Llobell A. 2001. An antifungal Exo-a-1,3-glucanase(AGN13.1) from the biocontrol agent *Trichoderma harzianum*. *Appl. Env. Microbiol*. 67:5833-5839.

137. Lorito M, Di Pietro A, Hayes CK, Woo SL, Harman CE. 1993a. Antifungal, synergistic interactions between chitinolytic enzymes from *Trichoderma harzianum* and *Enterobacter cloacae*. *Phytopathology* 83:721-728.

138. Di Pietro A, Lorito M, Hayes CK, Broadway RM, Harman GE. 1993. Endochitinase from *Gliocladium virens*: Isolation, characterization, and synergistic antifungal activity in combination with gliotoxin. *Phytopathology* 83:308-313.

139. 139Baek JM, Howell CR, Kenerley CM, 1999. The role of extracellular chitinases from *Trichoderma virens* Gv 29-8 in the biocontrol *of Rhizoctonia solani*. *Curr. Genet*. 35:41-50.

140. Woo SL, Donzelli B, Scala F, Mach R, Harman GE, Kubicek CP, Del Sorbo G, Lorito M. 1999. Disruption of the ech42 (endochitinase-encoding) gene affects biocontrol activity in *Trichoderma harzianum*. Pl. *Mol. Plant-Microbe Interact*. 12:419-429.

141. Lorito M, Woo SL, Fernandez IG, Collucci G, Harman GE, PintorToros JA, Filippone E, Muccifora S, Lawrence CB, Zoina A, Tuzun S, Scala F. 1998. Genes from mycoparasitic funge as a source for improving plant resistance to fungal pathogens. *Proc. Natl. Cad. Sci*. USA 95:7860-7865.

142. Bolar JP, Norilli JL, Wong KW, Hayes CK, Harman GE, Aldwinckle HS. 2000. Expression of endochitinase from *Trichoderma harzianum* in transgenic apple increases

resistance to apple scab and reduces vigor. *Phytopathology* 90:72-77.

143. Ghisalberti EL, Sivasithamparam K. 1991. Antifungal antibiotics produced by *Trichoderma* spp. *Soil Biol. Biochem*. 213:1011–1020.

144. Sivasithamparam K, Ghisalberti EL. 1998. Secondary metabolism in *Trichoderma* and *Gliocladium*. In: Harman GE, Kubicek CP. (Eds.), *Trichoderma* and *Gliocladium*, Vol. 1. Taylor and Francis Ltd., London, pp. 139–191.

145. Demain AL, Fang A. 2000. The natural functions of secondary metabolites. *Adv. Biochem. Engin. / Biotechnol.* 69:1–39.

146. Ghisalberti EL, Narbey MJ, Dewan MM, Sivasithamparam K. 1990. Variability among strains of *Trichoderma harzianum* in their ability to reduce take-all and to produce pyrones. *Plant Soil* 121:287–291.

147. Rebuffat S, El Hajji M, Hennig, P., Davoust, D., Bodo, B., 1989. Isolation, sequence and conformation of seven trichorzianines B from *Trichoderma harzianum*. International Journal of Peptide and Protein Research 34, 200–210.

148. Vinale F, Marra R, Scala F, Ghisalberti EL, Lorito M, Sivasithamparam K. 2006. Major secondary metabolites produced by two commercial *Trichoderma* strains active against different phytopathogens. *Letters Appl. Microbiol.* 43, 143–148.

149. Howell CR, Stipanovic RD, Lumsden RD. 1993. Antibiotic production by strains of Gliocladium virens and its relation to the biocontrol of cotton seedling diseases. *Biol. Sci. Technol.* 3:435–441.

150. Howell CR, Stipanovic RD. 1983. Gliovirin, a new antibiotic from Gliocladium virens, and its role in the biological control of Pythium ultimum. *Canadian J. Microbiol.* 29:321–324.

151. Chet I, Inbar J, Hadar I. 1997. Fungal antagonists and mycoparasites. In: Wicklow, D.T., So¨ derstro¨ m, B. (Eds.), The Mycota IV: Environmental and Microbial Relationships. Springer, Berlin, pp. 165–184.

152. Howell CR. 1998. The role of antibiosis in biocontrol. In: Harman GE, Kubicek CP. (Eds.), *Trichoderma* and *Gliocladium*. Enzymes, Biological Control and Commercial Application, vol. 2. Taylor and Francis Ltd., London, pp. 173–183.

153. Howell CR. 2002. Cotton seedling preemergence damping-off incited by *Rhizopus oryzae* and *Pythium* spp and its biological control with *Trichoderma* spp. *Phytopathology* 92:177–180.

154. Howell CR, Hanson LE, Stipanovic RD, Puckhaber LS. 2000. Induction of terpenoid synthesis in cotton roots and control of *Rhizoctonia solani* by seed treatment with *Trichoderma virens*. *Phytopathology* 90:248-252.

155. Hanson LE, Howell CR. 2004. Elicitors of plant defense responses from biocontrol strains of *Trichoderma virens*. *Phytopathology* 94:171–176.

156. Lorito M, Farkas V, Rebuffat S, Bodo B, Kubicek CP. 1996b. Cell wall synthesis is a major target of mycoparasitic antagonism by *Trichoderma harzianum*. J. Bacteriol. 178:6382–6385.

157. Fogliano V, Ballio A, Gallo M, Woo SL, Scala F, Lorito M. 2002. *Pseudomonas* lipodepsipeptides and fungal cell wall-degrading enzymes act synergistically in biological control. *Mol. Plant–Microbe Inter.* 15:323–333.

158. Celar F. 2003. Competition for ammonium and nitrate forms of nitrogen between some phytopathogenic and antagonistic soil fungi. *Biol. Control* 28:19–24.

159. Blakeman JP. 1978. Microbial competition for nutrients and germination of fungal spores. *Ann. Appl. Biol.* 89:151–155.

160. Sivan A, Chet I. 1986. Biological control of *Fusarium* spp. in cotton, wheat and muskmelon by *Trichoderma harzianum*. *J. Phytopathol.* 116:39–47.

161. Srinivasan U, Staines HJ, Bruce A. 1992. Influence of media type on antagonistic modes of *Trichoderma* spp. against wood decay basidiomycetes. Mater. Org. 27:301–321.

162. Brewer MT, Larkin RP. 2005. Efficacy of several potential biocontrol organisms against *Rhizoctonia solani* on potato, *Crop Prot*. 24:939–950.

163. Hutchinson CM. 1999. *Trichoderma virens*-inoculated composted chicken manure for biological weed control. *Biol. Control* 16:217–222.

164. Hanson LE, Howell CR. 2002. Biocontrol efficacy and other characteristics of protoplast fusants between *Trichoderma koningii* and *T. virens*. *Mycol. Res.* 106:321–328.

165. Schirmb¨ock M, Lorito M, Wang YL, Hayes CK, Arisan-Atac I, Scala F, Harman GE, Kubicek CP. Parallel forma-

tion and synergism of hydrolytic enzymes and peptaibol antibiotics, molecular mechanisms involved in the antagonistic action of *Trichoderma harzianum* against phytopathogenic fungi, 1994, pp. 4364–4370.

166. Gullino ML. 1992. Control of *Botrytis* rot of grapes and vegetables with *Trichoderma* spp. In: Tjamos EC, Papavizas GC, Cook RJ. (Eds.), Biological Control of Plant Diseases, Progress and Challenges for the Future. Plenum Press, New York, pp. 125–132.

167. Sivan A, Chet I. 1989. The possible role of competition between *Trichoderma harzianum* and *Fusarium oxysporum* on rhizosphere colonization. *Phytopathology* 79:198–203.

168. Bae YS, Knudsen GR. 2005. Soil microbial biomass influence on growth and biocontrol efficacy of *Trichoderma harzianum*. *Biol. Cont.* 32:236–242.

169. de Boer W, Verheggen P, Klein Gunnewiek PJ, Kowalchuk GA, van Veen JA. 2003. Microbial community composition affects soil fungistasis. *App. Environ. Microbiol.* 69, 835–844.

170. Mukherjee PK, Latha J, Hadar R, Howeitz BA. 2004. Role of two G-protein alpha subunits, TgaA and TgaB, in the antagonism of plant pathogens by *Trichoderma virens*. *Appl. Env.Microbiol.* 70:542-549.

171. Leong J. 1986. Siderophores: Their biochemistry and possible role in the biocontrol of plant pathogens. *Annu. Rev. Phytopathol.* 24:187-209.

172. Weindling R. 1941. Experimental consideration of the mould toxin of *Gliocladium* and *Trichoderma*. *Phytopathology* 31:991-1003.

173. Rocha-Ramirez V, Omero C, Chet I, Horwitz BA, Hererra-Estrella A. 2002. *Trichoderma atroviride* G-protein alpha subunit gene tga1 is involved in mycoparasitic coiling and conidiation. *Eukaryot Cell* 1:594-605.

174. Reithner B, Brunner K, Schumacher R, Peissl I, Seidl V, Krska R, Zeilinger S. 2005. The G protein alpha subunit Tga1 of *Trichoderma atroviride* is involved in chitinase formation and differential production of antifungal metabolites. *Fungal Genet. Biol.* 42:749-760.

175. Mendoza-Mendoza A, Pozo MJ, Grzegorski D, Martinez P, Garcia JM, Olmedo-Monfil V, Cortes C, Kenerley C, Herrera-Estrella A. 2003. Enhanced biocontrol activity of Trichoderma through ina-

ctivation of a mitogen-activated protein kinase. *Proc. Natl. Acad. Sci.* USA 100:15965-15970.

176. Mukherjee M, Mukherjee PK, Kale SP. 2007 cAMP signaling is involved in growth, germination, mycoparasitism and secondary metabolism in *Trichoderma virens. Microbiology* 153:1734-1742.

177. Reithner B, Brunner K, Schumacher R, Stoppacher N, Pucher M, Brunner K, Zeilinger S. 2007. Signaling via the *Trichoderma atroviride* mitogen-activated protein kinase Tmk1 differentially affects mycoparasitism and plant protection. *Fungal Genet. Biol.* 44:1123-1133; doi:101016/jfgb20704001.

178. Yedidia I, Benhamou N, Chet I. 1999. Induction of defense responses in cucumber plants *(Cusumis sativus* L.) by the biocontrol agent *Trichoderma harzianum Appl. Environ. Microbiol.* 65:1061-1070.

179. Vinale F, D'Ambrosio G, Abadi K, Scala F, Marra R, Turra` D, Woo SL, Lorito M. 2004. Application of *Trichoderma harzianum* (T22) and *Trichoderma atroviride* (P1) as plant growth promoters, and their compatibility with copper oxychloride. *J. Zhejiang Uni. Sci.* 30, 2–8.

180. Cutler, HG, Himmetsbach DS, Arrendale RF, Cole PD, Cox RH. 1989. Koninginin A: a novel plant regulator from *Trichoderma koningii. Agric. Biol. Chem.* 53:2605–2611.

181. Brenner ML. 1981. Modern methods for plant growth substance analysis. *Ann. Rev. Pl. Physiol.* 32:511–538.

182. Chaube HS, Pundhir VS. 2005. Crop diseases and their management. Prentice-Hall of India, New Delhi pp. 233-264.

183. Kalra A, Singh HB, Pandey R, Patra NK, Katiyar N, Jupta ML, Dhawan OP, Kumar S.2002. Strains of *Trichoderma harzianum* useful as nematode inhibitor, fungicide and plant growth promoter and a process for the isolation thereof.(USPatent No.6,475,772).

184. Yedidia I, Srivastava AK, Kapulnik Y, Chet I. 2001. Effect of *Trichoderma harzianum* on micro element concentration and increased growth of cucumber plants. *Plant Soil* 235:235-242.

185. Samuels GJ, Lieckfeldt E, Nirenberg HI. 1999. Description of *Trichoderma asperellum* sp. Nov. and comparison of *T. viride. Sydowia* 51:71-88.

186. Alfano G, Lewis Ivey ML, Cakir C, Bos JIB, Miller SA, Madden LV, Kamoun S,

Hoitink HAJ. 2007. Systemic modulation of gene expression in tomato by *Trichoderma hamatum* 382. *Phytopathology* 97:429–437.

187. Viterbo A, Harel M, Horwitz BA, Chet I, Mukherjee PK. 2005. *Trichoderma* MAP-kinase signaling is involved in induction of plant systemic resistance. *Appl. Environ. Microbiol.* 71:6241-6246.

188. Zeinlinger S, Galhaup C, Payer K, Woo SL, Mach RL, Fekete C, Lorito M, Kubicek CP. 1999. Chitinase gene expression during mycoparasitic interaction of *Trichoderma harzianum* with its host. *Fungal Genet. Biol.* 26:131-140.

189. Walters DR, Newton AC, Lyon GD. 2005. Induced resistance: helping plants to help themselves. Biologist 52:28-33.

190. De Meyer G, Bigirimana J, Dlad Y, Hofte M. 1998. Induced systemic resistance in *Trichoderma harzianum* biocontrol of *Botrytis cineria, Eur.J. Plant Pathol.*, 107:279-286.

191. Lanzuise S, Ruocco YF, Sallenave-Namont C, et al. 2004. Cloning of fist ABC transpoter encoding gene from *Trichoderma* spp. and its expression during stress and mycoparasitism 11th SIPAV Annual Metteing Milan, September 29-October 1st 2004, Abst. *J Plant Pathol.* 86:322.

192. Lavania M, Chauhan P.S, Chauhan SVS, Singh HB, Nautiyal CS. 2006. Induction of plant defense enzymes and phenolics by treatments with plant growth promoting rhizobacteria *Serratia marcescens* NBRI 1213. *Curr. Microbiol.* 52:363-368.

193. Liu SD, 1988. The uses of benomyl rasistance mutant of *Trichoderma koningii* as a biocontrol agent against root rot disease of chrysanthemum and adjuki bean. *Trichoderma News Letter*, 4:5.

194. Mukherjee PK, Haware MP, Raghu P. 1997. Induction and evalutation of benomyltolerant mutants of *Trichoderma viride* for biological control of botrytis gray mould of chickpea. *Indian Phytopathol.* **50**:485-489.

195. Troutman JL, Matezka JC. 1978. Induced tolerance of *Trichoderma viride* to benomyl. *Phytopathological News,* 12:131(abstr.)

196. Papavizas GC. 1987. Genetic manipulation to improve the effectiveness of biocontrol fungi for plant disease control. In: *Innovative Approaches to Plant Desease Control.* ed.

Chet, I.) Wiley Interscience, N.Y.

197. Mukherjee PK, Mukhopadhyay AN. 1993. Induction of stable mutants of *Gliocladium virens* by gamm-irradiation. *Indian Phytopathol.* **46**:393-397.

198. Rajappan K.1997. Induction of mutants in *Trichoderma viride* by UV-irradiation. *Plant Dis. Res.* 12(1):1-5.

199. Kumar A, Gupta JP. 2000. Variation in the enzymatic activity of tebuconazole tolerant biotupes of *Trichoderma viride. In:* Proceeding of Indian Phytopathological Society-Golden Jubilee, International Conference on *"Integrated Plant Disease Management for sustainable Agriculture"*, pp391-392.

200. Stasz TE, Harman GE, Weeden NF. 1988. Protoplast preparation and fusion in two biocontrol strains of *Trichoderma harzianum. Mycologia* 80:141-150.

201. Paulitz TC, Belanger RR. 2001. Biological control in greenhouse systems. *Ann. Rev. Phytopathol.* **39**:103-135.

202. De la Cruz, J, Pinor-Toro JA, Benitez T, Llobell A, Romero LC. 1995. A novel endo-*b*-1, 3-glucanase, BGN 13.1. Involved in the mycoparasitism of *Trichoderma harzianum. J. Bacteriol.*, 177:6937-6945

203. Limon MC, Lora JM, Garcia I, De la Cruz, J, Llobell A, Benifez T., Pintor-Toro JA. 1995. Primary structure and expression of 33kDa chitinase gene from the mycoparasitic fungus *Trichoderma harzianum. Curr. Genet.* 28:478-483.

204. Harman GE, Bjoorjman T. 1998. Potential and existing uses of *Trichoderma* and *Gliocladium* for plant disease control and plant growth enhancement. In: *Trichoderma* and *Gliocladium*, Vol.2. (eds, G.E. Harman and C.K.Kubicek,) Taylor and Francis Ltd., London. pp 229-265.

205. Howell CR. 2003. Mechanism employed by *Trichoderma* species in the biological control of plant diseases: The history and evolution of current concepts. *Plant Desease,* 87:4-12.

206. Flores A, Chet I, Herrera-Estrella, A. 1997. Improved Biocontrol activity of *Trichoderma harzianum* by overexpression of the proteinase encoding gene *prb 1. Curr Genet.* 31:30-37.

207. Baker KF, Cook RJ. 1982. Biological control of plant pathogens. APS Press, St. Paul, Minn. Beagle-Risataino JE and Papavizas GC (1985). Biological control of *Rhizoc-*

tonia stem canker and black scruf of protein. *Phytopathology* 75:560-564.

208. Singh A, Srivastava S, Singh HB. 2007a. Effect of substrates on growth and shelf life of *Trichoderma harzianum* and its use in biocontrol of diseases. *Biores. Technol.* 98:470-473.

209. Papavizas GC. 1985. *Trichoderma* and *Gliocladium*: Biology and potential for biological control. *Annu. Rev. Phytopathol.*, 23:23-54

210. Jin X, Harman GE, Taylor AG. 1991. Conidial biomass and desiccation tolerance of *Trichoderma harzianum* produced at different medium water potentials. *Biol. Control* 1:237-243.

211. Lewis JA, Papavizas GC. 1983. Production of chlamydospores and conidia by *Trichoderma* spp. in liquid and solid growth media. *Soil Biol. Biochem.* 15:351-357.

212. Elad Y, Kirshner B. 1993. Survival in the phylloplane of an introduced biocontrol agent (*Trichoderma harzianum*) and populations of the plant pathogen *Botrytis cinerea* as modified by abiotic conditions. *Phytoparasitica* 21(4):303-313.

213. Krishna C. 2005. Solid state fermentation systems-an over view. *Crit. Rev. Biotechnol.* 25(1-2):1-30.

214. Tewari AK. 1996. Biological control of chick pea wilt-complex using different formulations of *Gliocladium virens* through seed treatment. Ph.D. Thesis, G.B. Pant University of Agric. & Tech., Pantnagar (India), p167

215. Mukherjee PK, Mukhopadhyay AN.1995. *In situ* mycoparasitism of *Gliocladium virens* on *Rhizoctonia solani*. *Indian Phytopathol.*, 48(1): 101-102.

216. Mishra DS, Singh US, Dwivedi TS. 2001. Comparative efficacy of normal seed treatment and seed bio-priming with commercial formulations of *Trichoderma* spp. In: 53rd annual meeting of Indian Phytopathological Society and National symposium on "Ecofriendly approaches for plant disease management" held at Chennai, 21-23, 2001.

217. Baby UI, Manibhushanrao K. 1996. Fungal antagonists and VA mycorrhizal fungi for biocontrol of *Rhizoctonia solani*, the rice sheath blight pathogen. In: *Recent Developments in Biocontrol of Plant Pathogens.* (Eds. K. Manibhushan Rao and A. Mahadevan).pp.1-9

218. Prasad RD, Rangeshwaran R.2000. Shelf life and bioeffic-

acy of *Trichoderma harzianum* formulated in various carrier materials. *Plant Dis. Res.*, 15:38-42.

219. Haware MP, Lenne JM, Mukherjee PK. 1996. Integrated manage-ment of *Botrytis* gray mold of chickpea. Proceedings of XI International *Botrytis* symposium 23-27 June, Wageningen pp. 98.

220. Janisiewicz WJ. 1988. Biocontrol of postharvest diseases of apples with antagonists mixtures. *Phytopathology,* 78:194-198

221. Nautiyal CS, Mehata S, Singh HB. 2006a. Biological control and plant growth promotion by *Bacillus* strains from milk. *J. Microbiol Biotechnol* 16(2):184-192

222. Nautiyal CS, Mehata S, Singh HB, Pushpangadan P. 2006b. Synergistic bioinoculant composition comprising bacteria strains of Accession Nos. NRRL B-30486, NRRL B - 30487 and NRRL B-30488 and a method of producing said composition thereof. USPatent No.71097,830B2.

223. Chaube HS, Singh US. 1991. Plant Disease Management: Principles and Practices. CRC Press, Boca Raton, USA.

224. Datnoff LE, Nemec S, Pernezny K. 1995. Biological control of *Fusarium* crown and root rot of tomato in Florida using *Trichoderma harzianum* and *Glomus* intraradices. *Biol. Control.*, 5:427-431.

225. Guetsky R, Shtienberg D, Elad Y, Dinoor A. 2001. Combining biocontrol agents to reduce the variability of biological control . *Phytopathology*, 91: 621 -627.

226. Dandurand LM, Knudsen GR.1993. Influence of *Pseudomonas fluorescens* on hyphal growth and biocontrol activity of *Trichoderma harzi-anum* in the spermosphere and rhizosphere of pea. *Phytopathology,* 83: 265-270.

227. Hayes CK, Klemsdal S, Lorito M, Di Pietro A, Peterbauer C, Nakasa JP, Tronsmo A, Harman GE. 1994. Isolation and sequence of an endochitinase-encoding gene from a cDNA library of *Trichoderma harzianum. Gene,* 138:143-148.

228. Lorito M, Woo SL, D¢ Ambrosio M, Harman GE, Hayes CK, Kubicek CP, Scala F. 1996. Synergistic interaction between cell wall degrading enzymes and membrane affecting compounds. *Mol Plant Microbe Interact* 9:206–213

229. Bolar JP, Norelli JL, Harman GE, Brown SK, Aldwinckle HS. 2001. Synergistic activity

of endochitinase and exochitinase from *Trichoderma atroviride* (*T. harzianum*) against the pathogenic fungus (*Venturia inaequalis*) in transgenic apple plants. *Transgenic Res.* 10:533–543.

230. Mora A, Earle ED. 2001. Combination of *Trichoderma harzianum* endochitinase and a membrane-affecting fungicide on control of *Alternaria* leaf spot in transgenic broccoli plants. *Appl. Microbiol. Biotechnol.* 55:306–310.

231. Liu PG, Yang Q. 2005. Identification of genes with a biocontrol function in *Trichoderma harzianum* mycelium using the expressed sequence tag approach. *Res. Microbiol.* 156:416-423

232. Emani C, Garcia JM, Lopata-Finch E, Pozo MJ, Uribe P, Kim DJ, Sunil kumar G, Cook DR, Kenerley CM, Rathore KS. 2003. Enhanced fungal resistance in transgenic cotton expressing an endochitinase gene from *Trichoderma virens*. *Plant Biotechnol. J.* 1:321-336.

233. Noe¨l A, Levasseur C, Le VQ, Seguin A. 2005. Enhanced resistance to fungal pathogens in forest trees by genetic transformation of black spruce and hybrid poplar with a *Trichoderma harzianum* endochitinase gene. *Physiol. Mol. Plant Pathol.* 67(2):92–99.

234. Samac DA, Tesfaye M, Dornbusch M, Dornbusch M, Saruul P. Temple SJ. 2004. A comparison of constitutive promoters for expression of transgenes in alfalfa (*Medicago sativa*). *Transgenic Res.* 13:349-361.

235. Peterbauer C, Brunner K, Mach RL, Kubicek CP. 2002a. Identification of the N-acetyl-D-glucosamine-inducible element in the promoter of the *Trichoderma atroviride* bag1 gene encoding N-acetyl-glucosamidase. Mol. Genet. Genomics 267:162-170.

236. Peterbauer C, Litscher D, Kubicek CP. 2002b. The *Trichoderma atroviride seb1* (stress response element binding) gene encodes an AGGGG-binding protein which is involved in the response to high osmolarity stress. *Mol. Genet. Genomics* 268:223-231.

237. Zeilinger S, Reithner B, Scala V, Piessl I, Lorito M, Mach R. 2005. Signal transduction by Tga3, a novel G protein alpha subunit of *Trichoderma atroviride. Appl. Environ. Microbiol.* 71:1591-1597.

238. Zhou X, Zu S, Liu L, Chen J. 2007. Degradation of cyanide by *Trichoderma* mutants constructed by restriction enzyme mediated integration (REMI). Bio. Res. Technol. 98:2958-2962.

239. Liu M, Zhu J, Sun Z. Xu T. 2007. Possible suppression of exogenous β-1,3-glucanase gene gluc 78 on rice transformation and growth. *Plant Sci*. 172:888-896.

240. Mukhopadhyay AN. 1994. Biological control of Soil borne fungal plant Pathogens-current status, future prospects & potential limitations. *Indian Phytopathol.,* 47:119-126.

241. Csino AS, Bel DK, Minton NA, Wells HD. (1983). Evaluation of *Trichoderma* spp., fungicides and chemical combinations for control of southern stem rot of peanuts. *Peanut Science* 10:75-79.

242. El-Mohamedy RSR. 2004. Integration between biological & chemical treatments to control *Fusarium* root rot of some citrus root stocks under saline soil conditions. *Ann. Agric Sci*. (Cairo) 49(1):357-375.

243. Kidwai MK, Vikas, Srivastava S, Singh HB. 2006. Compatibility of *Trichoderma harzianum* to selected fungicides *J. Eco-fried. Agric.* 1(2): 156-161.

244. Kidwai MK, Vikas, Srivastava S, Srivastava SC, Singh HB 2007. Compatibility of *Trichoderma harzianum* to selected insecticides. *J. Eco-fried Agric.* 2(2): 170-174.

AGROBACTERIUM TUMEFACIENS - PRACTICAL IMPLICATIONS IN AGRICULTURE

Anju Bala , T.S. Thind, S. Gowswami and S. Raheja

Department of Plant Pathology, Punjab Agricultural University, Ludhiana 141 004 E-mail: anjuchani@yahoomail.co.in

KEY WORDS: Agrobacterium, Genetic transformation, Phytohormones, T- DNA

ABSTRACT

Agrobacterium tumefaciens is a unique, highly sophisticated pathogen which causes disease by the transfer of a particular DNA segment (T-DNA) into the nucleus of infected cells where it is suitably integrated into host genome and transcribed. The genes of T- DNA code for production of cytokinins, indole acetic acid, synthesis and release of opines and agrocinopines. The physical transfer of the t-DNA from *Agrobacterium* into the plant cell and integration into the host genome are mediated by approximately 20 virulence genes. Phenolic compounds produced by plants are perceived by *Agrobacteium* and stimulate the functioning of *vir* A genes which leads to synthesis of complimentary strand of T-DNA. *A. tumefaciens* naturally infects only dicotylednous plants, thus *Agrobacterium* mediated genetic transformation of monocotyledonous plant species are focused on the use of super-binary vector systems and use of tissue that consists of actively dividing embryonic cells such as calli induced from scutella. Genetic transformation of filamentous fungi and yeast has also become possible using structures like LBA 4404. Recent progress of *A. tumefaciens* mediated transfer includes production of edible vaccines in banana and vitamin A & iron content in rice grains. *Agrobacterium* can serve as an important tool in the modern agricultural biotechnology for creating transgenic plants with useful traits.

In nature, many strains of bacteria exchange genetic information during a process known as conjugation, and the new genetic information is passed on to subsequent generations. The advantage of using bacteria relates to the single-celled nature of the organisms. Only one cell needs to be changed in order to integrate the new genetic information and allow for its transmission to the next generation. This natural exchange of genetic information allows the organism an increased capacity to adapt to its environment.

Transformation is usually more difficult with multicellular organisms, however, and can be particularly challenging with some important plant species like cereals. The multicellular nature of most plants introduces the complication of transforming each cell of the plant in order to fully integrate the new information. Often the approach taken in the transformation of higher organisms, such as plants, is to transform an individual plant cell and then regenerate it into a whole organism.

Genetic transformation of plants and other organisms occurs naturally. Bacteria carry this out routinely. Soil bacteria, such as *Agrobacterium tumefaciens* and *Agrobacterium rhizogenes* are great examples of natural transformation systems, causing crown gall disease and 'hairy root syndrome' respectively. *Agrobacterium* can transfer a section of its own DNA, known as the T-DNA or 'transfer DNA' into plant cells. For *A. tumefaciens* this normally occurs at wound sites and the transferred DNA causes the plant to develop crown gall disease. The story does not end there. Along with causing a gall, the *Agrobacterium* harnesses the plant's machinery to produce unique sugars called opines that the bacterium uses as a nutrient supply.

The wonder bacterium

Agrobacterium tumefaciens is a Gram-negative, non-sporing, motile, rod shaped bacterium which is commonly found in the rhizosphere of plants where it survives in root exudates. It is a unique, highly sophisticated pathogen which causes disease by the transfer of a particular DNA segment (T-DNA) into the nucleus of infected cells where it is then suitably integrated into the host genome and transcribed, which leads to increased production of phytohormones causing tumor or gall formation and release of unique plant metabolites such as opines and agrocinopines (3,45).

Virulent strains of *A. tumefaciens* have a large plasmid in their cells called Ti-plasmid which is approximately 200 kb in size, having T-DNA region, virulence region (*Vir* gene) and opine catabolism region. T-DNA carries tumorogenesis genes. Certain genes of Ti plasmid are necessary for T-DNA transfer, other genes play role in pathogenesis (32,34,63).

The T-DNA is flanked by two imperfect 25 bp repeat motifs at either end, called the left border (LB) and right border (RB). Importantly, the LB and RB, together with small surrounding elements, are the only two DNA sequence elements that define which piece of DNA is transferable and which is not. Therefore, any piece of DNA inserted between the two borders can be transferred from *Agrobacterium* to the plant. These well established facts helped in construction of the first vector and bacterial strain system for plant transformation (20,28,32,55).

Only a small part of the plasmid (the T-DNA) enters the plants, the rest of the plasmid remains in the bacterium to serve further roles. When integrated into the plant genome, the genes on the T-DNA code for production of cytokinins, indoleacetic acid, synthesis and release of novel plant metabolites- the opines and agrocinopines (58). Thus the bacterium creates for itself a marvelous nutrients rich environment.

Synthesis of phytohormones and plant metabolites

Agrobacterium genes are involved in transformation by elevation of phytohormones. Different genes are required for the synthesis of auxins and cytokinins. For auxin synthesis two genes are involved.

First iaaM gene product helps to synthesise Indole 3 acetamide from tryptophan which is then converted to indoleacetic acid (IAA) with the help of iaa H gene product where as for the synthesis of cytokinins only one gene product called ipt gene product is required which synthesises isopentenyl-AMP (a cytokinin precursor) from isopentenyl – pyrophosphate and AMP.

Plant metabolites produced by *Agrobacterium* include the opines and agrocinopines. The opines are unique amino acid derivatives, different from normal plant products, and the agrocinopines are unique phosphorylated sugar derivatives. The bacterium can use them as carbon and energy source. Two types of opines are present which include nopalines and octopines. Nopalines are produced by the action of nopaline synthase gene of nopaline strains of *Agrobacterium* and octopines are produced by the action of octopine synthase gene of octopine strains of *Agrobacterium.* These plant metabolites are uniquely taken up by *Agrobacterium,* thus provide a competitive advantage for the pathogen because no other microorganism can metabolize these compounds.

Bacterial infection

The infection process is composed of chemotaxis and initial binding at the site of wounding and development of a stable attachment. Initiation of t-DNA excision starts by the action of virulence genes. Bacterial DNA is then transferred to plant cell through construction of trans membrane channel. Now t-DNA to nucleus integrates in the plant genome and starts transcription of bacterial genes.

The physical transfer of the t-DNA from *Agrobacterium* into the plant cell and its subsequent integration into the host plant nuclear genome are complex process that are still not entirely understood at the molecular and cellular level. However much of the process is mediated by the large group of *Agrobacterium* virulence (*Vir*) genes which are about twenty in number (59). The 30 kb virulence (*Vir*) region is organized in six operons that are essential for the T- DNA transfer (*Vir* A, *Vir* B, *Vir*D and *Vir* G) or for increasing transfer efficiency (*Vir* C and *Vir* E) (32,63). In natural *Agrobacterium* strains all of the 20 *Vir* genes reside on the Ti plasmid.

However, as explained in more detail below, *Vir* genes and T-DNA can be separated into two different DNA molecules, which have many advantages for the molecular biologist.

Functioning of *Vir* genes

T-DNA transfer to plant cell is regulated by a series of virulence genes known as *Vir* genes which functions as follows:

Signalling molecules (phenolic compounds, e.g. acetosyringone) produced by the plants are perceived by *Agrobacterium*. Level of *Vir* A protein increases dramatically upon stimulation with plant phenolic inducer molecules. *Vir* A is a transmembrane dimeric sensor protein that detects signal molecules, mainly small phenolic compounds, released from wounded plants (9,46). The signals for *Vir* A activation include acidic pH, phenolic compounds such as acetosyringone (60) and certain class of monosachharides which acts synergistically with phenolic compounds (2,23,49). Activated *Vir* A has the capacity to transfer its phosphate to a conserved aspartate residue of the cytoplasmic DNA binding protein *Vir* G (35,36,46). The activation of *Vir* system also depends on external factors like temperature and pH. At temperatures greater than 32°C, the *Vir* genes are not expressed because of a confirmational change in the folding of *Vir* A induce the inactivation of its properties (37). The effect of temperature on *Vir* A is suppressed by a mutant form of *Vir* G (*Vir* G^{C}), which activates the constitutive expression of the *Vir* genes. *Vir* G activates transcription of other *Vir* genes like *Vir* D, *Vir*E and *Vir* B. *Vir* D proteins catalyse endonucleolytic cleavage of the Ti plasmid and creates a single stranded nick for attachment of external DNA to the plant cell (16,50, 63). *Vir* E proteins act as DNA binding proteins that bind the external DNA to the plasmid (34). When the naked ss T-DNA-*VIR* D complex is inside the plant cell, it is coated by *Vir* E (4,40). *Vir* B proteins stimulate the bacterium to produce filaments resembling the pilli, which act as conjugation bridge for interbacteial conjugation (25,48,62). Plant DNA polymerase produces complementary strand to the single strand of T- DNA that is transferred before the DNA gets integrated into the plant genome. Inside the plant cell, the ssT-DNA complex is targeted to the nucleus crossing the nuclear membrane. Two *Vir* proteins have been found to be important in this step: *Vir* D2 and *Vir* E2, which are most important, and probably *Vir* F, which has a minor contribution to this process (32,52). Complex repeats of the T-DNA can be generated at the integration site, and T-DNA can integrate at more than one locus per genome. Once the introduction of T- strand in the 3'-5' strand of the plant DNA is completed, a tortion followed by a nick into the opposite plant DNA strand is produced. This situation activates the repair mechanism of the plant cell and the complementary strand is synthesized using the early inserted T- DNA strand as a template (54).

Genetic transformation of plant using *A. tumefaciens*

The first record of plant transformation by using Ti plasmid vector for introduction of foreign genes into plant cells without alteration of their normal regeneration capacity appeared in 1983 for genetic transformation of tobacco (31,61). Since that time, a great progress has been achieved in understanding the *Agrobacterium* mediated gene transfer to plant cells. However, *A. tumefaciens* naturally infects only dicotyledonous plants and many economically important plants, including cereals were inaccessible for genetic manipulation using *Agrobacterium*. Thus, alternative direct transformation methods have been developed such as polyethyleneglycol-mediated transfer, microinjection, protoplasm and intact cell electroporation and gene gun technology. However, *Agrobacterium* – mediated transformation has remarkable advantages over direct transformation methods. It reduces the copy number of the transgene potentially leading to fewer problems (29,39). Transformation has been used for genetic manipulation of more than 120 species of at least 35 families, including the major economic crops, vegetables, ornamentals, fruit trees and medicinal plants. Using these procedures, crop plants like soybean, canola and cotton have been developed that are protected from glyphosate herbicides (Roundup ready), have increased resistance against insects (Bt gene) or diseases e.g cotton, potato and maize, or have improved characteristics of products e.g enhanced vitamin A level (Table 1). Transgenic tomatoes with increased shelf life, vine-ripened flavour and better consistency of tomato paste was developed by Calgene and Zeneca companies by using these technologies during the year 1994 & 1995 respectively. Some of the important strains and helper plasmids of *A. tumefaciens* used in genetic transformation programme are listed in Table 2.

Table 1: Examples of transgenic crop plants resulting from *Agrobacteirum* mediated gene transfer

Transgenic phenotype	Crop plants
Herbicide resistance	
Phosphinothricin	Tobacco, tomato, potato, oil seed rape, sugar beet, alfalfa, poplar
Glyphosate	Tobacco, tomato oilseed rape, flax, sugar beet, alfalfa
Bromoxycil	Tobacco, tomato, cotton
Sulphonurea	Tobacco, tomato, potato, flax
Atrazine	Tobacco
Insect resistance	
Insecticidal BT proteins	Tobacco, tomato potato, cotton
Proteinase inhibitors	Tobacco
Virus resistance	
Alfalfa mosaic virus	Tobacco, alfalfa
Tobacco mosaic virus	Tobacco
Tomato mosaic virus	Tomato
Potato virus X	Potato
Potato virus Y	Potato
Rhizomania virus	Sugar beet
Cucumber mosaic virus	Cucumber
Other characters	
Heavy metal resistance	Tobacco
Firm fruit	Tomato
Thaumation production	Potato
Modified seed storage protein	Oil seed rape

Source: Plant Biotech- recent advances (56).

Genetic transformation of cereal plants

Efficient methodologies of *Agrobacterium*- mediated gene transfer have been established mainly for dicotyledonous plants. To develop methodologies for monocotyledonous plants it is important to take into account the critical aspects in the *A. tumefaciens* – plant interaction, the cellular and tissue culture methodologies developed for that species. The suitable genetic materials (bacterial strains, binary vectors, reporter, marker genes and promoters) and molecular biology techniques available in the laboratory, are necessary for selection of the DNA to be introduced (38). This DNA must be able to be expressed in plants making possible the identification of transformed plants in selectable medium and using molecular biology techniques to test and characterize the transformation events (5). Extensive research aimed at understanding and improving the molecular machinery of *Agrobacterium* responsible for the generation and transport of the bacterial DNA into the host cell has resulted in the establishment of many recombinant *Agrobacterium* strains, plasmids and technologies currently used for the successful transformation of numerous plant species (42,57). Recent works on *Agrobacterium* – mediated genetic transformation of monocotyledonous plant species are focused on the use of "superbinary" vector systems, i.e. binary vectors carrying a DNA fragment from the *A. tumefaciens* virulence region (55).

Table 2: Some *Agrobacterium tumefaciens* strains and their helper plasmids used in genetic transformation program

Agrobacterium strain	Helper plasmid
A205	pGV3850::pA41003, TIT37ASE::pMON200
	pTi201, pGV2298, pGV2441, pGV2774, V3304, GV3850
A281	pTiBo542
A479	pAL4404
Ach5	pTiAch5
EHA101	pTiAch5 , pAL1050
EHA105	pEHA105
LBA1050	pA11050
LBA4301	pRAL4404
LBA4401	pTiAch5
LBA4404	pAL4404
LBA4434	pAL1050
LBA4434	pAL4404

GV3101	pGV3850::pCAT4
GV3101	pGV3850::pLGV1103
GV3111SE	pTiB6S3::pMON120
T37	pTiT37

Source: Fillipone and Penza (1992) (24)

Therefore, cereal crops such as rice, wheat, maize and barley which could not be transformed historically using this system are now being transformed because significant progress has been made recently on the identification of more efficient *Agrobacterium* strains that can enhance the efficiency of *Agrobacterium* mediated transformation in cereals crops. Different cereal crops have different protocols for *Agrobacterium* mediated genetic transfer in different crops but the key point for those protocols is the use of tissue that consists of actively dividing embryonic cells such as calli induced from scutella (14,19). Such cells are cocultivated with *Agrobacterium* in the presence of acetosyringone (potent inducer of *Vir* genes that are involved in transfer of DNA). By following these protocols and using super virulent/more efficient strains of *Agrobacterium,* crops like rice, wheat, barley and maize are being transformed for antibiotic and herbicide resistance (26, 47,55) (Table 3).

Among cereals *Agrobacterium* mediated transformation was first reported in Japonica cultivars of rice (11) whereas efficient transformation of maize using *Agrobacterium* strains was first reported by 1996 (33). Barley is the third cereal where *Agrobacterium* mediated transfer is a method of choice for the production of transgenic plants. First report of successful transformation of barley and wheat through *Agrobacterium* came in 1997 (12,53). Improved quality of wheat could be obtained by transfer of gene glutenin, which is helpful for modification of gluten polymer in the endosperm. In rice stilbene synthase has been transferred from grapevine which promotes biosynthesis of phytoalexin thus imparting resistance to *Pyricularia oryzae* causing rice blast (41). Similarly chimeric chitinase gene responsible for production of chitinase enzyme was transferred to rice for imparting resistance against *Rhizoctonia solani* (18). Transfer of Cry I A (b) and Cry I A (c) genes in japonica rice imparted resistance to yellow stem borer (13,44). Quality improvement in rice has been achieved by the transfer of phytoene synthase gene from daffodils which regulates the bio-synthesis of Provitamin A (8). Quality of maize was also enhanced by transfer of α zein mutant gene for production of soft and starchy endosperm (17). Different rice cultivars were transformed by using different strains of *Agrobacterium* and found that three cultivars mainly Basmati 122, Tulsi and Vaidehi were positive for chitinase gene which impart resistance to sheath blight (18).

Table 3: Recent advances in *Agrobacterium* gene transfer technology for the transformation of cereals

Plant	Agrobacterium strain	Vector name	Target tissue
Rice	EHA 101, EHA 105, LBA 4404	PIG 121P TOK 233	Immature embryos, callus, suspension cells
Maize	LBA 4404	PTOK 233	Immature embryo
Barley	AGL 1	PDM 805	Immature embryo
Wheat	C58 (AB1)	PMON 18365	Immature embryo, embryogenic callus

Genetic transformation of filamentous fungi and Yeasts

Agrobacterium can transform fungi including yeast (6,7), Ascomycetes, Basidiomycetes and Fungi imperfecti (1,27). Various workers have reported genetic transformation of filamentous fungi and yeast using different *A. tumefaciens* strains.

Different workers have achieved genetic transfer of filamentous fungi. *Penicillium chrysogenum* was transformed for increased production of Penicillin by using strain LBA 4404 for transfer of gene Vgb and vitreoscilla hemogloebin gene (51). A gene for hygromycin resistance was introduced in *Beauveria bassiana* by using LBA 4404 strain (22). Fungi like *Aspergillus awamori, A. niger, Trichoderma rosei, Fusarium venenatum, Colletotrichum gloeosporioides* and *Agaricus bisporus* have been transformed by using LBA 1100 strain of *Agrobacterium* (10,43).

Challenges and limitations

The major challenge is to use *Agrobacterium* for site directed or homologous recombination which occurs at 10^{-5} frequency of illegitimate recombination. So we need an *Agrobacterium* mediated transfer system that delivers T-DNA to the plant nucleus efficiently but is deficient in random T-DNA integration. Stable and predictable transgene expression in plants is the other important aspect. We need to understand the role of position effects, chromatin effects and T-DNA integration pattern in transcriptional and post-transcriptional gene silencing. Transformation of recalcitrant plant species and medically and agronomically important fungi and plastid genetic transformation instead of nucleus are the other challenges.

The use of transgenic crops or genetically modified organisms (GMOs) caused a major debate on consumers' acceptance in Europe, and it is becoming a significant agricultural trade issue with other countries.

Future prospects

Recent progress includes the production of edible vaccines in bananas and vitamin A and iron content in rice grains. There are many other research groups working on modifying almost every aspect of plant development and physiology. Scientists at Texas are involved in finding ways to make plants more resistant to environmental stresses such as extreme temperatures and water deficit. While these techniques provide exciting new tools for the development of crop plants with new characteristics, they will not replace traditional plant breeding. Rather the integration of transgenic plants and traditional breeding techniques will be necessary to effectively incorporate these new technologies into useful varieties.

Genetic transformation of human and animal cells is the next objective of scientists. Human gene therapy using *Agrobacterium* has also been suggested by researchers (30). According to this report gene transfer to human Hela cells have common methodology as followed for plant transformation.

Conclusions

Agrobacterium mediated genetic transformation is the dominant technology used for the production of genetically modified transgenic plants because it is a natural gene transfer system and a poor man's vector system requiring no high tech. microinjection equipment or protoplast regeneration skills. It is a clean system introducing only a desired DNA package (15,21). It is a predominant route used to create transgenic dicotyledonous plants but T-DNA transfer to monocotyledonous plants has only recently become possible. Extensive research aimed at understan-ding and improving the molecular machinery of *Agrobacterium* responsible for the generation and transport of the bacterial DNA into the host cell has resulted in the establishment of many recombinant *Agrobacterium* strains, plasmids and technologies currently used for the successful transformation of numerous plant species. Unlike the role of bacterial proteins, the role of host factors in the transformation process has remained obscure for nearly a century of *Agrobacterium* research, and only recently have we begun to understand how *Agrobacterium* hijacks host factors and cellular processes during the transformation process. The identification of such factors and studies of these process hold great promise for the future of biotechnology and plant genetic engineering, as they might help in the development of concept-ually new techniques and approaches needed today to expand the host range of *Agrobacterium* and to control the transformation process and its outcome during the production of transgenic plants. Modern agricultural biotechnology is heavily dependent on using *Agrobacterium* to create transgenic plants. It evolved millions of years ago to genetically transform a wide range of organisms and is doing what molecular biologist are now striving to do. It is now upto scientists to harness the natural ability of this bacterium.

REFERENCES

1. Abuodeh RO, Orbach MJ, Mandel MA, Das A and Galgiani JN. 2000. Genetic transformation of *Coccidioides immitis* facilitated by *Agrobacterium tumefaciens. J Infect Dis* 181: 2106–2110.
2. Ankenbauer RG and EW Nester. 1990. Sugar-mediated induction of *Agrobacterium tumefaciens* virulence genes: structural specificity and activities of monosaccharides. *J Bacteriol* 172: 6442-6446.
3. Binns AN and Thomashow M F. 1988. Cell biology of *Agrobacterium* infection and transformation of plants. *Ann Rev of Microbiol* 42: 575-606
4. Binns AN, Beaupre CE and Dale M. 1995. Inhibition of Vir B-mediated transfer of diverse substrate from *Agrobacterium tumefaciens* by

the IncQ plasmid RSF1010. *J Bacteriol* 177: 4890-4899.

5. Birch RG. 1997. Plant transformation problems and strategies for practical application. *Pl Mol Biol* 48: 297-326.

6. Bundock P and Hooykaas PJJ. 1996. Integration of *Agrobacterium tumefaciens* T-DNA in the *Saccharomyces cerevisiae* genome by illegitimate recombination. *Proc Natl. Acad Sci* USA 93:12172–12175.

7. Bundock P, Ras A D, Beijersbergen A and Hooykaas PJJ. 1995. Trans-kingdom T-DNA transfer from *Agrobacterium tumefaciens* to *Saccharomyces cerevisiae. EMBO J* 14: 3206–3214.

8. Burkhardt PK, Beyer P, Wtinn J, Kliiti A, Armstrong GA, Schledz M, lintig JV and Potrykus I. 1997. Transgenic rice (*Oryza sativa*) endosperm expressing daffodil (*Narcissus pseudonarcissus)* phytoene synthase accumulates phytoene, a key intermediate of provitamin A biosynthesis. *Plant J* 11: 1071-l 078.

9. Cangelosi GA, Ankenbauer RG and Nester EW. 1990. Sugars induce the *Agrobacterium virulence* genes through a periplasmic binding protein and a transmembrane signal protein. *Proc Natl Acad Sci USA* 87: 6708-6712.

10. Caroline B M, Arthur F J R, Paul J J H and Cee A M J J. 2004. *Agrobacterium*-mediated Transformation of *Aspergillus awamori* in the absence of full length Vir D_2, Vir C_2 and Vir E_2 leads to insertion of Aberrant T-DNA structures. *J Bacteriol* 186 (7): 2038-2045.

11. Chan MT, Lee TM and Chang HH. 1992. Transformation of Indica rice (*Oryza sativa* L.) mediated by *Agrobacterium tumefaciens. Plant Cell Physiol.* 33: 577.

12. Cheng M, Fry JE, Pang SZ, Zhou HP, Hironaka CM, Duncan DR, Conner TW and Wan YC. 1997. Genetic transformation of wheat mediated by *Agrobacterium tumefaciens. Plant Physiol* 115: 971-980.

13. Cheng XY, Sardana R, Kaplan H. and Altosaar I. 1998. *Agrobacterium-transformed* rice expressing synthetic cry1Ab and cry1Ac genes are higly toxic to striped stem borer and yellow stem borer. *Proc. Natl. Acad. Sci., USA* 95: 2767-2772.

14. Cheng M, Fry JE, Pang SZ, Zhou HP, Hironaka CM, Duncan DR, Conner W. and Wan YC. 1997. Genetic transformation of wheat mediated by *Agrobacterium tumefaci-*

ens. Plant Physiology 115: 971-980.

15. Chilton MD. 1993. *Agrobacterium* gene transfer: Progress on a "Poor Man's Vector" for maize. *Proc. Natl. Acad. Sci.* 90: 3119-3120.
16. Christie PJ. 1997. *Agrobacterium tumefaciens* T-complex transport apparatus: a paradigm for a new family of multifunctional transporters in Eubacteria. *J Bacteriol.* 179: 3085-3094.
17. Coleman CE, Clore AM, Ranch JP, Higgins R, Lopes MA and Larkins BA. 1997. Expression of a mutant a-zein creates the *floury* phenotype in transgenic maize. *Proc Natl Acad Sci USA* 94:7094-7097.
18. Datta K, Koukolikova ZN, Baisakh NO and Datta SK. 2000. *Agrobacterium* mediated engineering for sheath blight resistance of indica rice cultivars from different ecosystems. *Theor Appl Genet* 100: 832-839.
19. De la Pena, Lorz A H. and Schell J. 1987. Transgenic rye plants obtained by injecting DNA into young floral tillers. *Nature* 325: 274-276.
20. Deblaere R, Bytebier B, De Greve, H, Deboeck F, Schell J, Van Montagu M and Leemans J. 1985. Efficient octopine Ti plasmid-derived vectors for *Agrobacterium*-mediated gene transfer to plants. *Nucleic Acid Res* 13: 4777-4788
21. Dehleen LS, Patricia AO and Bleehl AE. 2001. Transgenic approaches to control Fusarium Head Blight in Wheat and Barley. *Crop Sci.* 41: 628-37.
22. Dos R M C, Pellegrineere F M H, Delgado D R T, Furlanetol and Furlaneto MC. 2004. *Agrobacterium tumeficiens* – medicated genetic transformation of the entamopathogenic fungus *Beauveria bassiana. J. Bacteriol.* 58 (2): 197-202.
23. Doty SL, Yu NC, Lundin JI, Heath JD. and Nester EW. 1996. Mutational analysis of the input domain of the Vir A protein of *Agrobacterium tumefaciens. J Bacteriol* 178: 961-970.
24. Fillippone E and Penza R. 1992. *Agrobacterium tumefaciens* – mediated gene transfer. In *Biotechnology: Enhancing research on Tropical Crops in Africa*. CTA/IITA Co-Publication –IITA, Ibadan, Nigeria. 376 pp.
25. Finberg KE, Muth TR, Young SP, Maken JB, Heitritter SM, Binns AN and Banta LM. 1995. Interactions of VirB9, -10 and -11 with the memb-

rane fraction of *Agrobacterium tumefaciens*: solubility studies provide evidence of tight associations. *J. Bacteriol.* 177: 4881-4889.

26. Gelvin BS. 2003. *Agrobacterium* - Mediated Plant transformation: the Biology behind the Gene- Jockeying tool. *Mol Biol Rev* 67: 16-37.

27. Groot D, Bundock MJA, Hooykaas, PJJ and Beijersbergen AGM. 1998. *Agrobacterium tumefaciens*-mediated transformation of filamentous fungi. *Nat Biotechnol* 16: 839–842.

28. Hamilton CM. 1997. A binary-BAC system for plant transformation with high-molecular-weight DNA. *Gene* 200: 107-116.

29. Hansen G, Shillito R.D and Chilton M.D. 1997. T-strand integration in maize protoplasts after codelivery of a T-DNA substrate and virulence genes. *Proc Natl Acad Sci USA* 94: 11726-11730.

30. Hanur V. 2004. Genetic transformation of human cells by a soil phytopathogen presents consume molecular strategies. *Curr Sci* 87 (7): 856-57.

31. Herrera-Estrella L. 1983. Transfer and expression of foreign genes in plants. Ph. D. Thesis. Laboratory of Genetics, Gent University, Belgium.

32. Hooykaas PJJ and Shilperoort RA. 1992. *Agrobacterium* and plant genetic engineering. *Pl Mol Biol* 19: 15-38.

33. Ishida Y, Saito H, Ohta S, Komari HYT and Kumashiro T. 1996. High efficiency transformation of maize (*Zea mays* L.) mediated by *Agrobacterium tumefaciens*. *Natl Biotechnol* 14: 745–750.

34. Jeon GA, Eum JS and Sim WS. 1998. The role of inverted repeat (IR) sequence of the *virE* gene expression in *Agrobacterium tumefaciens* pTiA6. *Molecules and Cells* 8: 49-53.

35. Jin S, Prusti RK, Roitsch T, Ankenbauer RG and Nester EW. 1990a. The Vir G protein of *Agrobacterium tumefaciens* is phosphorylated by the autophosphorylated Vir A protein and this is essential for its biological activity. *J Bacteriol* 172: 4945-4950.

36. Jin S, Roitisch T, Christie PJ and Nester EW. 1990b. The regulatory Vir G protein specifically binds to a *cis* acting regulatory sequence involved in transcriptional activation of *Agrobacterium tumefaciens* virulence genes. *J Bacteriol* 172: 531-562.

37. Jin S, Song Y, Pan S and Nester EW. 1993. Characterization of a *vir G* mutation that confers constitutive virulence gene expression in *Agrobacterium tumefaciens. Molecular Microbiology* 7: 555–5.

38. Komari T, Hiei Y, Ishida Y, Kumashiro T and Kubo T. 1998. Advances in cereal gene transfer *Current opinion in Pl Biol* 1: 161-65.

39. Koncz C, Németh K, Redei GP and Scell J. 1994. In: p. 167-189, J Paszkowski (ed.). Homologous recombination and gene silencing in plants. Kluwer, Dordrecht, The Netherlands.

40. Lessl M and Lanka E .1994. Common mechanisms in bacterial conjugation and Ti-mediated T-DNA transfer to plant cells. *Cell* 77: 321-324.

41. Lorenzen SP, Nelke B, Hanbler G, Muhbach HP, Thomzik JE. 1997. Transfer of a grapevine stilbene synthase gene to rice (*Oryza sativa* L). *Pl Cell Rep* 16: 668-673.

42. Lörz H, Baker B and Schell J. 1985. Gene transfer to cereal cells mediated by protoplast transformation. *Molecular General Genetics* 199:473-497.

43. Marcel JA, Groot D, Paul B, Paul JJH and Aliee GMB. 1998. *Agrobacterium* mediated genetic transformation of filamentous fungi. *Nature Biotech* 16: 839-42.

44. Nayak P, Basu D, Das S, Basu A, Ghosh D, Ramakrishnan NA, Ghosh M and Sen SK. 1997.Transgenic elite lndica rice plants expre-ssing CrylAc delta-endotoxin of *Bacillus thuringiensis* are resistant against yellow stem borer (*Scirpophage incertulas) Proc Natl. Acad. Sci. USA*, 94: 211l -2116.

45. Nester EW, Gordon MP, Amasino RM and Yanofsky MF. 1984. Crown gall: a molecular and physiological analysis. *Ann Rev. Pl Physiol* 35: 387-413.

46. Pan SQ, Charles T, Jin S, Wu ZL and Nester EW. 1993. Preformed dimeric state of the sensor protein VirA is involved in plant-Agrobacterium signal transduction. *Proc Natl Acad Sci USA* 90: 9939-9943.

47. Rashid H, Yokoi S, Toriyama K and Hinata K. 1996. Transgenic plant production mediated by *Agrobacterium* in Indica rice. *Plant Cell Reports* 15: 727-730.

48. Rashkova S, Spudlich GM and Christie PJ. 1997. Characterization of membrane and protein interaction determinants of the *Agrobacterium tumefaciens* VirB11 ATPase. *J Bacteriol* 179: 583-591.

49. Shimoda N, Toyoda-Yamamoto A, Nagamine J, Usami S, Katayama M, Sakagami Y and Machida Y. 1990. Control of expression of *Agrobacterium tumefaciens* genes by synergistic actions of phenolic signal molecules and monocaccharides. *Proc Natl Acad Sci USA* 87: 6684-6688.

50. Stephens KM, Roush C and Nester EW. 1995. *Agrobacterium tumefaciens* VirB11 protein requires a consensus nucleotide binding site for function in virulence. *J Bacteriol* 177: 27-36.

51. Sun CB, Kong GL, Xu WS. 2002. Efficient transformation of *Penicillium chrysogenum* mediated by *Agrobacterium tumefaciens* LBA 4404 for clonning *vitreoscilla* hemoglobin gene. *Electronic J Biotech* ISSN.5: 0717-3458.

52. Thomashow MF, Karlinsey JE, Marks JR and Hurlbert RE. 1987. Identification of a new virulence locus in *Agrobacterium tumefaciens* that affects polysaccharide composition and plant attachment. *J Bacteriol* 169: 3209-3216.

53. Tingay S, McElroy D, Kalla R, Fieg S, Wang M, Thornton S and Brettell R. 1997. *Agrobacterium tumefaciens*-mediated barley transformation. *Plant J.* 11: 1369–1376

54. Tinland B, Schoumacher F, Gloeckler V, Bravo AM, Angel M and Hohn B. 1995. The *Agrobacterium tumefaciens* virulence D2 protein is responsible for precise integration of T-DNA into the plant genome. *EMBO Journal* 14: 3585-3595.

55. Torisky RS, Kovacs L, Avdiushko S, Newman JD, Hunt AG and Collins GB. 1997. Development of a binary vector system for plant transformation based on supervirulent *Agrobacterium tumefaciens* strain Chry5. *Plant Cell Reports* 17: 102-108.

56. Trivedi P C. 2003. *Plant Biotech – Recent Advances* pg 11, Panima Publishing Corporation, New Delhi.

57. Tzfira and Citovsky. 2006. *Agrobacterium* –mediated genetic transformation of plants: Biology and biotechnology. *Curr Opin Biotechnol* 17 (2): 147-154.

58. Walden R, Reiss B , Knocz C and Schell J .1997.The impact of T_i plasmid –derived gene vectors on the study of the mechanism of action of phytohormones. *Ann Rev Phytopathol* 35: 45-66.

59. Walt R. 1989. *Agrobacterium tumifaciens* and inter kingdom genetic exchange. *Annu.*

Rev. Phytopathol. 27: 583-618.

60. Winans SC. 1992. Two-way chemical signaling in *Agrobacterium*-plant interac-tions. *Microbiol Rev* 56: 12-31.

61. Zambryski P, Joos PH, Genetello C, Leemans J, Montagu MV and Schell J. 1983. Ti plasmid vector for the introduction of DNA into plant cells without alteration of their normal regeneration capacity. *EMBO J*. 2: 2143-2150.

62. Zhou XR and Christie PJ. 1997. Suppression of mutant phenotypes of the *Agrobacterium tumefaciens Vir* B11 ATPase by overproduction of *Vir* B proteins. *J Bacteriol 179*: 5835-5842.

63. Zupan JR and Zambryski PC. 1995. Transfer of T-DNA from *Agrobacterium* to the plant cell. *Plant Physiology* 107: 1041.1047.

ROLE OF PSEUDOMONADS IN PLANT DISEASE CONTROL

K.V.B.R. Tilak* and G. Bagyanarayana

Department of Botany, Osmania University, Hyderabad-500 007, A.P.
*Corresponding author-mail : *tilakkvbr@yahoo.com*

KEY WORDS: *Pseudomonas* spp., plant diseases, suppression, mechanisms

ABSTRACT

Biological control should be justified on its own merits without compromising its perceived importance to chemical controls. In this context, the microorganisms that harbour in the root region are the most befitting candidates for use as biocontrol agents since the rhizosphere provides the frontline defense for root against attack by pathogens. Pathogens face a complicated phenomena of antagonism from roots. They also compete with each other for site of colonization on the root surface. The ideal biocontrol agent introduces and/or promotes the antagonists only, whenever required and is most effective in minimizing the wasteful application of inoculum to non-targets.

Many genera belonging to bacteria, fungi, actinomycetes and viruses are used as biocontrol agents to combat several important plant diseases. The present review focuses on the use of bacterial antagonists, particularly pseudomonads, as biocontrol agents of fungal diseases of plants, their mechanism of action, molecular and genetic basis of antagonistic property and possible genetic manipulations to bring about desirable changes.

Biological control of plant diseases can be defined as the reduction of inoculum density or disease producing activities of a pathogen or a parasite in its active or dormant stage by one or more organisms accomplished naturally or through manipulation of the environment, host or antagonist or by mass introduction of one or more antagonists (7). In nature, biological control can occur without any external aid. Cross protection by viral infection, phage attack on plant pathogenic bacteria, fungi parasitizing on plant pathogenic fungi, etc. are a few examples (101).

Pseudomonads as biological control agents of plant pathogens

Fluorescent pseudomonads have revolutionized the field of biological control of soil-borne plant pathogenic fungi. Most of them fall either in *fluorescens* or *putida* group. During the last three decades, they have emerged as the largest potentially most promising group of plant growth promoting rhizobacteria involved in the biocontrol of plant diseases (8,23,24,45,77,78,84,106,116). Fluorescent pseudomonads have received the most attention for several compelling reasons. First, they readily colonize roots in nature, where they are frequently the most common among microorganisms (117). The simple nutritional requirement and the ability to use many carbon sources that exude from roots and to compete with indigenous microflora, may explain their ability to colonize the rhizosphere (73). Additionally, pseudomonads are amenable to genetic manipulation. These characteristics make them useful vehicle for the delivery of antimicrobial and insecticidal compounds and plant hormones to the rhizosphere. O'Sullivan and O'Gara (82) reviewed the traits of fluorescent pseudomonads such as production of antibiotics, hydrogen cyanide, siderophore which are involved in suppression of plant root pathogens.

There are numerous examples of biocontrol of several devastating fungal plant pathogens of important crops by fluorescent pseudomonads and has been reviewed from time to time (23,33,66,68,82,84,88,95,107,118).

Natural disease suppression involving pseudomonads has been reported by many workers. In disease suppressive soils, though the pathogen initially causes disease, on continued cropping it fails to do so. Naturally occurring soil pseudomonads are important elements in these soils suppressive to diseases.

A numbér *of Pseudomonas* strains have been used as biological control agents in green house and field conditions against an array of plant pathogens (Table 1). In many cases they not only help in suppressing the pathogens, but also improvd the plant yield by acting as plant growth promoters (82,32). Eventhough there are problems like variable results obtained in different soil types and inadequate survival on seeds prior to planting, *Pseudomonas* spp. have great potential in biological control of plant pathogens (88).

Table 1. Introduced pseudomonads as biological control agents

Pseudomonad	Pathogen suppressed	Host Plant	Reference
P.fluorescens 89 B-61	*Fusarium* wilt	Cotton	(21)
P.fluorescens 89 B-61	Anthracnose, Bacterial angular leaf spot	Cucumber	(116)
P. glumae EM 85	Damping-off Charcoal rot, *Fusarium* wilt	Cotton, Maize	(87)
P. glumae EM 85	Damping-off	Cotton	(3)
P.fluorescens NBRI 1303	*Fusarium oxysporum* f.sp.*ciceri*	Chickpea	(76)
P.fluorescens NBRI 1303	*Macrophomina phaseolina*	Chickpea	(76)
P.fluorescens NBRI 1303	*Pythium*	Chickpea	(76)
P.fluorescens WCS 374	*Fusarium* wilt	Radish	(69)
P. putida WCS 358	*Fusarium* wilt	Carnation	(36)
P.fluorescens GL-92	*Sclerotinia* wilt	Sunflower	(38)
P.putida FR-38	*Rhizoctonia solani*		(45)

P.fluorescens CH 31	Pythium root rot	Cucumber	(75)
P.fluorescens CR-20	*Rhizoctonia solani*	Cowpea	(8)
P. aureofaciens PA 147-2	Root rot	*Asparagus*	(18)
P.fluorescens	Damping-off	Cotton	(53)
P. cepacia 55B	Stem rot	*Poinsettia*	(19)
P. cepacia AMMD	*Aphanomyces*	pea	(61)
P. chlororaphis	*Pythium aphanidermatum, P. dissotocum* root rot	Pepper	(20)
Pseudomonas spp.	Take-all disease	Wheat	(84, 34)
Pseudomonas spp.	Grey mold		(112)
Pseudomonas spp.	Anrhracnose	Sorghum	(74)
Pseudomonas spp.	Seedling blight	Cucumber	(120)
Pseudomonas spp.	Sheath blight	Rice	(44)
P. putida	Sheath blight	Rice	(67)
P. fluorescens PfAIR2	Sheath Blight	Rice	(107)
P .maltophila	*Rhizoctonia solani*	Potato	(43)
P. fluorescens	Sheath blight	Rice	(100)
P.fluorescens P16	*Fusarium* wilt	Tomato & chilli	(102)
P. fluorescens PFBC 25 PFBC 26	Root rot	Chickpea	(62)

Mechanisms involved in biological control by pseudomonads

Soil pseudomonads involved in plant disease suppression produce a number of metabolites. It is important to know which metabolite is involved in disease suppression in individual cases to use the strain efficiently and for further genetic modifications. Biocontrol activity is mediated by production of allelochemicals including siderophores, antibiotics, biocontrol volatiles, lytic enzymes and detoxification enzymes (5,42,49).

There are two approaches to know the exact mechanism of disease suppression by pseudomonads. One requires a purified compound of known structure. The disease suppressive ability of the compound is compared with that of the producer strain in a conducive soil. If the compound mimics the disease suppression by the disease controling strain, it is assumed that the metabolite has a role to play in disease suppression (64). The second approach involves generation of mutants from the *Pseudomonas* strain, either by random transposon mutagenesis or site directed mutagenesis, having no more production of an extracellular antifungal metabolite. The effectiveness of these mutants in suppressing the disease is compared with that of the wild type strain. The loss of ability to suppress a disease by a mutant defective in production of a particular metabolite indirectly suggests the role of that metabolite in disease control.

The possible mechanisms of disease suppression by pseudomonads as suggested by several workers (82,32) are as follows :

(i) Competition for Fe^{3+} by producing siderophores

(ii) Producing volatile or diffusible metabolites : antibiotics, HCN, ammonia

(iii) Induction of systemic resistance in plants

(iv) Successful root colonization

Siderophore production and disease suppression

Though iron is the fourth abundant element on the earth's crust, it is almost unavailable to living organisms due to its insoluble nature. In aerated and oxidized soils iron is present as Ferric (Fe^3) form which is less soluble. Its solubility in water is very less (10^{-18}M), which is too low to support microbial requirements. Since iron is a very essential component in the metabolic processes, its unavailability acts as a limiting factor for growth of all organisms. Rhizosphere is a region where there is a very high competition for nutrients among the heterogeneous microflora present. Under these conditions some microorganisms secrete siderophores, which are iron binding ligands and sequester iron in the ferric form and outgrow other microflora (79,13). Using a very specific receptor on the outer membrane, these organisms take up the siderophore-iron complex which other organisms cannot use (82).

Fungal siderophores generally have a lower affinity than do the siderophores produced by plant growth promoting rhizobacteria (PGPR) (71,89,119), hence, they are better competitors for available iron.

Siderophores produced by fluorescent pseudomonads mediated iron regulated antagonism against a variety of phytopathogens in culture (33,64) and in soil (98). The fluorescent siderophores termed pyoverdines, pyochelin and pseudobactin represent only one class of sideropores produced by fluorescent *Pseudomonas* spp. (23). Pyoverdines produced by *Pseudomonas* spps. May deplete the medium of available iron, thereby inhibiting the growth of an indicator strain, while ferric complexes have no effect. Unlike microbial phytopathogens, plants are not adversely affected by localized depletion of iron caused by PGPR because plants can grow at much lower concentration than microorganisms (82). Also certain plants use microbial Fe (III) siderophores such as ferrioxamine B (25), the ferrichromes (81), rhodotorulic acid (Crowley et al 1988). The mechanism is that the plants transport Fe from microbial Fe (III) siderophores by reduction to Fe (II) and ligand exchange with phytosiderophores respectively, or by the direct uptake of the Fe (III) siderophores (9,111,113).

Although the concentrations of pyoverdines produced in the rhizosphere by *Pseudomonas* spp. are adequate, a number of factors like host plant, the target phytopathogen, edaphic and other environmental parameters and the affinity of the specific siderophore for iron determine its affectivity as a biocontrol agent.

Antibiotic mediated suppression

In vitro inhibition of the growth of pathogen by the biocontrol agent is the first evidence of role of antibiotic production in the disease suppression. The approaches to confirm the involvement of an antibiotic in the disease suppression include;

(a) Application of the purified compound in the place of the biocontrol agent in *in vivo* conditions and comparing with the efficiency of biocontrol agent, provided the chemical structure of the compound is known.

(b) Isolation of the inhibitory compound from the rhizosphere.

(c) Isolation of mutants defective in the production of inhibitory compound and comparison of disease control ability of the mutant with that of wild type.

The polyketide antibiotic 2,4-diacetylpholoroglucinol (DAPG) has received particular attention as it plays a key role in the ability of introduced *Pseudomonas fluorescens* strains to suppress a broad spectrum of crop Diseases (29,33,49). Pseudomonads produce a wide range of antibiotics such as phenazine carboxylic acid, pyoluteorin, pyrolnitrin, aceto phloroglucinols, salicylic acid, oomycin etc. (26,2977,78,90).

Table 2 enlists examples of certain antibiotics produced by pseudomonads and their effect on disease suppression. The antibiotics pyoluteorin (Plt), pyrrolnitrin (Prn), Phenazine-1-carboxylic acid (PCA), and 2-4-diacetyl phloroglucinol (Phl) are currently a major focus of research in biological control. The polyketide antibiotic 2,4- diacetylpholoroglucinol (DAPG) has received particular attention as it plays a key role in the ability of introduced *P. fluorescens* strains to suppress a broad spectrum of crop diseases.

Table 2. Antibiotics produced by Pseudomonas spp involved in disease control

Antibiotic(s)	Crop	Disease suppressed	Reference
Pyrolnitrin	Rice		(92)
Phenazine 1-carboxylic acid			
Pyocyanine Pyoluteorin	Cotton	Damping-off	(56)
Pyrolnitrin	Vegetable crops	Wilt	(54)
	Vegetable crops	*Pythium ultimum*	(47)
		Rhizoctonia solani	(57)
		Verticillium dahliae	
2-4-diacetyl-phloroglucinol	Tobacco	Black root rot	(27, 60)
	Wheat	Take-all	(51,59,110)

	Sugarbeet	Damping-off	(99)
Oomycin A	Cotton	Damping-off	(48, 57)
Phenazine 1- carboxylic acid and its derivatives	Wheat	Take all	(14, 46, 50, 83,104)
	Wheat	Spikeled blotch	(39)
Phenazine-1-Carboxylic acid		*Rhizoctonia* root rot	(58)
		Pythium aphanidermatum	(20)
		P. dissotocum	
Pyrolnitrin		*Fusarium sambucinum*	(16)

Conservation of *phl* genes for biosynthesis of DAPG among ecologically and geographically diverse antagonistic pseudomonads further supports the global importance of DAPG production in biocontrol (61,114). Keel et al (61) reported that DAPG producers carry *hcn* genes for producing broad spectrum hydrogen cyanide (HCN) indicating a relationship between the two metabolites.

Genes for Antifungal Toxins

Pseudomonas glumae strain EM 85 was found to produce antifungal antibiotic, the gene for which was found to be located on a 23 kb fragment of chromosomal DNA (2).

Burkhead et al (16) reported that *Pseudomonas cepacia* B 37 W produced pyrrolnitrin antibiotic inhibitory to *Fusarium. sambucinum.* Michereff et al (74) could correlate the *in vitro* inhibition of *Pythium* and *Rhizoctonia* by 2,4-diacetyl pholoroglucinol, an antibiotic produced by *P. fluorescens* PF 5 and *in vivo* control of sorghum anthracnose caused by *Colletotrichum gramminicola. P.fluorescens* F113 lac zy produces an antifungal pholoroglucinols which protects sugarbeet against *Pythium* mediated damping-off. The work of Carruthers et al (18) conclusively implicated the role of antibiotic in disease suppression. *Pseudomonas aureofaciens* PA 147-2 produces an antibiotic (AF++) which inhibits the growth of phytopathogens *in vitro* (18). They developed Tn5 antibiotic deficient mutant PA109 and tested along with the wild type for their ability to suppress root rot of *Asparagus officinalis*. Seedlings coinoculated with the pathogen and wild type strain showed a significantly reduced level of infection and disease severity compared to seedlings inoculated with the pathogen alone. However, all seedlings treated with Af – mutant PA 109 were diseased (18).

Enhanced antibiotics pyoluteorin and 2,4-diacetylphloroglucinol production in mutants of *P. fluorescens* strain CHAO resulted in corresponding improvement in suppression of *Pythium ultimum* induced diseases of cucumber plants. Introduction of cosmid pME3090 carrying 22kb of CHAO DNA into strain CHAO enhanced antibiotic pyoluteorin (Plt) and 2,4-diacetyl phloroglucinol (Phl) production by 3-5 fold *in vitro* and in an artificial wheat rhizosphere (72). The difference between the amounts of plt and phl produced by two strains was greater in presence than in absence of *P.ultimum*. Presence of phytopathogen induces the host plant to produce more nutrients in root exudates which in turn may stimulate *Pseudomonas fluorescens* CHAO/pME 3090 to overproduce the antibiotics Plt and Phl. Differential protection offered by mutants to different plants show that bacterial antibiotic production varies in the rhizosphere of different plants due to difference in root exudates (72). The correlation between six to seven fold over production of antibiotic phloroglucinol *in vitro* and in wheat rhizosphere environment with a corresponding enhancement in the biocontrol activity was observed activity (11,72).

Complementation of specific mutants defective in antifungal property have been employed for identification of gene(s) responsible for antifungal toxin biosynthesis (6,18,52). Genetic analysis show that the production of 2-4 diacetylphloroglucinol by *Pseudomonas aureofaciens* Q2-87 is coded by 4.8 kb DNA fragment and in *P.fluorescens* Q2-87 by a 5 kb DNA segment (6).

Anith et al (3) while studying the analysis of mutation affecting antifungal property of a fluorescent *Pseudomonas* sp during cotton-*Rhizoctonia* interaction observed that the mutant AN 21 generated from the wild type strain by chemical mutagenesis showed no antifungal property. The mutant was unable to produce an antifungal antibiotic and failed to inhibit the fungal growth. *In vivo* experiment also established that antifungal antibiotic was responsible for the disease control ability of the strain.

A number of antifungal metabolites produced by pseudomonads appears to be regulated by a global regulator (gac A) (41). This leads to difficulty in identification of structural genes for individual antifungal factors by mutagenesis, since regulatory mutants affected in the global regulation of all antifungal factors are commonly found (52).

Another strategy to genetically engineered *Pseudomonas* spp. with improved biocontrol activity is through increasing the amount of antibiotic produced by a given strain or by transferring biosynthetic loci into a heterologous strain (105). In another study, inactivation of pqq genes, responsible for biosynthesis of pyrroloquinoline quinine, a co-factor of different hydrogenases in *P. fluorescens* CHAO stimulated the production of antibiotic pyoluteorin (97). It is proposed that the possible mechanism of stimulation of antibiotics production is through shunting metabolites from other metabolic pathways to pyoluteorin biosynthesis pathway.

HCN Production

In the late exponential phase of growth many pseudomonads produce HCN (4). Antagonistic effect of *P. fluorescens* strain CHA0 on *Thielaviopsis basicola* in the rhizoplane of tobacco appears to be the result of HCN production (103,109). It has also been proposed that HCN may induce plant defense mechanism .

An HCN deficient mutant *P. fluorescens* derived from HCN producer strain CHAO, is less effective than the wild type of suppressing tobacco black root rot caused by *Thielaviopsis basicola* (103). This deficiency can be restored by a plasmid carrying a 5.0 kb genomic fragment from strain CHAO. The fragment contains three continuous open reading frames designated as ABC which encode cyanide synthase (108). Three genes lem A, gac A and anr have shown to be involved in the regulation of cyanogenesis in *Pseudomonas* spps. The lem and gac A genes globally control antibiotic regulator of arginine deiminase pathway and cyanogenesis (108,122). *Pseudomonas putida* strain BK 8661 which produces siderophore(s), antibiotics(s) and low levels of hydrogen cyanide (HCN) suppressed growth of *Septoria tritici* and *Puccinia recondita* f. sp. *tritici in vitro* and on wheat leaves. HCN over producing derivatives, the pleiotropic mutant of strain BK 8661 are deficient in siderophore and antibiotics production. HCN produced by the over producing bacterial strains resulted in a small but significant increase in the suppression of symptoms caused by *S. tritici* and *P. recondita* f. sp. *tritici* of wheat seedling leaves.

Lytic Enzymes

Production of lytic enzymes like chitinases and β-1,3-glucanases by certain bacteria forms the basis of control of plant-pathogenic fungi in the rhizosphere. Koby et al (65) reported the introduction of Tn7 based chiA gene into *P. fluorescens* and the construct could improve the biocontrol activity against *Rhizoctonia solani*.

Enzyme β 1,3 glucanase produced by *Pseudomonas stutzeri* YPL-1 could suppress *Fusarium solani, Sclerotium rolfsii* and *Pythium ultimum* by 85, 48 and 71% respectively (40). Production of hydrogen cyanide and ammonia has been reported as a mechanism of disease suppression by few bacteria (103).

Induction of systemic resistance

Plants develop systemic resistance to a variety of pathogens when inoculated previously with a pathogen or with inducing chemical compounds (66,70,91,93, 115). It is speculated that the aggressive colonization by the beneficial microorganisms, which itself is not a pathogen, may generate some signals that induce plant defence response.

Induced systemic resistance increases the activity of chitinases, β-1, 3-glucanases, protease, lipases, peroxidases and other pathogensis related proteins (PR). These PR proteins show antimicrobial activity *in vitro*. Some biocontrol PGPB strains have been found to produce chitinase, β-1,3-glucanase, protease and lipase that aids pathogen cell lysis (22). *P. stutzeri* secreting extracellular chitinase and laminarinase could digest and lyse *F. solani* mycelia resulting in suppression of root rot disease. Similarly suppression of diseases caused by *Rhizoctonia solani, S. rolfsii* and *P. ultimum* was mediated by β-1,3 glucanase producing strain of *Pseudomonas cepacia* (40). Transfer of chitinase gene (Chi A) into a PGPR *P. fluorescens* resulted in gene expression, secretion of chitinase enzyme and biocontrol of *R. solani* (65). PGPR induced ISR in *Arabidopsis* occurs without induction of transcripts of PRs (85). Similarly in radish, *Pseudomonas fluorcscens* strain WCS 417 mediated ISR is not associated with accumulation of PRs (51).

Ryu et al (93) opined that bacterial volatiles induced systemic resistance in *Arabidopsis thaliana.* They further demonstrated that the plant growth promoting bacteria induced systemic aquired resistance (SAR) in *A. thaliana* against cucumber mosaic virus by a salicylic acid and NPRI-independent and jasmonic acid-dependent signaling pathway(94).

Role of Ethylene

One mechanism by which plants respond to fungal phytopa-thogen infection is by synthesizing "stress" ethylene. Ethylene can induce some of the PR proteins e.g. β-1-3-glucanase and chitinase (1). Structural reinforcements of cell wall such as lignification and accumulation of hydroxyproline rich cell wall proteins are also enhanced by ethylene (10). However, reports exist that ethylene is not the signal involved in induction of systematic resistance (SAR) because SAR gene expression in ethylene-insensitive mutant of *Arabidopsis* is similar to the wild type plants (17). Majority of plant growth promoting bacteria appear to trigger induced systemic resistance via a SA-independent pathway involving jasmonate and ethylene signals (86).

Root Colonization and Competition

Extensive colonization of root surface by pseudomonads is essential for any efficient biocontrol activity because unsuccessful biocontrol field experiments are often correlated with poor colonization of the root system by the biocontrol bacteria. A positive correlation was observed between population size and biocontrol activity of *P. fluorescens* strain 2-79 against *Gaeumannomyces graminis* var. *tritici*, the take - all pathogen of wheat (15). The application of an increased population of *Pseudomonas putida* on seed pieces of potato to control the pre-emergent decay caused by *Erwinia carotovora* sub sp. *atroseptica* had a positive effect as reported by Xu and Gross (121). Similar observations were reported in case of *Pythium* damping off disease of cucumber controlled by *Pseudomonas cepacia* isolate 808 (37). Mutational studies to elucidate the molecular basis of rhizospheric colonization show that presence of flagella, the presence of O-antigen of lipopolysaccharides (LPS) and ability to synthesize amino acids are important colonization traits (30).

The process of root colonization involve migration of the introduced bacterial strain towards the plant root, attachment, distribution along the elongating root and proliferation. Though *Pseudomonas* strains have been shown to migrate towards soybean seeds in soil, no correlation between chemotactic ability and root colonization has been reported (96). Transfer of site-specific recombinase gene for a rhizosphere–competent *P. fluorescens* into a rhizosphere-incompetent *Pseudomonas* strain enhanced the ability to colonize the root tips (28).

Attachment of the bacterial cells to the root surface by agglutination has been suggested as an important aspect in the process of successful disease suppressive ability of the introduced strain. de Wegar ct al (31) reported that lipopolysaccharides play a major role in the agglutination process of pseudomonads to the root surface.

Brand et al (12) reported competition with respect to nutrient utilization in the rhizosphere as a major attribute required by the biocontrol agent, which is brought about by the root colonization ability and establishment.

Genetic studies on antagonistic pseudomonads with respect to disease suppression

Disease suppression by pseudomonads being a complicated process, mutant analysis has been used to establish the relative role of each mechanism exhibited by a particular bio-control agent. Two important requirements as far as this approach is concerned are :

(i) The production of mutants lacking ability to produce the antagonistic metabolite which is supposed to be responsible for biocontrol activity, and

(ii) A system which measures the disease suppression quantitatively in terms of plant growth and yield.

(iii) Genetic analysis and localization of genes responsible for the production of various antifungal metabolites produced by *Pseudomonas* sp. has been done by many workers (2,6,18,110).

The mutants generated should be impaired only in the antifungal metabolite production and otherwise isogenic to the wild type strain. Once the desired mutants are obtained either by chemical mutagenesis or by biological mutagenesis, the next step is to identify the DNA fragment which has been mutated. This could be achieved through a complementation approach, in which a cosmid or plasmid- based genomic library of the wild type strain is constructed and mobilized into the defective mutant and looked for complementation of the lost character.

A genomic library of an organism contains the whole genome of it, cloned to a specific vector molecule in an overlapping fashion of DNA fragments, and established and maintained in a bacterial host, usually *Escherichia coli).* Overlapping genomic DNA fragm-ents of the whole genome are generated by partial restriction of the genomic DNA with a specific restriction enzyme followed by size fractionation using ultra centrifugation. The fragments thus obtained are then cloned to a specific purpose vector linearized with the same restriction enzyme used for generation of the genomic fragments. The cloned molecules are then mobilized and stored in *E. coli* and maintained (80).

Cosmids are usually used as cloning vectors for establishment of gene libraries. They combine the characters of plasmids and bacteriophage vectors. They have the genetic information necessary for *in vitro* packaging of DNA into 1 phage heads in *cis* configuration. They are capable of carrying large fragments of DNA, and can be propagated in *E. coli* and handled as plasmids.

The molecular basis of antifungal toxin production by fluorescent *Pseudomonas* strain EM 85 was studied by Anith et al (3). The bacterial strain showed *in vitro* inhibition of growth of *R. solani.* NTC mutagenesias of the wild type strain helped in the isolation of an antifungal-toxin-defective mutant AN 21. A genomic library of the wild type strain was constructed (2) in the cosmid vector PlAFR 1 and maintained in an *E. coli* background. Complementation analysis with cosmid library resulted in the isolation of a cosmid clone which complemented the defective character in the mutant AN 21. The size of the complementing DNA fragment was found to be 23.5 kb.

Acknowledgement

One of the authors (KVBRT) is grateful to The National Academy of Sciences, India for providing financial support as Senior Scientist Platinum Jubilee Fellowship and to Dr. (Mrs) I.K. Kunwar, Department of Botany, Osmania University, Hyderabad, A.P. for critically going through the manuscript and suggestions.

REFERENCES

1. Abeles FB, Morgan PW and Saltveit ME Jr. 1992. Regulation of ethylene production by internal, environmental and stress factors. In. *Ethylene in Plant Biology*: II edn Academic Press. pp. 56-119.
2. Anith KN, Tilak KVBR, Khanuja SPS and Saxena AK. 1998. Cloning of genes involved in the antifungal activity of a fluorescent *Pseudomonas* sp. *World J Microbiol Biotech* 14: 939-941.
3. Anith KN, Tilak KVBR, Khanuja SPS and Saxena AK. 1999. Molecular basis of antifungal toxin production by fluorescent *Pseudomonas* sp. strain EM 85- A biological control agent. *Curr Sci* 77(5): 671-677.
4. Askeland RA and Morrison SM. 1983. Cyanide production by *Pseudomonas fluorescens* and *Pseudomonas aeruginosa*. *Appl Environ Microbiol* 45: 1802-1807.
5. Bais HP, Park SW, Weir TL, Callaway RM and Vivanco JM. 2004. How plants communicate using the underground information superhighway. *Trends Plant Sci* 9: 26-32.
6. Bangera MG and Thomashow LS. 1996. Characterization of a genomic locus required for synthesis of the antibiotic 2,4-diacetylphloroglucinol by the biological agent *Pseudomonas fluorescens* Q2-87. *Mol Plant-Microbe Interact* 9: 83-90.
7. Baker KF and Cook RJ. 1983. Nature and practice of biological control of plant pathogens. APS Press, Minnesota, USA.
8. Barbosa MAG, Michereff SJ, Mariano RLR and Maranthao E. 1995. Biocontrol of *Rhizoctonia solani* in cowpea by seed treatment with fluorescent *Pseudomonas* spp. *Summa-Phytopath*. 21: 151-157.
9. Bar-Ness E, Hadar Y, Shanzer A and Libman J. 1992. Iron uptake by plants from microbial siderophores, A study with 70-nitrobenz-2 oxa-1,3-diazole-des-ferrioxa-

mine as fluorescent ferrioxamine B analog. *Plant Physiol* 99: 1329-1335.

10. Boller T. 1990. Ethylene and plant-pathogen interactions. *Curr Top Plant Physiol* 5: 138-145.

11. Bonsall RF, Weller DM and Thomasshow LS 1997. Quantification of 2,4-Diacetylphloroglucinol produced by fluorescent *Pseudomonas* spp. *in vitro* and in the rhizosphere of wheat. *Appl Environ Microbiol* 63: 951-955.

12. Brand I, Luttenberg BJJ, Glandorf DCM, Baker PAMH, Schippers B and de Weger LA. 1991. Isolation and characterization of a superior potato root colonizing *Pseudomonas* strain. In : *Plant growth promoting rhizobacteria – Progress and Prospects*. Keel C et al (Eds.) IOBC/ WPRS Bulletin pp.350-354.

13. Briat JF. 1992. Iron assimilation and storage in Prokaryotes, *J Gen Microbiol* 138: 2475-2483.

14. Brisbane PG and Rovira AD. 1988. Mechanisms of inhibition of *Gaeumannomyces graminis* var. *tritici* by fluorescent pseudomonads. *Plant Pathol* 37: 104-111.

15. Bull CT, Weller DM and Thomashow LS. 1991. Relationship between colonization and suppression of *Gaeumannomyces graminis* var. *tritci* by *Psedomonas fluorescens* strain 2-79. Phytopathol. 81: 954-959.

16. Burkhead KD, Schisler DA and Slininger PJ. 1994. Pyrrolnitrin production by biological control agent *Pseudomonas cepacia* B. 37w in culture and in colonized wounds of potatoes. *Appl Environ Microbiol* 60 : 2031-2039.

17. Chang C, Kwok SF, Blucker AB and Meyerowitz EM. 1993. *Arabidopsis* ethylene – response gene ETR-1 similarity of product to two-component regulators. *Science* 262: 539-544.

18. Carruthers FL, Shum-Thomas T, Conner AJ and Mahanty HK. 1995. The significance of antibiotic production by *Pseudomonas aureofaciens* PA 147-2 for biological control of *Phytophthora megasperma* root rot of *Asparagus*. *Plant Soil* 170: 339-344.

19. Cartwright DK, Chiton WS and Benson DM 1995. Pyrrolnitriin and Phenazine production by *Pseudomonas cepacia*, strain 5.5B, a biocontrol agent of *Rhizoctonia solani*. *Appl Microbial Biotechnol* 433: 211-216.

20. Chatterton S, Sutton JC and Boland GJ. 2004. Timing

Pseudomonas chlororaphis applications to control *Pythium aphanidermatum, Pythium dissotocum* and root rot in hydroponic peppers. *Biol Control* 30: 360-373.

21. Chen C, Bauske EM, Musson G, Rodriguez-Kabana R and Kloepper JW. 1995. Biological control of *Fusarium* wilt on cotton by use of endophytic bacteria. *Biol Control* 5: 83-91.

22. Chet I and Inbar J 1994. Biological control of fungal pathogens. *Appl Biochem Biotechnol* 48: 37-43.

23. Compant S, Duffy B, Nowak J, Clement C and Essaid Ait B. 2005. Use of plant growth-promoting bacteria for biocontrol of plant diseases, mechanisms of action and future prospects. *Appl Environ Microbiol* 71(9): 4951-4959.

24. Cook RJ. 1993. Making greater use of introduced microorganisms for biological control of plant pathogens. *Ann Rev Phytopath* 31: 53-80.

25. Crowley DE, Reid CPP and Szaniszlo PJ. 1988. Utilization of microbial siderophores in iron acquisition by oat. *Plant Physiol* 87: 680-695.

26. Defago G. 1993. 2,4-diacetylphloroglucinol, a promising compound in biocontrol. *Pl Path* 42: 311-312.

27. Defago G, Berling CH, Burger U, Hass D, Kahr G, Keel C, Voisard C, Writhner P and Wuthrich B. 1990. Suppression of black rot of tobacco and other root disease by strains of *Pseudomonas fluorescens* : Potential applications and mechanisms. In: *Biological control of soil borne plant pathogens.* Hornby D. (Ed). Oxon CAB international pp. 93-108.

28. Dekkers LC, van der Bij AJ , Mulders IHM, Phoelich CC, Wentwood RAR, Glandorf DCM, Wijffelman A and Lugtenberg BJJ. 1998. Role of the O-antigen of lipopolysaccharide, and possible roles of growth rate and of NADH ubiquinone oxidoredutase (*nuo*) in competitive tomato root-tip colonization by *Pseudomonas fluorescens* WCS 365. *Mol Plant-Microbe Interact* 11: 763-771.

29. de Souza JT, de Boer M, de Waard P, van Beek TA and Raaijmakers JM. 2003. Biochemical, genetic and zoosporicidal properties of cyclic lipopeptide surfactants produced by *Pseudomonas fluorescens. Appl Environ Microbiol* 69: 7161-7172.

30. de Weger LA, Jann B, Jann K and Lutenberg B. 1987.

Lipopolysaccharides of *Pseudomonas* spp. that stimulate plant growth : composition and use for strain identification. *J Bacteriol* 169: 1441-1446.

31. de Weger LA, Van Loosdrecht MCM, Klaasen HE and Lugtenberg B. 1989. Mutational changes in physicochemical surface properties of plant growth stimulating *Pseudomonas* sp. do not influence the attachment properties of the cells. *J Bacteriol* 171: 2756-2761.

32. Dowling DN and O'Gara F. 1994. Metabolites of *Pseudomonas* involved in biocontrol of plant diseases. *TIBTECH* 12: 133-141.

33. Duffy B, Keel C and Defago G. 2004. Potential role of pathogen signaling in multitrophic plant-microbe interactions involved in disease protection. *Appl Environ Microbiol* 70: 1836-1842.

34. Duffy BK and Weller DM 1996. Biological control of take-all of wheat in the *pacific northwest of the USA using hypovirulent Gaeumannomyces graminis* var. *tritici* and fluorescent pseudomonads. *J Phytopath* 144: 585-590.

35. Duijff BJ, Gianinazzi-Pearson V and Lemanceau P. 1997. Involvement of the outer membrane lipolysacharides in the endophytic colonization of tomato roots by biocontrol *Pseudomonas fluorescens* strain WCS 417r. *New Phytol* 135 : 325-334.

36. Duijff BJ, Meijer JW, Bakker PAHM and Schippers B. 1993. Siderophore-mediated competition for iron and induced resistance in the suppression of *Fusarium* wilt of carnation by fluorescent *Psudomonas* spp *Neth J Plant Pathol* 99: 277-289.

37. Elad Y and Chet I. 1987 Possible role of competition for nutrition in biocontrol of *Pythium* damping off by bacteria. *Phytopathol* 77: 190-195.

38. Expert JM and Digat B. 1995 Biocontrol of *Sclerotium* wilt of sunflower by *Pseudomonas fluorescens* and *Pseudomonas putida* strains. *Can J Microbiol* 41 : 685-691.

39. Flaishman M, Eyal Z, Voisard C and Hass D. 1990 Suppression of *Septoria tritici* by phenazine or siderophore deficient mutants of *Pseudomonas*. *Curr Microbiol* 20: 121-124.

40. Fridlender M, Inbar J Chet I 1993. Biological control of soil borne plant pathogens by a β 1,3-glucanase producing

Pseudomonas cepacia. Soil Boil Biochem 25: 1211-1221.

41. Gaffney T, Friedrich L, Vermooij B, Negrotto B, Nye G, Ukness S, Ward E, Kessmann H and Ryals J. 1993. Requirement of salicyclic acid for the induction of systemic acquired resistance. *Science* 261: 754-756.

42. Glick BR. 2003. Plant growth promoting bacteria. In: *Molecular Biotechnology – Principles and Applications of Tecombinant DNA.* Glick BR and Pasternak JJ (Eds.) ASM Press Washington DC, USA pp. 436-454.

43. Gandhi SK and Kumar P. 2007. Antagonism of fungal and bacterial isolates against *Rhizoctonia solani. J Mycol Pl Pathol* 37(1); 178.

44. Gnanamanickam SS, Candole BL and Mew TW. 1992. Influence of soil factors and cultural practice on biological control of sheath blight of rice with antagonistic bacteria. *Pl Soil* 144: 67-75.

45. Gomes AMA, Peixoto AR, Mariano RLR and Michereff SJ. 1966. Effect of bean seed treatment with fluorescent *Pseudomonas* spp. on *Rhizoctonia solani* control. *Aruivas-de-Biologiae-Technologia* 39: 537-545.

46. Gurusiddaiah S, Weller DM, Sarkar A and Cook RJ. 1986. Characterization of an antibiotic produced by a strain of *Pseudomonas fluorescens* inhibitory to *Gaeumannomyces graminis* var. *tritici* and *Pythium* spp. *Antimicrob Agents Chemother* 29: 488-495.

47. Gutterson I. 1990. Microbial fungicides : Recent approaches to elucidating mechanisms. *Crit Rev Biotechnol* 10: 69-91.

48. Gutterson NI, Layton TJ, Ziegle JS and Warren GJ. 1986. Molecular cloning of genetic determinants for inhibition of fungal growth by a fluorescent pseudomonad. *J Bacteriol* 165: 696-703.

49. Haas D, Blumer C and Keel C. 2000. Biocontrol activity of fluorescent pseudomonads genetically dissected: Importance of positive feedback regulation. *Curr Opinion Biotechnol* II: 290-297.

50. Hamdan H, Weller DM and Thomashow LS. 1991. Relative importance of fluorescent siderophores and other factors in biological control of *Gaeumannomyces graminis* var. *tritici* by *Pseudomonas fluorescens* 2-79 and M 4-80R. *Appl Environ Microbiol* 57: 3270-3277.

51. Harrison LA, Letendre L, Kovacevih, Pierson EA and Weller DM. 1993. Purification

of an antibiotic effective against *Gaeumannomyces graminis* var. *tritici* produced by a biocontrol agent *Pseudomonas aureofaciens. Soil Biol Chem* 25: 215-221.

52. Hill SE, Hammer P and Ligon J. 1997. The role of antifungal metabolites in biological control of plant disease. In: *Technology Transfer of Plant Biotechnology*. Greshoff PM (Ed.) CRC Press Inc, Boca Raton, Florida.

53. Hill DS, Stein JI, Torkewitz NR, Morse AM, Howell CR, Pachlatko JD and Ligon JM. 1994. Cloning of genes involved in the synthesis of pyrrolnitrin from *Pseudomonas fluorescens* and role of pyrrolnitrin synthesis in biological control of plant disease. *Appl Environ Microbiol* 60: 78-85.

54. Homma Y and Suzui T. 1989. Role of antibiotic production in suppression of raddish damping off by seed bacterization with *Pseudomonas cepacia. Ann Phytopathol Soc* Japan 55: 643-652.

55. Howell CR and Stipanovic RD. 1979. Control of *Rhizoctonia solani* on cotton seedlings with *Pseudomonas fluorescens* and with an antibiotic produced by the bacterium. *Phytopathol* 69: 480-482.

56. Howell CR and Stipanovic RD. 1980. Suppression of *Pythium ultimum* induced damping off of cotton seedlings by *Pseudomonas fluorescens* and its antibiotic pyoluteorin. *Phytopathol* 70: 712-715.

57. Howie WJ and Suslow T. 1991. Role of antibiotic biosynthesis in the inhibition of *Pythium ultimum* in the cotton spermosphere and rhizosphere by *Pseudomonas fluorescens. Mol Plant-Microbe Interact* 4: 393-399.

58. Huang Z, Bonsall RF, Mavrodi DV, Weller DM and Thomashow LS. 2004. Transformation of *Pseudomonas fluorescens* with genes for biosynthesis of phenazine-1- carboxylic acid improves biocontrol of *Rhiz-octonia* root rot and *in situ* antibiotic production. *FEMS Microbiol Ecol* 49: 243-251..

59. Keel C, Schinder U, Maurhofer M, Voisard C Laville J, Murger U, Wirthner P, Hass D and Defago G. 1992. Suppression of root disease by *Pseudomonas fluorescens* CHAO. Importance of bacterial secondary metabolite 2,4-Diacetyl phloroglucinol. *Mol Plant Microbe Interact* 5: 4-13,

60. Keel C, Voisard C, Berling CH, Khar G and Defago G. 1989. Iron sufficiency, a prerequisite for suppression

of tobacco root rot by *Pseudomonas fluorescens* strain CHA0 under gnotobiotic conditions. *Phytopathol* 79: 584-589.

61. Keel C, Weller DM, Natsch A, Defago G, Cook RJ and Thomashow LS. 1996. Conversion of the 2,4-diacetylphloroglucinol biosynthesis locus among fluorescent *Pseudomonas* strains from diverse geographic locations. *Appl Environ Microbiol* 65: 552-563.

62. Khan MA and Gangopadhyay S. 2007. Efficacy of *Pseudomonas fluorescens* in controlling root rot of chickpea caused by *Macrophomina phaseolina*. *J Mycol Plant Pathol* 37(1): 200.

63. King EB and Parke JL. 1993. Biocontrol of *Aphanomyces* root and *Pythium* damping-off by *Pseudomonas cepacia* AMMD on four pea cultivars. *Plant Dis* 77: 1185-1188.

64. Kloepper JW, Leong J, Teintze M and Schroth MN. 1980. *Pseudomonas* siderophores : A mechanism explaining disease suppression in soils. *Curr Microbiol* 4: 317-330.

65. Koby S, Schickler H, Chet I and Oppenheim AB. 1994. The chitinase encoding Tn7-based chi gene endows *Pseudomonas fluorescens* with the capacity to control plant pathogens in soil. *Gene* 147: 81-83.

66. Krishnamurthy K and Gnanamanickam SS. 1997. Biological control of sheath blight of rice : Induction of systemic resistance in rice plant associated *Pseudomonas* spp. *Curr Sci* 72: 331-334.

67. Krishnamurthy K and Gnanamanickam SS. 1998. Biocontrol of rice sheath blight with formulated *Pseudomonas putida*. *Indian Phytopath* 51: 233-236.

68. Kumar BSD and Dube HC. 1992. Seed bacterization with a fluorescent *Pseudomonas* for enhanced plant growth, yield and disease control. *Soil Biol Biochem* 24: 539-542.

69. Leeman M, Pelt J A-van, Hendrickx MJ and Scheffer RJ. 1995. Biocontrol of *Fusarium* wilt of radish in commercial greenhouse trails by seed treatment with *Pseudomonas fluorescens* WCS 374. *Phytopath* 85: 1301-1305.

70. Liu L, Kloepper JW and Tuzan S. 1995. Induction of systemic resistance in cucumber against *Fusarium* wilt by growth promoting rhizobacteria. *Phytopathol* 85: 695-698.

71. Loper JE and Henkels MD. 1999. Utilization of heterologous siderophores enhances levels of iron available to *Pseudomonas putida* in the rhizosphere. *Appl Environ Microbiol* 65: 5357-5363.

72. Maurhofer M, Keel C, Haas D and Defago G. 1995. Influence of plant species on disease suppression by *Pseudomonas fluorescens* strain CHAO with enhanced antibiotic production. *Plant Pathol* 44: 40-50.

73. Mazzola M and Cook RJ. 1991. Effects of fungal root pathogens on the population dynamics of biocontrol strains of fluorescent pseudomonads in the wheat rhizosphere. *Appl Environ Microbiol* 57: 2171-2178.

74. Michereff SJ, Silveira NSS and Mariano RLR. 1994. Antagonism of bacteria to *Colletotrichum graminicola* and potential for biocontrol of sorghum anthracnose. *Fitopatologia Brasileira* 19 : 541-54.

75. Moulin F, Lemanceau P and Alabouvette C. 1996. Suppression of *Pythium* root of cucumber by a fluorescent pseudomonad is related to reduced root colonization by *Pythium apahanidermatum. J Phytopath* 144: 125-129.

76. Nautiyal CS. 1997. Selection of chickpea-rhizosphere-competent *Pseudomonas fluorescens* NBRI 1303 antagonistic to *Fusarium oxysporum* f.sp. ciceri. Curr *Microbiol* 35 : 52-58.

77. Nautiyal CS. 2006. Biological cnrol of plant diseases by natural and genetically engineered fluorescent *Pseudomnas* spp. In:*Microbial biotechnology in horticulture. Science Publishers.* Ray RC and Owen PW (Eds.) Science Publishers, (NH),USA, pp.125-162.

78. Nautiyal, C. S.,Tilak, K. V. B. R. 2009. Exploitation of rhizobacteria to improve soil, plamt, human and the environment health. In: *Recent advances in biopesticides biotechnlogcal applications.* Johri JK (Ed). New India Publ. Agency, New Delhi, pp.117-153.

79. Neilands JB and Leong SA. 1986. Siderophores in relation to plant growth and disease. *Ann Rev Plant Physiol* 37: 187-208.

80. Old RN and Primorse SB. 1989. *Principles of gene manipulation* Blackwell Scientific Publications, Oxford.

81. Orlando JA and Neilands JB. 1982. Ferrichrome compounds as a source of iron for higher plants. In: *Chemistry and Biology of Hydroxamic Acids,* Kehl H and Karger S (Eds.) Basel, Switzerland.

82. O'Sulivan DJ and O'Gara F. 1992. Traits of fluorescent *Pseudomonas* spp. involved in suppression of plant root pathogens. *Microbiol Rev* 56: 662-676.

83. Ownley BH, Weller DM and Thomashow LS. 1992. Influence of *in situ* and *in vitro* pH on suppression of *Gaeumannomyces graminis* var. *tritici* and non pathogenic *Fusarium oxysporum*. *Phytopathol* 78: 190-194.

84. Pierson EA and Weller DM. 1994. Use of mixtures of fluorescent pseudomonads to suppress Take-all and improve the growth of wheat, *Phytopathol* 84: 940-947.

85. Pieterse CMJ, van Wees M, Hoffland E, Van Pelt JA and Van Loon LC. 1996. Systemic resistance in *Arabidoposis* induced by biocontrol bacteria is independent of salicyclic acid accumulation and pathogenesis related gene expression. *Plant Cell* 84: 940-947.

86. Pieterse CMJ, van Wees M, Van Pelt JA, Knoester M, Laan R, Gerrits H, Weisbeek PJ and van Loon C. 1998. A novel signaling pathway controlling induced systemic resistance in *Arabidopsis*. *Plant Cell* 10: 1571-1580.

87. Pal KK, Tilak KVBR, Saxena AK, Dey R and Singh CS. 2000. Antifungal characteristics of a fluorescent *Pseudomonas* strain involved in the biological control of *Rhizoctonia solani*. *Microbiol Res* 155: 233-242.

88. Pal KK, Tilak KVBR, Saxena AK, Dey R and Singh CS. 2001. Suppression of maize root diseases caused by *Macrophomina phaseolina, Fusarium moniliforme* and *Fusarium graminearum* by plant growth promoting rhizobacteria. *Microbiol Res* 156: 209-223.

89. Raaijmakers JM, Vandersluis I, Koster M, Bakker PAHM, Weisbeek PA and Schippers B. 1995. Utilization of heterologous siderophores and rhizosphere competence of fluorescent *Pseudomonas* spp. *Can J Microbiol* 41: 126-135.

90. Raaijmakers JM, Vlami M and de Souza JT. 2002. Antibiotic production by bacterial biocontrol agents. *Antonie Leeuwenhoek* 81: 537-547.

91. Ramamoorthy V, Viswanathan R, Raghuchander T, Prakasam V. and Samaiyappan R. 2001. Induction of systemic resistance by plant growth-promoting rhizobacteria in crop plants against pests and diseases. *Crop Prot* 20: 1-11.

92. Rosales AM, Thomashow L, Cook RJ and Mew TW. 1995. Isolation and identification of

antifungal metabolites produced by rice-associated antagonistic *Pseudomonas* spp. *Phytopathol* 85: 1028-1032.

93. Ryu CM, Farag MA, Hu CH, Reddy MS, Kloepper JW and Pare PW. 2004 a. Bacterial volatiles induce systemic resistance in *Arabidopsis. Pl Physiol* 134: 1017-1026.

94. Ryu CM, Murphy JF, Mysore KS and Kloepper JW. 2004 b. Plant growth promoting rhizobacteria systemically protect *Arabidopsis thaliana* against cucumber mosaic virus by a salicylic acid and NPRI-independent and jasmonic acid –dependent signaling pathway. *The Plant J* 39: 381-392.

95. Saxena AK, Pal KK and Tilak KVBR. 2000. Bacterial biocontrol agents and their role in plant disease management. In *BioControl Potential and its Exploitation in Sustainable Agriculture. Vol.1. Crop Diseases, Weeds and Nematodes.* Upadhyay RS, Mukerji KG and Chamola BP (Eds.) Kluwer Academic/Plenum Publishers, New York, pp. 25-37.

96. Scher FM, Kloepper JW, Singleton CA, Zaleska I and Laliberte M. 1988. Colonization of soybean roots by *Pseudomonas* and *Serrratia* species : relationship to bacterial motility, chemotaxis and generation time. *Phytopathol* 78 : 1055-1059.

97. Schnider U, Keel C, Voisard C, Defago G and Hass D. 1995. Tn5-directed cloning of pqq genes from *Pseudomonas fluorescens* CHAO : Mutational inactivation of the genes results in over production of antibiotic pyluteorin, *Appl Environ Microbiol* 61: 3856-3864.

98. Schipper B, Bakker AW and Bakker PAHM. 1987. Interactions of deleterious and beneficial rhizosphere microorganisms and the effect of cropping practices. *Ann Rev Phytopathol* 25: 339-358.

99. Shanahan P, Borro A, O'Gara F and Glennon JD. 1992. Isolation, trace enrichment and liquid chromatographic analysis of diacetyl phloroglucinol in culture and amperometric detection. *J Chromatogr* 606: 171-177.

100. Singh R and Singh AP. 2007. Management of sheath blight of rice with *Pseudomonas fluorescens*. *J Mycol Pl Pathol* 37(1): 18-21.

101. Snyder WC, Watts GW and Smith SN. 1976. Biological control of plant

102. pathogens. In: *Theory and practice of biological control.*

Huffaker CBand Messenger PS, Academic press, NY pp. 521-535.

103. Srivastava R, Khalid A, Sharma AK and Singh US. 2007. Management of *Fusarium* wilt of chilli and tomato using biointensive integrated crop management. *J Mycol Pl Pathol* 37(1): 203.

104. Stutz EM, Defago G, and Kerri H. 1986. Naturally occurring fluorescent pseudomonads involved in suppression of black root rot of tobacco, *Phytopath* 76 : 181-185.

105. Thomashow LS and Weller DM. 1988. Role of a phenazine antibiotic from *Pseudomonas fluorescens* in biological control of *Gaeumannomyces graminis* var. *tritici*. *J Bacteriol* 170: 3499-3508.

106. Thomashow LS and Weller DM. 1996. Current concepts in the use of introduced bacteria for biological disease control: mechanisms and antifungal metabolites. In: *Plant-Microbe Interactions* Vol.1 Stacey G and Keen N (Eds.) Champman and Hall, New York. pp.187-235.

107. Tilak K.V.B.R., Pal. K.K. and Dey, R. 2010. *Microbes for sustainable agriculture*. I.K. International Publ (India) Pvt. Ltd. New Dlhi, India, pp. 200.

108. Vidhyasekharan P and Muthamilan M. 1995. Development of formulations of *Pseudomonas fluorescens* for control of chickpea wilt. *Plant Dis* 79: 782-786.

109. Voisard C, Bull CT, Keel C, Laville J, Maurhofer M, Schinider U, Defago G and Hass D. 1994. Biocontrol of root diseases by *Pseudomonas fluorescens* CHAO: Current concepts and experimental approaches. In: *Molecular Ecology of Rhizosphere Microorganisms*. O'Gara F, Dowling DN and Boesten B (Eds.) VCH Weinhaim, Germany pp. 67-89.

110. Voisard C, Keel C, Hass D. and Defago G. 1989. Cyanide production by *Pseudomonas fluorescens* helps suppress black rot of tobacco under gnotobiotic system. *EMBO J* 8: 351-358.

111. Vincent MV, Harrison LA, Brackin JM, Kovacevic PA, Mukerji P, Weller DM and Pierson EA. 1991. Genetic analysis of the antifungal activity of a soilborne *Pseudomonas aureofaciens* strain. *Appl Environ Microbiol* 57: 2928-2934.

112. 111 .Visca P, Ciervo A, Sanfilippo V and Orsi N. 1993. Iron regulated salicylate synthesis by *Pseudom-*

onas spp. *J Gen Microbiol* 139: 1995-2001.

113. Walker R, Emslie KA and Allan EJ. 1996. Bioassay methods for the detection of antifungal activity by *Pseudomonas antimicrobica* against the grey mould pathogen *Botrytis cinerea. J Appl Biotechnol* 81: 531-537.

114. Wang J, Budde AD and Leong SA. 1989. Analysis of ferrichrome biosynthesis in the phytopathogenic fungus *Ustilago maydis* : cloning of an ornithine-N-Oxygenase gene. *J Bacteriol* 171: 2811-2818.

115. Wang CN, Ramette A, Panjasamernwong P, Zala M, Nalseh A, Moeme-Loecoz Y and Defago G. 2001. Cosmopolitian distribution of phlD-containing diacetylphloroglucinol crop associated pseudomonads of worldwide origin. *FEMS Microbiol Ecol* 37: 105-116.

116. Ward ER, Uknees SJ, Williams SC, Dincher SS, Weiderhold DL, Alexander DC, Ahl-Goy P, Metraux JP and Ryals JA 1991. Coordinate gene activity in response to agents that induce systemic acquired resistance. *Plant Cell* 3: 1085-1094.

117. Wei G, Kloepper JW and Tuzun S. 1996. Induced systemic resistance to cucumber diseases and increased plant growth by plant growth promoting rhizobacteria under field conditions. *Phytopath* 86: 221-224.

118. Weller DM. 1988. Biological control of soil borne plant pathogens in the rhizosphere with bacteria. *Ann Rev Phytopathol* 26: 379-407.

119. Weller DM and Thomashow LS. 1993. Use of rhizobacteria for biocontrol. *Curr Opinion Biotechnol* 4: 306-311.

120. Whipps JM. 1997. Developments in the biological control of soil-borne plant pathogens. *Adv Bot Res* 26: 1-133.

121. Wu-Ws 1995. Biocontrol of seedling blight of cucumber and poinsettia. *Pl Path Bull* 4 : 97-105.

122. Xu GW and Gross DC. 1986. Field evaluation of interactions among fluorescent pseudomonads, *Erwinia carotovora* and potato yields. *Phytopathol* 76: 423-430.

123. Zimmermann A, Reimmann C, Galimand M and Haas D 1991. Anaerobic growth and cyanide synthesis of *P. aeruginosa* depend on *anr*, a regulatory gene homologous with *fnr*. of *E. coli. Mol Microbiol* 5: 1483-1490.

PLANT DISEASES AND CHANGING WAYS OF THEIR MANAGEMENT

Muralidharan, K.[1,2] *and Raghu, P.*[3]

[1]Formerly Project Director (i/c), Principal Scientist & Head, Crop Protection, Directorate of Rice Research (ICAR), Hyderabad;
[2]Presently: Director (Technical), Sowbhagya Biotech (P.) Ltd., #79 Industrial Development Area II, Cherlapally, Hyderabad 500 051 India
[3]Department of Botany, Osmania University 500 007, India.
E-mail: muralidharan_km@yahoo.com, raghugenes@yahoo.com

KEY WORDS: Plant disease, resistance, biological control, chemical control, genomics

ABSTRACT

As the world population is marching towards 8 billion and is expected to be 10 billion by 2050, famine is still a very real threat to most of the human populace. Dwindling land, water and nutrient resources limit human capacity to produce food and other plant products. Additionally, plant diseases take a heavy toll on whatever is produced. It is a primary necessity to prevent yield losses due to plant diseases. Plant diseases cause never-ending problems. Sound management of plant diseases requires knowledge and judgment because the more we know about the pathogen, the more chance we have of producing a healthy crop. Research is focused to gather information and knowledge about plant pests, their capabilities, and resources, about crop plants and plant resistance, and about the environment in which plant diseases occur. The object is to provide crop plants with every possible advantage against pathogens, to manage diseases at the least possible cost in monetary terms and cost to the environment, and to minimize losses. Successful disease management costs money. Therefore, farmers must know how to expend their resources wisely. If they do not, costs and losses may be intolerably high. Crop yield may be lost, and then the money spent for disease management is wasted. Every rupee spent in pest management should return three to four rupees to farmers.

Here we discuss in details on the pathogen, disease, resistance, and biological and chemical control methods as well as what the new developments in genomics can offer for an efficient and cost-effective control of the plant diseases.

INTRODUCTION

Continuous pest management is basic to efficient production. Plant pests are extremely resourceful and constantly challenge the judgment, skills, and resources of the farmer. Hence, farmers must know not only the current capabilities of the pathogen at any given time and place, but also their own ability to apply corrective measures. Farmers must make suitable preparations in advance of the outbreaks of disease that are bound to come. Sometimes the pathogens can infect plants without warning; plant resistance breaks down, and epidemics occur. A successful strategy to manage plant diseases involves two main objectives: to reduce the initial amount of the pathogen present and to reduce the rate of pathogen increase and spread. Suitable management systems must be developed. Machinery and resources must be available in adequate amounts at the correct time and place to implement management strategies. Management practices must be safe, efficient, and effective. Farmers need to know what each management practice is for, how and when it is to be employed, and how effective it is. They should use chemicals wisely and avoid contaminating or polluting the environment. Overall disease management practices may be complex, especially where more than one plant pest is involved. Successful management may require the use of various combinations of measures that assist and strengthen each other. Sometimes diseases occur unexpectedly, and large yield losses occur. Such occurrences require a re-evaluation of the present management strategy and the development and employment of new methods designed to prevent recurrent disasters.

Successful management of plant diseases must not be neglected if a hungry world is to be fed.

CAUSAL AGENTS IN PLANT DISEASES

A plant disease is defined as any disturbance that prevents normal development of plant and reduces its economic or aesthetic value. A disease interferes with the normal function of some part of the plant resulting in lower yields or reduced quality. Diseases reduce income to farmers, availability of quality produce in adequate quantities to consumers and lower natural beauty while increasing maintenance costs. Disease are caused by living agents such as microorganisms (pathogens) and parasitic plants (biotic agents) and non-living agents like environmental factors, faulty nutrition and toxic chemicals (abiotic agents). A plant disease develops over a period of time while from insect feeding or damage from equipment, machinery and chemical use occur immediately or within a short period of time. Usually a disease begins several days before the appearance of visual symptoms. Microorganisms that cause plant disease include nematodes, fungi, bacteria and mycoplasma. As viruses and viroids must have a living cell for reproduction and are composed of nucleic acid and protein, they can also be classified as biotic agents. Seed producing higher plants that are parasitic on other plants are also pathogens. Each type of causal agent is discussed briefly.

Nematodes

Since the beginnings of agriculture, a potential enemy has lurked in every farmland and garden soils. It is called the nematode. There are thousands of kinds of nematodes. Probably every living thing-animal or plant-bigger than a nematode is attacked or parasitized by them. Nematodes that feed on animals, however, will not or cannot feed on plants; likewise nematodes that feed on plants do not attack animals. This is a general rule in biology: Plant parasites do not attack animals; animal parasites will not attack plants. Other reasons these remarkable creatures escape notice are that they are microscopic in size or barely visible to the naked eye, and almost transparent. They spend their lives hidden in the tissues of host plants or in the soil near the roots. Another unusual thing about plant-parasitic nematodes is that they neither feed nor reproduce except on living plants. To survive, each individual nematode, after hatching, must reach a plant on which it can feed before its reserve food supply is exhausted. That means the nematode must swim through the soil in a film of water in search of food. Nearly 11% of the production of the 21 most important crops is lost worldwide to nematodes each year. Plant pathogenic species of nematodes are of tremendous importance as disease agents. All plants are parasitized by nema-todes, but only a few hundred species of nematodes are known to be plant parasites. Nematodes constitute a significant part of that threat in tropical countries, where they are a more serious and complex problem. Warm, sandy soils are favorable to nematode infection, especially in irrigated areas used for continuous crop production. Perennial crops, and annual crops grown in the same fields year after year, often are so seriously attacked by nematodes that such crops can barely survive. The people however, depend on perennial crops and non-rotated crops for their food supply.

Kingdom Animalia - Phylum Nematoda consists of 197 Genera, 4300 species of phytoparasites They are obligate parasites and cannot culture axenically, but can be cultured monoxenically. Most damaging plant parasitic nematodes are *Meloidogyne* (causing root-knot), *Pratylenchus* (lesion nematode), and *Globodera* (cyst nematodes), *Ditylenchus* (stem and bulb nematode). *Tylenchulus* (citrus nematode), *Xiphinema* (dagger nematode), *Radopholus* (burrowing nematode), *Rotylenchulus* (reniform nematode), *Helicotylenchus* (spiral nematode) and *Belonolaimus* (sting nematode). *Anguina tritici* is a common parasite of wheat on the plateaus. *Aphelenchoides besseyi* (white tip disease of rice), is responsible for major losses of this cereal. *Aphelenchoides ritzemabosi* is widespread in hilly areas of northern India on *zinnia* and *A. agarici* seems to be a destructive pest of cultivated mushrooms in India. The stem nematode, *Ditylenchus dipsaci*, attacks several crops and causes major losses of onion, garlic, broad bean and strawberry crops. *Helicotylenchus indicus* causes some damage to vegetable crops, but only when at high population densities and *H. multicinctus* occurs wherever bananas are cultivated. *Pratylenchus coffeae* is an important pest of coffee, banana, cowpea and citrus. *Radopholus similis* is a destructive parasite of banana and coconut palms. *Hemicriconemoides mangiferae* may be the most widely associated nematode species with mango. *Tylenchorhynchus brassicae* is the most frequently occurring in the rhizosphere of crops. Other common species are *T. claytoni* and *T. indicus*. *Rotylenchulus reniformis* is widespread and causes economic damage to vegetable crops, fruit-trees and staple and fibre crops. The citrus nematode, *T. semipenetrans* is invariably associated with citrus groves. *Paralongidorus citri* has been consistently found associated with declining groundnut crops. *Xiphinema elongatum* is a parasite of sugarcane.

Fungi

Fungi are microorganisms that contain no chlorophyll and usually reproduce by producing spores. Although they have sometimes been classified as simple plants, they have no roots, stems or leaves. True fungi are considered as kingdom Eumycota, which includes Chytridiomycota, Zygomycota, Ascomycota and Basidiomycota. The Fungi Imperfecti or Deuteromycotina being anamorphic fungi whose perfect states lie either in Ascomycotina or Basidiomycotina has not been included as one of the phyla. Kingdom Chromista includes Hyphochytridiomycetes and Omycetes. The slime molds, protostelids and Plasmodiophoromycota belong to the kingdom Protozoa (15, 18). The fungi are a very old group of microorganisms that have lived on earth for more than two billion years. Fossils of fungi have been found. Approximately 100,000 species of fungi have been described, and more are being found each year. Mushrooms, toadstools, molds, and mildews are fungi known by most everyone, but many fungi are so small that it takes an expert to find and see them through a microscope. The body or vegetative stage of a fungus consists of branching, delicate, threadlike structures called hyphae. A single strand is known as a hypha. Many hyphae together form an interlacing tangle known as a mycelium (plural mycelia). The hyphae of many fungi are divided by crosswalls or septae (singular septum) into definite cells full of protoplasm. The protoplasm is held within a semipermeable cytoplasmic membrane containing one or more nuclei. Hyphae of other fungi possess no septae or crosswalls and are essentially long tubes containing protoplasm; the streaming back and forth of the active liquid protoplasm is often visible under the microscope. Mycelia may form on or inside the host plant or on decaying organic matter and have different branching habits and structures, which help identify the fungus. Together with bacteria, fungi are principal agents of decay, and through the decomposition of organic matter they play an essential role in recycling nutrients and in the nutrition of plants. All fungi are capable of breaking down complex organic foods into simple substances and using these substances as sources of energy. The fungi that can live exclusively on decaying organic matter are called saprophytes. Fungi decompose all sorts of materials including books, cloth, food, fruits, leather, meat, nuts, paper stored seed, twine, vegetables, and wood. A symbiotic association of beneficial fungi is formed with plant roots. These fungi form mycorrhizae and live primarily in (endomycorrhizae) or on (ectomycorrhizae) roots. The fungi obtain needed nutrients from the plant roots, but also aid the roots in the absorption of certain nutrients such as phosphorus. Some mycorrhizae also may help protect roots against invasion by harmful or pathogenic fungi.

Some harmful fungi cause diseases of humans, other animals and especially plants. About 10000 species of fungi can cause plant diseases, and all plants are attacked by some fungi. Some plant parasitic fungi can attack many species of plants; others attack only one type of plant. Some fungi can grow and multiply only on a living host and are called obligate parasites. Others can grow and multiply on dead organic matter as well as on living plants and are called facultative or non-obligate parasites. In general, fungi that cause plant diseases will not attack humans, and fungi that attack animals also do not cause plant diseases.

A few substances produced by plant-parasitic fungi in food called mycotoxins, are toxic to animals. When cereal grains such as sorghum, wheat, maize, rice or millets, or foods such as groundnuts are placed in storage, certain fungi can grow on the stored seeds if adequate moisture is present. Some infections of the seeds occur in the field, some in storage. Growth of the *Aspergillus* fungi results in the production of aflatoxin. If the level of aflatoxin is sufficiently high, it is poisonous to poultry, swine cattle, and, in some cases, humans. Certain species of *Fusarium* also can grow on cereal grains and produce a range of dangerous toxic substances. Ergot of rye (caused by the fungus *Claviceps purpurea* also results in the contamination of the grain with sclerotia or survival structures of the fungus that contain several highly toxic and often hallucinogenic chemicals harmful to livestock and humans. Plant diseases manifest themselves in various symptoms such as wilts, rots, antracnose, leaf spots, blu moulds, etc. Crop losses occur yearly due to plant diseases and post-harvest spoilage of food caused by fungi throughout the world. A portion of farmers' investment in crop production is spent annually for chemicals fungicides to combat pathogenic fungi. More amounts are spent in breeding resistant plants throughout the world.

Bacteria

Bacteria are essential to human existence. Bacteria are the humankind's most useful servants as well as the most destructive masters. Bacteria produce useful food and chemicals, increase the fertility of soil, fix nitrogen from the air and aid in digestion. Together with fungi, bacteria decompose the complex bodies of dead plants and animals, and salvage the vital chemical elements and compounds for re-use. Bacteria have caused more deaths than all wars. Bacteria have defeated armies and destroyed nations. Yet most bacteria are beneficial, not harmful. In 1876 Robert Koch showed that a specific bacterium was the cause of specific disease-anthrax-in animals. A few years later, in 1882, Louis Pasteur identified bacteria as the cause of fermentation. Many plant parts, when deprived of free oxygen, still continue to give off carbon dioxide and often produce alcohol and other compounds. This process is called anaerobic respiration or fermentation. The theory that living agents cause fermentation became the basis of modem bacteriology and laid the foundation for aseptic surgery, the control of epidemics, preservation of foods, purification of water and milk (pasteurization), and sewage disposal.

The two prokaryotic domains (or empires), Bacteria (or Eubacteria) and Archaea (or Archaeobacteria), are subdivided into 30 phyla (or divisions): 27 phyla in the domain Bacteria, and 3 phyla in the domain Archaea. The class Ktedonobacteria is currently not classified in a phylum (http://www.bacterio.cict.fr/classifphyla.htm). When bacteria fill the water-conducting cells of a plant and kill plant tissues by means of toxins, the plant suffers a wilt or blight disease. If the bacteria kill leaf cells, leaf spots may develop. When fruits, tubers, or roots are infected, soft rots may develop, which result in a slimy, foul-smelling mass. Other bacteria stimulate tissue growth to form galls or tumors that resemble cancerous tissues of animals. In addition to killing plant tissues, bacteria can stunt the growth of plants, change plant color from green to brown or black, or cause distortion of leaves, stems, flowers, or fruits.

Plant Viruses and Viroids

The discovery of viruses as causal agents of disease is major event in understanding of the life sciences and diseases. We now know that all living things animals, plants, even fungi and bacteria-are attacked by viruses. The invention of the light microscope permitted us to see and study bacteria and helped produce the germ theory of disease. But until about 1935 many human diseases - such as chicken pox, colds, infantile paralysis, influenza, measles, mumps, smallpox, scarlet fever, and yellow fever-for which no causal organism could be observed, were believed to be caused by ultra-microscopic organisms that were beyond the range of the microscope. Similarly, some plant diseases confounded the best-trained and most efficient workers. Some of these diseases were known to early workers by symptom names as chlorosis, dwarf, etch, leaf curl, leaf roll, mosaic, mottling, vein banding, and yellows. A few diseases were readily transmitted by grafting a diseased plant to a healthy one or by rubbing sap of a diseased plant onto a healthy plant. Other similar diseases were rarely or never transmitted to healthy plants by either method. Many of these puzzling diseases caused a partial yellowing of plants, and some have been shown to be caused by mycoplasmas.

Viruses are very small particles, usually less than 300 nm in length, composed of nucleic acid and protein. They can be seen only under an electron microscope at magnifications of 2000 to 3000 times. Viruses that infect plants have several different shapes: some are spherical; some are, comparatively speaking, long rigid rods; some are long flexuous rods; and some are short thick rods. Nucleic acid (usually ribonucleic acid or RNA) composes the center of the particles and is surrounded by and connected to a protein coat. The RNA in viruses directs certain enzyme systems in plant cells to manufacture components used to produce more virus particles. Viruses cause diseases in plants by diverting energy and structural components, normally used in plant growth, into reproductive processes for the virus. Viruses are spread from one plant to another by insect or nematode vectors or by humans. Another major means of spread is in infected plant parts. Viroids are smaller than viruses. They act like viruses, but the infectious particle is simply a strand of RNA and contains no protein. Therapeutic chemicals that inactivate plant viruses and viroids are not yet available.

In the 1890s, scientists determined that tobacco mosaic was caused by a contagious living fluid that did not contain bacteria. In 1935 W. M. Stanley isolated and purified some tiny white crystals from the leaves of mosaic-infected tobacco plants. He had crystallized the proteinaceous part of tobacco mosaic virus. He wrote: "*Tobacco mosaic virus is regarded as an autocatalytic protein, which, for the present, may be assumed to require the presence of living cells for multiplication.*" He was correct that a protein was involved and that living cells were essential for virus multiplication. For this important work Stanley received the Nobel Prize for chemistry in 1946. However, Stanley did not have a complete understanding of the nature of plant viruses. F. C. Bawden and N. W. Pirie, of England, demonstrated that a virus is really a nucleoprotein and, therefore, contains both protein and nucleic acid. Bawden also found later that the nucleic acid was required for infection of plants, and that the protein served only to protect the nucleic acid component. About 1945 the electron microscope was invented. With the aid of this new tool, scientists were able to see virus particles, which are so small that they are invisible under the highest power of ordinary microscopes. Since the 1940s, developments in virology have been rapid. In the early 1950s, F. H. Crick and J. D. Watson, working in England, set forth their ideas on the double helix structure of deoxyribonucleic acid (DNA) and its replication. The work of Crick and Watson permitted us to visualize how viruses reproduce and, more than that, how genes and chromosomes replicate.

In 1971, T O. Diener described a new class of infectious units called viroids, composed only of nucleic acids. Viroids are plant pathogens that consist of a short stretch (a few hundred nucleobases) of highly complementary, circular, single-stranded RNA without the protein coat that is typical for viruses. The smallest discovered is a 220 nucleobase scRNA (small cytoplasmic RNA) associated with the rice yellow mottle sobemovirus (RYMV) (22). In comparison, the genomes of the smallest known viruses capable of causing an infection by themselves are around 2 kilobases in size. Viroid RNA does not code for any protein. The replication mechanism involves RNA polymerase II, an enzyme normally associated with synthesis of messenger RNA from DNA, which instead catalyzes rolling circle synthesis of new RNA using the viroid's RNA as template. Some viroids are ribozymes, having catalytic properties which allow self-cleavage and ligation of unit-size genomes from larger replication intermediates (109). The first viroid to be identified was the potato spindle tuber viroid (PSTVd). Some 33 species have been identified and primary and secondary structure of the PSTVd viroid is now known. There has long been confusion over how viroids are able to induce symptoms in plants without encoding any protein products within their sequences. Evidence now suggests that RNA silencing is involved in the process. First, changes to the viroid genome can dramatically alter its virulence. This reflects the fact that any siRNAs produced would have less complementary base pairing with target messenger RNA. Secondly, siRNAs corresponding to sequences from viroid genomes have been isolated from infected plants. Finally, transgenic expression of the non-infectious hpRNA of potato spindle tuber viroid develops all the corresponding viroid like symptoms (119). This evidence indicates that when viroids replicate via a double stranded intermediate RNA, they are targeted by a dicer enzyme and cleaved into siRNAs that are then loaded onto the RNA-induced silencing complex. The viroid siRNAs actually contain sequences capable of complementary base pairing with the plant's own messenger RNAs and induction of degradation or inhibition of translation is what causes the classic viroid symptoms.

The symptoms of plant disease caused by different viruses often are quite similar and difficult to distinguish. Symptoms vary widely, depending upon virus strain, host, cultivar, plant age, nutrition, weather, and infection by more than one virus. Accuracy in field diagnosis frequently is improved with knowledge of such factors as the pattern of infected plants in a field and the cultivar being grown. Identification of a particular disease may require isolation of the causal virus from infected plants and its identification. Various characteristics of isolated virus particles may be used to pinpoint their identity, including their serological specificity, size and shape as determined by electron microscopy, reactions to different hosts, and chemical composition. The effectiveness of measures for managing virus diseases often depends upon accurate diagnosis. Although some plant viruses can be identified by the visible symptoms and general behavior of the diseases they cause, the identity of many viruses can be determined only by trained personnel using specialized serological, EM and molecular techniques.

The Plant Virus Subcommittee established in 1966 by the Executive Committee of the International Committee on Nomenclature recommended a classification of viruses into 16 subgroups, including last four monotypic groups. These groups are Tobravirus, Tobamovirus, Potexvirus, Carlavirus, Potyvirus, Cucumovirus, Tymovirus, Comovirus, Nepovirus, Bromovirus, Tombusvirus, Caulimovirus, Alfalfa mosaic virus, Pea enation mosaic virus, Tobacco necrosis virus and Tomato spotted wilt virus. Over 800 plant viruses have been recognized and characterized. The genomes of most of them, such as the *tobacco mosaic virus* (TMV), are infective single-stranded RNAs; some RNA viruses have double-stranded RNA genomes. *Cauliflower mosaic virus* and *bean golden mosaic virus* are examples of viruses having double-stranded and single-stranded deoxyribonucleic acid (DNA), respectively. The genome of many plant viruses is a single polynucleotide and is contained in a single particle, whereas the genomes of brome mosaic and some other viruses are segmented and distributed between several particles. There are also several low-molecular-weight RNAs (satellite RNAs) which depend on helper viruses for their replication. The classification of plant virus is carried out by the plant virus sectional committee of International Committee on Taxonomy of Viruses. 750 viruses are classified into 20 families and 88 genera. They are roughly classified according to that genome is DNA or RNA, single strand or double strand, and plus strand or minus strand of single strand RNA, and are classified into families according to the size of particle, form, and basic structure of genome, and into genus according to host range, infection style, and gene sequence. Specific name and notation are also explained.

Mycoplasmas

First shown to cause plant diseases in the 1960s, mycoplasmas are small single-celled organisms that occur inside plant cells. Mycoplasmas do not have rigid cell walls; each is enclosed by a triple-layer cell membrane. These cell membranes can be easily ruptured if the cells are not in a favorable environment. Many diseases that were earlier thought to be caused by viruses, and were called yellows diseases, are caused by mycoplasmas. These organisms are spread from one plant to another by a vector, usually an insect such as a leafhopper. Mycoplasmas can also be spread in infected plant parts. Tetracyclines are antibiotic substances that have been used to kill some mycoplasmas in plants. By electron microscope and light microscope observations, it is difficult to separate mycoplasmas from bacterial L-forms. Molecular weight and the quantum of genetic information in mycoplasmas are between the pox viruses and L-form bacteria. Plant pathogenic mycoplasmas were discovered by electron microscopy, in 1967, long after the discovery and culture in 1898 of the first pathogenic mycoplasma of animal origin, Mycoplasma mycoides. Mycoplasmas are Eubacteria of the class Mollicutes, a group of organisms phylogenetically related to Gram-positive bacteria. Their more characteristic features reside in the small size of their genomes, the low guanine (G) plus cytosine (C) content of their genomic DNA and the lack of a cell wall. Plant pathogenic mycoplasmas are responsible for several hundred diseases and belong to two groups: the phytoplasmas and the spiroplasmas. The phytoplasmas (previously called MLOs, for mycoplasma like organisms) were discovered first; they are pleiomorphic, and have so far resisted in vitro cultivation. Phytoplasmas represent the largest group of plant pathogenic Mollicutes. Only three plant pathogenic spiroplasmas are known today. *Spiroplasma citri*, the agent of citrus stubborn was discovered and cultured in 1970 and shown to be helical and motile. *S. kunkelii* is the causal agent of corn stunt. *S. phoeniceum*, responsible for periwinkle yellows, was discovered in Syria. There are many other spiroplasmas associated with insects and ticks. Plant pathogenic mycoplasmas are restricted to the phloem sieve tubes in which, circulates the photosynthetically-enriched sap, the food for many phloem-feeding insects (aphids, leafhoppers, psyllids, etc.). Interestingly, phytopathogenic mycoplasmas are very specifically transmitted by leafhoppers or psyllid species (34).

For a long time a group of diseases distinguished by stunting, chlorosis, yellowing of leaves, changing of flowers to leaflike structures, the production of numerous leaves, or combinations of these symptoms were thought to be caused by viruses. Some common diseases of this type are aster yellows, big bud of tomato and tobacco, com stunt, and lethal yellowing of coconut palm. However, these diseases are now known to be caused by mycoplasmas. In the absence of a cell membrane, these organisms are soft, fragile, and subject to distortion under slight shifts in osmotic pressure. They assume many different shapes and reproduce by forming chains, budding, or fission. Mycoplasmas are transmitted from plant to plant by leafhoppers. An unusual property of mycoplasmas is their ability to multiply in leafhoppers, in some cases with harmful effects such as decreased hatchability of insects from infected eggs and decreased life span of infected adults. Maize stunt is caused by a spiral mycoplasma or spiroplasma transmitted by leafhoppers. Mycoplasmal diseases of expensive perennial ornamentals and other valuable plants can be managed or delayed by treatment with injections of certain antibiotics such as tetracyclines. In some cases, management of the leafhopper vector helps delay spread of the disease.

Plant Parasites

Usually we think of parasites as lower forms of plant life such as bacteria, fungi or nematodes. However, a number of flowering or seed plants are parasites on other plants. Parasitic seed plants vary widely in their dependence on a host. The more independent ones, called half-parasites or hemiparasites, have chlorophyll and roots and can manufacture their own food; but they depend on their host for dissolved minerals and organic substances. Others, such as mistletoe or *Loranthus*, have chlorophyll but no roots and depend on their host for minerals and water. Some other parasitic seed plants, such as dodder (*Cuscuta*), have neither chlorophyll nor true roots, and depend entirely on their host for their existence. Parasitic seed plants cause disease principally by depriving their hosts of water and nutrients and by disturbing the balance of growth regulators. The full spectrum of diseases caused by parasitic seed plants still is not known.

Abiotic Disease Causing Agents

Some plant diseases are not caused by living agents, but are due to an upset in the environment usually too much or too little of something necessary for growth or the presence of some harmful substance in contact with the plant. These diseases are called as abiotic diseases or non-infectious diseases that occur in the absence of pathogens and thus cannot be transmitted from one plant to another. These diseases can occur at any stage from seed to mature plant; they may cause damage in the field, during storage or transit, or at the market. Symptoms may range from slight to severe, and affected plants may be killed. Like every other living thing, plants require specific environmental conditions to prosper. Some plants thrive under moist conditions; others do well in dry atmospheres. Some plants grow fastest in bright sunshine; others need shade. Some require fertile soils; others do not. Whenever the proper balance of temperature, soil moisture, light, purity of air, nutrients, and pH is upset to a smaller or a larger degree, this imbalance is reflected in the growth of plants. Moreover, cultivated plants frequently are grown in artificial environments or subjected to cultural practices (e.g., fertilization, irrigation, pesticide, or hormone treatments) that may affect their growth considerably.

DISEASE COMPLEXES

Frequently plant diseases, especially those induced by soilborne pathogens, involve more than one pathogen. This joint infection leads to increased damage or death of plant. A plant may be able to resist the attack of one parasite, but when two or more parasites attack the same plant, it may be overwhelmed. Environmental extremes-such as too much or too little water, temperature, or relative humidity (RH)-often affect disease development; nutritional excesses or deficiencies may increase the susceptibility of plants to some diseases. Insects, nematodes, fungi, weeds, and even people often transport pathogens from one area to another. Satisfactory management programs for disease complexes require accurate diagnosis of the problem and the combined use of various practices if the affected crop is to be grown profitably.

OPTIONS FOR MANAGEMENT OF PLANT DISEASES

Once the causal agent of a disease has been correctly identified, it is possible to develop plans to manage the disease. Research has been conducted extensively from the ancient times on pathogens, diseases, and management methods. This vast store of knowledge helps us in our efforts to manage plant pests. Intelligent plant disease management is an economic necessity. Together with weed and insect management, it is a form of insurance that helps guarantee the farmer a profitable yield. It helps prevent wide fluctuations in yield. It helps prevent disastrous epidemics and famines. There are currently in farmers use five categories for the management of plant diseases (79):

- Avoidance of pathogens by exclusion of the pathogen from a geographical area, either voluntarily or by legislation, and by evasion by the host plant, to prevent the pathogen from coming into contact with the host.
- Eradication or removal of the pathogen from the host, soil, from other reservoirs of inoculum, or from a geographical area. The actual result may be a reduction of inoculum rather than the total eradication of the pathogen.
- Protection of plants by environmental changes, biological methods, or physical methods that produce conditions less favorable for disease development or by the use of pesticides.
- Use of cultivars with genetic resistance that prevent the pathogen from causing disease
- Chemotherapy with systemic chemicals to kill pathogens that are already in the plant.

Predicting Disease Epidemics

An epidemic depends on the coincidence of available inoculum of a virulent pathogen, susceptible host tissue, and a favorable environment (79). All of these factors can cause a complicated balance of factors to shift in favor of disease or health. Plant disease cycles represent pathogen biology as a series of interconnected stages of development including dormancy, reproduction, dispersal, and pathogenesis. The progression through these stages is determined by a continuous sequence of interactions among host, pathogen, and environment. The stages of the disease cycle form the basis of many plant disease prediction models that help to take appropriate control measures. The relationship of temperature and moisture to disease development and pathogen reproduction serve as the basis for most contemporary plant disease prediction systems (78). Pathogen dormancy and inoculum dispersal are also considered. Advances in information technology have stimulated innovations in model application. This trend must accelerate to provide the disease management strategies needed to maintain food supplies.

Economical Management Options

Cultural control and rotations will remain the cornerstones of safe organic production, but recent developments in understanding interactions between soil micro-organisms, plants, pathogens and the environment offer the opportunity to manage specific disease problems should they occur. The use of composts and soil amendments or cover crops appears to offer good prospects for improving disease control. The use of covers and mulches should provide a range of benefits particularly if used for successive crops. Biological control agents either natural occurring or introduced may be advantageous against particularly damaging or persistent problems. Similarly, plant extracts require further evaluation for specific uses. Disease problems occur throughout the production cycle from seed to propagation to the growing crop and its harvest and storage. Approaches for disease control are considered at each stage. Organic growers will almost certainly require stringent standards for seed and propagation to ensure that diseases are not introduced into their systems. A long term commitment to strategic studies of organic systems will be needed to ensure that the planned expansion in production can meet the challenges it undoubtedly faces.

Disease prevention and sanitation are especially important in greenhouses, which are designed to provide optimum conditions for plant growth. Their subtropical environment of moderate temperature, high RH, and adequate water is also conducive to the growth of many disease organisms. In addition, continuous culture of the same or related crops year-round increases disease problems by providing an ever-present host for pathogens. Sometimes it is impracticable or impossible to exclude or eradicate a pathogen from a given area. In such cases the crop must be protected from it by either physical or chemical means. Physical means include solarization, moisture management, drying, and radiation. Bulbs, roots, seeds, and tubers are usually treated with chemicals to prevent decay after planting from seedborne pathogens or those occurring in the soil. Chemicals are applied as dusts or slurries, or by soaking the plant material in water solutions. Many seed merchants routinely treat seed before sale (100).

Biological Control

Farmers have observed for centuries that there is less plant disease on certain soil types, called suppressive soils, whereas on other soil types (conducive soils), plant diseases cause significantly more damage. They also knew that plowing under green cover crops or adding manure or lime to soil, rotating crops, or flooding the land often would decrease disease and increase yields. These age-old observations together with attempts through the years to increase crop production by changing agronomic practices altered the microbial balance in soil and probably decreased plant disease through biological control - the control of one organism by another. All living things have enemies. A parasite that feeds off a living host does not preclude it from being attacked in turn by other parasites and predators. Soil is the natural habitat of countless hordes of microorganisms, consisting of many kinds and types, living in a constantly changing environment. It is an important medium in nature for the growth, multiplication, survival and other activities of microbes which live in soil not in pure culture but as complex populations wherein various forms of antagonistic relationships are operative. Usually their existence produces a healthy soil and a more favourable environment for plant growth. The role of microorganisms in natural processes and in maintaining the biological balance in nature is among the most significant ones.

Sir Alexander Fleming, an English scientist, noticed in 1928 that a fungus, a *Penicillium* species, produced a toxin that killed bacteria. A decade later, this amazing antibiotic was purified and named penicillin. Shortly after Fleming's discovery in 1932, R. Weindling published a report stating that a fungus, a *Trichoderma* species that lives in the soil, parasitized other fungi and produced antibiotics that inhibited or suppressed the growth of several soil pathogens that caused plant diseases. Since these early discoveries, many scientists have demonstrated that one organism can suppress the growth of another with a subsequent decrease in the amount of disease. The severity of a disease depends upon the interactions of the pathogens, antagonists, hosts, and the environment. The many antagonistic organisms that live in the soil may produce antibiotics harmful to the pathogen, compete for food in the soil, or parasitize the pathogen. Basically the mechanisms that interfere with the growth of soilborne pathogens work in two ways: by reduction of the amount of inoculum produced by the pathogen, and by prevention or reduction of the rate of spread of the pathogen.

Fungi such as *Trichoderma* or *Penicillium* or bacteria such as *Pseudomonas* or *Bacillus,* when added to soils, can suppress the amount of disease. We now know that some organisms in suppressive soils produce chemicals that are harmful to the soilborne pathogens that cause seed rots, damping off, or other diseases. For example, penicillin interferes with bacterial cell wall formation, streptomycin interferes with bacterial protein synthesis, bacteriocins prevent the growth of species closely related to the producer, and siderophores inhibit absorption of certain necessary ions (iron) needed for growth by pathogens. Marigold (*Tagetes* sp.) roots exude terthienyls that are toxic to some nematodes and fungi. Seeds can be protected from soil pathogens by coating the seed with spores of antagonistic fungi that prevent pathogens from harming the seed or germinating seedlings. Undoubtedly many additional antagonistic compounds and organisms will be discovered with further investigations on biological control mechanisms. Numerous attempts have been made to identify the organisms that produce suppressive chemicals and the mechanisms involved. A successful pathogen needs to overcome antagonistic effects of potential competitors. For an antagonist or biological control agent to be successful, it will require a growth promoting or neutral relationship with other potential antagonists in the pathogen's environs while neutralizing the defensive mechanisms of the pathogen. We can expect great progress in the future as investigators combine biological control with sensible chemical management, IPM and forecast to make farming more profitable, safe, and environmentally sound.

Rhizosphere Microorganisms in Biocontrol

The health of a flowering plant is determined by countless biotic and abiotic interactions in the soil and on the root surfaces (95). As early as in 1687, Hooke first suggested the existence of a relationship between root exudates and the soil environment (25). Later, the activity of microorganisms was observed to be stimulated in the soil volume close to roots of a plant and this part of soil volume was denoted as rhizosphere. Since then, some remarkable advances have been made on the importance of the rhizosphere microorganisms on the plant growth. The rhizosphere is heavily populated by different microorganisms that exert both beneficial and deleterious effects on plant growth. Many strains of plant growth promoting rhizobacteria (PGPR) were discovered in the rhizosphere. These bacteria were shown to survive in the rhizosphere. The new opportunities to apply biotechnological developments provided the impetus to exploit commercially the key beneficial microorganisms that inhabit plant roots. Seed bacterization which introduced and established beneficial microorganisms in the rhizosphere was identified as an excellent way for improving plant growth. The natural defense provided by microorganisms in the rhizosphere deserves much more scientific attention, both because of its importance in plant ecology and because of its potential as a practical means to improve the health and productivity of crop (23).

It is evident from the innumerable research reports over the years that several rhizosphere bacteria have the potential to suppress several plant pathogens and to promote plant growth (95). However, in spite of these extensive research efforts, the use of biocontrol in commercial agriculture is still limited. The inconsistency in performance of these biocontrol agents over time and space in comparison to chemical pesticides is a major impediment to their large scale use in commercial agriculture. Thus the use of biocontrol agents is being tried in a system where chemical fungicides provide a better and more economical result *i.e.*, the biocontrol agents are tried to fit in to a chemical paradigm (43). Therefore, efforts are needed to develop widely adapted biocontrol agents with broad-spectrum activity. Clear understa-nding of different molecular mechanisms of antagonism and root colonization are essential so that desirable traits can be pulled together genetically in one biocontrol agent. Improving known biological agents, with genetic manipulation, can enhance their biological control attributes and expand their spectrum of activity. Cook (24) advocated an approach to biological control that uses combinations of biocontrol agents that suppresses diseases by complementary mechanisms. This may also allow long term control since individual components colonize different crops and are adapted to different environments. Similarly, timely and augmented releases of biocontrol agents have been suggested to make biocontrol more effective. Another aspect which needs attention is the wide spread and multilocation testing of the biocontrol agents against specific diseases at the respective hot spot locations (77). Research should also be directed to focus on shelf life of the biocontrol products. Thus it is hoped that biocontrol will form an important component in the integrated disease management in near future.

PGPRs in Biocontrol

Modern varieties have been selected in homogenous, high fertility systems with a primary focus on yield. This has contributed to the loss of plant genes associated with efficient nutrient acquisition and adaptation to soil-related biotic and abiotic stresses. It will be crucial to re-introduce desirable rhizosphere-related traits into elite cultivars. Breeding are to be targeted at selecting traits that offer possibilities for improving nutrient acquisition capacity, plant–microbe interactions and tolerance to abiotic and biotic soil stresses. Modern molecular tools and novel approaches can help to make a significant contribution to breeding for rhizosphere related traits (121). Rhizosphere is the habitat in which several biologically important processes and interactions takes place and the microorganisms that grow in the rhizosphere are ideal for use either as biological control agents or plant growth promoters. Rhizosphere is the zone of soil where microorganisms are stimulated by plant roots. The rhizosphere bacteria those help plant growth in various ways are designated as plant growth promoting rhizobactera (PGPR). They support seed germination, increase plant growth, produce nodules on legumes, enhance nitrogen fixation and also suppress crop diseases. Simultaneously, the use of pesticides has also become an integral and economically essential part of agriculture.

Biocontrol through the use of PGPRs provides disease control besides being relatively harmless to human and non-polluting. Their use improves soil health and the sustainability of agriculture. PGPRs are subjected to variation in soil characteristics, soil temperature, pH, rainfall affecting their adaptability to a specific environment and edaphic conditions. The demand for organic crop produce in the world market is very high, and is increasing. Hence biofertilizers and bioprotectants can be the best alternative to these chemicals as a component of integrated management of plants from the ecological point of view.

Since the rhizosphere is expected to be a rich source of microorganisms, bacteria from plant rhizosphere itself could be exploited for use as biofertilizer and bioprotectant capable of improving the growth of plant, either by suppression of pathogenic soil fungi or by growth promotion. These bacteria appear completely harmless, do not cause any symptoms and yet induce substantial resistance against different pathogens (117). These beneficial bacteria have been applied to a wide range of agricultural species for the purpose of growth enhancement, including increased seed emergence, weight, crop yields and disease control (37). The biological control of *Ralstonia solanecearum* wilt in tomato with *Serretia* sp, *P. fluorescens* or *Bacillus* has been reported by Guo *et al.*, (41). They have also found increased yield of tomato with these PGPRs under green house and field conditions. *B. subtilis*, *P. fluorescens* and *Bacillus* sp. have been reported inhibit *X. campestris* pv. *malvacearum* and also increasing germination of treated cotton seeds (by 13 %) and seedling growth (by 23 %) (69).

The mechanisms by which PGPRs influence plant growth may differ from species to species and from strain to strain. Symbiotic plant colonizers such as rhizobia mostly contribute to plant growth by nitrogen fixation. Free living rhizobacteria usually do not rely on single mechanism of promoting plant growth. There are several determinants for mechanisms of growth promotion that include bacterial synthesis of plant hormones like indole-3- acetic acid (IAA), cytokinin and gibberellin, production of 1-aminocyclopropane-1- carboxylate (ACC) deaminase, breakdown of plant induced ethylene and increased mineral and N availability in the soil (37, 53). *Bacillus megaterium* TRS-4 isolated from rhizosphere of healthy tea plants, was found to be antagonistic to several root pathogens *viz, Fomes lamaoensis, Poria hypobrumea, Sphaerostilbe repens, Sclerotium rolfsii* and *Sclerotinia sclerotiorum* in vitro. Rhizobacteria were also found to exhibit other plant growth promoting and disease suppressing characteristics such as phosphate solubilization and production of IAA and siderophores (9). Several indirect possible mechanisms have also been proposed that reduces the stress by limiting the damage caused by phytopathogens. These mechanisms may include suppression of diseases (107), through competition with pathogenic microorganism by colonizing root. PGPR mediated induced systemic resistance (ISR) results in reinforcement of plant cell wall by lignin, callose and phenolic compounds, alteration of physiological and biological reactions of plant cells, and production of antimicrobial substances, such as pathogenesis related proteins (PR) and phytoalexins (94).

PGPRs represent a wide variety of soil bacteria which, when grown in association with a host plant, results in stimulation of growth of their host. Rhizosphere bacteria are present in large numbers on the root surface, where nutrients are provided by plant exudates and lysates. Plant growth promoting rhizobacteria induced systemic resistance (ISR) is dependent on colonization of the root system by sufficient numbers of PGPR and this has been achieved by pouring a suspension of the bacteria on, or mixing it with autoclaved soil, dipping the roots of seedlings in a bacterial suspension at transplanting, or coating seeds with high numbers of bacteria before sowing (54, 117). Once natural plant-resistance mechanisms are activated, increase defensive capacity is maintained for prolonged period against multiple pathogens. The interaction between the pathogen and host plant induces some changes in cell metabolism, primarily in the enzyme activities including that of phenylalanine ammonia lyase (PAL), peroxidase (POX), polyphenol oxidase (PPO) and lipoxygenase (LOX). The increased level of defence related enzymes during ISR are known to play a crucial role upon the degree of host resistance (93).

Phosphate Solubilizing Microbes

Phosphorus is one of the macro-elements required as a plant nutrient and is second only to nitrogen in importance. About 95% of phosphorus in the soil is in an insoluble state and cannot be utilized as such by plants. Further, any application of phosphorus as a chemical fertilizer quickly accumulates as an insoluble form. Availability of phosphorus to crops is subject to chemical fixation in soil and other metal cations depending on soil pH. Soil contains only 0.02 – 0.5% of total P, which exists in poorly soluble form. The scantily accessible P typically comprises of calcium phosphate ($Ca_3(PO_4)_2$), ferric phosphate ($FePO_4$) and aluminium phosphate ($AlPO_4$). Indian soils in general have low to medium available P status (35,49). Phosphate solubilizing microorganisms have been reported to occur in different environmental niches and are abundant in rhizosphere region (81, 17). Rhizosphere is a we recognized and specialized ecological niche for fungi, bacteria, actinomycetes and other microbes. It is the narrow zone of soil around roots supporting diversified microbes and is influenced by root exudates, soil factors, age and type of the host plant and molecular signals. Phosphate solubilization by fungi, bacteria and other microbes is often considered as one of the parameters related to plant growth promotion. There are many reports that document plant growth promotion due to inoculation of phosphate solubilizing fungi and bacteria (16, 20, 21, 97).

Using Vesicular-Arbuscular Mycorrhiza

The ectomycorrhizas are characterized by an extracellular fungal growth in the root cortex. The ectomycorrhizas are characterized by an extracellular fungal growth in the root cortex. forming specific fungal structures, referred to as vesicles and arbuscles. This characteristic growth gives the endomycorrhiza the alternate name, vesicular arbuscular mycorrhiza (VAM). It is the most widely distributed association in plants. About 80% of all terrestrial plant species form this type of symbiosis (103) and 95% of the world's present species of vascular plants belong to families that are charcteristically mycorrhizal. Arbuscular mycorrhizal fungi (AMF) are most ubiquitious and form mutualistic relationship with more than 80 percent of major group of vascular plants. AM fungi are able to confer bioprotection against a number of soil borne pathogens causing root diseases of various plants. In mycorrhizal plants, there are improvements in nutrition, root biomass and growth which increases plant tolerance and compensate for root damage caused by a pathogen, changes in rot system morphology, modifications of antagonistic microbial populations, and competition with pathogenic fungi to colonize root tissue.

Arbuscular mycorrhizal fungi (order Glomales, division Glomeromycota) form a symbiotic relationship with most plant species (4). The effect of mycorrhizal fungi on plant performance has been proposed to have far reaching ecological consequences, from increasing productivity of ecosystems to enhancing plant diversity (52, 116). This symbiosis is increasingly being recognized as an important and integral part of natural ecosystems throughout the world. The AM fungus-plant association is a mutually beneficial event: the plant supplies the fungus with carbon (from its fixed photosynthates) while the fungus assists the plant in its uptake of phosphate and other mineral nutrients from the soil (103). This bidirectional exchange of nutrients takes place through extensively branched haustoria, termed arbuscules. In addition to increased nutrition, mycorrhizal plants also show increased resistance to root pathogens and tolerance to drought stress, and their hormonal balance is altered. AM fungi inhabit various ecosystems with a wide range of host plant species. However, as these fungi are obligate symbionts with living roots, the hosts play an important role in mycorrhizal development, spore formation and distribution of AM fungi. The community of AM fungal species in the rhizosphere may vary with host species. The phenomenon of AMF protecting plants from root pathogens is known from studies involving root-infecting pathogens e.g. *Phytophthora parasitica* or *Fusarium sp.*, root-invading nematodes (31) and horticulutural and agricultural species such as tomato (*Lycopersicum esculentum*), alfalfa (*Medicago sativa*) (28) and in grasses (83). *G. mosseae* induced local and systemic resistance to *P. parasitica* and was effective in reducing symptoms produced by this pathogen (91).

Plant Disease Management with Chemicals

Agricultural scientists agree that pesticides can be hazardous, especially when used by ignorant or careless applicators. Chemicals are, however, essential in agriculture today. Without them our food supply would be threatened, and it would be impossible to grow certain crops profitably. Moreover, chemical application is the only economical means for managing some plant diseases. Many plant scientists and chemists are working diligently to accumulate information to make chemical pesticides safer and even more effective. Agribusiness is a technical, skillfully organized industry devoted to providing food for an exploding world population, dedicated to the safe and effective use of pesticides. Each day researchers learn more about chemicals and their mode of action. New and safer chemicals are being developed. Pesticides will be used widely until effective substitutes are discovered. The use of chemicals for plant disease management is merely one tool in managing plant pests, and it should be used in conjunction with the other components of IPM. When used improperly, chemicals are dangerous and expensive, and often will not reduce pest losses. The biggest problem with agricultural chemicals is the people who apply them. Many pesticide applicators make significant application errors because of inaccurate calibration, incorrect mixing, worn equipment, and failure to read the product label. The increased possibilities of harm to the crop, applicators, farm workers, consumers, and the environment are self-evident. The application of pesticides to large acreages with little or no regard for their deleterious side effects or impact on non-target species are no longer tolerated.

Chemicals either kill or inhibit germination, growth, and multiplication of the pathogen. Some pesticides are non-selective, that is, toxic to many pathogens; others are selective, affecting only a few or one kind of pathogen. For effective results, a disease must be correctly identified and the correct chemical selected and applied at the correct time, in correct amounts, and in the correct way. Most of the older disease management chemicals can only protect plants from subsequent infection and cannot cure a disease after it has started. Such chemicals are called protectants. These chemicals protect only that part of the plant that was covered before the pathogen came in contact with the plant. Protectant chemicals are not absorbed into or translocated through the plant, but form a chemical barrier between the plant and the pathogen. Thus, when the plant makes additional growth and puts out new leaves, or rain washes off the chemicals, another application must be made.

Several of the newer chemicals in use today have therapeutic, eradicant or curative properties. These chemicals are absorbed and spread internally in the sap stream through the plant and are called systemic pesticides. The systemic fungicide benomyl (benzimidazole) and dimethrimol (amino pyrimidine) in the early 1970's have site-specific mode of actions. Another new type of pesticide with some curative properties is the sterol inhibitors. These chemicals inhibit ergosterol biosynthesis in certain fungi. An antibiotic is a substance produced by one microorganism that is toxic to another. Most antibiotics used today in managing plant diseases are absorbed and translocated systemically by the plant. Streptomycin, a well-known antibiotic, is used against several bacterial pathogens that cause blights, rots, spots, or wilts. Tetracyclines are also active against many bacteria and mycoplasmas. Cycloheximide is effective against several fungi. It is effective against powdery mildews. However, cycloheximide injures some plants, so its use is limited. Fungicide research has produced a diverse range of products with novel modes of action during the last few years. More novel compounds with different modes of action (notably strobilurins, phenylpyrroles, anilinopyrimidines, spiroxamines, quinolines and phenylpyridylanines) have been developed and they possess more potent action against diverse fungal pathogens at much lower rates compared to most of the earlier fungicides. These new classes of fungicides which include anilinopyrimidines, phenoxyquinolines, oxazolidinediones, spiroketalamines, scytalone dehydratase inhibitors, phenylpyrroles, strobilurins, and activators of systemic acquired resistance are expected to have a significant impact on disease control (40). The most recent results in fungicide discovery originated from different approaches: traditional empirical screening of synthetic chemicals; traditional screening of natural products; biochemically directed screening of chemicals to discover; and non-conventional agents stimulating plant defense.

When a plant detects an invading micro-organism a complex series of events unfolds within minutes. First, elicitor molecules of either plant or pathogen origin are released at the site of infection which, when recognised by the plant, trigger the production of signals. These signals then travel within the plant and are responsible for triggering a complex array of defence responses. When recognition occurs quickly, the plant can effectively resist pathogen attack. However, if there is a failure in recognition and defences are triggered too late, the plant succumbs to pathogen invasion and disease develops. It is possible to elicit resistance using either elicitor (Dinaker 30) or signal molecules, or synthetic derivatives or mimics.

CONVENTIONAL METHODS OF CROP IMPROVEMENT

Throughout recorded history the domestication, breeding and improvement of animals and plants has depended on the fundamental observation that living organisms both resemble and differ from their parents. Within the past century researchers have gained an increasing understanding of heredity and change among living things – the science of genetics, which has developed and expanded rapidly. It is now established that each inherited trait or function of an organism depends upon a unit of inheritance, a specific gene, which directs its development. In microorganisms such as bacteria that commonly reproduce asexually and whose single cells lack a defined nucleus, a single gene controls each specific trait. In higher plants and animals, two genes exist with respect to each trait: one inherited from each parent.

Many plants have 10,000 or more genes. Each gene is a tiny segment of a threadlike component of the cell called a chromosome. The chromosomes, located in the nucleus, are composed of deoxyribonucleic acid (DNA) and protein. DNA is a wondrous chemical that can reproduce exact copies of itself. Many genes make up one chromosome. Each gene is a tiny piece of DNA that regulates the reproduction of a specific protein, which in turn governs the composition and development of a specific trait. Thus, the fidelity and the diversity of inherited traits are locked within the DNA of chromosomes. The cytoplasm also contains other organelles, such as chloroplasts and mitochondria, floating as tiny blobs in the cytoplasmic soup. Some genetic material is located in the cytoplasm. The inheritance of the following important traits such as male sterility, flower morphology, resistance to fungal toxins, organelle morphology, albinism and variegation and Fraction 1 protein production-nitrate growth response has been linked with cytoplasmic genes. By using the accumulated knowledge of the laws of genetics and the techniques of controlled crossbreeding and selection, plant pathologists and breeders have greatly improved the usefulness of many farm crops. Development of disease-resistant cultivars accounts for a major portion of the yield increases. Each major agricultural crop has numerous relatives, both wild and cultivated, that contain many desirable characteristics. If some way could be found to transfer immunity found in one species and incorporate the resistance into another crop species, then crops would be protected from attack by devastating pathogens. However, until the last few decades, the transfer of desirable traits from one species of plant to another was possible only if they could be mated by cross-pollination. It still is impossible to crossbreed or hybridize unrelated plants such as soybeans and tomatoes by cross-pollinating them. In addition, many plant species are difficult to manipulate genetically by conventional methods. Therefore, more rapid and convenient ways to improve plants are being sought. Classical or conventional genetics will continue to play the major role in plant breeding, but new genetic techniques will assume ever-increasing importance.

Our efforts have been aimed at understanding and combating diseases, primarily through host plant resistance breeding with selected donors. In many cultivars, genes for resistance to a particular pathogen have been identified. In some plant species, near isogenic line carrying these resistance genes were developed as in rice. Multi-environment tests demonstrated that durable resistance may be achieved in rice cultivars with single gene (*Pi-1, Pi-2* or *Pi-4*) for one disease (blast) and two or more resistance genes (*Xa4*, *xa5*, *xa13* and *Xa21*) to another disease (bacterial leaf blight) (72, 73, 77).

BIOTECHNOLOGY AND PLANT DISEASE MANAGEMENT

Generally, biotechnology includes any technique that is used to make or modify the products of living organisms to improve plants or animals, or to develop useful microorganisms. In modern terms, biotechnology has come to mean the use of cell and tissue culture, cell fusion, molecular biology, and in particular, recombinant deoxyribonucleic acid (DNA) technology to generate unique organisms with new traits or organisms that have the potential to produce specific products. Biotechnology is also a collection of techniques, tools, and procedures. It is bringing about profound changes in plant pathology. The potential impact of these changes on our understanding of biological systems and in the successful production of food and fiber is immense. In the arena of genetic engineering alone the stable introduction of foreign genes into plants represents one of the most significant developments in agricultural technology. Genetic engineering of plants and micro-organisms, protoplast and tissue culture, the sequencing of RNA and DNA, the use of monoclonal and polyclonal antibodies and molecular probes, are some of the frontiers explored intensively. These frontiers represent new avenues of research that could not previously be pursued and, as a result, our horizons expand and our view of what may be possible in managing plant diseases changes.

Since 1960, scientists have made great progress in finding ways to genetically change and improve plants, including bacteria and fungi. This new technology, a form of biotechnology, deals with ways to change and/or grow plants in test tubes in the laboratory, rather than by use of conventional plant breeding techniques that involve crossbreeding. As a result, the development of superior strains of food crops and trees can proceed at a faster pace. Researchers accomplish within a few years in the laboratory what usually takes 10 to 40 years to do by traditional methods. Biotechnology includes genetic engineering, by which the hereditary apparatus (DNA) of an organism is altered so that the cell can produce more or different chemicals or perform completely new functions. These altered cells can then be used in industrial production. When genes from other organisms, for example, the gene for producing the coat protein of a virus, are inserted into plants, the plants are called transgenic plants (70).

Biotechnology is already having a direct impact in disease management of crops, beginning with rice and this will become more significant in the next few years. Current research achievements include genetic modification of crops to confer pest and disease resistance; study the structure and function of genes and molecular markers to breed disease resistant crops; detection and identification pathogens and those exhibiting in-sensitivity to fungicides; and production of genetically improved biological control agents. Advances in automated technology enabled a new approach in marker-assisted breeding, termed as breeding by design particularly in rice in India. IR 64 lines with bacterial blight resistance genes (*xa5, xa13* and *Xa21*) developed at the Central Rice Research Institute, Cuttack through marker-assisted selection showed high disease resistance and yield potential (76). PCR based molecular markers were used in a backcross-breeding program to introgress three major bacterial blight resistance genes (*Xa21, xa13* and *xa5*) into Samba Mashuri from a donor line (SS1113) in which all the three genes are present in a homozygous condition (106). Using conventional plant breeding approach integrated with MAS and two bacterial blight resistance genes X*a13* and *Xa21* incorporated in Pusa Basmati-1 (39). They were released for commercial cultivation.

Genetic modification is a tool that is now in routine use to help understand biological process and produce new products. This tool has been used in concert with traditional methods of plant breeding based on hybridization and selection. Genetic modification of organisms (GMO) involves the incorporation of cloned genes or other sequences into the genetic makeup of the organism by the process of genetic transformation. So far no rice culture has been found to be resistant to *R.solani* infection. Transgenic IR 64 rice plants over-expressing rice chitinase, a pathogen-related protein, have been generated (26). Transgenic rice with cloned gene *Xa 21* in IR 72 was developed at the International Rice Research Institute, Philippines using particle bombardment. In wheat, defense response genes have been expressed to control powdery mildew (11) and rust (84) infection. Wheat lines expressing ß-1,3- glucanase, thaumatin-like protein1 (tlp-1), ribosome inactivating protein (RIP), -1-purothionin, and *Arabidopsis NPR1* (At*NPR1*) transgenes exhibited enhanced resistance against *F. graminearum* in greenhouse and field trials (61). Approaches to transgenic nematode control can be organized as acting on nematode targets, at the nematode-plant interface, and in plant response. Degrees of resistance have been demonstrated through recombinant expression of protease inhibitors, dsRNAs, and cloned plant resistance genes (63).

NEW ERA OF GENOMICS

Genomics is the study of all of the nucleotide sequences, including structural genes, regulatory sequences, and non-coding DNA segments, in the chromosomes of an organism. The term DNA sequencing refers to sequencing methods for determining the order of the nucleotide bases adenine, guanine, cytosine, and thymine in a molecule of DNA. The transition to a new era of biological research is underway, and both the public and private sectors are moving to exploit the new tools and opportunities presented by genomics. In response to the promise of both fundamental advances and profitable applications, there has been an infusion of funding that is enabling large-scale experimentation and rapid progress. Genomic approaches are already beginning to impact fundamental and applied plant biology. A transitory change is occurring from hypothesis-driven research to a period of descriptive study involving the generation of large amounts of data. Subsequently, hypotheses will often be derived from *in silico* analyses of databases, and testing may initially involve computer simulations prior to actual experimentation. *In silico* is an expression used to mean performed on computer or via computer simulation. Testing of hypotheses will still require detailed phenotyping, but experimental studies will access a broad range of new tools capable of global analyses of RNAs, proteins, and metabolites rather than a gene-by-gene or protein-by-protein approach. We are more concerned with the trends in genomics and their current and potential impact on disease resistance, particularly in resistance genes and the genes that they regulate (66).

Genomics is the discovery and study of many genes simultaneously on a genome-wide scale. Three inter-related areas have been variously described: structural genomics, which is primarily concerned with the determination of genome structure at the sequence level; comparative genomics, which involves the molecular basis of differences between organisms at a variety of taxonomic levels; and, functional genomics, which focuses on the function of genes. Structural genomics is the discovery engine for the other areas. Comparative genomics provides the allelic variation for functional genomics. Proteomics is an outgrowth of functional genomics that involves global studies of gene expression at the protein level. Bioinformatics is the acquisition, curation and interrogation of large collections of complex biological data. Developments in plant genomics include the following:

- Sequencing: *De novo* generation, rice, *Arabidopsis* and large portions of other crop species sequenced and re-sequencing of allelic variants.
- Genetic mapping: Based on segregation analysis, low density maps for many species, high density maps for a few, hybridization to contiguous arrays of genomic clones and extensive inferences between species based on macro- and micro-synteny.
- Gene expression: Predominantly at mRNA level, sequencing of random cDNAs, microarray analysis beginning of proteomics in its infancy, quantitative catalogs of all expressed genes for many species and situations, routine global analyzes using DNA chips and protein arrays and/or other proteomics.
- Gene discovery: Gene-by-gene basis, high throughput phenotyping of mutants beginning, candidate gene approaches by relating phenotypes to sequenced genomes, bulk discovery relating differentially expressed sequences to phenotypes.
- Comparisons between homologs: Emphasize similarities for functional inferences and analyses of allelic differences to explain variation in function.

- Focus of research: Basic biology of model species, transfer of paradigms from model species to crops and study of crop problems in model species
- Traits characterized: Mostly simple Mendelian traits, complex traits and genotype x environment interactions
- Experimental design: Initially based on empirical observations Often derived from *in silico* analyses, refined through practical experience, simulated *in silico* and refined before being performed

Plant Genome Sequencing

In plants, the genomes of *Arabidopsis thaliana* (108), *Oryza sativa* (45), grapes (46) poplar (111) and papaya (68) have been completely sequenced. However, now that they have been completed and their impact is being felt in basic research, attempts are on to sequence of other major crops such as *Zea mays*, *Glycine max*, *Triticum aestivum*, *Brassica repa*, *Gossypium hirsutum*, *Hordeum vulgare*, *Lotus japonicus*, *Medicago truncatula*, *Physcomite carpa*, *Sorghum vulgare* and *Solanum lycopersicum*.

There are major differences between crop plant genomics and the human genome project. For one thing, human genome research contributes to a trillion dollar biomedical research and development. Crop plant genome research also underlies enormously important industries in food, feed, energy, and fiber, but here the analogy ends. First, several species must be targeted to cover agriculturally important plants, rather than one genome in the case of biomedical research. Second, the seed industry operates on far lower margins than the pharmaceutical industry, and has raised public concerns over food safety and security. Finally, the genetic information available to plant breeders is usually thought to be far less extensive than the vast array of epidemiological data collected by the biomedical community, making the sequence less useful. Plant genomes pose problems to sequencing technology because of their size and repetitive content, as well as the number of different species required. This issue has been addressed by taking advantage of the observation that most methylation in plant genomes is restricted to transposons and high copy. By sequencing only the un-methylated portion of the genome, or by subtracting repeats, costs can drop by 10 fold or more, making the sequence of multiple large genomes practical (92). Shotgun sequences can be linked to the genetic and physical maps by bacterial artificial chromosome (BAC) fingerprinting, and then ordered and oriented. Unfortunately, anchoring methods such as BAC-end sequencing, which has been widely employed in animals, is of limited use in crop plants because of the high proportion of long identical repeats. Instead, skimming methodologies can be used to anchor sequence islands to the physical map (62).

Controlled pedigrees are un-available in human populations. Therefore, the value of the collective knowledge gathered by plant breeding has been underestimated. Once genes underlying individual traits are known, the basis for disease resistance and stress tolerance is likely to emerge as it has in model organisms, allowing more precise diagnosis in breeding programs as well as genetic modification. The sequence can also be used to detect epigenetic, as well as genetic variation, which likely contributes to traits such as flowering time, perennialism, apomixis, and heterotic performance. New pesticides and herbicides will also emerge from comparative genomics of crops and their pests, just as new antiviral and antimicrobial drugs are emerging in the pharmaceutical industry by selecting pathogen targets that are not found in the human genome and are less likely to be toxic. Finally, the economic value of agriculture to a growing world population must not be underestimated. While the pharmaceutical industry is successful because of aging western populations, agriculture is the priority for crowded, youthful and hungry nations that make up the rest of the world. Though genomics cannot replace the breeder – pathologist combination, it can open new areas of research un-imagined by conventional plant breeding. Imagine, for example, small, non-toxic molecules that delay flowering or trigger apomixis. Given the importance of agricultural research for food security, energy conservation, and the environment, as well as the rich genetic resources available, genomics may yet make a greater impact on plant breeding than the human genome has had in biomedical research (62).

Resistance genes and defense response like genes which can identified by genome sequence of plant species and validated by using functional genomics approaches. Genomes of several pathogens have also been sequenced: *Agrobacterium tumefaciens* - 5.67 mb (38, 122); *onion yellows phytoplasma* - 0.86 mb (86); *Pseudom-*

onas syringae pv. tomato - 6.4 mb (13); *Ralstonia solanacearum* - 5.8 mb (Genoscope/INRA 99); *Xanthomonas axonopodis pv. citri* – 5.2 mb and *Xanthomonas campestris pv. campestris* - 5.1 mb), *Xylella fastidiosa* - 2.7 mb, *Xylella fastidiosa* - 2.6 mb (113). *Magnoporthe greisea* - 40 mb (27). Many of the following species are also sequenced: *Burkholderia cenocepacia* 8.1 mb, *Clavibacter michiganensis subsp. sepedonicus* - ~2.6 mb and *Erwinia carotovora subsp. atroseptica* SCRI1043 ~5.0 mb (Sanger Institute); *Clavibacter michiganensis subsp. michiganensis* ~3.3 mb, *Xanthomonas campestris pv. campestris* B100 and *Xanthomonas campestris pv. vesicatoria* - 5.2 mb (The Competence Network); *Erwinia chrysanthemi* - 3.7 mb (TIGR/Univ. of Wisconsin); *Leifsonia xyli subsp. xyli* - ~3.0 mb (AEG); *Pseudomonas syringae pv. phaseolicola* 1448A ~6.0 mb (TIGR); *Pseudomonas syringae pv. syringae* 5.6 mb, *Xylella fastidiosa –almond* Dixon 2.6 mb and *Xylella fastidiosa–oleander* Ann1 2.6 mb (JGI); *Spiroplasma kunkelii* - 1.48 mb (Univ. Oklahoma); *Xanthomonas axonopodis pv. aurantifolii* (Sao Paulo State Brazil Consortium); *P. ultimum* genome 42.8 mb (56b) and *Xanthomonas campestris pv.campestris* 8004 ~5.0 mb (Chinese National Human Genome Center). *Phytophthora* genome sequences uncover evolutionary origins and mechanisms of pathogenesis. *Phytophthora* genome sequences have uncovered evolutionary origins and mechanisms of pathogenesis (110b)

Structural Genomics

Structural genomics or structural bioinformatics refers to the analysis of macromolecular structure particularly proteins, using computational tools and theoretical frameworks. One of the goals of structural genomics is the extension of idea of genomics, to obtain accurate three-dimensional structural models for all known protein families, protein domains or protein folds. Structural alignment is a tool of structural genomics. Ability to generate and analyze immense amounts of DNA sequence data has revolutionized by the advances in sequencing chemistries and automation as well as computational power and algorithms. Technologies currently under development will probably increase this capacity yet further through massively parallel sequencing and microfluidic processing (65). The complete genomic sequence of a variety of microorganisms and an increasing number of model organisms are being determined as those of *Arabidopsis* and rice

Large portions of the genomes of crop species will also be sequenced as technologies improve and costs decrease; this will be facilitated by the identification of and focusing on gene-rich islands of the genome (7, 88). Sequencing of resistance gene clusters is an objective of newly funded projects. The nucleotide binding site (NBS) is a protein motif that is present N-terminal to a leucine-rich repeat region (LRR) in predicted proteins encoded by the majority of resistance genes cloned from a variety of species (67; 114). These NBS can be grouped into two distinct classes: those with toll/interleukin-1 receptor homology N-terminal to the NBS (i.e. the toll-interleukin receptor i.e. TIR class) and those without (i.e. the non-TIR class). The TIR class includes proteins encoded by the resistance genes *N* from tobacco (59), *M* and *L6* from flax (101), and *RPP5* from *Arabidopsis*. The non-TIR class includes proteins encoded by the resistance genes *RPS2* and *RPM1* from *Arabidopsis*, and *I2, Mi* and *Prf* from tomato (confer resistance to root knot nematodes) as well as *Dm3* from lettuce (confers resistance to the oomycete downy mildew fungus *Bremia lactucae*).

Genomic sequencing of *Arabidopsis* and rice has already yielded interesting insights into the numbers and organization of disease resistance genes. Analysis of 1.9 million base-pairs (Mb) of *Arabidopsis* suggested that ~14% of the genes are potentially involved in disease resistance, encoding either signaling components or antimicrobial proteins (10). Analysis of ~67 Mb representing >50% of the *Arabidopsis* genome, detected 120 predicted gene products with similarity to the NBS domain encoded by plant R-genes (65) Assuming a similar distribution of genes in the remaining part of the genome, ~200 NBS-encoding genes are present in *Arabidopsis* which include ~150 encoding NBS of the TIR- and ~50 of the non-TIR types. This would represent close to 1% of all *Arabidopsis* genes. NBS-encoding sequences tend to be clustered in the *Arabidopsis* genome (1,12,104) numerous phenotypically defined resistance loci map to the clusters of NBS-encoding sequences on chromosomes IV and V (44, 56) Analysis of the currently available BAC end-sequences, which represent ~5% of the rice genome, suggests that there are probably 750–1500 NBS-encoding genes in rice (65); this estimate is several times greater than the number that would be predicted for rice on the basis of the representation of NBS-encoding genes in *Arabidopsis*.

All NBS-encoding genes in rice encode non-TIR-type NBS; TIR-type genes have not been detected in genomic or expressed sequence tag (EST) sequences from any grass species (65, 87).

From the sequence information (www.tigr.org), we have deduced the putative candidate genes for resistance in the mapped regions to rice tungro virus disease (55, 82). Tungro virus epidemics occur frequently in some parts of the country Muralidharan et al. (75). These epidemics could cause a maximum production loss of 53% in a district, 23% in a state and 2% in this country. Muralidharan *et al.* (74) showed tungro to be primarily caused by RTSV. The genetic basis of resistance to RTSV was investigated in these two cultivars by using QTL analysis. Two mapping populations were developed using susceptible cultivar TN1 (score 9) x Vikramarya, and TN1 x Utri Rajapan. For Utri Rajapan, genes in chr 7 between RM21576 and RM21640 comprised of two NB-ARC domains, one LRR domain and one protease inhibitor (~1.4 Mb), and in chr 2 between RM13530 and RM13608 comprised of domains - one NBS LRR, six LRR and one serine threonine kinase. For Vikramarya, genes in chr 7 comprised of one LRR domain, one NB-ARC domain and two protease inhibitors (~1.3Mb) between RM21135 and RM21205, and in chr 1 between RM10123 and RM10133 comprised of three LRR and one NB-ARC genes (~0.3 Mb) (82). Further fine mapping of candidate genes is underway.

Comparative Genomics

Comparative genomics is the study of relationships between the genomes of different species. These studies of whole genomes help researchers understand what parts of the genome in one organism are similar to those in another, how the overall structures of genes and genomes have evolved, and what these findings tell us about gene expression, gene regulation, and how to interfere with these events in model organisms or humans. Comparative genomics is also a critical enabling field for functional genomics, because it gives researchers an indication of which model organisms are most useful for which studies.

During the evolution of angiosperm plants, only a finite number of chromosomal re-arrangements have occurred. Significant blocks of genetic material may therefore be collinear (i.e. syntenic) among genomes of related species (65). Macrosynteny based on linkage analysis is becoming increasingly well-documented among monocot species as well as among *Brassica* species and *Arabidopsis*; nevertheless, preliminary data indicate only limited synteny between monocot and dicot species (29). Also, comparison of orthologous regions at the sequence level reveals that the level of microsynteny is variable. As several plant species, particularly rice and *Arabidopsis*, are sequenced and once the extent and pattern of synteny has been established for a particular species, it will become possible to predict the position of some, but not all, of the genes in each part of the genome. One of the challenges in comparative genomics is to distinguish orthologs (i.e. homologous genes with a common ancestor that have been separated by a speciation event) from paralogs (i.e. homologs resulting from a gene duplication event. This is particularly problematic within large diverse multigene families such as the resistance genes. Resistance genes may be located in less a stable region of the genome in which synteny is poorly preserved as in grass species (57). Resistance-gene homologs are located in syntenic positions within the Solanaceae but the resistance specificities encoded by these genes are not conserved (http://www.intlpag.org). In several species, resistance genes seem to be either telomeric or close to the centromere.

PCR with degenerate oligonucleotide primers are used due to sequence similarity between cloned resistance genes for the cloning of large numbers of resistance-gene candidate sequences from diverse species. These sequences often map to regions containing known disease resistance genes. Over 130 NBS-encoding sequences similar to those of known resistance genes currently in public databases have been identified by PCR (65, 87; http://www.ncgr.org/rgenes). This number will continue to increase as this approach is applied to an increasing number of species and new combinations of primers are used to amplify different subsets of sequences (87). As such work progresses, the likelihood that a candidate sequence will be available when a new resistance gene is genetically mapped will greatly increase (65).

Candidate sequences, with the exception of those from *Arabidopsis* are mapped by analysis of segregating progeny. In *Arabidopsis*, resistance-gene homologs were mapped relative to known resistance genes by hybridization to an ordered array of yeast artificial chromosome (YAC) clones (104). Mapping by hybridization to such an array will be superceded as the genomic sequence approaches completion. Arrays of contiguous genomic BAC clones will become available for rice and maize, and then soybean. Hybridization to such contigs will provide a rapid and accurate method for mapping cloned sequences and will replace segregation analysis. Such a hybridization strategy has the added advantage that it does not require polymorphism between the parents of a mapping population. High-throughput genotyping will allow the high-resolution mapping of phenotypic resistance genes relative to PCR-based markers or using DNA chips (19). Together, these technologies will facilitate the isolation of many resistance genes of known specificity. High-throughput genotyping will also facilitate the genetic analysis of populations that are large enough to allow the accurate mapping of quantitative trait loci (QTLs) determining quantitative disease resistance and the dissection of genotype x environment interactions. Integration of QTL mapping with genomic sequence data and information on allelic differences will provide the basis for candidate-gene approaches to cloning the QTLs for disease resistance. Another application of high-throughput genotyping will be in monitoring the dynamics of allelic variation at resistance loci in wild populations.

A variety of genetic mechanisms, including point mutation, recombination, un-equal crossing- over and gene conversion, generate diversity in resistance-gene clusters and may prevent a reliable designation of homologs within a cluster as orthologs or paralogs (67). The relative contributions of each of these mechanisms to generating diversity in resistance-gene specificities remain to be determined. Comparative genomics also provides allelic variation for research into the molecular basis of specificity. So far, only a few domain-swap experiments have been reported (33) and their results suggest that the LRR region of NBSLRR-encoding genes is an important, but not the only, determinant of specificity. As the efficiency of sequencing improves, libraries of resistance-gene-related sequences will be generated from such studies. These sequences will also act as templates for gene-shuffling experiments (16) for the generation of new resistance-gene specificities.

Functional Genomics

Understanding the function of genes and other parts of the genome is known as functional genomics. Functional genomics is a field of molecular biology that is attempting to make use of the vast wealth of data produced by genome sequencing projects to describe genome function. Functional genomics uses high-throughput techniques like DNA microarrays, proteomics, metabolomics and mutation analysis to describe the function and interactions of genes. Proteomics is the large-scale study of proteins, particularly their structures and functions. This term was coined to make an analogy with genomics, and is often viewed as the "next step", but proteomics is much more complicated than genomics. Most importantly, while the genome is a rather constant entity, the proteome is constantly changing through its biochemical interactions with the genome and the environment. One organism will have radically different protein expression in different parts of its body, in different stages of its life cycle and in different environmental conditions. A DNA microarray (also DNA chip or gene chip in common speech) is a piece of glass or plastic on which pieces of DNA have been affixed in a microscopic array. Scientists use such chips to screen a biological sample for the presence of many genetic sequences at once. The affixed DNA segments are known as probes. Thousands of identical probe molecules are affixed at each point in the array to make the chips effective detectors. Many microarrays use PCR products, genomic DNAs, BACs, plasmids, or longer oligos (35 to 70 bases). Microarrays may be made by any number of technologies including printing with fine-pointed pins onto glass slides, photolithography using pre-made masks, photolithography using dynamic micromirror devices, or ink-jet printers (http://genomebiology.com/2004/5/8/R58). Proteomics, or the analysis of the protein complement of the genome, provides experimental continuity between genome sequence information and the protein profile in a specific tissue, cell or cellular compartment during standard growth or different treatment conditions. Although the genome defines potential contributions to cellular function, the expressed proteome represents actual contributions. Moreover, by using proteomic approaches, differences in the abundance of proteins actually present at the time of sampling can be distinguished and different forms of the same protein can be resolved. The analysis of proteomes from organisms has been performed extensively by exploring the high resolution of two-dimensional electrophoresis (2DE) coupled with mass spectroscopy (MS). These data, when complemented by de novo sequencing, allow the unequivocal identification of proteins involved in different biological functions.

The proteomic approach is a fundamental method by which we can obtain an understanding and identification of the functions of proteins expressed in a given condition. Because of the large quantity of data and the desire to be able to find patterns and predict gene functions and interactions bioinformatics is crucial to this type of analysis. Comparative genomic analyses can also help determine gene function. Molecular biology is the study of biology at a molecular level. The field overlaps with other areas of biology particularly genetics and biochemistry. Molecular biology chiefly concerns itself with understanding the interactions between the various systems of a cell, including the interrelationship of DNA, RNA and protein synthesis and learning how these interactions are regulated.

A variety of methods for global analyses of gene expression combined with predictions from DNA-sequence data are greatly increasing our ability to make inferences on gene and protein function (6). Methods for global analysis of protein profiles and cataloging protein–protein interactions on a genome-wide scale are technically more difficult but improving rapidly, although they have yet to be applied extensively to plants. Genetic stocks encompassing insertions or deletions in nearly every potential gene will become available for the analysis of phenotypes in model species such as *Arabidopsis*. Catalogs of genes expressed under a range of different conditions, in different organs, or in different individuals will become available within a few years. The full potential of global analysis of plant gene expression is still far from being realized. Both the technology and the algorithms for collecting, displaying and analyzing the vast amounts of quantitative expression data are still being developed (6). Careful standardization and replication are required to provide robust data sets and to allow comparisons within and between experiments. Genes that have altered expression in compatible and incompatible plant–pathogen interactions have been targeted for characterization by microarray analysis (http://www.unil.ch/ibpv/docs/WWWPR/Docs/collaboration.htm). These analyses will provide comprehensive data on expression profiles, both for genes already implicated in plant–pathogen interactions as well as for many genes that were not previously known to be involved in resistance or susceptibility.

Data generated by expression profiling may imply the function of a particular gene but function will still have to be confirmed on a gene-by-gene basis (65). This confirmation will be aided by existing data for genes and proteins that are induced in compatible and incompatible interactions; although many proteins are known to be induced, few have been shown to be causal in resistance (50). Candidate gene approaches that map phenotypes onto sequenced regions will complement gene profiling data; they will not, however, have sufficient resolution to un-ambiguously identify individual genes. For testing gene function, high-throughput reverse genetics approaches are required. Potentially powerful approaches include viral-induced gene silencing (VIGS) or viral over-expression as well as the use of gene knock-out libraries and promoter-trap strategies. In addition to testing the function of individual genes, all three of these strategies can also be used with libraries of anonymous sequences for *de novo* gene discovery. It is likely that each of these approaches will succesfully demonstrate the function of some but not all genes: a combination of approaches may be required to overcome gene redundancy or lethality associated with manipulation of some genes. Virus-induced gene silencing is a technology that exploits an RNA-mediated antiviral defense mechanism. In plants infected with unmodified viruses the mechanism is specifically targeted against the viral genome. However, with virus vectors carrying inserts derived from host genes the process can be additionally targeted against the corresponding mRNAs. VIGS has been used widely in plants for analysis of gene function and has been adapted for high-throughput functional genomics. Until now most applications of VIGS have been in *Nicotiana benthamiana* (60). However, new vector systems and methods are being developed that could be used in other plants,

Research demonstrated that host resistance to *Fusarium verticillioides*, *Blumeria graminis* and *Puccinia striiformis f. sp. tritici* can be enhanced by the expression of fungal genes in transgenic host plants (83b).

Over the years, several studies have been performed to analyse plant–pathogen interactions. Recently, functional genomic strategies, including proteomics and transcriptomics, have contributed to the effort of defining gene and protein function and expression profiles. Using these genomic approaches, pathogenicity- and defence-related genes and proteins expressed during phytopathogen infections have been identified and enormous datasets have been accumulated. However, the understanding of molecular plant–pathogen interactions is still an intriguing area of investigation. Proteomics has dramatically evolved in the pursuit of large-scale functional assignment of candidate proteins and, by using this approach, several proteins expressed during phytopathogenic interactions have been identified (64). An understanding of how plants and pathogens recognize each other and differentiate to establish either a successful or an unsuccessful relationship is crucial in this field of investigation. The recognition and signaling events that occur in plant cells in response to microorganism challenge need to be extremely rapid, reliable and specific, and are part of the strategy evolved by plants to survive attacks. Therefore, successful pathogens must evolve mechanisms to interfere with or suppress basal defence provided by resistance to colonize the host and develop disease. Superimposed on the basal defence, some plant varieties express resistance proteins that guard against this interference and trigger a specific genetically defined hypersensitive response and subsequent programmed cell death. The function of the hypersensitive response is to contain the pathogen, and it is typified by various biochemical perturbations, known as generic plant responses including changes in ion fluxes, lipid hyperperoxidation, protein phosphorylation, nitric oxide generation and a burst of reactive oxygen species and antimicrobial compounds. This rapid incompatibility response effectively puts an end to pathogen invasion and prevents further disease development (3). In the case of pathogens several of the proteins involved in pathogenicity are secretion proteins, which were observed in bacteria, fungi and nematodes Proteomics has dramatically evolved in the pursuit of large-scal functional assignment of candidate proteins and, by using thi approach, several proteins expressed during phytopathogeni interactions have been identified. Although the involvement c defence-related proteins are PR proteins, including thaumatins glucanases, peroxidases and chitinases, has been described i several pathosystems, their direct role in resistance enhancemen still needs to be demonstrated. The examples reviewed her demonstrate the complex cellular network that exists in differen plant–pathogen interactions. The first step in the understanding of disease resistance is currently being met with the identification of the proteins expressed during plant–pathogen interactions. The next step will be to determine which proteins confer pathogenicity and disease resistance, and the mechanisms by which they do so.

Cloning Resistance Genes

For cloning of disease resistance genes, different approaches are used (33). Since protein products of disease resistance genes are not known, map based cloning approach has been predominantly used for cloning disease resistance genes in different host-pathogen systems. Disease resistance gene *Hm1* was the first gene cloned in maize by using transposon tagging approach (48). However, more than 80% of the resistance genes have been cloned by using the preferred map based cloning approach as it does not require any information of the gene product, and each step can be monitored and is very systematic. It is labor intensive and slow technique. It cannot be used to clone a gene residing in the complex locus of the genome. A blast resistant gene *Pi-kh* has been mapped in India on chromosome 11 in Tetep using DNA markers (102) and subsequently cloned. This gene has been later renamed as *Pi54* (102b). More than 70 disease resistance genes have been cloned from different crops (58). Of these 94% of the genes have been cloned from seven plant species like, *Arabidopsis*, rice, tomato, wheat, flax, barley and potato. The information published on plant genes cloned till date was further analyzed to find out as to which category of the pathogens drew more attention for molecular cloning of resistance genes. This analysis indicated that 63% of the genes cloned from different plant species were resistant to various fungal diseases, and 23% to bacterial diseases, while 7% of the genes were resistant to nematode and viral pathogens (47).

Resistance-gene discovery will become much faster and less expensive as resistance phenotypes are matched to candidate sequences identified by genomic sequencing or PCR using degenerate oligonucleotides. The rate-limiting step will be the confirmation of the function of candidate genes. Antisense inhibition or sense suppression can be used to demonstrate whether a member of a multigene family encodes a particular specificity. The antisense concept was first conceived in 1978 and has been through various technical refinements. Antisense compounds are designed to have the right nucleotide sequence to bind specifically to and interfere with its associated mRNA, the instructions for the production of a particular protein. To create antisense compounds, special chemically stabilized nucleotides are synthetically linked together in short chains of about 12-30 nucleotides (called oligonucleotides). Each antisense compound is designed with the right complementary genetic code to bind to a specific sequence of nucleotides in its mRNA target to form a short area of double strands. This double stranded region can inhibit the production of protein by a number of mechanisms. They include stopping the ribosome from reading the message, or by leading to the destruction of the mRNA by an enzyme already in the cells called RNase H. RNase H present both inside and outside of the nucleus, destroys any such double-stranded nucleotides. The identification of genes with individual specificities is still going to require careful experimentation involving a combination of mutation and transgenic analyses. Nevertheless, caution is required with this approach as ectopic expression of resistance genes may result in non-specific resistance (85). Genes encoding NBS-LRR-containing proteins are one of the most prevalent classes in plant genomes as discussed earlier but little is known of their function. Their sequence motifs indicate that they are involved at the beginning of signaling pathways. So far, the only demonstrated role for NBS-LRR-encoding genes is in disease or pest resistance.

Characterization of Disease Resistance Genes

Plants apparently recognize and respond to diseases. The gene for gene theory, suggests for example, two genes contribute to rust resistance – one in the plant and one in the rust. Little is known about how *Avr* and *R* gene products determine this incompatibility. If *R* or *Avr* genes are absent from either partner, the plant does not recognize the pathogen and disease ensues (42). The plant must have the correct rust resistance gene to fight the rust, and the rust must have a gene that identifies it to the plant as an unwanted intruder. The avirulence gene apparently makes a protein that is injected into the plant cells to help the pathogen infect them. But, unfortunately for the pathogen, this protein also gives it away to the plant's immune system. Avirulence genes are originally defined by their negative impact on the ability of a pathogen to infect their host plant. Many avirulence genes are now known to represent a subset of virulence factors involved in the mediation of the host–pathogen interaction. Characterization of avirulence genes has revealed that they encode an amazing assortment of proteins and belong to several gene families. Although the biochemical functions of the avirulence gene products are unknown, studies are beginning to reveal the features and interesting relationships between the avirulence and virulence activities of the proteins. Identification of critical virulence factors and elucidation of their functions promises to provide insight into plant defense mechanisms, and new and improved strategies for the control of plant disease (120).

A limited number of genes associated with infection-related morphogenesis have been characterized. For example, *Mpg*1 and *Mhp*1 for surface sensing and attachment, Pth11 is for G-protein signaling, *Mac*1, *PkA* and *CpkA* for cAMP signaling, *Atg*1 and *Atg*8 for autophagy, *Nima*, *Hex*1, *Nox*1, *Nox*2, *NoxR*, *Pex*6, *Pth*2, Tps for appressorial development, *Pmk*1, *Mst*50, *Mst*11, *Mst*7 are for MAPK signaling, and genes like *Mps*1, *Mig*1, *Mck*1, *Pls*1, *Pde*1, *Mst*12 are needed for invasive growth. In addition, cytochrome P450, laccases, and ABC transporters are also possibly being utilized to elude and suppress plant defence systems (98).

Rusts are biotrophic pathogens that attack many plant species but are particularly destructive on cereal crops. The stem rusts (caused by *Puccinia graminis*) have historically caused severe crop losses and continue to threaten production today. Barley (*Hordeum vulgare*) breeders have controlled major stem rust epidemics since the 1940s with a single durable resistance gene *Rpg1*. As new epidemics have threatened, additional resistance genes were identified to counter new rust races, such as the *rpg4*/*Rpg5* complex locus against races QCCJ and TTKSK. To understand how these genes work, we initiated research to clone and characterize them. The *Rpg1* gene encodes a unique protein kinase with dual kinase domains, an active kinase, and a pseudo-kinase. Function of both domains is essential to confer resistance. The *rpg4* and *Rpg5* genes are closely linked and function coordinately to confer resistance to several wheat (*Triticum aestivum* L.) stem rust races, including the race TTKSK (also called Ug99) that threatens the world's barley and wheat crops. The *Rpg5* gene encodes typical resistance gene domains NBS, LRR, and protein kinase but is unique in that all three domains reside in a single gene, a previously unknown structure among plant disease resistance genes. The *rpg4* gene encodes an actin de-polymerizing factor that functions in cytoskeleton re-arrangement (51).

A complete genetic resistance to white mold of soybean is not available. RNA from 15 day-old inoculated plants was labeled and hybridized to soybean cDNA microarrays. ANOVA identified 1270 significant genes from the comparison between time points and 105 genes from the comparison between genotypes. Selected genes were classified into functional categories. The analyses identified changes in cell-wall composition and signaling pathways, as well as suggesting a role for anthocyanin and anthocyanidin synthesis in the defense against *S. sclerotiorum*. In-silico mapping of both the differentially expressed transcripts and of public markers associated with partial resistance to white mold, provided evidence of several differentially expressed genes being closely positioned to white mold resistance markers, with the two most promising genes encoding a *PR-5* and anthocyanidin synthase (14).

With the availability of a reference genome sequence, research in rice genomics has now shifted to functional and applied genomics strategies. A novel methodology called genomics-based breeding approach has also started to emerge (110). In this strategy, many agronomically important phenotypic traits are located on the rice genome in nucleotide-level resolution using the genome-wide DNA markers generated by sequence polymorphism such as SNPs (single nucleotide polymorphism) and Indels (INsertion/DELetions). Indels are defined as either an insertion or deletion, or as a mutation resulting in a co-localized insertion and deletion and a net gain or loss in nucleotides. Indels can be used as genetic markers in natural populations, especially in phylogenetic studies. SNP markers are by far the most common form of DNA polymorphism in a genome. The DNA marker methodology has several advantages over classical phenotypic markers used in conventional breeding. Basically, genotyping could be performed even in the seedling stage, and confining the responsible portion within a narrow genomic region could reduce the chance of dragging totally unrelated genes. Finding the genome-wide polymorphism for any rice variety is unrealistic and impractical because this virtually means re-sequencing entire genomes. The cost of sequencing entire genomes has remained a limiting factor. However, if the ongoing revolution of sequencing technology could reduce the cost of total genome sequencing to a few thousands or even hundreds of rupees, sequencing will not be limited to reference genomes. Instead, revealing the nucleotide sequences of specific genetic analysis materials such as recombinant lines and their parent cultivars for example, will become a reality. This will allow breeders or researchers to immediately recognize the origin of genomic segments anywhere in the genome for every recombinant plant. Eventually, this will contribute to accelerating many aspects of agricultural research, particularly in introducing specific chromosome regions from one cultivar to another. Such a target-specific approach in genetic modification will benefit many breeding goals to increase yield, incorporate resistance to biotic and abiotic diseases, and develop cultivars that could grow well in adverse environmental conditions (45, 110).

Thiamine, in addition to its nutritional value, induces systemic acquired resistance (SAR) in plants (2). Thiamine-treated rice, *Arabidopsis*, and vegetable crop plants showed resistance to fungal, bacterial, and viral infections. Thiamine activates SAR-related genes in rice, tobacco, tomato (*Lycopersicon esculentum*), cucumber (*Cucumis sativus*), and *Arabidopsis* and prevents several diseases caused by semibiotrophic and biotrophic pathogens. Thiamine confers disease resistance through the priming of several plant defense responses, leading to a restriction of pathogen growth in planta and suppressed propagation of the inoculum. Thiamine treatment induces the transient expression of pathogenesis-related (PR) genes in rice and other plants. In addition, thiamine treatment potentiates stronger and more rapid PR gene expression and the up-regulation of protein kinase C activity. Treatment of *Arabidopsis* plants with thiamine resulted in the activation of *PR-1* but not *PDF1.2*.The maintenance of the resistance mimic status for a long period indicates that thiamine is a good candidate as a plant defense activation agent. Along with conventional antibiotics, previously developed plant defense activators, biocontrol organisms, and improved seed varieties, thiamine should provide novel disease control strategies that satisfy environmental regulations (2). A novel function of riboflavin (vitamin B2) in disease resistance has also been described (32). Treatment with riboflavin protects tobacco (*Nicotiana tabacum*) and *Arabidopsis* plants from fungal and bacterial infections without inhibiting pathogen growth.

Bioinformatics

Bioinformatics is the field of science in which biology, computer science, and information technology merge into a single discipline. There are three important sub-disciplines within bioinformatics: the development of new algorithms and statistics with which to assess relationships among members of large data sets; the analysis and interpretation of various types of data including nucleotide and amino acid sequences, protein domains, and protein structures; and the development and implementation of tools that enable efficient access and management of different types of information. Bioinformatics is critical to structural, comparative and functional genomics. There is already a vast amount of DNA genomic and EST sequence data in the public domain and an even greater amount in private databases. The amount of quantitative expression data becoming available in the public and private sectors will increase exponentially and rapidly dwarf the DNA sequence data; quantitative expression data will be considerably more difficult to manage and exploit. Improved and preferably semi-intelligent algorithms are therefore required to acquire, curate and query the data. A specialized, thematic database of plant NBS-encoding sequences has been developed at the National Center for Genomic Research (http://www.ncgr.org/rgenes). This internet database includes links to the underlying database records, data source, BLAST (i.e. basic local alignment search tool; National Center for Biotechnology Information) scores relating the NBS encoding sequences to known R-genes, organism, map positions in *Arabidopsis*, and graphic descriptions of motif organization. The existing databases can be used to search for homologs of known molecules from other signaling pathways and defense responses in other organisms. The increased power of algorithms such as PSI-BLAST (i.e. position specific iterated BLAST) (5) affords the opportunity for identification of distant homologies (e.g. for caspases) that can be the basis of functional testing of plant genes. Homologies with genes of known function in other organisms can predict the function of newly identified plant genes and provide opportunities for utilization of model systems to demonstrate function. For example, the sequence of the cloned *Rar1*, a gene required for Mla-1 activity in barley, implicated its involvement in a cell death pathway (90). Likewise, similarities between the NBS region of plant resistance proteins and CED-4 and Apaf-1 in nematodes and mammals, respectively, have led to inferences of functional parallels between the cell death pathways of the hypersensitive resistance response in plants and apoptosis in animals (18, 115).

CONCLUSIONS

Plant Pathology is experiencing a period of rapid change. We already have un-imagined tools and capabilities compared with those available few years ago. There is a shift away from the identification and manipulation of individual genes to the global characterization of resistance phenotypes. Ideally, disease resistance genes will control pathogens at a low metabolic cost by inducing defense responses only in those cells that are challenged by the pathogen. In the future, resistance genes will be designed that recognize essential components of pathogens. Pathogen propagules increase their population size rapidly and in a short period un-like their hosts. Consequently there is a greater opportunity for virulent pathogen races to arise than for corresponding new host resistance specificities. The pathogen evolution from avirulence to virulence usually results from loss-of-function mutation. The corresponding gain of function, that is resistance in the host, is unlikely to occur by simple mutation. It is unlikely that plants have a mechanism to generate new R gene specificities and keep pace with the evolutionary progress of pathogens. Research challenges include the determination of the mechanisms by which R gene products recognize pathogen elicitors and the plant defence response that blocks pathogen growth. The basic knowledge obtained from this research will undoubtedly help to produce novel forms of durable disease resistance and will lead to a decline in the use of environmentally damaging pesticides. Newer techniques in genome will unleash tremendous possibilities of achieving increased production levels and improved quality in crop commodities, especially in food crop species. It is also our hope that benefits from research will be in the public domain to enable use by small and marginal farmers from the developing countries.

REFERENCES

1. Aarts MGM, Hekkert BL, Holub EB, Beynon JL, Stiekema WJ and Pereira A. 1998. Identification of R-gene homologous DNA fragments genetically linked to disease

resistance loci in *Arabidopsis thaliana*. *Mol Plant–Microbe Interact* 11:251-258.

2. Ahn I, Kim S, and Lee YH. 2005. Vitamin B1 functions as an activator of plant disease resistance. *Plant Physiol* 138: 1505–1515

3. Alfano JR and Collmer A. 2004. Type III secretion system effector proteins: double agents in bacterial disease and plant defence. *Annu Rev Physiol* 42, 385–414.

4. Allen MF. 1991. *The ecology of mycorrhizae*. Cambridge University Press, Cambridge.

5. Altschul SF and Koonin EV. 1998. Iterated profile searches with PSI-BLAST — a tool for discovery in protein databases. *Trends Biochem Sci* 23: 444-447.

6. Baldwin D, Crane V and Rice D. 1999. A comparison of gel-based, nylon filter and microarray techniques to detect differential RNA expression in plants. *Curr Opin Plant Biol* 2:96-103.

7. Barakat A, Matassi G and Bernardi G. 1998. Distribution of genes in the genome of *Arabidopsis thaliana* and its implications for the genome organization of plants. *Proc Natl Acad Sci USA* 95:10044-10049.

8. Barr DJS. 1992. Evolution and kingdoms of organisms from the perspective of mycologist. *Mycologia* 84(1): 1-11.

9. Basnet M, Chakraborty BN and Chakraborty U. 2009. Exploitation of tea rhizosphere microorganism for improvement of plant health status. *J Mycol Pl Pathol* 39(1):1-13.

10. Bevan M, Bancroft I, Bent E, Love I, Goodman H, Dean C, Bergkamp R, Dirkse W, Van Staveren M, Stiekema W *et al.* 1998. Analysis of 1.9 Mb of contiguous sequence from chromosome 4 of *Arabidopsis thaliana*. *Nature* 391:485-488.

11. Bieri S, Potrykus I, and Futterer J. 2003. Effects of combined expression of antifungal barley seed proteins in transgenic wheat on powdery mildew infection. *Mol Breed* 11:37–48.

12. Botella MA, Coleman MJ, Hughes DE, Nishimura MT, Jones JDG and Somerville SC. 1997. Map positions of 47 *Arabidopsis* sequences with sequence similarity to disease resistance genes. *Plant J* 12:1197-1211.

13. Buell CR et al. 2003. The complete genome sequence of the Arabidopsis and tomato pathogen *Pseudomonas syringae* pv. tomato DC3000. *Proc.*

Natl. Acad. Sci. USA 100:10181–10186.

14. Calla B, Vuong T, Radwan O, Hartman GL and Clough SJ. 2009. Gene Expression Profiling Soybean Stem Tissue early response to *Sclerotinia sclerotiorum* and *in silico* mapping in relation to resistance markers. *Plant genome* 2(2):149-166

15. Cavalier-Smith T. 1989. The kingdom Chromista. In: *The Chromophyte Algae: problems and perspectives.* Green JC, Leadbeater BSC and Diver WL (eds), Systematics Association. Clarendon Press, Oxford, pp381-407.

16. Chang CC, Chen TT, Cox BW, Dawes GN, Stemmer WP, Punnonen J, Patten PA. 1999. Evolution of a cytokine using DNA family shuffling. *Nat Biotechnol* 17:793-797.

17. Chen X, Tang J, Fang Z, and Shuijin HU. 2002. Phosphate-solubilizing microbes in rhizosphere soils of 19 weeds in southeastern China. *J Zhejiang Univ Sci* 3(3): 355-361.

18. Chimmaiyan AM, Chaudhary D, O'Rouke K, Koonin EV and Dixit VM. 1997. Role of CED-4 in the activation of CED-3. *Nature* 338:728-729.

19. Cho RJ, Mindrinos M, Richards DR, Sapolsky RJ, Anderson M, Drenkard E, Dewdney J, Reuber TL, Stammers M, Federspiel N *et al.* 1999. Genome-wide mapping with biallelic markers in *Arabidopsis thaliana. Nature Genet* 23:203-207.

20. Chuang CC, Kuo YL, Chao CC and Chao WL. 2007. Solubilization of inorganic phosphates and plant growth promotion by *Aspergillus niger. Bio Fertil Soils* 43: 575-584.

21. Chuang CC, Park M, Madhiyan M, Seshadri, S, Song J, Cho I and Sa. 2005. Isolation and characterization of phosphate solubilizing bacteria from rhizosphere of crop of Korea. *Soil Biol Biochem* 37:1970-1974.

22. Collins RF, Gellatly DL, Sehgal OP, Abouhaidar MG. 1998. Self-cleaving circular RNA associated with rice yellow mottle virus is the smallest viroid-like RNA. *Virology* 241(2): 269–75.

23. Cook RJ, Thomashow LS, Weller DM, Fuzimoto D, Mazzola M, Bangera G and Kim DS. 1995. Molecular mechanisms of defense by rhizobacteria against root disease. Proc. National Acad. Sci., USA 92: 4197-4201.

24. Cook RJ. 1993. Making greater use of introduced microorganisms for biological

control of plant pathogens. *Annu Rev Phytopath* 31:53-80.

25. Curl EA and Truelove B. 1986. The rhizosphere (Advanced series in Agric. Science 15). Springer Verlag, Berlin, Heidelberg.

26. Datta K, Tu J, Oliva N, Ona I, Velazhahan R, Mew TW Muthukrishnana S and Datta SK. 2001. Enahanced resistance to sheath blight by constitutive expression of infection-related rice chitinase in transgenic elite *indica* rice cultivars. *Plant Sci* 160:405-414.

27. Dean RA, Talbot NJ, Ebbole DJ, Farman ML, Mitchell TK, Orbach MJ, Thon M, Kulkarni R, Xu J, Pan H, Read ND, Lee Y, Carbone I, Brown D, Oh YY, Donofrio N, Jeong JS, Soanes DM, Djonovic S, Kolomiets E, Rehmeyer C, Li W, Harding M, Kim S, Lebrun M, Bohnert H, Coughlan S, Butler J, Calvo S, Ma L, Nicol R, Purcell S, Nusbaum C,. Galagan JE and Birren BW. 2005. The genome sequence of the rice blast fungus *Magnaporthe grisea. Nature* 434, 980-986. http://www.nature.com/nature/journal/v434/n7036/abs/nature03449.html - a3#a3

28. Dehne HW and Schonebeck F. 1979. Untersuchungen zum einfluss der endotrophen mycorrhiza auf planzenkrankheiten. I. Ausbreitung von *Fusarium oxysporum* f. sp.*lycopersici* in tomaten. *Phyotol Z* 95:104-110.

29. DeVos K, Beales J, Nagamura Y and Sasaki T. 1999. *Arabidopsis*–rice: will colinearity allow gene prediction across the eudicot-monocot divide? *Genome Res* 1999, 9:825-829.

30. Dinaker C and Muralidharan K. 2006. Spore production, germination and appressorium formation by *Pyricularia grisea* as influenced by nutrients and leaf components of resistant and susceptible rice cultivars. *J Myco Pl Pathol* 36(2): 226-233.

31. Dodd JC. 2000. The role of arbuscular mycorrhizal fungi in agro-and natural ecossystems. *Outlook Agric* 29 (1):55-62.

32. Dong H and Beer SV. 2000. Riboflavin induces disease resistance in plants by activating a novel signal transduction pathway. *Phytopathology* 90: 801–811

33. Ellis JG, Lawrence GJ, Luck JE and Dodds PN. 1999. Identification of regions in alleles of the flax rust resistance gene *L* that determine differences in gene-for-gene specificity. *Plant Cell* 1999, 11:495-506.

34. Garnier M, Foissac X, Gaurivaud P, Laigret F, Renaudin J, Saillard C and Bové JM. 2001. Mycoplasmas, plants, insect vectors: a matrimonial triangle. *C R Acad Sci III* 324(10):923-8.

35. Ghosh AB, and Hasan R. 1979. Phosphorus fertility status of soils of India. *Bull Indian Soc Soil Sci* 12: 1-8.

36. Ghosh AB, Bajaj BC, Rehanul Hasan and Dhayan Singh. 1983. In: *Soil and water testing methods*. Indian Agriculture Research Institute, New Delhi., 79pp.

37. Glick, B. R. 1995. The enhancement of plant growth by free-living bacteria. *Can J Microbiol* 41:109-117.

38. Goodner B et al. 2001. genome sequence of the plant pathogen and biotechnology agent *Agrobacterium tumefaciens* C58. *Science* 294:2323 – 2328

39. Gopalakrishnan S, Sharma RK, Rajkumar KA, Joseph M,. Singh MP, Singh AK, Bhat KV, Singh NK and Mohapatra T. 2008. Integrating marker assisted background analysis with foreground selection for identification of superior bacterial blight resistant recombinants in Basmati rice. *Pl Breed* 127: 131-139.

40. Gullino ML, Leroux P and Smith CM. 2000. Uses and challenges of novel compounds for plant disease control. *Crop Prot* 19(1):1-11.

41. Guo, J. H., H. Y. Qi, Y. H. Guo, H. L. Ge, L. Y. Gong, L X. Ziang and P. H. Sun. 2004. Biocontrol of tomato wilt by plant –growth rhizobacteria. *Biol Contl* 29: 66-72.

42. Hammond-Kosack KE, Harrison K and Jones DGJ. 1994. Developmentally regulated cell death on expression of the fungal avirulence gene Avr9 in tomato seedlings carrying the disease-resistance gene *Cf-9*. *Proc Natl Acad Sci USA* 91:10445-10449.

43. Harman GE. 2000. Myths and dogmas of biocontrol. Changes in perceptions derived from research on *Trichoderma harzianum* T22. *Plant Dis* 84:377-393.

44. Holub EB. 1997 Organization of resistance genes in *Arabidopsis*. In *The Gene-for-Gene Relationship in Host–Parasite Interactions.* Crute IR, Holub EB and Burdon J. Ed. Wallingford, UK: CAB International pp 5-26.

45. IRGSP. 2005. The map based sequence of rice genome. *Nature* 436: 793-800.

46. Jaillon O, Jean-Marc A, and Noel B. 2007. The grapevine genome sequence suggests ancestral hexaploidization in major angiosperm phyla. *Nature* 449: 463-467.

47. Jalali BL. 2008. Molecular plant pathology: where do we stand? *J Mycol Pl Pathol 38(3)410-429.*

48. Johal GS, and Briggs SP. 1992. Reductase activity encoded by the HM1 disease resistance gene in maize. *Science* 258: 985-987.

49. Kapoor KK, Mishra MM and Kukreja K. 1989. Phosphate solubili-zation by soil microorganisms. *Indian J Microbiol* 29: 119-127.

50. Kitajima S and Sato F. 1999. Plant pathogenesis-related proteins: molecular mechanisms of gene expression and protein function. *J Biochem (Tokyo)* 125:1-8.

51. Kleinhofs A, Brueggeman R, Nirmala J, Zhang L, Mirlohi A, Druka A, Rostoks N and Steffenson JB. 2009. Barley stem rust resistance genes: structure and function. *Pl Genome* 2(2):109-120.

52. Klironomos JN, McCune J, Hart M and Neville J. 2000. The influence of arbuscular mycorrhizae on the relationship between plant diversity and productivity. *Ecol. Lett.* 3: 137-141.

53. Kloepper JW. 1992. Plant growth promoting rhizobacteria as biological control agents. In: *Soil microbiol ecology: applications in agricultural and environmental management.* Metting, FB Jr. Ed Marcel Dekker Inc. NY, USA. pp 255-274.

54. Kloepper JW. 1996. Host specificity in microbemicrobe interactions. *Bioscience* 46:406–409.

55. Krishnaveni D, Neeraja CN, Prasad GSV and Muralidharan K. 2006. Identification of microsatellite markers linked to rice tungro resistance. In *Int Symp Management of Vector-Borne Viruses*, Feb 7-10, 2006, Int Crops Res Inst Semi-Arid Tropics (ICRISAT), Patancheru 502 324, Hyderabad, Andhra Pradesh, India

56. Kunkel BN. 1996. A useful weed put to work - genetic analysis of disease resistance in *Arabidopsis thaliana. Trends Genet* 12:63-69.

56.b. Lévesque CA. 2010. Genome sequence of the necrotrophic plant pathogen Pythium ultimum reveals original pathogenicity mechanisms and effector repertoire. *Genome Biol.* http://genome-biology.com/2010/11/7/R73

57. Liester D, Kurth J, Laurie DA, Yano M, Sasaki T, Graner A, Schulz-Lefert P. 1998. Rapid reorganization of resistance gene homologues in cereal genomes. *Proc Natl Acad Sci USA* 95:370-375.

58. Liu J, Liu X, Dai L and Wang G. 2007. Recent progress in elucidating the structure, function and evolution of disease resistance genes in plants. *J Genet Genomic* 34:765-776.

59. Liu Y, Schiff M, Marathe R, Dinesh-Kumar SP. 2002. Tobacco *Rar1, EDS1* and *NPR1/NIM1* like genes are required for N-mediated resistance to *tobacco mosaic virus. Plant J.* 30(4):415-29.Lu

60. R, Martin-Hernandez AM, Peart JR, Malcuit I, Baulcombe DC. 2003. Virus-induced gene silencing in plants. *Methods* 30(4):296-303.

61. Makandar R, Essig JS, Schapaugh MA, Trick HN and Shah J. 2006. Genetically engineered resistance to *Fusarium* head blight in wheat by expression of *Arabidopsis NPR1. Mol Pl Microbe Interact* 19:123–129.

62. Martienssen RA, Rabinowicz PD, O'Shaughnessy A, and McCombie WR. 2004. Sequencing the maize genome. *Curr Opin Plant Biol* 7:102–107.

63. McCarter JP. Molecular approaches towards resistance to plant-parasitic nematodes. In: *Cell biology of plant nematode parasitism.* Berg RH and Taylor CG. Eds. Springer-Verlag, pp239-259.

64. Mehta et al. 2008. Plant–pathogen interactions: what is proteomics telling us? *FEBS J* 275:3731–3746.

65. Meyers BC, Dickerman AW, Michelmore RW, Sivaramakrishnan S, Sobral BWa nd Young ND. 1999. Plant disease resistance genes encode members of an ancient and diverse protein family within the nucleotide-binding superfamily. *Plant J* 20:317-332.

66. Michelmore R. 2000. Genomic approaches to plant disease resistance. *Curr Opin Pl Biol* 3:125–131

67. Michelmore RW and Meyers BC. 1998. Clusters of resistance genes in plants evolve by divergent selection and a birth-and-death process. *Genome Res* 1998, 8:1113-1130.

68. Ming, RS, Hou S and Feng Y. 2008. The draft genome of the transgenic tropical fruit tree papaya (*Carica papaya* Linnaeus). *Nature* 452: 991-996.

69. Mondal, K, Singh RP and Verma JP. 1999. Benificial effects of indigenous cotton rhizobacteria on seed germination, growth promotion and suppression of bacterial blight disease. *Indian Phytopath.* 52: 228–235.

70. Muralidharan K and Siddiq EA. Eds. 1993. *New frontiers in rice research.* Directorate of Rice Research, Hyderabad 500 030, India. 400p.

71. Muralidharan K, Krishnaveni D, Laha GS, Reddy CS, Srinivasprasad M and Sridhar R. 2004a. Performance of rice blast resistance genes in multi-environment test in India. *Indian Phytopath* 57:260-266.

72. Muralidharan K, Krishnaveni D, Laha GS, Reddy CS, Srinivasprasad M and Sridhar R. 2003a Evaluation of blast resistance genes in India. *Rice Genet Newsl* 20:94-96.

73. Muralidharan K, Krishnaveni D, Laha GS, Reddy CS, Srinivasprasad M and Sridhar R. 2003b. Appraisal of bacterial leaf blight resistance genes in India. *Rice Genet Newsl* 20:96-98.

74. Muralidharan K, Krishnaveni D, Neeraja CN and Prasad GSV. 2003c. Reflections on the complexity of rice tungro virus disease. *Annu Rev Plant Pathol* 2:379-436.

75. Muralidharan K, Krishnaveni D, Rajarajeswari NVL and Prasad ASR. 2003d. Tungro epidemics and yield losses in paddy fields in India. *Curr Sci.* 85(8):1-5.

76. Muralidharan K, Laha GS, Krishnaveni D, Reddy CS, Reddy JN and Sridhar R. 2005. Field performance of genotypes generated by molecular-assisted breeding to bacterial leaf blight in India. *Rice Genet Newsl* 21: 29-31.

77. Muralidharan K, Reddy CS, Krishanaveni D and Laha GS. 2004c. Field application of fluorescent *Pseudomonas* products to control blast and sheath blight diseases in rice. *J Mycol Pl Pathol* 34: 411-414.

78. Muralidharan K. 2000. Modeling for rice blast disease development. In: *Epidemiological principles in plant disease control.* Dang JK. and Rishi N. Eds. Department of Plant Pathology, CCS Haryana Agricultural University, Hissar, India & Indian Council of Agricultural Research, New Delhi.

79. Nagarajan S and Muralidharan K. 1995. *Dynamics of plant diseases.* Allied Publishers, New Delhi.

80. National Science Foundation: Plant Genome Research Program Collaborative Research and Infrastructure Projects.http://www.nsf.gov/bio/pubs/awards/genome

81. Nautiyal CS, Bhaduria S, Kumar P, Lal H, Mondal R, and Verma D. 2000. Stress induced phosphate solubiliz-

ation in bacteria isolated from alkaline soils. *FEMS Microbiol Lett* 18: 291-296.

82. Neeraja CN, Krishnaveni D, Saivishnupriya K, Prasad GSV and Muralidharan K. 2007. Molecular mapping of resistance to tungro virus in rice cultivars Vikramarya and Utri Rajapan. *Rice Genet Newsl* 23:80-83,2007.

83. Newsham KK, Fitter AH and Watkison AR. 1995. Arbuscular mycorrhiza protect an annual grass from root pathogenic fungi in the field. *J Ecol* 83:991-1000.

83b. Nunes CC and Dean RA. 2012. Host induced gene silencing: a tool for understanding fungal interaction and for developing novel disease control strategies. *Mol Plant Pathol.* 13:519-529.

84. Oldach KH, Becker D and L¨orz H. 2001. Heterologous expression of genes mediating enhanced fungal resistance in transgenic wheat. *Mol Pl Microbe Interact* 14:832–838.

85. Oldroyd GED and Staskawicz BJ. 1998. Genetically engineered broadspectrum disease resistance in tomato. *Proc Natl Acad Sci USA*, 95:10300-10305

86. Oshima K, Kakizawa S, Nishigawa H, Jung HY, Wei W, et al. 2004. Reductive evolution suggested from the complete genome sequence of a plant-pathogenic phytoplasma. *Nat Genet* 27-9.

87. Pan Q, Wendel J and Fluhr R. 2000. Divergent evolution of plant NBS-LRR resistance gene homologues in dicot and cereal genomes. *J Mol Evol.* 50:203-213.

88. Panstruga R, Buschges R, Piffanelli P, Schulze-Lefert P. 1998. A contiguous 60 kb genomic stretch from barley reveals molecular evidence for gene islands in a monocot genome. *Nucleic Acids Res* 26:1056-1062.

89. Pichard B and Thouvenot D. 1999. Effect of *Bacillus polymyxa* seed treatment on control of black rot and damping off cauliflower. *Seed Sci Tech* 27:455–465.

90. Piffanelli P, Devoto A and Schultz-Liefert P. 1999. Defense signaling pathways in cereals. *Curr Opin Plant Biol* 2:295-300.

91. Pozo MJ, Cordier C and Dumas-Gaudot E. 2002. Localized versus systemic effect of arbuscular mycorrhizal fungi on defence responses to Phytophthora infection in tomato plants. *J Exp Bot* 53(368):525-534.

92. Rabinowicz PD, McCombie WR and Martienssen RA.

2003. Gene-enrichment in plant genomic shotgun libraries. *Curr Opin Plant Biol* 6:150–156.

93. Ramamoorthy V, Raguchander T and Samiyappan R. 2002. Enhancing resistance of tomato and hot pepper to *Pythium* diseases by seed treatment with fluorescent pseudomonads. *Eur J Plant Pathol* 108:429-441.

94. Ramamoorthy V, Viswanathan R, Raghuchander T, Prakasam V and Samiyappam R. 2001. Induction of systemic resistance by plant growth promoting rhizobacteria in crop plants against pest and diseases. *Crop Prot* 20: 1-11.

95. Reddy CS, Laha GS and Muralidharan K. 2005. Interactions between rhizosphere microorganisms and plant pathogens. In: *Frontiers in Plant Sciences*. KG Mukerji, KVBR Tilak, SM Reddy, LV Gangwane, P Prakash, IK Kuwar. Eds. I.K. International Pvt. Ltd., New Delhi pp 163-218

96. Ren TY, Zhichao MA, Weiliang C and Bao LD. 1999. Purification and partial characterization of antagonistic proteins from *Bacillus subtilis* B034. *Acta Microbil Sinica* 39: 339–343.

97. Reyes I, Valery A and Valduz. 2006. Phosphatesolubilizing microorganisms isolated from rhizospheric and bulk soils of colonizer plants at an abandoned rock phosphate mine. *Plant Soil* 287: 69-75.

98. Roy-Barman S and Chattoo B. 2011. Molecular basis of rice blast disease. In: Muralidharan K and Siddiq EA, eds. 2011. Genomics and crop improvement: Relevance and reservations. Institute of Biotechnology, Acharya NG Ranga Agricultural University, Rajendranagar, Hyderabad 500030, India, pp 104-124.

99. Salanoubat M. et al. 2002. Genome sequence of the plant pathogen Ralstonia solanacearum, *Nature* 415:497-502.

100. Sastry MVS, Muralidharan K. and Prasad GSV. 2004. *Production and marketing of seed and seed material.* 440p. Padma Publishers, B11/2, HIGH II, Baghlingampally, Hyderabad 500044, India

101. Schmidt SA, Williams SJ, Wang CA, Sornaraj P, James B, Kobe B, Dodds PN, Ellis JG and Anderson PA. 2007. Purification of the M flax-rust resistance protein expressed in *Pichia pastoris. Plant J* (56) : 1107 – 1117 .

102. Sharma TR, Madhav MS, Singh BK, Shanker P, Jana TK, Dalal V, Pandit, Singh A, Gaikwad K, Upreti HC and Singh NK. 2005. High-

resolution mapping, cloning and molecular characterization of the gene of rice, which confers resistance to rice blast. *Mol Genet Genomics* 274:569-578.

102b.Sharma TR, Rai AK, Gupta SK and Singh NK. 2010 Broad spectrum blast resistance gene *Pi-k*^{h} designated as *Pi-54. J Plant Biochem Biotechnol* 19:987-989.

103. Smith SE and Read DJ. 1997. *Mycorrhizal symbiosis.* Ed 2. Academic Press, San Diego.

104. Speulman E, Bouchez D, Holub EB, Beynon JL. 1998. Disease resistance gene homologs correlate with disease resistance loci of *Arabidopsis thaliana. Plant J* 14:467-474.

105. Staskawicz BJ, Ausubel FM, Baker BJ, Ellis JG and Jones JDG. 1995. Molecular genetics of plant disease resistance? *Science* 268:661-667.

106. Sundaram RM., Manne R, Vishnupriya R, Biradar SK, Laha GS, Reddy GA, Rani NS, Sarma NP and Sonti RV. 2008. Marker assisted introgression of bacterial blight resistance in Samba Mahsuri, an elite *indica* rice variety. *Euphytica* 160:411–422.

107. Suslow TV and Schroth MN. 1982. Rhizobacteria of sugar beets, effects of seed application & root colonization on yield. *Phytopathology* 72:199-206.

108. TAGI. 2001. Analysis of the genome sequence of the flowering plant *Arabidopsis thaliana. Nature* 410:299.

109. Tsagris EM, de Alba AE, Gozmanova M, and Kalantidis K 2008. Viroids. *Cell. Microbiol.* 10 (11): 2168.

110. Tuberosa R and Salvi S. 2006. Genomics-based approaches to improve drought tolerance of crops. *Trend Pl Sci* 11:405–412.

110b. Tyler BM, Tripathy S, Zhang X, et al. 2006 *Phytophthora* genome sequences uncover evolutionary origins and mechanisms of pathogenesis. *Science* 313:1261–1266.

111. Tuskan GA, DiFazio S and Jansson S. 2006. The genome of black cottonwood, *Populus trichocarpa* (Torr. & Gray). *Science* 313:1596-1604.

112. Upreti HC and Singh NK. 2005a. High-resolution mapping, cloning and molecular characterization of the gene of rice, which confers resistance to rice blast. *Mol Genet Genomics* 274:569-578.

113. Van Sluys MA, Monteiro-Vitorello CB, Camargo LEA, Menck CFM, Da Silva ACR, Ferro JA, Oliveira MC, Setubal JC, Kitajima JP & Simpson AJ. 2002. Comp-

arative g^nomic analysis of plant-associated bacteria. *Annu Rev Phytol* 40:169–189.

114. vanderBiezen EA and Jones JDG. 1998a. Plant disease resistance proteins and the gene-for-gene concept. *Trends Biochem Sci* 23:454-456.

115. VanderBiezen EA and Jones JDG. 1998b. The NB-ARC domain: a novel signaling motif shared by plant disease resistance gene products and regulators of cell death in animals. *Curr Biol* 1998, 8:R226-R227.

116. vanderHeijden MGA, Klironomos JN, Ursic M, Moutoglis P, Streitwolf-Engel R, Boller T, Wiemken A and Sanders IR. 1998. Mycorrhizal fungal diversity determines plant biodiversity, ecosystem variability and productivity. *Nature* 396: 69–72.

117. VanLoon LC, Bakker PAHM and Pieterse CMJ. 1998. Systemic resistance induced by rhizosphere bacteria. *Annu Rev Phytopathol* 36: 453-483.

118. vanLoon LC. 1997. Induced resistance in plants and the role of pathogenesis related proteins. *European J Pl Pathol* 103: 753-765.

119. Wang MB, Bian XY, Wu LM, et al. 2004. On the role of RNA silencing in the pathogenicity and evolution of viroids and viral satellites". *Proc Natl Acad Sci USA* 101 (9): 3275–80.

120. White FF, Yang B and Johnson LB. 2000. Prospects for understanding avirulence gene function. *Curr Opin Pl Biol* 3(4):291-298.

121. Wissuwa M, Mazzola M and Picard C. 2009. Novel approaches in plant breeding for rhizosphere-related traits *Plant Soil* 321(1-2): 409-430

122. Wood DW. et al. 2001. The genome of the natural genetic engineer *Agrobacterium tumefaciens* C58. *Science* 294 (5550), 2317

HERBICIDAL POTENTIAL OF SOIL MICROFLORA: CURRENT STATUS AND FUTURE PROSPECTS

A.K. Pandey[1], Rekha Shukla, Divya Bajpai*, Mithilesh Jaiswal* Nikita Benerjee* and Sneha Pandey**

[1]MP Private University Regulatory Commission,
Bhoj University Campus, Bhopal (M.P.) India
*Department of Biological Science, R. D. University,
Jabalpur-482001 (M.P.) India E-mail: akpmycol@yahoo.co.in

KEY WORDS: Soil microflora, herbicidal potential, bacteria, actinomycetes, fungi.

ABSTRACT

Soil is considered as a natural medium which provide a conducible condition for survival and multiplication of a large and diverse group of microbial populations. Some of them are highly plant pathogenic while few of them are although saprophytes but able to produces highly phytotoxic metabolites. These characteristics make them highly suitable for development of cheaper, effective and ecofriendly bioherbicides and herbicidal agrochemicals. Several soil microbes have been discovered, evaluated and few of them have shown significant herbicidal potential against several weeds. Many compounds derived from soil microflora have been directly used as herbicides or provide a basis for the development of new synthetic herbicide which can be exploited by industry – led discovery programmes. The present review is focused on to highlights the potential of soil microflora in weed management.

Introduction

Soil is considered as a dynamic and ever changing system which provides a natural home to a large and diverse group of saprophytic and pathogenic micro-organisms specially fungi, bacteria and actinomycetes. They are involved in innumerable positive and negative interactions in soil. Plant pathogenic micro-organisms are capable of infecting underground parts of the hosts and incite many diseases viz., root-rot, seed-rot, girdling of tissues, wilt, collar-rot, blight or damping-off of seedlings in various plants including weeds. They can reduce weed populations through decay of seeds before emergence or cause death of seedlings just after emergence. In addition, soil micro-organisms are also responsible for synthesis of a variety of highly phytotoxic metabolites. These characteristics make such micro-organisms as novel agents for effective and efficient management of weeds. The potential of these organisms in weed control have been reviewed in many publications (14,25,33-34, 48,53,79,82, 101-103, 111). Therefore, the objective of this article is not to discuss all the aspects, but to highlight the current status and future prospects of exploitation of soil micro-organisms viz., fungi, bacteria and actinomycetes in the management of ever shifting problems of weeds.

A. Bacteria: Bacteria are the most important and dominant group of microorganisms in soil which probably represent total half of the microbial biomass present there. Their population ranges from 100,000 to several hundred million per gram of soil depending upon the physiochemical and biochemical conditions of soil. There are several reports where bacterial strains were directly or indirectly responsible for considerable losses in agriculture, forestry and horticultural crops. However, their role in weed management has not yet been properly investigated. Studies carried out on herbicidal potential of these are as follows:

a. Phytopathogenic bacteria: Several devastating plant diseases viz., dwarfing, wilts, lesions, blight, tumors, galls, scab, distortion and discoloration have been recorded by plant pathologists all over the world (87). Such enormous pathogenic potential of bacteria have attracted the attention of weed pathologists to exploit them as bioherbicides. Dodd (32) successfully controlled *Opuntia* sp. employing secondary soft rotting bacteria. Substantial control of *Solanum viarum* by two strains of *Pseudomonas* *solanacearum* has also obtained. *P. syringae* pv. *tagetis* when applied @PST+ 109 cfu/ml with Silwet L-77 (0.3% v/v) successfully controlled many weeds including *Cirsium arvense* and *Xanthium strumarium* (5). Johnson *et al.* (62) extensively reviewed the role of phytopathogenic bacteria in weed management and stated that herbicidal potential of *P. syringae* against Canada thistle can be increased by use of suitable surfactant viz., polyalkylene oxide modified polydimethyl siloxane. Kishan and Chand (70) reported significant herbicidal potential of another isolate of *P. solanacearum* against *Parthenium hysterophorus, Ralstonia solancearum* incites severe root rot and wilt in *Parthenium* (121). Chand *et al.* (21) and Ovies and Larring (96) recovered few strains of *Xanthomonas campestris* pv. *parthenii* incite severe blight in *Parthenium.* Another strain of *X. campestris* pv. *poae* has been registered in Japan as Camperico® an effective bioherbicide to control Blue grass (*Poa annua*) (59,12). There are several reports where phytopathogenic bacteria perpetuate on weeds. However, herbicidal potential of these are still not properly evaluated.

Application of bacterial pathogens to weed requires special formulation because they are not able to penetrate plants directly. Their entry to plants requires artificial facilitation. This has been done in two ways. The first is through the use of non ionic organosilicone surfactant which lowers the surface tension of aqueous solutions to the point that stomata are penetrated. Excellent example of this method is infection incite by *P. syringae* pv. *phaseolicola* (164). Mechanical wounding is another method for bacterial pathogens. *P. campestris* pv. *poanura* has been commercially applied in golf course greens in Japan to control annual blue grass in conjunction with mowing which provide a mode of entry for bacteria (60). In contrast, some bacteria *viz.* rhizobacteria do not require facilitated infection (69). Technological, biological and economical feasibility of exploitation of bacteria in weed management have been extensively discussed (62, 98).

b. Exopathogenic bacteria: Naturally occurring non-pathogenic and root colonizing bacteria that impose reverse allelopathic harmful effect on plants are termed as Deleterious Rhizobacteria (DRB). Mechanism by which rhizobacteria suppress plant growth are not fully understood but may include over production of indoleacetic acid (123), production of siderophores (80), extracellular polysaccharides and hydrogen cyanide (74,97,

126). Presence of any or all of these effectively reduced plant growth. Screening of several thousand bacterial strains by scientists against many annual grasses clearly support the above hypothesis. They inhibit root or shoot growth by producing phytotoxic or plant suppressive compounds that are absorbed by plant roots (12, 72). They impair seed germination and delay root and shoot elongation and subsequent development. Volatile metabolites released from cultures of DRB can inhibit plant growth. They are specific in nature. DRB include strains of *Enterobacter, Klebsiella, Citrobacter, Arthrobacter* and Fluorescent Pseudomonads such as *Pseudomonas cichorii, P. fluorescence* (137). Herbicidal potential of DRB have been critically evaluated against several weeds (12, 14, 69). Charrington and Elliot (24) successfully controlled downy brome (*Brome tectorum* # BROTE) and other broad leaved weed seedlings by DRB strains. Out of 15 isolates of rhizobacteria screened against *Parthenium.* Pandey *et al.* (106) reported high reduction in overall growth of the weed by *Pseudomonas* RB#108(RD), RB#109, RB#112(Sh) and *Xanthomonas* RB#116 (RDU). Partially purified toxin from *P. fluorescens* D7 consists of polypeptides, fatty acids, esters and lipopolysaccharides (44) which reduced root elongation and caused root discolouration (145). D7 toxin directly affects lipid synthesis and membrane integrity while there is no effect on cell division, respiration or protein synthesis (144). Significant inhibition in root length of field bindweed due to overproduction of auxin by bacterial strains in the media containing L-tryptophan as the precursor have also been reported (73, 123). Pre emergence application of DRB strains reported to cause more significant effect on growth and biomass production of many weeds (124). According to Johson *et al* (61) soil factors especially temperature plays a very impor-tant role in efficacy of DRB. Significant suppression in weed population where DRB was applied in combination with herbicides has also been reported by some workers (47). Rhizobacteria have been encapsulated into sodium alginate granules, which have been shown to be a suitable medium to ensure the survival and distribution of microbial inocula in soil environment (93). Wide spectrum activity of many DRB sometimes creates doubt about their suitability as bioherbicide. However, there are several reports of host specificity in bacterial strains (24, 69).

Apart from the effect of phytotoxin the destruction of cortical cells and root hairs by DRB invasion is likely to contribute to impaired root function. DRB have been directly applied to soil or regulative residues to attack germinating seeds and seedlings. Efficacy of DRB can be characterized by its ability to colonize over root and seed, extent and rate of weed suppression and adaptability to pesticide application method. Potential of *Xanthomonas* for synthesis of phytotoxins is well documented (146). *Pseudomonas syringae* pv. *tagetis* (pst) produced a non-host specific phytotoxin, Tagetitoxin, which is translocated to emerging leaves where it inhibits RNA Polymerase III which finally prevents chlorophyll biosynthesis. Significant herbicidal potential of DRB against leafy spurge (*Euphorbia esula*) have been clearly demonstrated by using suspended cell and callus tissue culture bioassay (75). Varghase (148) screened several rhizobacterial strains of *Pseudomonas* sp., *Xanthomonas* sp. and *Rhizobium* sp. and reported very high herbicidal activity against *Parthenium*. Kremer and Kennedy (72) identified several strains of rhizobacteria effective against many weeds.

DRB do not rely on complete killing of weed but on reducing the competitive ability of weeds growing with crops. According to Kennedy *et al.* (69) DRB agents reduced competitive ability, density, seed production and inhibition in biomass accumulation. Release of deleterious metabolites by bacteria may enhance the permeability of the plasma lemma or damage plasma lemma protein function in the root. The effect is highly dependent on environmental conditions such as water, nutrition, bacterial density, soil structure and texture. Boyetchko and Holmstrom-Ruddick (13) reported that specific bacterial strains and their metabolites exhibiting suppressive properties to downy brome did not inhibit growth of spring and winter cereal crops. In some instances, plant growth promotion was observed and was dependent on the bacterial strains and crop cultivar. However, Kremer and Kennedy (72) suggested that researchers should recognize the risk of utilizing DRB with a wide host range because of potential detrimental effects on crop plants. The bacterial weed pathogen may have significant potential as biological control agents due to several traits. They generally grow rapidly in liquid culture, viability can be preserved in frozen or dried formulation and they can be genetically manipulated and screened for mutant selection with relative ease.

B. Actinomycetes: They constitute a very important component of soil micro-biota and are considered as one of the most prolific source for the production of phytotoxic metabolites. Several chemical herbicides have been synthesized, patented and commercialized on the basis of the natural products of this group of organism (79). Out of twenty two isolates of *Nocardiopsis* recovered four strains inhibited the germination of cress seed and two affected barnyard grass seeds (92). Anismycin recovered from culture filtrate of *Streptomyces toyocaensis* was the first microbial phytotoxin to be used as herbicide and provided a basis for the development of a new synthetic herbicide methexyphenon or NK049 (54, 55). Anismycins recovered from *S. actinomycetes* is a kind of growth inhibitor for annual grassy weed and it destroys the plant chlorophyll (163). Bialaphos produced by *S. hygroscopicus* was successfully developed as a commercial herbicide for barnyard millet, which is being converted into Phosphinothrincin by plants (127). It has been marketed first in Japan and subsequently in Germany. It has a very short half-life i.e. 2-3 days in soil. Application with nitrogenous fertilizer enhanced its activity (33). Bialaphos produced by *S. viridochromogenes* is highly effective against a wide range of weeds including perennial weeds (28). A peptide analog of Bialaphos known as Phosalacine produced by *Kitasatosporia phosalarinea* showed very high herbicidal activity (31). Gulfosinate or HOE- 39866 is the synthetic version of phosphoinothricin a potent inhibitor of glutamine synthetase produced by *S. viridochromogens* and *S. hygroscopicus*. Glufosinate (PPT) is a nonselective, postemergence herbicide used for weed control. Herbicidin and herbimycin produced by *S. saganonensis* are more commonly used to control grassy weeds in paddy fields and both mono and dicot weeds respectively (135).

Pandey *et al.* (107) reported maximum phytotoxicity in benzene fraction of filtrate of *Streptomyces* WC#150 against *Parthenium.* CFCF of the test strain was also responsible for severe phytotoxicity against *Lantana camara* (65). Pandey *et al* .(109) reported high herbicidal properties in CFCF obtained from 21 days old fermented broth of *Streptomyces* FGCC#280 against *Hyptis suaveolens*. Bajpai and Pandey (11) also reported high phytotoxic activity of CFCF of *Streptomyces* sp. DPN # 20 against *Parthenium*. Hydantocin originally extracted from *S. hygroscopicus* SANK 13584 has been patented to control a wide spectrum of mono and dicotyledonous weeds (138). In addition, pathoxazolin and homoalanosin obtained from *Streptomyces* can control several weeds. Research on isolation and screening of microbial metabolites to discover new herbicidal compound has witnessed an upsurge as evidenced by the number of reviews published (2, 53). Phytotoxic metabolites produced by actino-mycetes which are highly effective against a wide range of weeds are listed in Table 1.

Table 1: Herbicidal Phytotoxins from Actinomycetes.

Source	Phytotoxin	Target Weed
Actinomyces sp.	Actinonin	Many monocotyledonous weeds
Kitasatosporia phosalacinea	Phosalacine	Alfa alfa
Streptomyces sp.	Methoxyhygromycin	Crabgrass, Barnyardgrass, *Digitaria sanguinalis*
Streptomyces sp.	Pyridazocidin	Velvet leaf, Ivyleaf, Morning glory, Barnyardgrass, Gaint fortail
Streptomyces sp.	Anismycin	Crabgrass, Barnyardgrass
Streptomyces sp.	2,5 - Dihydrophenylalanine	Duckweed, *Arabdopsis thaliana*, *Echinochloa crusgalli*, *Abutilon theophrasti.*
Streptomyces sp.	α – Methylene – β - alanine	*Arabdopsis thaliana*
Streptomyces sp.	Oxetin	Alfa alfa
S. viridochromogenes *S. hygroscopicus*	Glufosinate/ Phosphothinothricin	*Imperata cylindrica*, *Paspalum comjugatum*, Johnson grass
S. hygroscopicus	Bialphos	Wide range of weeds including Perennials
S. hygroscopicus	Hydantocidin	Barmudagrass, Quackgrass, Field Bindweed
S. toyocaensis	Anisomycin	Echinochloa crusgalli, Digitaria adscendense, Horsenettle, Purple nutsedge, Yellow nutsedge.
S. chromofuscus	Herboxidiene	Wild buckwheat, Rape, Morning glory.

S. sioyaensis	Altemicidin	*Sesbania exallata, Ipomea* sp., *Stellaria media*
S. filipenensis	Filpin	Beet seeds
S. gougertotti	Gougerotin	Many weeds
S. galilaeus	Homoalanosine	*E. crusgalli, Polygonum perricaria*
S. platensis	Resormycin	*Digitaria adscendens, Amaranthus viridis, Polygonum lapthifolium, Bidens pilosa.*
S. hygroscopicus	Herbimycin A	*E. crusgalli, Digitaria adscendens, Cyperus microiria, Chenopodium album.*
S. hygroscopicus	Herbimycin B	*E. crusgalli, Digitaria adscendens, Cyperus microiria, Chenopodium album*
S. saganonensis	Herbimycin A & B	Many monocotyledonous and dicotyledonous weeds
S. cinnamonensis	Monensin	*Velvet leaf, Prickly sida, Sicklepod, Johnson grass*
S. griseus	Naramycin B	Many weeds

C. Fungi: The potential of soil borne fungi to incite diseases in crop plants are well known. Several epiphytotics in plant populations have been reported in the past. This potential fungus has also been exploited in weed management. Fungi that are virulent, host specific and genetically stable, but constrained naturally by low inoculum production and poor dissemination are probably the best candidate for development of mycoherbicide. Efforts made by the weed pathologists reached to a point where four major strategies i.e. Classical, Mycoherbicidal, Biorational and integrated weed management approaches have been clearly defined (2, 33, 53, 66, 101-104, 110). Few successful examples of mycoherbicides and toxin are given in Table 2 and 3. Some of the well-known examples of fungal bioagents are as follows:

Phytophthora, a well known soil borne fungus is responsible for several devastating diseases in plants. The first registered commercially available microbial herbicide is the soil borne fungus, *P. palmivora* marketed as DeVine™ applied to soil around citrus trees for the control of Stranglervine (*Morrenia odorata*

Lindl) in Florida. Market product contained wet formulation of chlamydospores (6.7 x 10^5/ml) with a shelf life of six weeks in refrigerated storage. It is an exceptionally effective soil-borne weed pathogen, and a single application of DeVine™ often gives 95-100% control of the weed for six years or more because the inoculum becomes established in the vine root debris (27, 68). According to Kennedy (68) improvement in formulation for longer shelf life at ambient temperatures would have been necessary.

Table 2: Fungal Pathogens including soil borne fungi developed/ under developed as Mycoherbicides

Pathogens	Target weeds
Alternaria cassiae	*Cassia obtusifolia*
A. tenuissima f. sp. *euphobiae*	*Euphobiae esula*
A. crassa	*Datura stromonium*
A. helianthi	*Xanthium stromarium*
A. zinniae	*Xanthium spinosum*
Colletotrichum gleosporioides f. sp. *aeschynomene*	*Aeschynomene virginica*
C. gleosporioides f. sp. *malvae*	*Malva pusilla*
C. gleosporiodes cuscutae	*Cuscuta chinensis, C. austalis*
C. orbiculare (Dar 48942)	*Xanthium spinosum*
C. coccodes	*Abutilon theophrasti*
C. dematium	*Ipomoea purpurea*
C. truncatum	*Desmodium tortuosum*
C. malvacearum	*Sida spinosa*
Phytopthora palmivora	*Morrena odorata*, Strangler wine
Rhizoctonia solani	Water hyacinth
Sclerotinia sclerotiorum	*Canada thistle*
Cephalosporium diospyri	*Diospyros virginiana*
Acremonium zonatum	*Eicchornia crassipes*
Cercospora rodmanni	*Ecchornia crassipes*
Bipolaris sorghicola	*Sorghum halepense*
Fusarium lateritium	*Sida spinosa*
F. solani	*Cucurbita texana*
F. oxysporum var. orthoceras	*Orobanche* sp.

F. lateritium	Prickly sida spurred anoda, Velvet leaf spurred anoda
F. oxysporum f. sp. *cannabis*	Hemp
F. oxysporum f. sp.*carthami*	Yellow netsedge
F. roseum	Hydrilla
F. solani f. sp. cucurbitae	Texas groud

(Source: Connick *et al.*, 1990; Pandey 1999; Chutia *et al.*, 2007)

Table 3: Fungal toxins with promising herbicidal properties.

Name of toxin	Source organism
Fumonisins (B, Fusaric acid, Moniliformin)	*Fusarium moniliforme*
Maculosin	*Alternaria alteranata*
Gliotoxin	*Rhizoctonia solani, Pythium ultimum* (ATCC 32393)
Monocerin	*Exserohilium turcicum*
Ophiobolins, I, A, C	*Drechslera maydis, D. oryzae*
Epianhydrophiobolin, 6- epi – ophiobolin	*Drechslera sorghicola*
Perhelminthoporol	*Drechslera sorokiniana, Bipolaris* sp.
Curvulins	*Drechslera indica*
Tryptophol	*Drechslera nodulosum*
Triticones	*Drechslera tritici – repentis, Curvularia clavata*
Cornexistin	*Paecilomyces variotii* SANK21086
Cyclocolorenone	*Pythium ultimum*
Acetylaranotin	*Aspergillus terreus*
Spiciferones A,B,C	*Cochilobolus spicifer*
Compound 1233A, (not named)	*Scopulariopsis candidus*
Viridol	*Gliocladium virens*
Eremophilans (PR – Toxin, Phenonmenon, Bipolorotin, Pestasol, Cigantenone, Ziuniol)	*Bipolaris cynodontis, Drechslera gigantean, Phoma destructiva, P. exigua, Penicillium roquefortie*
Stemphyltotin, Stemphyperylenol	*Stemphylium botryosum* var *lactatum*

Botrydienal	*Botrytis cinerea*
Chaetoglobosin (Cytochalasin B, H)	*Chaetomium globosum*
Citreoviridin	*Penicillium charlesii*
Cyclocarbamide A, B	*Streptoverticilllium* sp.
Desmethoxyviridol	*Nodulosporium hinnuleum*
Dihydropergillin	*Aspergillus ustus*
Nigerrazine A, B	*Aspergillus glaucus, A. niger*
Orlandin, Kotanon	*Aspergillus niger*
Terphenyllin, Hydroxy terphenyllin	*Aspergillus candidus*
Mevinolin	*Aspergillus terreus*
Fusicocin	*Fusicoccum amygdale*
Macroliodes (Cladosporin, Recifeilide, Cladospolide A, B)	*Cephalosporium recifei, Aspergillus flavus, Cladosporium sladosporioides*
Oosprorein	*Chaetomium trilaterale*
Trichothecones	*Myrothecium verrucaria, M. roridum*
Corneristin hydroxycornexistin	*Paecilomyces variotii*
(-) - Dihydropyrenophorin	*Drechslera avenae*
Curuulin, o–methyleurnulinic acid	*Drechslera indica*
Gigantenone, Petasole phenomenon	*Drechslera gigantea*
B – nitropropionic acid	*Septoria cirsii*
Bipolaroxin	*Bipolaris cynodotis*
Cyperine	*Ascochyta cyperine*

(Source: Sugawara and Strobel, 1986; Pandey, 1999).

Cephalosporium diospyri Crandell incite wilt in *Diospyros virginiana* L. (Persimmon) and is responsible for substantial control of weed in Oklahoma (43). The fungal spores are provided in suspension to co-operating growers by the Nobal Foundation, Ardmore, Oklahoma (141). Another wilt inducing strain of *Cephalosporium* is used in control of Kolomona weed (*Casia surrattensis)* a woody plant (147). Requirement of hand inoculations is one of the major constraints in commercialization of this agent.

Chondrostereum purpureum, a wound infecting Basidiomycete when applied even at very small dose killed *Prunus serotina* Ehrh. (Blackcherry) effectively in the forest of Netherlands (154). Initially it acts as saprophyte on wounded hard wood and becomes pathogenic by invading the cambium in the injured areas thus killing resprouts of perennial weeds (46). It is marketed as 'Biocon' by Kopport Biological Systems (Berkel en Rodenrijs, Netherlands).

Sclerotium rolfsii a well known necrotrophic fungus incites severe collar rot diseases in many plants including *Parthenium* (100, 102-103,119). Mishra *et al.* (88) reported significant level of host specificity in this pathogen which further was proved by (131, 132). According to Mishra *et al.* (89) it can be servive for a longer duration at lower temperature and moisture in host debris. The pathogen can infect host plant by production of oxalic acid in advance and produce several sclerotia at advanced stage (90). Physicochemical requirements of two mycoherbicidal strains of *S. rolfsii* have been standarized by Mishra *et al.* (91). The pathogen significantly kills *Parthenium* seedlings when applied @ 90 sclerotia/100ml with sterilized distilled water (105). Actively growing mycelium when used as inoculum provides better control than sclerotia. Shukla and Pandey (131) reported significant pathogenic diversity in various isolates of *S. rolfsii* effective against *Parthenium.* Various aspects of mycoherbicidal potential of *S. rolfsii* against *Parthenium* have been extensively reviewed (102, 106, 131, 132). Shukla and Pandey (130) also reported significant level of mycoherbicidal activity in *S. rolfsii* isolate against *Hyptis suaveolens*. Various strains of *Sclerotium* spp. synthesize oxalic acid which has a reported to cause severe phytotoxicity and are responsible for pathogenesis (31, 32). Significant herbicidal activities in crude culture filtrate of *Sclerotium rolfsii* strain containing oxalic acid against weeds have been reported (130). Shukla and Pandey (32) reported very significant control through different formulation of bioactive molecules of *S. rolfsii*.

Sclerotinia sclerotiorum, appears to be among the most non- specific, omnivorous and successful soil borne plant pathogen (115), incites severe stem rot in *Parthenium* (41). An isolate of *S. sclerotiorum* has been evaluated as bioherbicide against Canada thistle (*Circium arvense*), spotted knapweed (*Centauria maculosa*) and Dandelion (*Taraxacum officinale*) (19). Repeated application of heat killed perennial ryegrass with *S. sclerotiorum* reduced 80-85% dandelion populations (120). 20-80% control of Canada thistle has been recorded when wheat grains or sclerotia of this pathogen were applied in the field (19). *Centauria diffusa* is also severely infected by this pathogen (156). The pathogen has a very wide host range, however, a genetic approach has been utilized to make the agent environmentally safe (139). *S. minor* have shown high herbicidal potential against some weeds (18)

Fusarium spp. has been tried for effective control of several weeds. *Fusarium pallidoroseum* is an important soil borne pathogen reported to be a potential mycoherbicidal agent against *Parthenium* (38, 81). Pandey *et al.* (100) reported very high mycoherbicidal potential of *Fusarium* #LC34 against *Lantana camara*. *F. oxysporum* and *F. moniliforme* incite severe seedling blight and are responsible for more than 90% inhibition in seed germination (99). *F. solani* is also reported as potential pathogen and responsible for significant damage to *Parthenium* (100). Farkya *et al.* (38) reported considerable difference in physiochemical condition suitable for *in vitro* growth and sporulation of two strain of *F. oxysporum*. According to Farkya *et al.* (39), *F. oxysporum* PR#12 and *F. solani* PR#13 showed maximum mycoherbicidal potential against *Parthenium* seedling in sandy soil at 25^{0}C, 70% moisture and high RH. The agent has shown significant host specificity against *Parthenium* (37). Abbas *et al.* (3) reported that an indigenus isolate of *F. moniliforme* caused very high damage to jimson weed and some other weed. *F. avenaceum* showed very high mycoherbicidal potential against spotted Knapweed (*Centaurea maculosa*) in Montana, U.S.A. (30). The agent significantly reduced the seed germination, overall plant weed growth. *F. oxysporum* and *F. lateritium* incite severe blight and have shown promising herbicidal activities against yellow nutsedge (*Cyperus esculentus*) in Georgia (113).

A strain of *F. oxysporum* f. sp. *orthoceras* (FOO) has shown promising result and is an effective mycoherbicide candidate for broomrapes (*Orobanche cumana*) in laboratory, green house as well as in field conditions (143). Indigenous isolates of *F. oxysporum* and *F. solani* reduced the number and weight of emerging *O. ramosa* shoots and tubercells by 60% and 70%. Similarly *F. camptoceras* and *F. chlamydosporum* were responsible for more than 50% control of this weed. Types of inocula and formulation greatly influenced the short life and efficacy of the bioherbicidal agent. Ciotola *et al.* (26) discovered an isolate of *F. oxysporum* with high potential against *Striga hermonthica.* Amsellem *et. al* (8) successfully controlled *Orobranche aegyptiaca* by a formulation containing conidia and mycelia of *F. oxysporum* 'Foxy' with absorbent starch, corn oil, sugar and silica. They reported that mycelia based formulation is better storable than conidia. Granular formulation of 'Pesta' containing propagules of *F. oxyporum* 'Foxy 2' and 'Foo' are also reported to provide significant control of *Striga hermonthica* and *O. cumana* (35, 76, 94, 128). Connick *et al.* (27) reported one year shelf life of *F. oxyporum* EN4-S stored with pesta granules at 35⁰C (0.12a_w) and 2 years at 25⁰C (0.12 & 0.33 a_w). Formulation of various strains of *F. oxysporum* has been extensively reviewed in many publications (35,36). Pesta being free flowing granules has several advantages as it is non toxic, relatively cost effective, easy to mass produce, convenient to store, simple to use, can be easily applied using agricultural machinery. *F. reseum* has been reported as one of the most lethal to hydrilla and is considered as safe agent to control this weed (23). *F. avenaceum* #1003 was responsible for 100% inhibition in seed germination of spotted knapweed (*Centaurea maculosa*) in Montana (30).

F. solani f. sp. *cucurbitae* effectively controlled *Cucurbita texana* (Texas gourd) in artificial conditions and in field tests (15 158). Yu and Templeton (162) reported that tank mixture of sequential applications of this agent and trifluraolin caused higher and rapid seedling mortality in the target weed. Boyette *et. al* (15 used oat seed infested with this pathogen to control Texas gourd *F. lateritium* reported as very effective pathogen against spurred anoda, prikly sida and velvet leaf by Walker (152). *F. lateritium* formulated with kaolin (without nutrient) and *F. solani* f.sp *cucurbitae* with Kaolin and CMC as nutrient provide significant control to *Abutillon theosphrasti* and *C. texan* respectively (140). Hildebrand and McCain (52) used wheat straw that was infested with *Fusarium oxysorum* f. sp. *cannabinensis* to control *Cannabis sativa* (Marijona) and Texas gourd. Efficacy of *F. lateritium* to control spurred anoda (*Anoda cristata*) enhanced significantly by prior application of *Alternaria macrospora* to the weed (29 Quimby (118) developed a dispensible spray formulation of *F. lateritium* by mixing macroconidia with hydrated silica powder peptone and starch. Boyette and Walker (16) recorded significant control of velvet leaf and prickly sida by *F. lateritium* alginate granules. It was considered as an excellent formulation for applying bioherbicide to soil where effective dissemination of conidia is less of a problem (27). Similarly tank mixture of *F. lateritium* and acifluorfen controlled 100% prickly sida seedling as compared to 8% in case of pathogen and 68% in case of herbicide respectively (118).

Fusarium spp. are also well known for their production of phytotoxic metabolites such as Fuminosins (1, 4), Fusaric acid (4), Moniliformin (3), Enniatin (20) and Trichothecene (6). Fumonsin B has been reported to cause severe phytotoxicity against many weeds including jimson weed (1, 3, 150). Pandey *et. al* (110) reported very high phytotoxicity of cell free culture filterate of *F. solani* FGCC#02 and *F. oxysporum* FGCC#6 against *Lantana camara*. Manickam *et al.* (83) reported 70% inhibition in seed germination of *Parthenium* treated with phytotoxic metabolite of *Fusarium moniliforme*.

Acremonium diospyri, *a* wilt inducing fungus, is not commercially available, but is routinely used as a mycoherbicide to control persimmon trees *(Diospyros virginiana)* in rangeland in South-Central Oklahoma. Wilson (159) reported that the fungus has been used since 1960 to control trees upto 10 cm diameter. Hand inoculation of conidial suspension provides 100% control within 3 years (43). It is provided free to local ranches by the Nobal Foundation near Ardmore, Oklahoma. Loss of virulence is a major problem with this fungus (142). *A. alternatum* recovered from rhizospheric soil caused 80% inhibition in seed germination and 70% seedling mortality in *Parthenium* (99, 119).

Rhizoctonia solani is a destructive, versatile, widespread, noxious soil borne pathogen incites severe diseases in many weeds *viz*; *Parthenium* (77) and *Cyperus rotundus* (113). It has high competitive saprophytic ability in soil but requires sufficient food base prior to colonisation of host plants (112). Haywood and Martin (49) reported significant pathogenic potential of *R. solani* against Centipede grass (*Eremochloa ophiuroides*) and St. Augustine grass (*Stenotaphrum secundatum*).

Aspergillus spp. viz. *A.flavus, A. parasiticus, A. nominis, A. nidulans* and *A. niger* are potential source of Aflatoxins which cause severe phytotoxicity in many plants including weeds (85). They are absorbed by the plants; translocated and distributed within specific cellular compartments (86).

Chaetomium globosum, a common saprophytic fungus colonized abundantly on seeds. More than 90% inhibition in seed germination of *Parthenium* has been reported (99, 119).

Drechslera, a common colonizer of plant residues, reduced weed population at pre-emergence stage. Chandramohan *et. al* (22) reported significant control of many weeds by combined application of *Drechslera gigantea*, *Exerohilum longirostratum* and *E. rostratum. D. indica* produce highly phytotoxic metobolites which cause severe necrosis in purslane and spiny amaranth (67). Similarly (-) - Dihydropyrenophorin, a noval and elective phytotoxin produced by *D. avenae* may have potential herbicidal properties (136).

Myrothecium spp. including *M. verrucaria and M. roridum* are common inhabitants of soil and facultative pathogens of various plants including weeds. Significat herbicidal potential of *M. verrucaria* has been recorded against *Cassia obtusifolia* (153), *Pueraria montana* (6), *Euphorbia esula* (161) and *Carduus acanthoides* (160). Anderson and Hallet (10) reported high herbicidal activity against metabolites produced by the fungus against *C. obtusifolia. Myrothecium roridum* is reported to kill *Parthenium* seedlings at both pre and post emergence (119). It inhibits more than 90% germination of seeds and responsible for 100% seedling mortality in this weed (99). Pandey *et al.* (100) standardized the inoculum doses for considerable level of control of the weed. Many other isolates of *Myrothecium* spp. viz., *M. roridium* # FO252 and *M. verrucaria* have also shown very high mycoherbicidal potential against several weeds (17, 56, 78).

Paecilomyces varioti a soil borne fungus has recently been reported as potential mycoherbicide against Horse purslane or Saranai (*Trianthema portulacastum*) responsible for more than 80% seedling mortality in the weed. Another strain of this pathogen i.e. SANK21086 produces some highly effective herbicidal compounds viz; Cornexitin and Hydroxycornestin against many weeds (40).

Phoma spp. are also known to synthesize several phytotoxic metabolites, however, herbicidal properties have not yet been explored properly. Herbicidal compound 3-nitro-1, 2 benzene carboxylic acid (3 nitrophthalic acid) synthesized by *Phoma herbarum* FGCC#75 causes severe damage in *Parthenium* (119, 151). Very high herbicidal activity in Cell Free Culture Filtrate obtained 21 days old fermented broth of *Phoma* sp. FGCC#54 against *Parthenium* have been recorded by some workers (116,117). Herbicidal potential of various spp. of this pathogen have extensively been reviewed (71).

Trichoderma spp viz; *T. viride* and *Trichoderma (Gliocladium) virens* are well known soil borne fungi *T. virens*, produces broad spectrum herbicidal compounds i.e. gliovirin, gliotoxin, viridian and a metabolic derivative of viridin, viridol which has been successfully used for control of many weeds (57, 63, 64). Very high herbicidal production of viridol was recorded on chicken manure (50, 58). Significant herbicidal potential of these fungi against *Parthenium* have also been recorded (149). Mode and mechanism of action of phytotoxic metabolites have been extensively discussed in many publications (125).

Ulocladium botrytis a very common fungus incites severe disease in *Orobanche crenata* and is responsible for significant control of this weed (95).

There are numerous reports of endophytic fungal parasites of annual and perennial wild plants that prevent their host form flowering or setting seed, and very often these have little visible effect on their host until flowering, particularly species of smut and bunts viz., *Ustilago, Sporisorium* and *Telletia*. In many cases the developing inflorescence is destroyed and a mass and smut spores is produce instead. In others sporulation is confined only to the ovaries, ovules and stamens. *Epichole* sp. also causes sterilization of host and completely suppresses the flowering. The key feature of these sterilizing fungi is that infection is systematic and the host is prevented from producing seed but instead produces spores of the pathogen. Some of such endophytes have been successfully applied for weed control. *Sporisorium ophiuri* and *Sphacelotheca holci* have been successfully controlled *Rottboellia cochincinesis* and *Sorghum halapense* respectively (84). Biology and feasibility of these as bioherbicides have also been reviewed (133,134). Several soil inhabiting microorganisms have been successfully exploited for microbial transformation of xenobiotic into potential herbicidal metabolites.

Factors affecting herbicidal potential

Soil inhabiting fungi are known to be influenced by several physical factors like location profile, soil texture, aeration, moisture, pH, light and temperature. Several biotic factors are also known to affect its potential and survival like colonization by mycoparasites, type of inoculum used and its application technology also significantly influence its survival and herbicidal efficiency (114,155). Difficulties in mass production, shelf life, market competition with synthetic chemicals etc. are the other constraints influencing bioherbicide development.

Genetic Engineering: Use of molecular approaches and genetic engineering to develop a weed control agent with enhanced activity has also been investigated (122). Applications of transgenic biocontrol agents (42,129) or transgenic plants containing pathogenic genes encoding enzymes and toxins to produce pest resistance (45) for enhancement of virulence, survival and efficacy have been employed. Transformed *Fusarium oxysporum* (FOXY) and *F. arthrosporioides* (FARTH) of *Orobanche aegyptica* with two genes of the indole-3-acetamide (IAM) pathway have shown more efficacy against *Orobanche* than natural one (9, 42). However, the creation of genetically engineered mycoherbicides must rely on detailed knowledge of the biology of the organisms, both agent and target host. Any attempt to induce genetical change into mycoherbicides without an extensive preliminary study may result in products that may cause agricultural and environmental risks.

Conclusion

The exploitation of microorganisms and their metabolic by-products has become a thrust area of research among scientists involved in weed management. It is a promising alternative for designing environmentally safe herbicides (125). Microbial toxins pose minimal to no risk to the environment due to their chemical make up and rapid degradation of the small amounts required for effective control. The known molecular sites of action of microbial phytotoxins differ in almost every case from the sites of commercial herbicides Microbial phytotoxins can be used as tools for envisaging new molecular sites of action not discovered by the traditional or commercial herbicides. Microbial products are less detrimental to the environment and less toxic to humans than

15. Boyette CD, Templeton GE and Oliver LR. 1984. Texas ground (*Cucurbita texana*) control with *Fusarium solani* f.sp. *Curcurbitae*. *Weed Sci* 32: 649 – 655.

16. Boyette CD and Walker HL. 1986. Evaluation of *Fusarium lateritum* as a biological herbicide for controlling velvet leaf (*Abutilon theosphrasti*) and prickly sida (*Sida spinosa*). *Weed Sci* 34: 106-109.

17. Boyette CD, Walker HL and Abbas HK. 1999. Biological control of Kudzu (*Puereria montana*) with an endemic fungal pathogen. *Proc South Weed Sci Soc* 52: 237.

18. Briere SC, Waton AK and Hallett SG. 2000. Oxalic acid production and mycelial biomass yields of *Sclerotinia minor* for formulation enhancement of a granular turf bioherbicide. *Biocontrol Sci Technol* 10: 281-289.

19. Brosten BS and Sands DC. 1986. Field trials of *Sclerotinia sclerotiorum* to control Canada thistle (*Cirsium arvense*). *Weed Sci* 34: 377- 380.

20. Burmeister HR and Plattner RD. 1987. Enniatin production by *Fusarium triticum* and its effect on germinating wheat seeds. *Phytopathology* 77: 1483-1487.

21. Chand R, Singh BD and Singh PN. 1995. *Xanthomonas campestris* pv. *parthenii* nov. incite of leaf blight of *Parthenium*. *Antonie Van Leeuwenhoek* 68: 161-164.

22. Chandramohan S, Charudattan R, Sonoda RM and Singh M. 2002. Field evaluation of a fungal pathogen mixture for the control of seven weedy grasses. *Weed Sci* 50: 204 – 213.

23. Charudattan R, Freeman TE, Cullen RE and Hofmeister FM. 1980. Evaluation of *Fusarium roseum "Culmorum"* as a biological control agent for *Hydrilla verticillata:* safety. In: Proc. Symp. Biol. Control Weeds. CSIRO Publication, Canberra, Australia. pp 307-323.

24. Cherrington CA and Elliott LF. 1987. Incidence of inhibitory pseudomonads in the Pacific Northwest. *Plant Soil* 101: 159-165.

25. Chutia M, Mahanta JJ, Bhattacharya N, Bhuyan M, Boruah P and Sarma TC. 2007. Microbial herbicides for weed management: Prospects, Progress and Constraints. *Plant Pathol J* 6: 210-218.

26. Ciotola M, Watson AK and Hallett SG. 1996. Discovery of an isolate of *Fusarium oxysporum* with potential to control *Striga hermonthica* in Africa. *Weed Res* 35: 303-305

27. Connick Jr W J, Daigle DJ, Pepperman AB, Hebbar KP, Lumsden RD, Andersen TW and Sands DC. 1998. Preparation of stable, granular formulation containing *Fusarium oxysporum* pathogenic to narcotic plants. *Biol Control* 13:79-84

28. Copping LJ. 1996. Crop protection agents from nature: natural products and analogues. The Royal Society of Chemistry, Cambridge, UK. p 501.

29. Crawley DK and Walker HL. 1983. Interaction of two fungal pathogens of spurred anod. *Proc South WSSA* 36: 136.

30. Czembor E and Strobel GA. 1997. Limitations of exotic and indigenous isolates of *Fusarium avenaceum* for biological control of spotted Knapweed- *Centaurea maculosa. World J Microbiol Biotechnol* 13: 119-123.

31. Daniel JT, Templeton GE, Smith Jr RS and Fox WT. 1973. Biological control of northerm jointvetch in rice with an endemic fungal disease. *Weed Science* 21: 303-307.

32. Dodd AP. 1927. The biological control of Prickly pear in Australia. *Aust Coun Sci Ind Res Bull* 34: 1-44.

33. Duke SO. 1986. Microbially produced phytotoxins as herbicides, a perspective. In: *The science of allelopathy* (eds. Putnam AR and Tang CS). New York, Wiley Interscience. pp 287-304.

34. Duke SO and Abaas HK. 1995. Natural products with potential use as herbicides In: *Allelopathy organisms, process and Applications* (eds. Inderjit, K.M.M. Dekshini) Washington, D.C, Am Chem Soc pp 348-362.

35. Elzein AEM. 2003. Department of granular mycoherbicidal formulation of *Fusarium oxysporum* 'Foxy 2' for the biological control of *Striga hermonthica*. In: *Tropical Agriculture-Advances in Crop Research* 12 (2). (Kros-chel J ed.) Margraf Verlag, Weikersheim, Germany, p 199 (Abst).

36. Elzein AEM and Kroschel J. 2006. Host range studies of *Fusarium oxysporum* 'Foxy 2': An evidence for a new forma specialis and it's implications for *Striga* control. *J Plant Diseases and Protection* (special issue), 20: 875-887.

37. Farkya S, Pandey AK and Rajak RC. 1994 Evaluation of *Fusarium* spp. as Mycoherbicides for *Parthenium* I: Host Specificity. *Nat. Acad Sci Letters* 17: 81- 82.

38. Farkya S, Pandey AK and Rajak RC. 1996. Mycoherbicidal potential of *Fusarium* spp. against *Parthenium*:

factors affecting in vitro growth and sporulation. *Bioved* 1: 1-9.

39. Farkya S, Pandey AK and Rajak RC. 2001. Effect of edaphic and climatic factors on the mycoherbicidal potential of *Fusarium* spp. against *Parthenium hysterophorus* L. *J Biocontrol* 15: 177-182.

40. Fields SC, Mireles L and Gerwick BC. 1996. Hydroxycornestin: a new phytotoxin from *Paecilomyces variotis. J Nat Proc* 59: 698-700.

41. Ghasolia RP and Asha Shivpuri. 2002. *Parthenium hysterophorus*-1 a new host record for *Sclerotinia sclerotiorum. J Mycol Pl Pathol* 34: 242-243.

42. Grassel J. 2001. Mitigating the spread and introgression of native and transgenic biological control agents. In: *Proc. of the NATO Advanced Research Workshop on Enhancing Biocontrol Agents and Handling Risks.* (eds. Maurizio, V., Grassel, J., Butt, T., Harman, G.E., Pilgeram, A., Leger, R.J. St. & Nuss, D.L.). 9-15 June 2001, Florence, Italy. NATO Science Series: Life and Behavioural Science- Vol. 339. IOS Press, Amsterdam, The Netherlands, pp 167-176.

43. Griffith CA. 1970. Persimmons wilt Research. Annual Report 1960-1970. Noble Foundation Agriculture Division. Ardmore, Oklahoma.

44. Gurusiddaiah S, Gealy DR, Kennedy AC and Ogg AG Jr. 1994. Isolation and characterization of metabolites from *Pseudomonas fluorescence.* D7 for control of downy brome (*Bromus tectorum* L.) *Weed Sci* 42: 492-501.

45. Harman GE and Donzelli BGG. 2001. Enhancing crop performance and pest resistance with genes from biological control agents. In: Proc of the NATO Advanced Research Workshop on Enhancing Biocontrol Agents and handling Risks. (Maurizio, V.J., Grassel, J., Butt, T., Harman, G.E., Pilgeram, A., Leger, R.J. St. & Nuss, D.L. eds.) 9-15 June 2001, Florence, Italy. NATO Science Series: Life and Behavioural Sciences. Vol. 39. IOS Press, Amsterdam, Netherlands. pp 114-126.

46. Harper GJ, Comeau PG, Hintz W, Wall RE, Prasad R and Becker EM. 1999. *Chondrostereum purpureum* as a biological control agent in forest vegetation management. II. Efficacy on Sitka alder and aspen in western Canada. *Can J For Res* 29: 852-858.

47. Harris PA and Stahlman PW. 1992. Biologcial weed control using deleterious

rhizobacteria. *Weed Sci Soc Am* 32: 50 (Abs).

48. Hasija S K, Rajak RC, and Pandey AK. 1994. Microbes in the management of obnoxious weeds. In: *Vistas in Seed Biology* (Singh, T., & Trivedi, P. C., eds). Print Well, Jaipur, pp 82-104.

49. Haywood RA and Martin SB. 1990. Characterization and pathogenicity of species of *Rhizoctonia* associated with centipede grass and St. Augustine grass in South Carolina. *Plant Dis* 74 (7): 510-513.

50. Heraux FMG, Hallett SG and Wellar SC. 2005. Combining *Trichoderma virens*-inoculated compost and a rye cover crop for weed control in transplanted vegetables. *Biol Control* 24: 21- 26.

51. Higgins, BB. 1927. Physiology and parasitism of *Sclerotium rolfsii*. *Phytopathol* 17: 417-448.

52. Hildebrand DC, McCain AM. 1978. The use of various substrates for large scale production of *Fusarium oxysporum* f. sp. *cannabis* inoculum. *Phytopathol* 68: 1099-1101.

53. Hoagland RE. 1990. Microbes and microbial products as herbicides: An overview. In: *Microbes and microbial products as herbicides* (Hoogland RE ed). *Am Chem Soc Symp* 439 ACS Books Washington DC.

54. Hoagland RE. 2001. The genus *Streptomyces*: A rich source of novel phytotoxins In: *Ecology of desert environment* (ed. Ishwar Prakash) Scientific Publishers, Jodhpur, India: pp 139-169

55. Hoagland RE and Cutler J. 2000. Plant and microbial compounds as herbicides In: *Allelopathy, Ecological Agriculture and Forestry* (eds. Narwal S.S., Hoagland R.E., Didly R.H. and Reigosa M.J.), Bostan, MA Kluwer Academic, pp 73-99.

56. Hoagland RE,Weaver MA and Boyette CD. 2007. *Myrothecium verrucaria* fungus: a bioherbicide and strategies to reduce its non-target risks. *Allelopathy J* 19: 179-192.

57. Howell CR and Stipanovic RD. 1984. Phytotoxicity to crop plants and herbicidal effects on weeds of viridiol produced by *Gliocladium virens*. *Phytopathol* 74: 1346-1349.

58. Hutchinson CM. 1999. *Trichoderma virens*-inocula-ted composted chicken manure for biological weed control. *Biol Control* 16: 217-222.

59. Imaizumi S, Nishino T, Miyabe K, Fujimori T and Yamada M. 1997. Biological control of annual bluegrass (*Poa annua* L.) with a Japanese isolate of *Xanthom-*

onas campestris pv. *poae* (JT-P482). *Bio Cntrl* 8: 7-14.

60. Johnson BJ. 1994. Biological control of annual bluegrass with *Xanthium campestris* pv. *poannura* in Bermuda grass. *Hort Sci* 29: 659-662.

61. Johnson BN, Kennedy AC and Ogg AG Jr. 1993. Suppression of downy brome growth by a rhizobacterium in controlled environments. *Soil Sci Soc. Am J* 57: 73-77.

62. Johnson D R, Wyse DL and Jones KJ. 1996. Controlling weeds with Phytopathogenic bacteria. *Weed Technol* 10: 621-624.

63. Jones RW and Hancock JG. 1987. Conversion of viridin to viridol by viridin producing fungi. *Can J Microbiol* 33: 963-966.

64. Jones RW and Hancock JG. 1990. Soil borne fungi for biological control of weeds. In: *Microbes and Microbial products as herbicides*. (ed. Hoagland RE). ACS Symp Ser No. 439 Am Chem Soc Washington DC, pp 276-286.

65. Joseph S, Lal S and Pandey A K. 2002. Preliminary evaluation of herbicidal potential of *Streptomyces* WC#150 against *Lantana camara*. *Ann Pl Protec Sci* 10: 134-136.

66. Kenfield DG, Bunkers GA, Strobel F and Sugawara. 1988. Potential new herbicides- Phytotoxins from plant pathogens. *Weed Technol* 2: 519-524.

67. Kenfield DG, Hallock Y, Clardy J and Strobel G. 1989. Curvulin and O methylcurvulinic acid: phytotoxic metabolites of *Drechslera indica* which cause necrosis in purslane and spiny amaranth. *Plant Sci* 60: 123-127.

68. Kennedy AC. 1996. Molecular biology of bacteria and fungi for biological control of weeds. In: *Molecular Biology of Biological control of pests and Diseases of plants*, (eds. Gunsakeran M and Webes DJ). CRC Press, Inc. pp 155-172

69. Kennedy AC, Elliott LF, Young, FL and Douglas CL. 1991. Rhizobacteria suppressive to the weed downy brome. *Soil Sci Soc Am J* 55: 722-727.

70. Kishan R and Chand R. 1987. New collateral hosts for *Pseudomonas solanacearum*. *Indian J Mycol Pl Pathol* 17: 237.

71. Kovics G J, Pandey AK and Rai MK 2005. *Phoma* Saccordo and related genera: some new perspectives in taxonomy and Biotechnology. In: *Biodiversity of Fungi: Their Role in Human Life* (Deshmukh SK & Rai MK eds), Science Publishers Inc. Enfield (NH), USA. Plymouth, UK. pp 129-153.

72. Kremer RJ and Kennedy AC. 1996. Rhizobacteria as Biocontrol agents of weeds. *Weed Technol* 10: 601-609.

73. Kremer RJ and Sarwar M. 1995. Microbial metabolites with potential application in weed management. *Proc Plant Growth Reg Soc Am* 21: 48-51.

74. Kremer RJ and Souissi T. 2001. Cyanide production by rhizobacteria and potential for suppression of weed seedling growth. *Curr Microbiol* 43: 182-186.

75. Kremer RJ, Souissi T and Stanley L. 1998. Characterization of rhizosphere microorganisms for potential as biocontrol agents of weeds. In: *Trans. 16th World Congress of Soil Science*, CD-ROM, International Soil Science Society, Montpellier, France.

76. Kroschel J, Muller-Strover D, Elzein A and Sauerborn J. 2000. The development of mycoherbicides for the management of parasitic weeds of the genus *Striga* and *Orobanche* – a review and recent results. In: *Proc. Xth Int. Symposium on Biological Control of weeds* (Spencer, N.R. ed.). 4-14 July 1999, Bozeman, Montana, USA. p 139.

77. Kumar S, Jayaraj S and Muthukrishnan TS. 1979. Natural enemies of *Parthenium hysterophorus*. *J Entomol Res* 3: 32-35.

78. Lee HB, Kim Jin-Chool, Hong, Kyang-Sik and Chang Jin Kim. 2008. Evaluation of fungus strain *Myrothecium roridum* EF#0252 as a bioherbicidal agent. *J Plant Pathol* 24(4): 453-460.

79. Li Y,Sun Z, Zhang X, Xu L, Chen S and Li M. 2003. Research progress on microbial herbicides. *Crop Protec* 247-252.

80. Loper JE and Buyer JS. 1991. Siderophores in microbial interactions on plant surfaces. *Mol Plant Microbe Interact* 4: 5-13.

81. Madhukeshwara SS, Jagdish PS, Methur SG and Kumar TB. 2002. *Fusarium pallidoroseum* – a new host for *Parthenium hysterophorus*. *Indian Phytopath* 55: 352.

82. Mallik MAB. 2001. Selective isolation and screening of soil microorganisms for metabolites with herbicidal potential. *J Crop Production* 4: 219-236.

83. Manickam K, Doraisamy S and Sankaran S. 1997. Inhibitory effect of *Fusarium moniliforme* on seed germination of *Parthenium hysterophorus* L. In: *Proc. Ist. Int. Conf. Parthenium Management* (eds. Mahadevappa M & Patil VC) UAS Dharwad, pp 90-91.

84. Massion, CL, Lindow, SE. 1986. Effects of *Sphacelotheca holci* infection on morphology and competitiveness of Johnsongrass (*Sorghum halepens*). Weed Sci., 34: 883 – 8.

85. McLean M. 1994. The phytotoxicity effects of aflatoxins B1: a review (1984-1994). *S Afr J Sci* 90: 385-390.

86. McLean M, Watt MP, Berjak P, Dutton MF and Smyman C. 1994. Effect of aflatoxin B1 on in vivo cultures of *Nicotiana tobacum* var. *samsum* in root and shoot development in tobacco plantlets. *Mycopathologia* 120: 107-117.

87. Mehrotra RS and Aggarwal A. 2003. Plant Pathology IInd edn. Tata Mc-Graw-Hill Publishing Co.

88. Mishra J, Pandey A K, Rajak R C and Hasija S K. 1994. Microbial management of *Parthenium*-I host specificity of *Sclerotium rolfsii* Sacc. *Nat Acad Sci Letters* 17: 169-173.

89. Mishra J, Pandey AK and Hasija SK. 1995. Evaluation of *Sclerotium rolfsii* Sacc as mycoherbicide for *Parthenium*: Factors affecting viability and virulence. *Indian Phytopathol* 48: 476-79

90. Mishra J, Pandey AK and Hasija SK. 1996a. Mycoherbicidal potential of *Sclerotium rolfsii* Sacc. against *Parthenium hysterophorus* L. histopathological studies. *Indian J Appl. Pure Biol* 11: 73-77.

91. Mishra J, Pandey AK and Hasija SK. 1996b. Mycoherbicidal potential of *Sclerotium rolfsii* Sacc. against *Parthenium*: Factors affecting *in vitro* growth and sclerotium formation. *J Phytol Res* 9: 19-24.

92. Mishra SK, Whitenack CJ and Putnam AR. 1988. Herbicidal metabolites from several genera of microorganisms. *Weed Sci* 36: 122-126.

93. Mooney HD, Boyetchko SM and Punja ZK. 1996. Development of application techniques for biological weed control using rhizobacteria. In: *IX Inetrnat. Symp Biological Control of weeds*. Stellenbosch, South Africa, pp 297-299.

94. Muller-Stover D. 2001. Possibilities of biological control of *Orobanche crenata* and *O. cumana* with *Ulocladium botrytis* and *Fusarium Oxysporum* f. sp. *orthoceras*. *Agroecology* 3. Apia Verlag, Laubach, Germany, pp 174.

95. Muller- Strover D and Kroschel J. 2005. The Potential of *Ulocladium botrytis* for biological control of *Orobanche* spp. *Biol Control* 33: 301-306.

96. Ovies J and Laringa L. 1988. Transmission of *Xanthomonas campestris* pv.phaseoli

by a wild host. *Ciencia Y Tecnica* en. *La Agricultura protection de plants* 11: 23-30

97. Owen, A and Zdor, R. 2001. Effect of cyanogenic rhizobacteria on the growth of velvetleaf (*Abutilon theophrasti*) and corn (*Zea mays*) in autoclaved soil and the influence of supplemented glycine. *Soil Biol Biochem* 33: 801-809.

98. Pandey AK. 1999. Herbicidal potential of microorganisms: Present status and future prospects. In: *Biotechnology for Susta-inable Development and Productivity*. (R.C. Rajak ed). Scientific Publishers, Jodhpur, pp 94-113.

99. Pandey, AK, Luka, RS, Hasija, SK & Rajak, RC. 1991. Pathogenicity of some fungi to *Parthenium* an obnoxious weed in Madhya Pradesh. *J Biol Control* 5: 133-115

100. Pandey AK, Farkya S and Rajak RC. 1992. A preliminary evaluation of *Fusarium* spp. for biological control of *Parthenium*. *J Indian Bot Soc* 71: 103-105.

101. Pandey AK, Rajak RC and Hasija SK. 1995. Management of Waterhyacinth (*Eicchornia crassipes* (Mart) Solms) through fungal pathogens in India: Present status and future prospects. *J Indian Bot Soc* 74A: 135-144.

102. Pandey AK, Mishra J, Rajak RC and Hasija SK. 1996a. Potential of indigenous strains of *Sclerotium rolfsii* Sacc. for the management of *Parthenium hysterophorus* L. A serious threat to biodiversity in India. In: *Herbal medicines, Biodiversity and Conservation Strategies* (Rajak, R.C. & Rai, M.K. eds). International Book distributors, Dehradun pp 104-138.

103. Pandey AK, Gaythri S, Rajak RC and Hasija SK. 1996b. Possibilities, problems and prospects of microbial management of *Parthenium hysterophorus* in India. In: *Perspectives in Biological Science* (Rai, V., Naik, M.L. & Manoharachary eds.). School of life Sciences, Pt. R.S. Shukla University, Raipur, pp 253-267.

104. Pandey AK, Gaythri S, Rajak RC and Hasija SK. 1997. Potential of indigenous *Colletotrichum* species for the management of weeds in India. In: *Achievements & Prospects in Mycology and Plant Pathology* (Chahal, S. S., Parashar, L. B., Randha, H. S. and. Arya, S. eds.). International Book Distributors, Dehradun pp 59-85.

105. Pandey AK, Mishra J and Hasija SK. 1998. Effect of inoculum on mycoherbicidal potential of *Sclerotium rolfsii*

against *Parthenium*. *J Mycol Pl Pathol* 28: 284-287

106. Pandey AK, Singh J and Varghese MA. 1999. Effect of Rhizobacteria on *Parthenium*. PCIN 31: 1-6.

107. Pandey AK, Lal S and Joseph S. 2000. Herbicidal activity of partially purified metabolites of *Streptomyces* sp. WC# 150 in *Parthenium*. PICN 32: 11-23.

108. Pandey A K, Rajak RC and Hasija SK 2001. Biotechnological development of ecofriendly mycoherbicides. In: *Innovative Approaches in Microbiology* (Maheshwari, D.K. and Dubey R.C. eds.) Bishen Singh, Mahendrapal Singh, Dehradun, India. pp 1-21.

109. Pandey AK, Singh AK, Quereshi S and Agrawal D. 2004. Herbicidal activities of secondary metabolites of *Strepomyces* sp. against *Hyptis suaveolens*. *J Basic Appl Mycol* 3: 95-97.

110. Pandey AK, Rajak RC and Hasija SK. 2002. Mycoherbicidal potential of *Colletotrichum*: present status and future prospects. In: *Biotechnology and plant pathology: current trends* (eds. Arora, D.K. and Trivedi P.C.), Print well Publishers and Distributors, Jaipur, pp 188-216.

111. Pandey, A K, Singh A K, Qureshi Sadaf and Wahab Seema. 2005. Microbial management of *Parthenium hysterophorus* in Madhya Pradesh: Current Status. In: *Proc. IInd Internat. Conf on Parthenium Management* (Ramchandra Prasad, T. et. al. eds.), UAS, Bangalore, pp.102-106.

112. Parmeter JR Jr. and Whitney HS. 1970. Taxonomy and nomenclature of the imperfect state in *Rhizoctonia solani*. *Biology and Pathology* (ed. J.R Parmeter Jr.) University of California Press, Berkeley, Los Angeles and London, pp 7-19.

113. Pathak SC, Callway MB and Vavrina CS. 1987. Biological control and its integration in weed management systems for purple and yellow nut sedge *(Cyperus rotundus* and *C. esculentus*). *Weed Technol* 1: 84-91.

114. Punja ZK and Grogan RG. 1981. Mycelial growth and infection without a food base by eruptively germinating sclerotia of *Sclerotium rolfsii*. *Phytopathol*. 71 : 1099-1103.

115. Purdy LH. 1979. *Sclerotinia sclerotiorum*: History, diseases and symtomatology, Host Range, Geographic Distribution and impact. *Phytopathol* 69: 875-850.

116. Quereshi S and Pandey AK. 2007. Evaluation of secondary metabolites of *Phoma* sp. test strain #54 (FGCCW)

for management of *Parthenium hysterophorus. Indian Phytopathol* 60 (4): 462-466.

117. Quereshi S, Pandey AK, Singh AK and Yadav KK. 2006. Herbicidal potential of *Phoma* sp. FGCC#54. A preliminary evaluation. *J Basic Appl Mycol* 5 (I & II): 60–61.

118. Quimby PC. 1985. Pathogenic control of prickly sida and velvet leaf: an altenative technique for producing and testing *Fusarium lateritium. Proc. South Weed Sci Soc* 38: 365-371.

119. Rajak RC, Farkya S, Hasija SK and Pandey, AK. 1990. Fungi associated with congress weed (*Parthenium hysterophorus* L.) *Proc Nat Acad Sci India* 60(II):165-168.

120. Riddle GE, Burpee LL and Boland CJ. 1991. Virulence of *Sclerotinia sclerotiorum* and *S. minor* on dandelion (*Taraxacum officinale*). *Weed Sci* 39: 109-118.

121. Samaddar KR, Chakarborty M and Kanjilal S. 1998. Identification of the race of *Pseudomonas solanacearum* causing wilt of Solanaceous vegetables in west Bengal and its survival. *J Mycopathol Res* 36: 51-58.

122. Sands DC, Pilgerma AL and Tiourebaev KS. 2001. Enhancing the efficacy of biocontrol agents against weeds. In: *Biocontrol Agents and handling Risks* (eds. Maurizio V., Grassel, J., Butt, T., Harman, G.E., Pilgerm, A., Leger, R.J.St. & Nuss, D.L.) Florence, Italy, NATO Science Series: Life and Behavioural Sciences-Vol. 339. IOS Press, Amsterdam, Netherlands, pp 3-13.

123. Sarwar M and Kremer RJ. 1995. Enhanced suppression of plant growth through production of L- Tryptophan –derived compounds by deleterious Rhizobacteria. *Plant Soil* 172: 261-269.

124. Sawchyn K, Nelson T, Gibson L, Leung S, and Boyetchko SM. 2000. Development of Rhizobacteria for biological control of grass weeds. In: *Proc. Third International Weed Science Congress*, Foz do Igaussu, Brazil, June-11 2000. p 183.

125. Saxena S and Pandey AK. 2001. Microbial metabolites as ecofriendly agro-chemicals for the next millennium. *Appl Microbial Biotechnol* 55: 395-403

126. Schippers B and Bakker AW. 1987. Interaction of deleterious and beneficial rhizosphere microorganisms and the effect of cropping practices. *Ann Rev Phytopathol* 25: 339-358.

127. Sekizawa Y and Takematsu T. 1983. How to discover new antibiotics for herbicidal use. In: *Pesticide Chemistry: Human Welfare and the*

environment. Vol. 2. *Natural Products* (Takahashi N. Yoshioka H. Misato T. and Matsunaka S. eds.). Oxford, U.K, Pergamon Press. pp 261-268.

128. Shabana YM, Muller-Stover D and Sauerborn J. 2003. Granular pesta formulation of *Fusarium oxysporum* f. sp. *orthoceras* for biological control of sunflower broomrape: efficacy and shelf-life. *Biol Control* 26: 189-201.

129. Sharon A, Barhoon S and Maor R. 2001. Genetic engineering of *C. gloeosporioides* f. sp. *aeschynomene*. In: *Proc. of the NATO Advanced Research Workshop on Enhancing Biocontrol Agents and handling Risks*. (Maurizio, V., Grassel, J., Butt, T., Harman, G.E, Pilgeram, A., Leger, R.J.st. and Nuss, D.L. eds.) 9-15 June 2001, Florence, Italy. Nato Science Series: Life and Behavioral Science- Vol. 33 IOS Press, Amsterdam, Netherlands, pp 240-247.

130. Shukla Rekha and Pandey AK. 2006. Maximisation of production of oxalic acid from *Sclerotium rolfsii*, a mycoherbicidal agent against *Parthenium hysterophorus* L. *Annl Plant Protec Sci* 14 (2): 202-205.

131. Shukla and Pandey AK. 2008a. Pathogenic diversity of *Sclerotium rolfsii* isolates, a potential biocontrol agent against *Parthenium hysterophorus* L. *Afr J Env Sc Tech* 2(4): 124-126

132. Shukla R and Pandey AK. 2008b. Formulation and evaluation of agrochemicals of *Sclerotium rolfsii* FGCC#02 against *Parthenium hysterophorus*. *J Pl Protec Res* 48(4): 487-494.

133. Smith MC and Holt J. 1996. Theoretical models for weed biocontrol: *Rottboellia cochinchinensis* infection with sterilising fungi. In: *Proceedings of the IX International Symposium on biological control of weeds*, (Moran VC, Hoffmann JH eds.) January 1996. Stellen-bosch, South Africa: University of Cape Town, pp 319-325.

134. Smith MC and Holt J. 1997. Analytical models of weed biocontrol with sterilizing fungi: the consequences of differences in weed and pathogen life-histories. *Plant Pathol* 46: 306-319.

135. Stephen OD and Lydon J. 1987. Herbicides from natural compounds. *Weed Technol* 1: 122-128.

136. Sugawara F and Strobel GA. 1986. (-)-Dihydropyrenophorin, a novel and selective phytotoxin produced by *Drechslera avenae*. *Plant Sci* 43: 1-5.

137. Suslow TV and Schroth MN. 1982. Role of deleterious

Rhizobacteria as minor pathogens in reducing crop growth. *Phytopathol* 72: 111-115.

138. Takahashi S, Nakajma N, Takeshi K, Haruyama H, Sugai S, Toyokuni H, Sato S and Haneishi T. 1994. Hydantocidin and cornexstin: new herbicidal antibiotics. *Am Chem Soc Symp Ser* 55: 74-84.

139. Te Beest DO. 1993. Biological control of weeds: Potential for genetically modified strains In: *Adavanced Engineered Pesticides* (Kim, L ed.). Marcel Dekker Inc New York, pp 147-163.

140. Templeton GE 1988. Biological control of weeds. *Am J Alt Agric* 2: 69-72.

141. Templeton GE and Greavest. 1984. Biological control of weeds with fungal pathogens. *Tropical Pest Management* 30: 333-338.

142. Templeton GE and Heiny DK. 1989. Improvement of fungi to enhance mycoherbicide potential In: *Biotechnology of Fungi for Improving Plant Growth* (Whipps JM & Lumsden RD eds.). CUP, Cambridge, UK, pp 127-152.

143. Thomas H, Sauerborn J, Muller-Stover, Zeigler A, Bedi JS and Kroschel J. 1998. The potential of *Fusarium oxyporum* f. sp. *orthoceras* as Bioocntrol agent for *Orobanche cumana* in sunflower. *Biol Control* 13: 41-48.

144. Tranel PJ, Gealy DR and Irzyk GP. 1993a. Physiological responses of downy brome (*Bromus tectorum*) roots to *Pseudomonas fluorescens* strain D7 phytotoxin. *Weed Sci* 41: 483-489.

145. Tranel PJ, Gealy DR and Kennedy AC. 1993b. Inhibition of downy brome (*Bromus tectorum*) root growth by a phytotoxin from *Pseudomonas fluorescens* strain D7. *Weed Technol* 7: 134-139.

146. Tripathi HHS, Vyas KM. 1986. Production of Toxins by Xanthomonas-Campestris Pathovar Betlicola Causing Leaf Spot Diseases of Piper-Betle L. *Acta Botanica Indica* 14: 220-225.

147. Trujillo EE and Obrero FP. 1976. *Cephalosporium* wilt of *Cassia surratencis* in Hawaii. In: *Proc. IVth Internat. Symp On Biological Control of Weeds* (ed. Freeman TE). University of Florida, Gainsville, Florida, pp 217-220.

Use in biological weed control. *J Gen Microbiol* 135: 2085–2095. Use in biological weed control. *J Gen. Microbial* 135: 2085-2095.

148. Varghese AM. 1998. Evaluation of byproducts (secondary metabolites) of some rhizobacterial strains

for the management of *P. hysterophorus* L. M.Sc. Dissertation, R.D University, Jabalpur.

149. Verma A, Kumar V, Upadhyay A and Kauraw LP. 2005. Study of the Biocontrol potential of *Trichoderma viride* and *T. virens* grown on different media against *Parthenium hysterophorus*. *J. Basic Appl Mycol* 4: 127-129.

150. Vesonder RF, Peterson RE, Labeda DPT and Abbas HK. 1992. Comparative phytotoxicity of fumonisins, AAl-toxins and yeast sphinzolipids in *Lemna minor* L. (duckweed). *Arch. Environ Contem Toxicol* 23: 464-467.

151. Vikrant P, Verma KK, Rajak RC and Pandey AK. 2006. Characterization of a phytotoxin from *Phoma herbarum* for management of *Parthenium hysterophorus* L. *J Phytopathol* 154: 1-8.

152. Walker, HL. 1981. *Fusarium lateritium*: A pathogen of spurred anoda (*Anoda cristata*), Prickly sida (*Sida spinosa*) and velvet leaf (*Abutilon theosphrasti*). *Weed Sci* 29: 629-631.

153. Walker HL and Tilley AM. 1997. Evaluation of an isolate of *Myrothecium verrucaria* from Sickle Pod (*Senna obtusifolia*) as potential mycoherbicide agent. *Biol Control* 10: 104-112.

154. Wall RE. 1990. The fungus, *Chondrostereum purpureum* as a silvicide to control stump sporouting in Landwoods. *North J Appl For* 7: 17-19.

155. Warcup, J.H. (1951). Soil-steaming: A selective method for the isolation of ascomycetes from soil. *Transactions of the British Mycological Society*, 34 (4): 515 – 18.

156. Watson AK, Copeman RJ and Renney AJ. 1974. A first record of *Sclerotinia sclerotiorum* and *Microsphaeropsis centaurea* on *Centaurea diffusa*. *Can J Bot* 52:2639-2640.

157. Weidemann, GJ. 1988. Effect of nutritional amendment for conidial production of *Fusarium solani* f.sp. *cucurbitae* on sodium alginate and conrol of texas gourd. *Plant Dis* 72: 36-38.

158. Weidemann GJ and Templeton GE. 1988. Control of texas gourd (*Cucurbita texana*) with *Fusarium solani* f.sp. *cucurbitae*. *Weed Technol* 21: 271-274.

159. Wilson CL. 1965. Consideration of the use of persimmon wilt as a silvicide for weed, persimmons. *Plant Dis Reptr* 49: 789-791.

160. Yang SM. 1984. Control of *Cardus acanthoides* with *Myrothecium verrucaria* in green house in the absence of dew. *Phytopathol* 58: 1136.

161. Yang SM, Dorlev WM and Jhonson DR. 1991. Comparison of methods for selecting fungi pathogenic to leafy spurge. *Plant Dis* 75: 1201-1203.

162. Yu SM and Templeton GE. 1983. The relationship of trifluoralin to collar rot of Texas gourd caused by *Fusarium solani* f.sp. *cucurbitae*. *Phytopathol* 73: 823.

163. Zhang J, Barko G, Wolfe S and Demain AL. 1987. Methionine induction of ACV Synthetase in *Cephalosporium acremonium*. *J Ind Microbiol* 2: 251-255

164. Zidack NK, and Backman PA. 1996. Biological control of Kudzu (*Pueraria lobata*) with the plant pathogen *Pseudomonas syringe* pv. *phaseolina*. *Weed Sci* 44: 645-649.

FOOT ROT AND BAKANAE OF RICE: AN OVERVIEW

Ram Singh and S. Sunder

CCS HAU Rice Research Station, Kaul - 136 021, Haryana, India

KEY WORDS: Foot rot, bakanae, Rice (*Oryza sativa* L.), *Fusarium moniliforme*, *Gibberella fujikuroi*

ABSTRACT

Foot rot and bakanae of rice caused by *Fusarium moniliforme* Sheldon [*Gibberella fujikuroi* (Sawada) Wollenweber] occurs in almost all the rice growing areas of the world and causes substantial yield losses. In this article, the information available on history and distribution; symptoms and economic importance of the disease; nomenclature, morphological characters, physiology and metabolic products of the pathogen; variation in sensitivity of *F. moniliforme* strains to different fungitoxicants; morphological, physiological and pathogenic variability; heterothallism, genetics, sexuality and heterokaryosis; host range, disease cycle, factors affecting disease development and disease management through host resistance, cultural practices, use of botanicals, biocontrol agents and fungitoxicants has been reviewed. The review contains published references for the period 1898-2009 and provides future lines of work.

HISTORY AND DISTRIBUTION

Foot rot and bakanae of rice caused by *Fusarium moniliforme* Sheldon [*Gibberella fujikuroi* (Sawada) Wollenweber] is said to have been known in Japan since 1828 (54). It is also known as root rot and white head disease (138), *Fusarium* blight, elongation disease, Fusariosis, white stalk in China, palay lalake (man rice) in British Guiana, foot rot in the Philippines, otoke nae (male seedling) in Japan, bakanae in USA, Africa and Australia, foot rot & bakanae in French Equitorial, Africa and Ceylon and foolish plant or foot rot in India (131).

The disease is widely distributed in almost all rice growing areas of the world including Africa, Australia, British Guiana, Cameroon, India, Italy, Ivory Coast, Japan, Kenya, Nigeria, Surinam, Tanzania, Thailand, Trinidad, Uganda, USA, Venezuela, Vietnam, Brazil, Ceylon, Indonesia, Philippines, Mexico, China, Malaysia, Taiwan, Pakistan, Iran, Spain, Bangladesh and Nepal. In India, it has been reported to occur in Andhra Pradesh, Uttar Pradesh, Tamil Nadu, Manipur, West Bengal, Assam and Haryana (131).

SYMPTOMS

Fusarium moniliforme causes different types of symptoms starting from pre-emergence seedling death to grain infection at maturity; elongation, elongation than normal growth, elongation than stunted growth, stunted growth, normal growth and no growth (119). The disease occurs both in main and ratoon crop, in seed-beds as well as in the field. Diseased seedlings are lanky, pale yellow and taller than the healthy ones. Such seedlings die either before or after transplanting. The ratio of gibberellic acid and fusaric acid is the deciding factor for elongation of rice plants (171). Tochinai and Ishizuka (179) found that presence of bakanae symptoms is limited to the actual period of contact with the toxin as the infected seedlings recovered after planting to an uninfected area.

Diseased plants show brown discolouration of tissues at lower nodes facilitating an easy separation of the root system at collar region resulting in foot rot symptoms (100,125). Often diseased plants have fewer tillers and their leaves dry one after the other, wilt and die within a few weeks. Plants surviving till maturity bear only empty panicles. Presence of white or pink mycelial growth, adventitious roots from the lower internodes (119) and wider leaf angles with stem (59) are the other diagnostic features of the disease. The leggy rice seedlings also showed an obvious phenomenon of symptoms hiding to the extent of 89.7 to 92.6% in different rice cultivars, which began in 13 days old seedlings and attained a peak at the full tillering stage. A part of hidden-symptom seedlings might retransform into leggy seedlings, particularly at early jointing stage. The leggy seedlings which did not show symptom hiding at 15 days after transplanting died before early earing. A part of leggy seedlings which entered into symptoms hiding state within 15 days could form ears (211). In the end of season, lower sheaths become blue and turn into black colour having small scattered perithecia under favourable conditions (158). In Japan, panicles are often infected and termed as pink panicles.

The fungus also causes necrotic lesions on leaves, stem and kernel, panicle rot with light brown spots on the glumes resulting in occasional twisting of panicles mostly within the sheath, withering of growing shoots resulting in dead heart or white ear head. The fungus is also known to cause abnormal profuse branching of the root system resulting in fasciculation (131) and sheath rot (33).

ECONOMIC IMPORTANCE

The disease is commonly seen in many parts of South East Asia but percentage of infection is usually small. However, under favourable conditions, it is known to cause 20-50% yield loss in Japan, 70% to almost complete loss in Australia, 3.7-14.7% in Thailand, 5-23% in Spain, 3.0-95.4% in India, 40% in Nepal (131), 6.7-58.0% in Pakistan (198) and up to 75% in Iran (138).

THE PATHOGEN

A. Nomenclature

Hori (54) first identified the causal organism as *Fusarium heterosporum* Nees. The development of ascigerous stage on host was first reported by Sawada (141) as *Lisea fujikuroi*. Wollenweber (191); Ito and Kimura (62) transferred *L. fujikuroi* to the genus *Gibberella* as *Gibberella fujikuroi* with *Fusarium moniliforme* Sheldon as its anamorph. Wineland (190) described *G. moniliformis* and proposed this name for the perithecial stage of *F. moniliforme*. It is, however, invalid according to the International Code of Botanical Nomenclature but continued to be used by some pathologists (119). Till 1945, it was identified as *G. moniliforme* (Sheld.) Snyder et Hansen. Finally, Sheldon named it as *F. moniliforme* (Sheld.) Snyder et Hansen. Wollenweber's (191) identification was, however, generally accepted as *F. moniliforme* Sheld. and perfect state as *G. fujikuroi* (Sawada) Wollenweber. The occurrence of perfect stage *G. fujikuroi* in nature has been reported in Japan, Taiwan, The United States, Australia, Mexico, Central Europe, Italy, Korea and India (131).

In Iran, foot rot is known to be caused by *F. proliferatum* var. *Proliferatum* (Matsushima) Nirenberg Nirenberg ex Gerlach & Nirenberg while in China *F. moniliforme* var. *zhejiangensis* has been identified as predominant pathogen associated with the disease (131).

B. Morphological characters

In host tissues, mycelium is hyaline, slender, richly branched, sparingly septate, 1-8 µm (mostly 3-5 µm) thick with swollen segm-ents in old cultures. Cultures from reverse appear typically dark violet but occasionally lilac or vinaceous. Surface is covered with rosy buff to vinaceous, floccose to felted mycelium, which has a powdery appearance due to formation of microconidia. Sporodochia on the plant surface are 200 µm or more in height, coral like, several times di- or trichotomously branched, consist of cells which are 9-24 x 5-9 µm in size, hyaline but pinkish in mass; the terminal sterigmata taper to a point (119).

Microconidia are abundant, hyaline, borne basipetally in chains of 15-32 spores on the host, pyriform, short, naviculate to obovate, aseptate, 5-12 x 3-3.5 µm, pale pinkish in mass on simple or verticillate conidiophores adhering to form chains. Macroconidia are delicate, awl-shaped, slightly sickle-shaped or somewhat straight, occasionally bent into a hook at the apex, distinctly or slightly foot-shaped, scattered, formed in sporodochia or pionnotes, in the mass clear, buff or salmon-orange, when dry carrot red or cinnamon-brown or rather pale, aseptate to 7 septate measuring 8.4-66 x 2.4-3.5 µm (119).

Perithecia on host plants are oval, bluish black to black, 200-400 x 150-400 µm, surface rugose, perithecial wall dark purplish, 24-54 µm (mostly 40 µm) in thickness, consist of several layers of cells (13-16 x 8-9 µm), asci and paraphyses emerging at the bottom. Asci are long, cylindrical, hyaline, attenuated to the base, slightly truncate at apex 66-129 x 7.4-11 µm, mostly 90-102 x 9-11 µm containing uniseriate 8-ascospores, sometimes only 4-6 spores arranged in one or two indistinct rows. Ascospores elliptical, mostly 1- (rarely 2) septate, 15-18 x 6.7 µm, slightly constricted at septa and contain several oil globules. Paraphyses obclavate, 3-5 septate, constricted at septa, hyaline, 120-135 x 15 µm (48).

Chlamydospores are absent both in mycelium and conidia. Sclerotia dark blue, spherical to irregularly globose, 80-100 µm, may or may not be present. Stroma is more or less plectenchymatous, yellowish, brownish or violet (119).

C. Physiology

The fungus can be easily cultured on various media. Maximum growth of the fungus was observed on potato dextrose agar (9,163) and sporulation on oat meal agar (182). Sultana and Khan (156) and Yu and Sun (207) reported better growth and sporulation in dark while Toledo et al (182) observed more sporulation under light. Deep freezing method was found to be most suitable for detection and isolation of *F. moniliforme* from rice seeds (73). The technique for selective isolation of *F. moniliforme* from seed, soil and plant tissues has been described by several workers (131). Manandhar (96) developed a selective medium (FMSM) that was more efficient in the detection and recovery of *G. fujikuroi* from rice seeds. Clear and Patrick (31) suggested a medium (CPK-20 %) for differentiating *F. moniliforme* from *F. proliferatum* and *F. subglutinans*. The trace elements such as boron, zinc and manganese increased the growth of the fungus, optimum level being 0.3 ppm (131) while phosphorus increased the conidial production in the medium and the amount of metachromatin in the cells. The volutin content fell during growth but raising the orthophosphate concentration in the medium increased it (209). Rice straw-soybean meal agar and Sach's agar plus rice straw supported the production of microconidia while those of macroconidia by glucose cosamino acid agar and rice straw cosamino acid agar. Xylose has been reported as the most suitable carbohydrate for macroconidia production (119). Other workers reported that the macroconidia production and growth of *F. moniliforme* were maximum on soluble starch or maltose while microconidia were produced most abundantly in 0.4-0.5 M KCl. Of various carbon and nitrogen sources, maximum growth and sporulation of the fungus was obtained by adding ammonium nitrate, lactose and sucrose to the basal medium (131). There was no sporulation on ammonium chloride, diethanolamine and L-asparagine (147,148).

The effect of temperature on mycelial growth has been studied by several workers, optimum being about 25-30°C, maximum 36-40°C and minimum 7-8°C. However, maximum sporulation has been reported at 30-35°C (9,119,163). Sung et al (166) observed that maximum growth and microconidia production was recorded at -14 bars osmotic water potential but was prevented at -84 and -94 bars, respectively. Macroconidia formation was best at -1.4 bars.

Moore (111) found best growth of *F. moniliforme* at pH 6.8-7.1. The fungus grew well over a pH range of 3-9 with an optimum at pH 5 (200) and pH 7 (9). However, Sunder et al. (163) reported that the *F. moniliforme* isolates grew and sporulated at pH 4-8, maximum being at pH 6. No significant effect of light and dark periods was observed on fungal growth; however, significant results were obtained when light was given for a continuous period of 24 h (9).

D. Metabolic Products

Culture filtrate of the fungus could produce bakanae symptoms on rice seedlings but it was not so with all the isolates (119). Hiroe and Nishimura (52) reported that only those strains of *F. moniliforme* produced gibberellin which caused bakanae of rice.

Two principal compounds viz; fusaric acid ($C_{10}H_{13}NO_2$) and gibberellin were isolated from culture filtrate of *G. fujikuroi*, which resulted in stunting and elongation of many species of plants beside rice (193). Subsequently, gibberellin A, B, C, dihydro-gibberellin A and other substances were separated in crystalline form by Yabuta et al in Japan (131). Unlike fusaric acid, gibberellic acid (GA_3) is produced at the later phase or cessation of the fungal growth (19,171). Ammonium sulphate, asparagine and peptone have been reported as good nitrogen sources for the production of GA_3 (131). Bruckner and Blechschmidt (20) reported that optimum production of GA_3 occurred at 19 mM ammonium sulphate and reduced drastically at 38 mM. Bilai et al (17) observed maximum accumulation of gibberellic acid at a C:N ratio of 56:1 in the culture. Potassium in the medium is essential for the formation of gibberellins and its production is inhibited by the omission of KH_2PO_4 or $MgSO_4$ from the culture solution (119). Maximum gibberellic acid formation was recorded at 5g/l KH_2PO_4 and inhibited at higher concentrations (209). Yabuta et al (194) found glycerol and glucose as good source of carbon for the production of gibberellins and fusaric acid, respectively. However, Seto (142) observed that glucose promoted gibberellic acid synthesis. Garcheva and Dimova (41) reported that gibberellin biosynthesis by *F. moniliforme* was maximum in a sucrose medium and it declined when sucrose was replaced by glucose, fructose or lactose. Zweig and DeVay (213) observed that high sugar concentration resulted in low carbon conversion to gibberellic acid. The optimum pH for gibberellin production was 3.4 while for fusaric acid it was 9 (119).

The GA_3 was found to be predominant in well-aerated shaky cultures (20,66,159) and in upper parts of the plants (154). Candau et al (21) reported that light had no effect on gibberellic acid production while some workers opined that light stimulated its production (66,103,213).

Stoll (153) reported that fusaric acid is produced most abundantly at 33°C in Richards' agar. The pathogen is also known to produce moniliformin, fumonisin (99,201), cellulolytic and pectolytic enzymes (131).

Growth retardants like Amo 1618, Cycocel and α-picoline have been found inhibitory to the biosynthesis of gibberellin (131) while rufianic acid inhibited pectic enzyme activity, production of fusaric acid reduced the disease in pots (170).

VARIABILITY IN PATHOGEN

A. Morphological, physiological and pathogenic variability

Several workers have observed a significant variability in strains of *F. moniliforme* with respect to production of gibberellins, fusaric acid, pectic enzymes and their ability to cause disease (21,

89,119,149,162,171,187). Matuo et al (102) characterized 56 isolates of *F. moniliforme* from different hosts with respect to mating, production of microconidia, macroconidia, chlamydospores, gibberellin and fusaric acid. Thakur (171) divided 48 variants of *F. moniliforme* into 4 groups on the basis of their variable growth rate while Sunder and Satyavir (162) categorized 28 isolates into five groups based on virulence and GA_3 production. A significant variation in pathogenic behaviour of *F. moniliforme* isolates has also been recorded by other workers (93,146). El-Bahrawi (36) found that the isolates from rice, maize and sugarcane produced gibberellin while those from cotton and wheat did not. Amoah et al (5) observed that DNA finger print detected as RFLP of ribosomal DNA and RAPDs separated the maize and rice isolates into discrete groups, which were generally host related. However, the pathogenicity of the isolates on rice, measured by their ability to cause bakanae symptoms, did not resolve the isolates into clearly defined groups suggested by the genetic studies. Yu and Sun (208) reported that isolates of *F. moniliforme* from rice were mainly sporodochial while those from other hosts produced few sporodochia. Kuhlman (82) described the four varieties of *G. fujikuroi* as *fujikuroi, subglutinans, moniliformis* and *intermedium* on the basis of mating groups and variation in ascospores, perithecial size, phialide type and microconidia formation of which var. *fujikuroi* has been found only on rice.

Wada et al (187) reported that isolates sensitive to ergosterol biosynthesis inhibitors (pefurazoate) and resistant to benomyl did not produce gibberellins while fusaric acid was produced by all the types of isolates. The isolates which were less sensitive to triflumizole produced smaller amount of gibberellin like substances and had reduced pathogenicity (46). Likewise, the carbendazim tolerant isolates in China had less mycelial growth reduced sporulation and pathogenicity (121).

Nisikado and Matsumoto (115) observed marked difference in pathogenicity of 66 strains of *G. fujikuroi* from rice as indicated by the degree of overgrowth. Chin (30) observed that some of the isolates caused dwarfing, some caused elongation and others had no effect on the size of rice seedlings. Sharma and Bagga (146) have also observed variation in symptoms production by four isolates of the pathogen. A positive correlation existed between GA_3 production in culture and bakanae (162) and pectic enzyme and foot rot development (171). Recently, Ma et al (95) also found a significant correlation between the length of seedlings treated with GA3 and bakanae injury.

B. Sensitivity to fungitoxicants

Several strains of *G. fujikuroi* resistant to benzimidazoles (benomyl, carbendazim) and triflumizole have been detected by various workers from Japan (46,60,116-118) and China (53), which could tolerate as high as 1000 ppm concentration of the former fungicides (199). For determining benomyl tolerance in *F. moniliforme*, use of Komada medium has been suggested by Yamashita et al (195) without the need of sterile operation. Ogawa (116) described the measures to combat appearance of benomyl tolerant strains in Japan. Holloman et al (53) observed that half the isolates of *F. moniliforme* from rice were carbendazim tolerant. The sensitive isolates caused high incidence of bakanae while less sensitive isolates produced smaller amount of gibberellin like substances resulting in reduced pathogenicity but produced similar amount of fusaric acid (46). Chen et al (28) reported stable resistance in *F. moniliforme* to carbendazim and prochloraz. The resistance to carbendazim was stable after these strains were 20 times transferred on carbendazim free PSA. However, the resistance to prochloraz decreased significantly after passing strains 5-20 times on fungicide free medium. No strain resistant to diisothiocyanatomethane was determined. Inoculation tests with mixed conidia of resistant and sensitive strains (1:1) revealed that the carbendazim resistant strains had stronger competitive ability than the sensitive ones on rice plants while resistant strains to prochloraz had lower competition than the sensitive ones. Iguchi and Takeuchi (57) reported cross-resistance in benomyl tolerant strains to thiophanate-methyl and thiabendazole. Other workers have also observed a considerable variation in sensitivity of strains of *F. moniliforme* to pefurazoate (187), benomyl and hygromycin B (196) and MEMC and carbendazim (159).

Heterothallism, Genetics, Sexuality and Heterokaryosis

F. moniliforme is a complex species that has been divided into several mating groups viz., A, B, C, D, E and F (56,70). Strains of group C are confined to rice while those of other groups have sugarcane, pine, rye, wheat, corn and sorghum as their major hosts (127). Of 37 *Fusarium* isolates collected from bakanae infected rice plants from three different provinces of China, 35 isolates belonged to A, D and F mating populations. Two isolates could not be crossed with any of the non-mating populations. In paring tests of 35 isolates for vegetative compatibility, 25 were heterokaryon and self compatibile and could be classified into 20 vegetative compatibility groups (VCGs). The genetic diversity for the members belonging to D, E and F mating populations were 0.7, 0.8 and 1.0, respectively (210).

Heterokaryosis (vegetative compatibility), responsible for generating and maintaining genetic variability in fungi, has been reported to occur in isolates of *F. moniliforme* from rice under laboratory (34) and field conditions (107) and was studied by using artificially induced auxotrophs (128) and *nit* mutants (150). Puhalla and Spieth (127) found that the vegetative compatibility was controlled by nuclear genes called 'het' or 'vic' genes and was heterogenic and allelic in all the four mating groups. There are at least four vic loci in group C strains (128). A total of nine genes were estimated for *G. fujikuro*i mating group A. Puhalla and Spieth (128) identified 12 VCGs among the isolates of group C (bakanae pathogen) in China and Taiwan while Sidhu (150) and Sunder and Satyavir (162) categorized 38 and 28 isolates of *F. moniliforme* in 13 and 10 VCGs, respectively using nit mutants. Isolates from different VCGs and also with in the same VCG varied considerably in virulence and GA3 production. Vegetative compatibility had no relationship with virulence pattern and GA3 production (162). *Nit* mutants lack the enzyme nitrate reductase and are recovered on chlorate containing media. The frequency at which spontaneous sectors (*nit* mutants) arise in culture on such media varied from strain to strain and was also affected by environmental conditions (81,150). The sectoring frequency on chlorate medium varied between 0.83-3.17 per colony in different isolates of *F. moniliforme* from rice (159) while it was reported to vary between 2-10 (150) and 0.9-1.3 (81) per colony in isolates of mating group A. Klittich et al (80) indicated that sectoring is controlled by many genes. They also reported that *nit* mutations occurred in 7 loci, a structural gene for nitrate reductase (*nit* 1), a regulatory gene specific for the nitrate-reduction pathway (*nit* 3) and five genes controlling the production of a molybdenum containing co-factor necessary for nitrate reductase activity (*nit* 2 *nit* 4, *nit* 5, *nit* 6, *nit* 7) commonly called Nit M. Phenotypic classes (*nit* 1, *nit* 3 or Nit M) of *nit* mutants are identified on media containing different nitrogen sources (81).

Hamamura et al (47) observed that triflumizole sensitive isolates from same geographical area generally formed a heterokaryon, whereas isolates from different areas were not complementary. Isolates less sensitive to triflumizole were frequently complementary even when obtained from different areas, but these were never complementary with sensitive isolates regardless of geographical origin. Only sensitive isolates could cause elongation symptoms. They also observed that the isolates that were less sensitive to triflumizole and carbendazim were not complementary to those which were sensitive to these fungicides. The less sensitive isolates to triflumizole could not easily acquire pathogenicity to rice through heterokaryon formation.

In Spain, Cerda et al (22) reviewed the research work done on development of genetic tools for studying *G. fujikuroi* and described the main features of the gibberellin biosynthesis pathway in *Gibberella*. Snyder and Sun (151) and Chang and Sun (25) reported the heterothallic nature of sexual reproduction of *G. fujikuroi* and identified sex factors viz. hermaphrodite, female, male and neuter. Two alleles A and a control the compatibility reactions. Min (106) confirmed the chromosome number (n = 8) in *F. moniliforme*. Sidhu (149) reported that inheritance of mating type, colony colour, morphology and growth rate are controlled by single nuclear gene. The gene for colony colour and growth rate are linked whereas, other genes showed independent assortment. Chaisrisook and Leslie (23) stated that nuclear gene controlling perithecial pigmentation in *G. fujikuroi* is expressed maternally.

HOST RANGE

F. moniliforme has a wide host range. Beside rice, the pathogen is known to attack wheat, mango, pineapple, cotton and pine, barley, *Panicum miliaceum*, sugarcane, maize, sorghum, fig, asparagus (131). Amoah et al (5) reported that maize and rice isolates were cross- pathogenic in Ghana. However, Sun (157) observed that isolates of the fungus from rice in Taiwan did not infect sugarcane, maize, sorghum or asparagus, nor did the isolates from these hosts infect rice. In China, *F. moniliforme* var. *zhejiangensis* has been found to infect the seedlings of wheat, barley, maize (93,188), sorghum, sudan grass, soybean (93), water melon and pea (188).

DISEASE CYCLE

The disease is both seed- and soil-borne but seed-borne infection is more important as soil-borne inoculum is rapidly reduced with passage of time (69). Hulls have been found to be the main site for infection. The recovery of pathogen was found to be more in hulled grains than in caryopsis. The highest recovery of the pathogen was found in lemma followed by palea, the basal glumes, embryo and endosperm. Embryo infection ranged from 2-41% in different cultivars. The isolation frequency of *G. fujikuroi* from infected, untreated rice seeds were 75, 25 and 15% from the hulls, endosperm and embryo, respectively (96). In Pakistan, the level of highest infestation of *F. moniliforme* in naturally infested samples was found to be 21.75% by Raza et al (133). The panicle infection is caused by secondary air-borne conidia and ascospores discharged from diseased plants from heading to harvest. The maximum discharge of ascospores from diseased plants in air has been observed in the night during 1:00-3:00 h or during rain even in day (205). The fungus grows intercellularly in stigma and anthers within 48 h and finally reaches and covers the ovary (67, 205). Kanjanasoon (69) found that the fungus remained viable in seeds and infected plants only for 4-10 months at room temperature. On the other hand, the survival of pathogen in seed has been reported for 22-26 months at 10°C, 40% relative humidity (96) for 10-28 months (35,110,161). Seed lot harvested during wet season had more *G. fujikuroi* infected seed than those harvested during the dry season regardless of cultivar and number of inoculations with conidial suspension during anthesis (96). Sowing of 100% infected seed produced 30% (205) and 68.4% (198) of diseased seedlings.

The fungus infects rice seedlings through roots and the base of the stem and becomes systemic multiplying within the infected tissues. Sun (157) observed that no vegetative parts above ground are sites for infection by the pathogen. The mycelium and microconidia of the pathogen concentrate in the vascular bundles especially in the large pitted vessels and lacunae of the xylem stem, leaf blades, sheaths and adventitious roots. The phloem and parenchyma were not invaded. The pathogen was found to be discontinuously distributed in infected seedlings (114,174) with a maximum recovery from the lower 3-4 inches of the stem (authors' observations). The fungus is also found in healthy looking seeds collected from an infested field. The frequency of isolation of pathogen was maximum (97%) from basal tissues of elongated seedlings followed by isolation from stunted seedling (63%) and minimum from healthy looking seedlings (32%). On sowing, healthy looking infected seeds bakanae symptoms were produced while stunted growth was observed by sowing reddish seeds (131). Chan et al (24) observed that sprouting period was the most effective period for disease occurrence. They observed positive correlations between disease incidence and soaking temperature (r = 0.8757), disease incidence and sprouting temperature (r = 0.9570) at 28-34°C. The most important sprouting temperature was 34°C and the most important inoculation time for soaking and sprouting period was 18 and 24 h, respectively. The correlation coefficient between disease incidence and soaking time in these two periods were 0.9620 and 0.8947, respectively. Bud phase was most suitable phase for inoculation of the pathogen.

Diseased debris also serves as a primary source of inoculum as the fungus has been reported to survive in plant debris for 10-28 months (161). Watanabe (189) also reported the role of soil-borne inoculum in disease development. The fungus survived in soil through conidia, ascospores, thick-walled hyphae (157) and sclerotia (145). Yu and Sun (208) reported the survival of *F. moniliforme* up to 280 days in dry soil [10% moisture holding capacity (MHC)], which decreased to 115 days at 45% MHC and only 45-70 days at 100% MHC. Mandal and Chaudhuri (97) and Sunder (159) also recorded longer survival of the pathogen in comparatively dry soil (5-10% MHC).

FACTORS AFFECTING DISEASE DEVELOPMENT

A. Soil moisture and temperature

Wollenweber and Reinking (192) stated that the fungus survived for at least 3 years under dry conditions in a building. Mandal and Chaudhuri (97) reported the survival of *F. moniliforme* for 210 days in soil. Soil moisture of 12.6% was more favourable to survival than dry (4.8%) or wet (48%) condition. Lower temperature (5-10°C) and relative humidity (30-35%) were found to be more favourable for the survival of the pathogen in infected grains and stubbles (69,90,97,159,208). More disease is reported in upland nurseries than in flooded conditions (6,48). In Thialand and Ceylon, disease was found to be more severe in hot season and at low water levels. Very wet soil inhibited the disease development and the infection was the least in flooded seedbeds (152). Seto (143,144) reported that elongated seedlings developed only under damp soil conditions while arrested growth and stunting occurred in dry soil. He also noted that bakanae symptoms did not appear at temperature lower than 20°C while Miah and Zaman (105) reported disease development even at 7-13°C. The optimum soil temperature for disease development has been reported to be 35°C (119). The grain infection is known to increase at high temperature during flowering (51,168).

B. Type of seed, cultivar and plant population

Rajagopalan and Bhuvaneswari (130) and Kanjanasoon (69) found more infection by sowing ungerminated seeds in infested soil in comparison to germinated seeds. The former workers reported that exudation of amino acids and sugars during 72 h after germination was critical for disease development as these acted as rich energy substrate for the fungus. The disease incidence has been observed to be more in scented cultivars particularly Pusa Basmati-1121 (Pusa-1121), CSR 30 and Taraori Basmati as compared to non-scented high yielding varieties and hybrids (authors' observation). The disease incidence has been observed more when planting density of the seeds was higher (119).

C. Planting time, crop stage and seedling methods

Bagga et al (12) observed maximum disease incidence (7.2%) in early transplanted (June 19) Basmati 385 followed by July 10 planting (4.3%) while the minimum disease was recorded in July 31 planted crop. These observations were attributed to prevalence of higher temperature during infection in early-planted crop. The disease development was found to be maximum 20 days after transplanting at maximum tillering stage and at boot stage. Zhang et al (211) reported three symptom display peaks at 3-leaf seedling stage, early jointing stage and early earing stage. The percentage of *F. moniliforme* infected seedlings was higher in nursery box than in protected semi-irrigated nursery regardless of seed treatment. In field, seedlings transplanted from nursery box had more bakanae than that of rice from protected nursery (165).

D. Fertilizers and organic amendments

Higher levels of nitrogen and potassium (KCl) have been reported to reduce the population of the fungus (97,160,202). The activity of pathogen was observed to be the best in soil amended with maltose, mannitol and lactose as carbon source and ammonium sulphate and sodium nitrate as nitrogen sources, respectively (208). Panneerselvam and Saravanamuthu (123) observed more suppression of pathogen in soil amended with neem cake than in groundnut cake.

The application of nitrogen to the soil stimulated disease development due to increased growth of the fungus rather than induced susceptibility in host and the effect was not modified by the addition of potash and phosphorus (125,176). The disease has been observed to be more serious in fertile and dry seedbeds than in standard seedbeds (27). Application of supplementary organic fertiliser increased the ear formation rate of diseased plants and decreased the yield loss (211).

E. Inoculum Potential

Vishwanath-Reddy (186) showed that different inoculum potentials produced both stimulating (bakanae) and suppressive growth effect on rice cultivars. Varying inoculum levels produced different kinds of symptoms. Seedling blight, dwarf seedlings and seedling yellows appeared in fairly high inoculum potential; bakanae in moderate propagule density while there was no infection in soils with very low propagule density (158,206).

DISEASE MANAGEMENT

1. Host Resistance

A. Screening Methods

Rajagopalan (129) tested the resistance of rice varieties to foot rot disease by growing seeds in infested soil. Aktas and Tunali (3) suggested the use of spore suspension for grain and soil inoculation for screening rice genotypes for resistance in pots. Amin and Reddy (4) and Wada et al (187) used spore suspension to inoculate the grains before sowing for screening varietal resistance and for testing fungicides, respectively. Haque et al (49) reported that incubation of just sprouted paddy grains inoculated with spore suspension for 7-10 days at room temperature in test tubes resulted in more bakanae infection than field sowing. Sukapanpotharam and Chantarasnit (155) developed a green house method to screen cultivars for resistance to bakanae disease. Surface sterilized seeds were inoculated with spore suspension for 15 min., sprouted and sown in aluminium trays. Gill et al (43) suggested addition of infected debris and fungal culture grown on wheat grains for screening rice genotypes for resistance against bakanae and foot rot of rice in Pakistan. Ahmed et al (2) standardized the test tube inoculation method which involved soaking of grains for 48 h and incubating 10 sprouted grains with 5 drops of spore suspension (125 x 10^3 spores/ml) in test tubes followed by incubation at room temperature for 7-10 days. A simple and easy method of inoculation called 'bud soaking method' has been developed for large scale screening against *G. fujikuroi* by Zheng et al (212). They advocated the use of mixed races from different areas for expanding disease spectrum.

B. Varietal resistance and Disease rating

Anonymous (7) suggested a 0-9 rating scale based on per cent infected plants. The genotypes showing 0, 0-1, 1-5, 5-25, 25-50 and >50% disease incidence were treated as highly resistant, resistant, moderately resistant, moderately susceptible and highly susceptible, respectively. Sukapanpotharam and Chantarasnit (155) described 1-9 scale for categorizing rice genotypes and considered entries with less than 15 % affected plants as resistant.

The use of resistant cultivars is the most effective, efficient, safe and practical mean of obviating loss in crop yield. Several genotypes have been found resistant to bakanae in India and abroad which can be utilized as donors in bakanae resistance breeding programme (Table 1).

Some genotypes behave uniformly at different crop growth stages. However, the reaction of others is known to vary with crop stage. In China, Lu (91) observed that rice genotype Longjiao 86074-6 was resistant at seedling stage but moderately susceptible at adult stage; Qingxi 96 was moderately resistant at seedling stage but resistant at adult stage while Zupei 7, Dongrong 84-21, G-6 and Sui 89-17 were moderately resistant at both the stages.

Table 1. Rice genotypes showing resistance to foot rot and bakanae

Country	Resistant sources	References
India	Wateribune, Aryan, GEB 24	175
	HR 98, HR 101, HR 105, Norin 18, Norin 20, Agami, Monthakab	129
	RP 5-3, RP 5-32, RP 5-48, RP 84-283, RP 176-6, CR 4-1048, CR 83-23	4
	Co 18, Co 22, Adt 8, PTB 7, GEB 24	131
	IR 20, IR 26, IR 32, IR 38, IR 44, IR 45	40
	C 4-64, Karjat X 13-21, BR 1067-84-1-3-2-1, BR 1257-31-1-1, BR 4363-8-11-4-9, IR 58109-109-1-1-3, AS 34011, BG 936, CN 3886, HKR 86-104, IR 39464-54-1-3-2-1-3, PR 106	159
	PAU 2343-60-2-3, PAU 2383-2-3-1-1, IR 67418-204-4-4-2, IR 67423-234-3-3-3, IR 67423-208-2-1-1, IR 67418-131-2-3-3, IR 58755-158-1-3-2, IR 64668-54-3-1, IR 66229-45-3-2, IR 67409-98-1-3-3-1, Double haploids 1, Double haploids 7, Double haploids 19	10
Pakistan	IR 6, IR 8, IR 9, Shadab, KS 282, PK 1339-12-1-1-0-6, PK 3717-9	74
	PK 3303-15-1	43
	DM 107-4 (Dwarf mutant of Basmati 370)	26
Bangladesh	Dharial, Surjamukhi, Charnock, Panbira, IR8	105
Japan	Shiroka, Akage No. 3, Kairyo-Mochi No. 1	62
Philippines	Macan Bino, Apostal, Macan Binan, Guinongang Strain 1	136
Thialand	MR 7, MN 62M	155
Iran	Binam, Kadous, Shafagh	138

Mechanism of resistance

Kanjanasoon (69) observed that unlike susceptible varieties, the fungus did not grow well on the surface of crown or the surrounding leaf sheath of resistant cultivars and did not form a mycelial mass dense enough to penetrate the host tissues. The content of total phenols was more in the resistant cultivar than susceptible. The phenol content increased sharply after infection in resistant but slightly in susceptible cultivar (172). A low nitrate reductase activity has been observed in roots and a high enzyme activity in the shoots of susceptible rice variety MTU 9 with infection while no such differences were noticed in resistant cultivar GEB 24 (173).

Inheritance of resistance

Khan et al (75) reported that resistance in rice against bakanae is monogenic, recessive in IR 6 and Dominant in KS 282. Two quantitative trait loci (QTLs) namely, QB 1 and QB 10 on chromosome 1 and 10 have been identified for bakanae resistance, which exhibited additive effect (197). Rice genotypes carrying dwarf gene like Sd 1 were not only sensitive to GA3 but also susceptible to bakanae while those carrying dwarf gene d 1 were insensitive to GA3 but susceptible to bakanae. All rice cultures carrying d 29, Sd 6 or Sdq (t) showed resistance to bakanae indicating that dwarf and semi-dwarf rice genotypes may be useful resources for improvement of bakanae resistance in rice breeding programme (95).

2. Cultural control

Sasaki (140) reported that the pathogen can be destroyed by keeping the diseased seeds under snow or in a cold room at 5°C. Kitamura (78) suggested selection of rice seeds by specific gravity. When the specific gravity was < 1.25, the percentage of non-viable seeds, incidence of bakanae and presence of pathogen in hulls and hulled rice seeds increased. The seeds with a specific gravity of >1.2 soaked in 250 ppm benomyl gave a disease control equal to that given by 2000 ppm. Selection of specific gravity did not eliminate all diseased seeds but was useful for increasing the efficacy of seed disinfection. Uprooting of rice nursery in standing water for transplanting has been found very effective in reducing the disease incidence (authors' observation). In Australia, bakanae infection could be avoided by postponing the sowing date of early maturing varieties from December to late January and by rotating rice with pasture (50).

3. Plant extracts

The conidial production, germination and mycelial growth of *F. moniliforme* was inhibited by root bark extract of *Plumbago zeylanica* (14); *Xanthium strumarium* (77); bulbils and tubers of wild variety of Yam (*Dioscoera bulbifera*) containing an alkaloid, dihydrodioscorine (1); root oil of *Curculigo orchioides* (63); onion, garlic, bird cherry, pelargonium, oak, juniper and tarragon (15, 44); extracts of *Sapindus trifoliatus*, *Leucaena leucocephala* and *Polyalthia longifolia* (104); garlic and *Sapindus trifoliatus* (44), volatile oil (cajeput oil) from leaves of *Melaleuca leucadendron* (108); neem oil (185); extract from leaves of tobacco or pawpaws infected with PVX and PVY (122). Mishra et al (109) reported that oil of *Hedychium spicatum* and *Acorus calamus* completely inhibited the growth of *F. moniliforme* at 1.0 x 10^3 ml/l and 0.5 x 10^3 ml/l, respectively and exhibited both fungistatic and fungicidal properties depending upon the concentration.

4. Biocontrol

Lakshmi Kumari et al (85) observed that thermolabile, ether soluble fungistatic substance produced by *Azotobacter* spp. inhibited growth and conidial germination of *F. moniliforme*. Seed treatment and soil incorporation of *Pseudomonas aureofaciens* and other antagonistic bacteria suppressed the growth of the pathogen (87) and reduced bakanae incidence by 71.7 to 96.3% (137). The isolates of *Chaetomium, Penicillium* and *Trichoderma* (180,181); *Aspergillus niger* (39); *Streptomyces fradiae* (13); *Pseudomonas* and *Bacillus* (113); *Penicillium chrysogenum, P. thomii* and *Stachybotrys atra* (98); *B. subtilis* and *B. megaterium* (27,88,94) were also found antagonistic to *F. moniliforme*. Kumar et al (83) reported that *Pseudomonas fluorescens* isolates PF-9, PF-13 and *Bacillus thuringiensis* isolate B-44 significantly reduced the fungal growth and bakanae incidence. The antagonists produced varying amounts of lytic enzymes [Chitinase and β-1,3-glucanase {endo-1, 3(4)- β-glucanase}], siderophores, salicylic acid and hydrogen cyanide. In green house experiments, B-44 was the most effective in the reduction of bakanae incidence. Of various antagonists, *B. subtilis*, *Trichoderma harzianum* and *T. virens* proved to be the most effective in reducing bakanae. However, these were inferior to thiophanate methyl thiram 80 WP at 2 g/l. Seed treatments with antagonists prior to inoculation with the pathogen proved better than their application after seed infection with the pathogen (32). Seed treatment has been found more effective than spraying antagonistic isolates and their efficacy was improved by combining the biological control agents (92). In China, Chen et al (29) observed that of 92 bacterial strains from rice seed, found antagonistic to *F. moniliforme*, two strains B-916 and P-91 gave good control of the disease. Kazempour et al (72) reported that molecular analysis using PCR based RAPD method is useful to differentiate strains of biocontrol agents effective against *F. moniliforme* at the intraspecific level. Motomura et al (112) compared the efficacy of seed treatment with metabolites produced by soil-borne bacteria with benomyl, triflumizole, pefurazoate and prochloraz and found that five microbial extracts efficiently inhibited bakanae development and were better than the fungicides.

5. Chemical control

A. *In vitro*

Phenyl mercury acetate and orthocide (37); Bavistin and Aliette (fosetyl aluminium) (68); benomyl and prochloraz (42); Derosal (16), Derosal and Benlate (93) have been found inhibitory to mycelial growth of *F. moniliforme*. Kutzner and Buchenauer (84) demonstrated that among four triazole fungicides, etaconazole was the most inhibitory followed by paclobutrazole, dichlobutrazole and bitertanol at 10^{-4} M. The fungicide also decreased the production of gibberellin like substances by 24-38%. Ferimzone, pefurazoate, carbendazim, propiconazole and MEMC have also been found to be very effective against *F. moniliforme*. The MIC values of pefurazoate recorded for various *F. moniliforme* isolates varied from 0.78-12.5 ppm (101,159,187). Carbendazim, carboxin and thiophanate-methyl at 5 ppm, pefurazoate and benomyl at 20 and 50 ppm completely inhibited the mycelial growth and provided effective disease control (64). Thiophanate methyl, benomyl (58,133), difenoconazole (Score 250) and penconazole (Topas-100) were found most inhibitory to *F. moniliforme* followed by tolclofos-methyl, thiram, oxadixyl, maneb + zineb, chlorothalonil, fosetyl, tebuconazole and thiabendazole. However, fosetyl and penconazole were phytotoxic when used as seed treatment (58).

B. *In vivo* and field testing

Seed soaking in 0.1% solution of organo-mercury compounds (acetate and chloride groups) for 16-24 h or 0.25% solution for 2 h is recommended in Japan. However, longer soaking for 3 h or more at higher concentration was phytotoxic, particularly, with the acetate group (55). Sunder (159) also reported an almost equal fungal recovery after 24 and 8 h of grain soaking at lower and higher concentration, respectively, of MEMC, carbendazim, propiconazole and flusilazole but the higher concentration of propiconazole was phytotoxic. Soaking of seeds in MEMC (0.5 g/l), MEMC (1 g/l) and propiconazole (2 ml/l) solution for 36, 24 and 24 h resulted in complete eradication of the pathogen from grains, respectively. However, Takeuchi (168) observed that the seed treatment with four organo-mercurials did not completely prevent Bakanae and wet seed treatment with these fungicides proved better than the dry ones. Kauraw (71) reported the control of *F. moniliforme* by immersing seeds in 0.1% Ceresan Wet (MEMC) for at least 48 h or in 0.2% solution for 36 h. Seed treatment with triforine (134), triflumizole (167), P 242 (164) and Ferimzone (101) have been found very effective in eradicating the seed-borne infection of the pathogen in addition to organomercurials (177). Higher doses of triforine (20 & 30 ml/Kg) were found phytotoxic as these affected emergence of rice seedlings (134). Sasaki (140) reported complete control of the disease by immersing the diseased seeds in 10 g/l benomyl for 10 min. Umehara and Akai (183) and Umehara and Oi (184) concluded that soaking of rice seeds in 0.1% benomyl and thiram solution for 16-24 h followed by drying before sowing gave maximum control by reducing the disease incidence at seedling and tillering stages. Effective disease control has also been reported by seed treatment with benomyl (12), carbendazim, MEMC and PCNB (12,35,133,139), propiconazole (57) and thiophanate-methyl (132), prochloraz (126), benomyl + thiram and thiophanate methyl + thiram (120), Shibaoke and Jinzhongling (29). Seed dressing or slurry treatment with prochloraz + carbendazim, mancozeb + benomyl, carboxin + thiram and iprodione gave best control of *F. moniliforme* (42). Seed treatment of Benlate T WP (203), thiophanate methyl, benomyl, carbendazim, kasugamycin (76) provided an effective disease control. However, Dithane S-60 WP at 4 or 2 μg/ml (203) and carboxin (76) enhanced the disease incidence. In Japan, EC formulation of triflumizole

(Trifmine) was more effective in controlling bakanae disease than its WP formulation because of higher amount of fungicide on the seed and early permeation of EC treatments into seeds through the husk than the WP treatment (167). Seed sorting by salt solution and drying treatment of seeds after disinfection improved the effect of chemical disinfection (57). Seed dressing with Chitosan S-II, widely used in medicine and food industry, did not affect rice growth but controlled bakanae by 91-98% at different growth stages (65). The chemical was found toxic to the pathogen and its efficacy as seed treatment increased with the increase in concentration of the chemical (204). Soluble chitin fragments released from fungal cell walls through the action of constitutive rice chitinases served as biotic elicitors of defense responses in rice (135). Tateishi et al (169) reported that seed treatments with WP containing 6% ipconazole protected rice against bakanae disease. The fungicide sufficiently permeated into the seeds during treatment conditions and successive water soaking. Racemic diastereomers 1, 2 and 4 of ipconazole showed greater fungicidal activity than 3 *in vitro* and *in vivo* against bakanae. *In vitro* fungicidal activities of (-) -1 and (-) -2 were higher than those of corresponding (+) -1 and (+) -2 (61).

Wada et al (187) observed that soaking of seeds at 500 ppm concentration of pefurazoate provided 90% disease control. The benzimidazole (benomyl) tolerant strains of *F. moniliforme*, frequently encountered in Japan, were also tolerant to thiophanate-methyl and thiobendazole but could be controlled by triflumizole, propiconazole, prochloraz, NF 30, pefurazoate (57, 187); captan + thiabendazole and triflumizole (38) and phenyl pyrrole fungicide CGA 173506 (86). In Iran, Bhalli et al (16) reported complete control of the disease both by treating rice seed with Derosal at 4 g/kg and by soil drenching in pots at 500 ppm in artificially inoculated soil.

Bagga and Sharma (11) reported that seedling treatment with Bavistin or benomyl at 0.1% for 6 and 8 h were very effective. Tilt 25 EC at 0.05% was the most effective treatment in controlling bakanae but showed signs of phytotoxicity.

Seed treatment followed by seedling dip treatment with carbendazim at 0.2% (124), soil drenching with carbendazim (8) and difenoconazole (58) have also been found effective in managing the disease. Foliar application of benzimidazoles (benomyl and carbendazim) at 0.1% significantly reduced the disease incidence and increased the grain yield while edifenphos, mancozeb and ziram proved ineffective (18). The benomyl and carbendazim (45,140) and propiconazole (124) protected the grains from infection and checked the spread of the disease. Kitamura (79) recommended that efficacy of seed disinfection with benomyl and thiophanate-methyl was significantly increased when it was followed by two sprays of 0.1% benomyl. Titone et al (178) reported the efficacy of thermal treatments and chemical treatments with carbendazim, mancozeb, ipodione + propiconazole and carboxin + thiram and found that the thermal (72°C for 5 min) and carbendazim treatment provided an average disease control of 98%.

CONCLUSION

Foot rot and bakanae of rice is widely distributed in almost all the rice growing countries of the world. Infected seedlings are lanky, pale yellow and taller than the normal seedlings, which ultimately die and result in substantial reduction in grain yield. A considerable progress has been made in research on various aspects of the disease in different countries. However, further research is necessary on pathogenic variability of the causal organism especially mapping of gene(s) for virulence; mechanism and genetics of resistance and mapping of resistant genes in the host. Biochemical and molecular aspects of pathogenesis and disease resistance also need attention. Besides, characterization of environmental conditions for disease epiphytotics; formation and maturation of perithecia; effect of pesticides, soil solarization and flooding on survival of the pathogen; role of internal seed-borne infection and soil-borne inoculum in annual recurrence and epidemiology of the disease need to be investigated along with development and commercialization of more effective strains of bio-control agents for disease management. Cultural practices need to be developed which can minimize the soil-borne inoculum of the pathogen and geographical areas need to be identified for production of disease free seed. New molecules need to be tested especially as seedbed drench or seedling dip or foliar application for effective disease management. An integrated approach involving development of resistant varieties, use of efficient biocontrol agents and chemicals for seedling dip treatment and foliar application is required for the management of foot rot an bakanae.

REFERENCES

1. Adeleye A, Ikotun T. 1989. Antifungal activity of Dihydrodioscorine extracted from a wild variety of *Dioscorea bulbifera* L. *J. Basic Microbiol.* 29: 265-67.
2. Ahmed HU, Mia MAT, Miah SA. 1986. Standardized test tube inoculation for bakanae disease (Bak). *Intern Rice Re. Newsl* 11(2): 21-22.
3. Aktas H, Tunali B. 1986. Determination of the reactions of promising rice cultivars and lines to *Pyricularia oryzae* Bri. and Cav., *Drechslera oryzae* Subram. & Jain and *Fusarium moniliforme* Sheld. in Turkey. *Bitki Koruma Bull.* 26: 41-58.
4. Amin KS, Reddy TB. 1976. Seedling and adult stage methods for screening rice to bakanae. *Indian Phytopath* 29: 46-49.
5. Amoah BK, Rezanoor HN, Nicholson P, Mac Donald MV. 1995. Variation in the *Fusarium* section *Liseola*: Pathogenicity and genetic studies of isolates of *Fusarium moniliforme* Sheldon from different hosts in Ghana. *Pl Pathol* 44: 563-72.
6. Anonymous. 1975. Cooper ative studies on control bakanae, brown spot an damping off of rice plant. *Re Chugoku Natn agric Exptl St* 6: 1-77.
7. Anonymous. 1996. *Standar Evaluation System for Ric* 4th edn. IRTP, IRRI, P.O. Bo 933, Manila, Philippines. 5 pp.
8. Aurangzeb M, Ahmed J, Ilya MB. 1998. Chemical control Bakanae disease of ric caused by *Fusarium monili orme*. *Pak J Phytopatho* 10(1): 14-17.
9. Aurangzeb M, Shafqat Bashir I, Gill MA. 2003. Physiological studies on *Fusariu moniliforme* Sheld., the caus organism of bakanae diseas of rice. *Mycopath* 1(1): 49-52.
10. Bagga PS, Kumar V. 200 Resistance to bakanae or fo disease in Basmati ric *Indian Phytopath*. 53: 321-22
11. Bagga PS, Sharma VK. 200 Evaluation of fungicides seedling treatment for contr lling bakanae/foot rot (*Fusa ium moniliforme*) disease Basmati rice. *Indian Phyt path* 59: 305-08.

12. Bagga PS, Sharma VK, Pannu PPS. 2007. Effect of transplanting dates and chemical seed treatments on foot rot disease of basmati rice caused by *Fusarium moniliforme*. *Pl Dis Res* 22: 60-62.

13. Balagopal C, Oblisami G. 1974. Inhibitory effect of *Streptomyces fradiae* on *Fusarium moniliforme* in the paddy rhizosphere (Eng.). *Curr Res* 3(9): 110-12.

14. Bambode RS, Shukla VN. 1974. Studies on fungitoxic properties of *Plumbago zeylanica* L. extract. *Punjabrao Krishi Vidyapeeth Res J* 2(2): 110-12.

15. Batikyan SG. 1975. Action of phytoncides on the spore germination and growth of some parasitic fungi of the genus *Fusarium*. *Biol Z Armenii* 28(3): 75-80.

16. Bhalli JA, Aurangzeb M, Ilyyas MB. 2001. Chemical control of bakanae disease of rice caused by *Fusarium moniliforme*. *On line J Biol Sci* 1(6): 483-84.

17. Bilai VI, Zakordonets LA, Bogomolova LA, Meteiko TYa. 1974. Effect of different ratios of carbon and nitrogen nutrient media on the growth and metabolism of *Fusarium moniliforme* Sheld. (strain 2801). *Mikol Fitopatol* 8(3): 177-84.

18. Biswas S, Das SN. 2002. Fungicidal spraying for control of bakanae disease of rice in field. *J Mycopathol Res* 40(2): 211-12.

19. Borrow A, Browm S, Jefferys EG, Kessell RHJ, Lloyd EC, Lloyd PB, Rothwell A, Rothwell B, Swait JC. 1964. Metabolism of *Gibberella fujikuroi* in stirred cultures. *Can J Microbiol* 10: 407-44.

20. Bruckner B, Blechschmidt D. 1991. Nitrogen regulation of gibberellin biosynthesis in *Gibberella fujikuroi*. *Appl Microbiol Biotechnol* 35: 646-50.

21. Candau R, Avalos J, Cerda-Olmedo E. 1991. Gibberellins and carotenoids in the wild type and mutants of *Gibberella fujikuroi*. *Appl Environ Microbiol* 57: 3378-82.

22. Cerda OE, Fernandez MR, Avalos J. 1994. Genetics and gibberellin production in *Gibberella fujikuroi*. *Antonie Van Leeuwenhoek* 65(3): 217-25.

23. Chaisrisook C, Leslie JF. 1990. A maternally expressed nuclear gene controlling perithecial pigmentation in *Gibberella fujikuroi* (*Fusarium moniliforme*). *J Hered* 81: 189-92.

24. Chan ZL, Ding KJ, Tan GJ, Zhu SJ, Chen Q, Su XY, Ma K, Wang AE. 2004. Epidemic

regularity of rice bakanae disease. *J Anhui agril Univ* 31(2): 139-42.

25. Chang YC, Sun SK. 1975. The perfect stage of *Fusarium moniliforme. J agric Res China* 24: 11-19.

26. Cheema AA, Awan MA, Ali Y. 1998. Screening of Basmati rice mutants against prevalent diseases in the Punjab province. *Pak J Phytopathol* 10(1): 39-41.

27. Chen CJ, Zhou MG, Ye ZY, Zhou ZY, Chen SL. 1997. Causes of rice bakanae (*Fusarium moniliforme*) occurrence in fertile and dry seed-bed. *J Nanjiang agric Univ* 20(2): 42-45.

28. Chen XJ, Lu GX, Tong YH, Ji Z, Xu JY. 2007. Resistance of *Fusarium moniliforme* rice bakanae disease pathogen to three fungicides for seed treatment and competition ability of resistant strains. *Acta Phytophylac Sin* 34: 425-30.

29. Chen ZY, Lu G, Zhao T.J, Yan DF, Wan HZ 1998. Effect of fungicides on antagonistic bacteria from rice seed surface. *Chinese J Biol Control* 14(2): 72-74.

30. Chin WF. 1940. A preliminary study on the physiological differentiation of *Fusarium fujikuroi* (Saw.) Wr. *Nanking J* 9: 305-21.

31. Clear RM, Patrick SK. 1992. A simple medium to aid the identification of *Fusarium moniliforme, F. proliferatum* and *F. subglutinans. J Food Prot* 55: 120-22.

32. Dehkaei FP, Jajaei SM, Rouhani H. 2004. Effects of paddy soil antagonistic microorganisms of Guilan on the causal agent of rice bakanae disease. *J Sci Technol agric Natural Resources* 8(1): 213-22.

33. Devi RKT, Singh NI. 1997. Distribution, pathogenicity and physiology of *Fusarium* spp. causing rice sheath rot in Manipur. In: *Himalayan Microbial Diversity, Part 1* (eds. Sati SC, Saxena J, Dubey RC), New Delhi, India, Today and Tommorrow's Printers & Publsihers pp. 295-303.

34. Dhillon TS, Garber ED, Wyttenbach EG. 1961. Genetics of pathogenic fungi. VI. Heterokaryons involving *Gibberella fujikuroi* and formae of *Fusarium oxysporum. Can J Bot* 39: 785-92.

35. Dodan DS, Ram Singh, Sunder S. 1994. Survival of *Fusarium moniliforme* Sheld. in infected rice grains and its chemical control. *Indian J Mycol Pl Pathol.*24: 135-38.

36. El-Bahrawi S. 1977. Survey of some *Fuasarium moniliforme* strains from different host plants for compounds possessing gibberellin-like activity. *Zentbl Bakt Parasitkde Infek Krank Hyg* 132: 178-83.

37. El-Hissy FT, Abdel-Kader MIA. 1980. The effect of five pesticides on the mycelial growth of some soil and pathogenic fungi. *Naturalia Monspeliensia* No. 37: 6 pp.

38. Furukawa M. 1990. Relationship between seed disinfection and occurrence of 'Bakanae' disease in paddy fields. *Bull Fukui agric Exptl Stn* 27: 21-25.

39. Gajbe DW, Lanjewar RD. 1991. Antagonism of *Aspergillus niger* on some seed borne fungi of two rice cultivars. *Punjabrao Krishi Vidyapeeth Res J* 15(2): 168-71.

40. Gangopadhyay S. 1983. Fusariosis. In: *Current Concepts on Fungal Diseases of Rice.* Today & Tomorrow's Printers and Publishers, New Delhi, pp. 193-204.

41. Garcheva V, Dimova TS. 1991. Influence of the carbon source on the biosynthesis of gibberellins. *Acta Microbiol Bulgarica* 27: 30-34.

42. Garibaldi A. 1985. Control of seed-borne *Drechslera oryzae* and *Fusarium moniliforme* on rice in Northern Italy. *Mededelingen Van de Faculteit Landouw Wetenschappen Rijksuniversiteit Gent.* 50(3b): 1251-57.

43. Gill MA, Yasin SI, Khan, TZ, Ahsanullah Q, Bhatt MA. 1993. Evaluation of resistance to Bakanae and foot rot disease. *Intern Rice Res Notes* 18(3): 14-15.

44. Gohil VP, Vala DG. 1996. Effects of extracts of medicinal plants on the growth of *Fusarium moniliforme. Indian J Mycol Pl Pathol* 26: 110-11.

45. Hajra KK, Ganguly LK, Khatua DC. 1994. Bakanae disease of rice in West Bengal. *J. Mycopathol Res* 32(2): 95-99.

46. Hamamura H, Kawahara M, Shimoda S. 1989. Some characteristics of *Gibberella fujikuroi* (*Fusarium moniliforme*) isolates less sensitive to triflumizole. *Annals Phytopathol Soc Japan* 55: 275-80.

47. Hamamura H, Kawahara M, Shimoda S, Kataoka M. 1991. Heterokaryosis of *Gibberella fujikuroi* (*Fusarium moniliforme*) isolates less sensitive to triflumizole. *Annals Phytopathol Soc Japan* 57: 387-93.

48. Hashioka Y. 1971. Rice Diseases in the World. VII. Diseases due to Hypocreales, Ascomy-

cetes (Fungal Diseases -5) *Riso* 20: 235-258.

49. Haque MA, Rahman MM, Miah SA. 1979. A rapid method of screening rice varieties for resistance to bakanae disease. *Intern Rice Comm Newsl* 28(2): 34-37.

50. Heaton JB., Morschel JR. 1965. A foot rot disease of rice variety Blue Bonnet, in Northern Territory Australia, caused by *Fusarium moniliforme* Sheldon. *Trop Sci* 7: 116-21.

51. Hino T, Furata T. 1968. Studies on the control of bakanae disease of rice plant, caused by *Gibberella fujikuroi*. 2. Influence of flowering season of rice plant on seed transmissibility through flower infection. *Bull. Chugoku-Shikoku agric. Exptl. Stn*. E2: 97-102.

52. Hiroe I, Nishimura M. 1961. The range of *Fusarium* species producing gibberellin. *Annals Phytopathol Soc Japan* 26: 50-51.

53. Hollomon DW, Zhou M, Wang J, Lu Y, Ju L. 1996. Selection for carbendazim resistance in *Fusarium* species on wheat and rice in China. In: *Brighton Crop Prot. Conf.: Pests and Diseases-1996, Vol. 2. Proc. Intern. Conf., Brighton, UK,* 18-21 Nov. 1996, Farnham, UK; British Crop Protection Council, pp. 707-12.

54. Hori S. 1898. Researches on bakanae disease of rice plant. *Nojishikenjyo Seiseki* 12: 110-19.

55. Hoshino Y. 1955. A study on rice-seed disinfection with special reference to the Bakanae disease. *Jubilee Publication in Commemoration of the 60th Birthdays of Prof. Y. Tochinai and T. Fukushi,* Sapporo, Japan. 290-299.

56. Hsieh WH, Smith SN, Snyder WC. 1977. Mating groups in *Fusarium moniliforme*. *Phytopathology* 67: 1041-1043.

57. Iguchi K, Takeuchi T. 1988. Occurrence and control of benomyl resistant strains of *Fusarium moniliforme*, the causal fungus of bakanae disease in Chiba Prefecture. *Bull Chiba-Ken agric Exptl Stn* 29: 115-23.

58. Ilyas MB, Iftikhar K. 1997. Screening of rice germplasm and fungitoxicants against bakanae disease of rice. *Pak J Phytopathol* 9(1): 67-73.

59. Imura J. 1940. On the angles between blade and culms in the accelerated rice seedlings caused by *Gibberella fujikuroi*. *Annals Phytopathol Soc Japan* 10: 45-48.

60. Ishii H, Takeda H. 1989. Differential binding of a N-phenyl formamidoxime compound in cell-free extract of

benzimidazole-resistant and sensitive isolates of *Venturia nashicole, Botrytis cinerea* and *Gibberella fujikuroi. Netherland J Pl Pathol* 95 (suppl. 1): 99-108.

61. Ito A, Saishoji T, Kumazawa S. 1997. Synthesis of stereoisomers of ipconazole and their fungicidal and plant growth inhibitory activities. *J Pestic Sci* 22(2): 119-25.

62. Ito S, Kimura J. 1931. Studies on the bakanae disease of the rice plant. *Rep Hokkaido Natn agric Exptl Stn* 27: 1-95 + 5.

63. Jaiswal S, Batra A, Mehta BK. 1984. The antimicrobial efficiency of root oil against human pathogenic bacteria and phytopathogenic fungi. *Phytopath Z* 109: 90-93.

64. Javaid MS, Ilyas MB. 1995. Effect of various fungicides on the control of bakanae disease of rice and rice yield. *Pak J Phytopathol* 7(1): 53-55.

65. Jiang HP, Kan LB, Ding LL, Chen YC, Sun JD. 1999. Seed dressing with Chitosan S-II can control bakanae disease of rice. *China Rice* 5: 29.

66. Johnson SW, Coolbaugh RC. 1990. Light-stimulated gibberellin biosynthesis in *Gibberella fujikuroi. Pl Physiol* 94: 1696-1701.

67. Kagiwata T. 1963. Studies on foot rot of the upland rice. *Kanagawa agric Exptl Stn Bull.* 101: 1-116.

68. Kakade DS, Khune NN. 1989. A study on the efficacy of some fungicides against selected phytopathogenic fungi. *J Maharashtra agril Univ* 14(2): 141-42.

69. Kanjanasoon P. 1965. *Studies on Bakanae Disease of Rice in Thailand*. Doc. agric. Thesis, Tokyo Univ., Japan.

70. Kathariou S. 1981. *Gene pool organization in Fusarium moniliforme*. Ph.D. Thesis, Univ. California, Berkeley.

71. Kauraw LP. 1981. Effect of length of treatment and fungicide concentration on seed germination and incidence of foot rot disease of rice. *Intern Rice Res Newsl* 6(6): 15.

72. Kazempour MN, Tabatabaei SM, Hassanzadeh N. 2007. Genetic diversity of antagonistic bacteria against sheath blight and bakanae rice disease by RAPD. *Intern J Biol Biotechnol* 4: 329-33.

73. Khan SAJ, Jalaluddin M, Ghaffar A, Khanzada AK. 1996. Detection of *Fusarium moniliforme* Sheld. the cause of seedling wilt of rice. *Pak J Bot* 28: 231-32.

74. Khan TZ, Gill MA, Bajwa KM, Khan MG. 1999. Screening of rice varieties/lines for resist-

ance against bakanae disease. *Pak J Phytopathol* 11: 166-68.

75. Khan TZ, Gill MA, Khan MG. 1999. Inheritance studies for bakanae disease of rice. *Pak J Phytopathol* 11: 181-82.

76. Khan TZ, Gill MA, Yasin SI, Khan MG. 1995. Effect of seed treatment on bakanae disease incidence and paddy yield. *Pak J Phytopathol* 7: 124-27.

77. Kishore N, Dubey NK, Tripathi RD, Singh SK. 1982. Fungitoxicity of the leaf extracts of some higher plants against *Fusarium moniliforme*. *Natn Acad Sci Letters* 5(2): 43-45.

78. Kitamura Y. 1975. Studies on the control of bakanae disease of rice plant I. Influence of the selection of rice seeds by specific gravity upon the control of bakanae disease. *Proc Kansai Pl Prot Soc* 17: 32-37.

79. Kitamura Y. 1976. Studies on the control of bakanae disease of rice plant II. Effect of benomyl and thiophanate-methyl by spraying at the flowering stage of rice plant on the control of bakanae disease. *Proc Kansai Pl Pro. Soc* 18: 23-27.

80. Klittich CJR, Correll JC, Leslie JF. 1988. Inheritance of sectoring frequency in *Fusarium moniliforme* (*Gibberella fujikuroi*). *Exptl Mycol* 12: 289-94.

81. Klittich CJR, Leslie JF. 1988. Nitrate reduction mutants of *Fusarium moniliforme* (*Gibberella fujikuroi*). *Genetics* 118: 417-23.

82. Kuhlman EG. 1982. Varieties of *Gibberella fujikuroi* with anamorphs in *Fusarium* section Liseola. *Mycologia* 74: 759-68.

83. Kumar MN, Laha GS, Reddy CS. 2007. Role of antagonistic bacteria in suppression of bakanae disease of rice caused by *Fusarium moniliforme* Sheld. *J Biol Control* 21(1): 97-104.

84. Kutzner B, Buchenauer H 1986. Effect of various triazole fungicides on growth and lipid metabolism of *Fusarium moniliforme* as well as on gibberellin contents in culture filtrates of the fungus. *Z Pflkrankh Pflschutz* 93: 597-607.

85. Lakshmi Kumari M Vijayalakshmi K, Subba Rao NS. 1972. Interaction between *Azotobacter* species and fungi I. *In vitro* studies with *Fusarium moniliforme* Sheld *Phytopath Z* 75: 27-30.

86. Leadbeater AJ, Nevill DJ Steck B, Nordmeyer D. 1990 CGA 173506: a novel fungicide for seed treatment. In *Brighton Crop Prot. Conf.*

Pests and Diseases - Vol. 2, Thorton Heath, U.K.; British Crop Prot. Council, pp. 825-830.

87. Lee YH, Shim GY, Lee EJ, Mew TW. 1990. Evaluation of biocontrol activity of fluorescent Pseudomonads against some rice fungal diseases *in vitro* and green house. *Korean J Pl Pathol* 6(1): 73-80.

88. Li B, Xie GL, Lu YL, Hao XJ, Luo JY, Liu, B, Li W. 2006. Community composition of gram positive bacteria associated with rice and their antagonists against the pathogens of sheath blight and bakanae disease of rice. *Chinese J Rice Sci* 20(1): 84-88.

89. Li DJ, Luo, K, Chen Z. 1993. Studies on resistance of rice varieties to bakanae disease and pathogenicity of pathogen (*Fusarium moniliforme*). *Acta Phytopath Sin* 23(4): 315-19.

90. Liddell CM, Burgess LW. 1985. Survival of *Fusarium moniliforme* at controlled temperature and relative humidity. *Trans Br mycol Soc* 84: 121-30.

91. Lu B. 1994. Preliminary report on evaluation of resistance to rice bakanae disease in rice germplasms. *Pl Prot* 20 (3): 20-21.

92. Lu F, Chen ZY, Chen YL, Qin JH, Din Z. 1998. Screening of biological control agents against *Fusarium moniliforme* and determination of their biological activity. *Jiangsu J agril Sci* 14(3): 149-53.

93. Luo JG. 1995. Studies on pathogenic *Fusarium* species of rice bakanae disease and strains. *Chinese J Rice Sci* 9(2): 119-22.

94. Luo JY, Xie GL, Li B, Luo YC, Zhao LH, Wang X, Liu B, Li W. 2005. Gram positive bacteria associated with rice in China and their antagonists against the pathogens of sheath blight and bakanae disease in rice. *Rice Sci* 12: 213-18.

95. Ma LY, Ji, ZJ, Bao JS, Zhu XD, Li XM, Zhuang JY, Yang CD, Xia YW. 2008. Responses of rice genotypes carrying different dwarf genes to *Fusarium moniliforme* and gibberellic acid. *Pl Prod Sci* 11(1): 134-38.

96. Manandhar J. 1999. *Fusarium moniliforme* in rice seeds: its infection, isolation and longevity. *Z Pflkrankh Pflschutz* 106: 598-607.

97. Mandal DN, Chaudhuri S. 1988. Survivality of *Fusarium moniliforme* Sheld. under different moisture regimes and soil conditions. *Intern J Trop Pl Dis* 6: 201-06.

98. Mangiarolti A, Frate GDel, Picco AM, Caretta G. 1987. Antagonistic activity *in vitro* of some saprophytic fungi occurring on the phylloplane of rice, wheat and maize. *Boletin Mycologico* 3(3): 183-189.

99. Marasas WFO, Thiel PG, Sydenham EW, Rabie CJ, Lubben A, Nelson PE. 1991. Toxicity and moniliformin production by four recently described species of *Fusarium* and two uncertain taxa. *Mycopathologia* 113: 191-197.

100. Martin-Sanchez JP, Jimenez-Diaz RM. 1982. Two new *Fusarium* species infecting rice in Southern Spain. *Pl Dis* 66: 332-34.

101. Matsuura K, Ishida K, Kuragano T, Konishi K. 1994. Development of a new fungicide, Ferimzone. *J Pestic Sci* 19(4): 5197-5207.

102. Matuo T, Endo T, Yoshii Y. 1976. Morphological and physiological characters of Japanese isolates of *Fusarium moniliforme* Sheldon. *Trans Mycol Soc Japan* 17: 295-305.

103. Mertz D, Henson W. 1967. Light stimulated biosynthesis of gibberellins in *Fusarium moniliforme. Nature* 214: 844-846.

104. Miah MAT, Ahmed HU, Sharma NR, Ali A, Miah SA. 1990. Antifungal activity of some plant extracts. *Bangladesh J Bot* 19: 5-10 (*Rice Abstr.* 14: 1838).

105. Miah SA, Zaman MQ. 1973. Study on reaction of high yielding dwarf rice to foot-rot and bakanae in Bangladesh. *Bangladesh J Bot* 2: 73-81.

106. Min BR. 1986. Chromosomal studies on the genus *Fusarium. Korean J Mycol* 14: 253-56.

107. Ming YN, Lin PC, Yu TF. 1966. Heterokaryosis in *Fusarium fujikuroi* (Saw.) Wr. *Sci Sin* 15: 1371-78.

108. Mishra D, Misra N. 1989. Antifungal activity of cajeput oil against *Aspergillus fumigatus* (EIDAM) Wint. (NRRL 1979) and *Fusarium moniliforme* Sheldon (NRRL 6398). *Indian Perfumer* 33: 151-55.

109. Mishra D, Samuel CO, Tripathi SC. 2003. Evaluation of some essential oils against seed-borne pathogens of rice. *Indian Phytopath* 56: 212-13.

110. Mishra JK, Gergon E, Mew TW. 1989. How long do fungal pathogens survive in/on rice seeds? *Rice Seed Health Newsl* 1(1): 6.

111. Moore WD. 1924. Corn root and ear rot studies. *44th Annu. Rep. New Jersey agric. Exptl. Stn. for year ending June 30.*

112. Motomura M, Suwa S, Hirooka EY. 1997. Biological control: microbial versus chemical fungicides on growth of rice infected with *Fusarium moniliforme*. *Revista de Microbiologia* 28(1): 11-16.

113. Nikogosyan VG, Shakhmuradyan SB. 1976. On the antagonism between oligonitrophils and other soil microorganisms. *Biol Z Armenii* 29(7): 52-58.

114. Nisikado Y, Kimura K. 1941. A contribution to the pathological anatomy of rice plants affected by *Gibberella fujikuroi* (Saw.) Wr. I. *Ber. Ohara Inst Landw Forsch* 8: 421-26.

115. Nisikado Y, Matsumoto H. 1933. Studies on the physiological specialization of *Gibberella fujikuroi* the causal fungus of rice bakanae disease. *Trans Tottori Soc agric Sci* 4: 200-211.

116. Ogawa K. 1988. Damage by bakanae disease and its chemical control. *Japan Pesticide Information* 52: 13-15.

117. Ogawa K, Takeda S. 1990. Population of benomyl-resistant rice "bakanae" fungus in paddy fields. *Annals Phytopathol Soc Japan* 56: 247-49.

118. Omatsu N, Izumi S, Shinyashiki I. 1990. Occurrence of benomyl resistant strains of *Gibberella fujikuroi* in Kagoshima Prefecture and control of bakanae disease of rice seed by seed disinfection. *Proc Assoc Pl Prot Kyushu* 36: 5-7.

119. Ou SH. 1985. *Rice Diseases*. 2nd edn. CMI, Kew, Surrey, UK, pp. 262-72.

120. Padasht F, Sharifi TA, Hedjaroude GA, Izadyar M, Okhovat M. 1996. Study on the efficacy of some fungicides in controlling rice bakanae disease in Gilan Province. *Iranian J Pl Pathol* 32 (3/4): 268-77 (Pe), 182-83 (En).

121. Pan Y, Wang ZY, Wu HZ. 1997. Mycclial growth, sporulation and pathogenicity of *Fusarium moniliforme* isolates with differential resistance to carbendazim (MBC). *Jiangsu J agric Sci* 13(2): 90-93.

122. Pandey AK, Shukla HS, Rao GP. 1989. Mycotoxic properties in leaf extracts of some virus infected plants. *Natn Acad Sci Letters* 12(1): 9-10.

123. Panneerselvam A, Saravanamuthu R. 1996. Studies on the saprophytic survival of *Fusarium moniliforme* Sheld. in soil under treatment of oil cakes. *Indian J agric Res* 30(1): 12-16.

124. Pannu PPS, Singh N, Rewal HS, Sabhiki HS, Raheja S. 2009. Integrated management of foot rot of Basmati rice.

Proc. National Symp. Plant Pathology in the Changing Global Scenario organised by Indian Society Plant Pathologists, Feb. 27-28, 2009, NBPGR, N. Delhi, p. 13/32.

125. Pavgi MS, Singh J. 1964. Bakanae and foot rot of rice in Uttar Pradesh, India. *Pl Dis Reptr* 48: 340-42.

126. Peng JS. 1995. Control effect of rice bakanae disease with Sportak. *J Jilin agril Univ* 17 (4): 49-52.

127. Puhalla JE, Spieth PT. 1983. Heterokaryosis in *Fusarium moniliforme. Exptl Mycol* 7: 328-35.

128. Puhalla JE, Spieth PT. 1985. A comparison of heterokaryosis and vegetative incompatibility among varieties of *Gibberella fujikuroi* (*Fusarium moniliforme*). *Exptl Mycol* 7: 39-47.

129. Rajagopalan K. 1961. Screening of rice varieties for resistance to foot rot disease. *Curr Sci* 30: 145-47.

130. Rajagopalan K, Bhuvaneswari K. 1964. Effect of germination of seeds and host exudations during germination on foot rot disease of rice. *Phytopath Z* 50: 221-226.

131. Ram Singh, Sunder S. 1997. Foot rot and bakanae of rice: retrospects and prospects. *Intern. J Trop Pl Dis* 15: 153-176.

132. Rathaiah Y, Das GR, Upendra Singh HK. 1991. Estimation of yield loss and chemical control of bakanae disease of rice. *Oryza* 28: 509-12.

133. Raza T, Ahmed MI, Nasir MA. 1993. Extent of infection of *Fusarium moniliforme* Sheld. in different rice seed samples and *in vitro* effect of certain fungicides on the fungus. *Sarhad J agric.* 9: 231-33.

134. Reddy OR. 1984. Evaluation of triforine (Saprol) against seed-borne pathogens of rice (*Oryza sativa*). *Intern J Trop Pl Dis* 2: 133-38.

135. Ren YY, West CA. 1992. Elicitation of diterpene biosynthesis in rice (*Oryza sativa* L.) by chitin. *Pl Physiology* 99(3): 1169-1178.

136. Reyes GM. 1939. Rice diseases and methods of control. *Philippine J agric* 10: 419-436.

137. Rosales AM, Mew TW. 1997. Suppression of *Fusarium moniliforme* in rice by rice-associated antagonistic bacteria. *Pl Dis* 81: 49-52.

138. Saremi H, Ammarellou A, Marefat A, Okhovat SM. 2008. Binam a rice cultivar, resistant for root rot disease on rice caused by *Fusarium*

moniliforme in North-West, Iran. *Intern J Bot* 4: 383-389.

139. Sarkar BB. 1986. Controlling bakanae (Bak) and foot rot disease with fungicides seed treatments *Intern Rice Res Newsl* 11(3): 18.

140. Sasaki T. 1987. Epidemiology and control of rice bakanae disease. *Bull. Tohoku Natn agric Exptl Stn.*74: 1-47.

141. Sawada K. 1917. Beitrage uber Formosas-Pilze no. 14. *Trans Natural History Soc Formosa* 31: 131-33

142. Seto F. 1932. Experimentelle Untersuchungen uber die hemmende und die beschleunigende Wirkung des Erregers der sogenannten `Bakanae'-krankheit, *Lisea fujikuroi* Sawada, auf das Wachstum der Reiskeimlinge. *Mem Coll agric Kyoto Univ* 18: 1-23.

143. Seto F. 1933. Untersuchungen uber die Bakanae- Krankheit der Reispflanze. III. Uber die Beziehungen Zwischen der Bodenfeuchtigkeit und dem Krankheitsbefall durch Bodeninfektion. *Forsch Gebiet Pflkrankh* 2: 125-37.

144. Seto F. 1933. Untersuchungen uber die Bakanae-Krankheit der Reispflanze. IV. Uber die Beziehungen Zwischen der Bodentemperaturen und dem Krankheitsbefall bei Bodeninfektion. *Forsch Gebiet Pflkrankh* 2:138-53.

145. Sharma RD, Singh RS. 1978. Survival of *Fusarium moniliforme* in soil. *Indian Phytopath* 31: 352-56.

146. Sharma VK, Bagga, PS. 2007. Pathogenic behaviour of *Fusarium moniliforme* isolates causing foot rot disease in Basmati rice. *Pl Dis Res* 22: 165-66.

147. Siddiqi SA, Mirza JH, Inam-ul-Haq M. 1995. Studies on the growth and sporulation of *Fusarium moniliforme* Sheld., the causal organism of bakanae disease of rice. I. Effect of nitrogen source. *Pak J Phytopath* 7(2): 145-47.

148. Siddiqi SA, Mirza JH, Inam-ul-Haq M. 1995. Studies on the growth and sporulation of *Fusarium moniliforme* Sheld., the causal organism of bakanae disease of rice. 2. Effect of carbon sources. *Pak J Phytopath* 7(2): 151-53.

149. Sidhu GS. 1983. Genetics of *Gibberella fujikuroi*. I. Inheritance of certain cultural traits. *Can J Genet Cytol* 25: 93-96.

150. Sidhu GS. 1986. Genetics of *Gibberella fujikuroi*. VIII. Vegetative compatibility groups. *Can J Bot* 64: 117-21.

151. Snyder WC, Sun SK. 1973. Heterothallism in *Fusarium*

moniliforme. *Rep. US-ROC Coop. Sci. Seminar Pl. Root Dis. Berkeley Univ., California.*

152. Stodola FH. 1958. Source book on gibberellin 1828-1957. Washington, USDA, 1958, pp. 138 + 426 unnumbered pages.

153. Stoll C. 1954. Ueber Stoffwe-schsel und biologisch wirksa-mestoff von *Gibberella fuji-kuroi* (Saw.) Woll., dem Erreger der Bakanae krank-heit. *Phytopath Z* 22: 233-274.

154. Subba Rao NV. 1957. *In vitro* detection of gibberellic acid in foot rot infected rice. *J Proc Indian Acad Sci B.* 45: 91-94.

155. Sukapanpotharam S, Chanta-rasnit AA. 1980. A screening method for bakanae resist-ance. *Intern Rice Res Newsl* 5(4): 6.

156. Sultana SN, Khan M. 1971. Effect of light on growth and sporulation of *Fusarium moniliforme* Sheld. *Sind Univ Res J* 5: 183-87.

157. Sun SK. 1975. The disease cycle of rice bakanae disease in Taiwan. *Proc Natn Sci Council* 8(2): 245-256.

158. Sun SK, Snyder WD. 1978. The bakanae disease of rice plant. *Sci Bull Taiwan* 10: (7) 2, (8) 4, (9) 4, (10) 4.

159. Sunder S. 1995. *Survival and Variability in Fusarium moniliforme Sheld. and Management of Bakanae Diseasc of Rice*. Ph.D. Thesis, CCS HAU, Hisar, 155 + XX pp.

160. Sunder S, Satyavir 1997. Survival of *Fusarium monilif-orme* in soil enriched with different nutrients and their combinations. *Indian Phytopath* 50: 474-81.

161. Sunder S, Satyavir 1997. Survival of *Fusarium moniliforme* in soil, grains and stubbles of paddy. *Indian Phytopath* 51: 47-50.

162. Sunder S, Satyavir 1998. Vegetative compatability, bio-synthesis of GA_3 and virulence of *Fusarium moniliforme* isolates from bakanae disease of rice. *Pl Pathol* 47: 767-772.

163. Sunder S, Satyavir, Goel SP. 1997. Morphological and cultural variability in *Fusarium moniliforme* from rice bakanae disease. *Indian J Pl Pathol* 15: 31-39.

164. Sung JM, Lee, EJ, Yang SS. 1985. Rice seed infection of *Fusarium* species at paddy fields and selection of seed disinfectants. *Korean J Mycol* 13(2): 99-103.

165. Sung JM, Yang SS. 1985. Occurrence of bakanae disease by *Fusarium moniliforme* associated with different seedling methods. *Korean J Mycol* 13: 23-26.

166. Sung JM, Yang SS, Lee EJ. 1984. Pathogenicity in nursery box and symptom appearance and yield damage in paddy field by each strain of *Fusarium moniliforme*. *Korean J Mycol* 12(2): 65-73.

167. Suzuki M, Hamamura H, Iwamori M. 1994. Relationship between formulations of triflumizole and their efficacy to bakanae disease in rice seed treatmant. *J Pestic Sci* 19(4): 251-56.

168. Takeuchi S. 1972. Climatic effect on seed infection of rice plant with bakanae disease and disinfection with organic mercury compounds. *Proc Kansai Pl Prot Soc* 14: 14-19.

169. Tateishi H, Saishoh T, Suzuki T, Chida T. 1998. Antifungal properties of the seed disinfectant ipconazole and its protection against bakanae and other diseases of rice. *Annals Phytopathol Soc Japan* 64(5): 443-50.

170. Thakur KSS. 1973. Effect of rufianic acid on the foot-rot disease of rice caused by *Fusarium moniliforme*. *Phytopath Z* 78: 81-85.

171. Thakur KSS. 1974. Role of gibberellic acid, fusaric acid and pectic enzymes in the foot-rot disease of rice. *Riso* 23: 191-207.

172. Thakur KSS, Ghosal M. 1976. Study of phenolic contents of resistant and susceptible varieties of rice in relation to foot rot disease. *Riso* 25: 87-91.

173. Thakur KSS, Subramanian S. 1966. Nitrate reductase of rice (*Oryza sativa* L.) under foot-rot disease. *Planta* 71: 125-29.

174. Thomas KM. 1931. A new paddy disease in Madras. *Madras agric* . 19: 34-36.

175. Thomas KM. 1933. The `foot rot' of paddy and its control. *Madras agric J* 21: 263-72.

176. Thomas KM. 1937. Administrative report of the Government Mycologist for the year 1936-1937. In: *Admn. Rep. agric. Chem. Entomol. Mycol. Madras, 1936-37*, p. 35-51.

177. Titone P, Polenghi L, Tamborini L, Garibaldi A. 2004. Chemicals and thermotherapy for rice seed dressing. In: *Proc. Conf. Challenges and Opportunities for Sustainable Rice Based Production Systems* (eds. Ferrero, A. and Vidotto, F.), Torino, Italy, 13-15 Sept., 2004.

178. Titone P, Tamborini L, Polenghi L, Garibaldi A. 2003. Efficacy of chemical and physical seed dressing against bakanae disease of rice: Comparison of fungicide treat-

ment and heat treatment. *Sementi Elette* 49(4): 23-27.

179. Tochinai Y, Ishizuka K. 1934. The after effect of the fungus filtrate of *Gibberella fujikuroi* on rice plants. *Trans Sapporo Natural History Soc* 13: 148-52.

180. Toledo ACDDe, Cardoso EJBN. 1975. Effect of natural soil microflora and soil amendments on the biological control of bakanae of rice. *Summa Phytopathol* 1(2): 81-86.

181. Toledo ACDDe, Cardoso EJBN. 1975. Antagonism of some soil fungi to *Fusarium moniliforme* Sheld. *Summa Phytopathol* 1(3): 164-68.

182. Toledo ACDDe, Kimati H, Cardoso CON. 1975. Effect of light and vitamins on conidia production in *Fusarium moniliforme* Sheld. *Phytopathologica* 1: 67-68.

183. Umehara Y, Akai K. 1974. On the control of the bakanae disease of rice plant by the seed disinfectants. 4. Relation between the preventive effect of benomyl, thiram, and temperature of solution (No. 2). *Proc Assoc Pl Prot Hokuriku* 22: 64-67.

184. Umehara Y, Oi J. 1978. Control of the bakanae disease of rice plant by the disinfecting of the seed. 9. Adherent property of seed disinfectant. *Proc Assoc Pl. Prot Hokuriku* 26: 31-35.

185. Vir D, Sharma RK. 1985. Efficacy of fungicide. XXXII. Evaluation of neem oil for control of plant pathogens. *Asian Farm Chemicals* 1(7/8): 23-24.

186. Viswanath-Reddy M. 1965. Inoculum potential and foot rot of rice (*Oryza sativa* L.). *Phytopath Z* 53: 197-200.

187. Wada T, Kuzuma S, Takenaka M. 1990. Sensitivity of *Fusarium moniliforme* isolates to pefurazoate. *Annals Phytopathol Soc Japan* 56: 449-56.

188. Wang GC, Chen HK, Xu PS, Bao JY. 1990. Studies on the pathogens of rice bakanae disease in Zhejiang. *Acta Phytopath Sin* 20(2): 93-97 (China).

189. Watanabe Y. 1974. The possibility of soil transmission of bakanae disease and the contamination of seeds with causal fungus during the hastening process of seed germination. *Bull Tokai-Kinki Natn Exptl Stn* 27: 35-41.

190. Wineland GO. 1924. An ascigerous stage and synonymy for *Fusarium moniliforme J agric Res* 28: 909-922.

191. Wollenweber HW. 1931. *Fusarium-Monographie*. Julius Springer, Berlin, 516 pp.

192. Wollenweber HW, Reinking O. 1935. *Die Fusarien*, Berlin, Paul Parey 355 + VIII pp.

193. Yabuta T, Hayashi T. 1940. Biochemical studies of bakanae fungus of rice. *J Imper agric Exptl Stn Japan* 3: 365-400.

194. Yabuta T, Sumiki Y, Uno SJ. 1939. Biochemical studies on `bakanae' fungus of rice. IV. The cultural conditions for producing gibberellin or fusaric acid. *J agric Chemical Soc Japan* 15: 1209-1220.

195. Yamashita T, Eguchi N, Saito Y. 1995. Straight forward methods for monitoring benomyl resistant strains of *Fusarium moniliforme*, the causal fungus of "Bakanae" disease. *Proc Kanto-Tosan Pl Prot Soc* 42: 23-25.

196. Yan KY, Dickman MB, Yu JR, Leslie JF. 1993. Sensitivity of field strains of *Gibberella fujikuroi* (*Fusarium* section *Liseola*) to benomyl and hygromycin B. *Mycologia* 85: 206-213.

197. Yang CD, Guo LB, Li XM, Ji ZJ, Ma LY, Qian Q. 2006. Analysis of QTLs for resistance to rice bakanae disease. *Chinese J Rice Sci* 20(6): 657-59.

198. Yasin SI, Khan TZ, Akhtar KM, Muhammad A, Mushtaq A. 2003. Economic evaluation of bakanae disease of rice. *Mycopath* 1(2): 115-17.

199. Yasuda Y. 1986. Present situation on benzimidazole resistance of *Gibberella fujikuroi* and *Botrytis cinerea* in Japan. *ISPP Chemical Control Newsl* 7: 8-9.

200. Yogeswari L. 1948. Trace element nutrition of fungi. 1. The effect of boron, zinc and manganese on *Fusarium* species. *Proc Indian Acad Sci.*, B 28: 177-201.

201. Yoshizawa T, Yamasaki M, Nanba N, Yamashita A, Ueda S, Shibata H. 1994. Fumonisin producibility of fungicide-resistant *Gibberella fujikuroi*, a pathogen of Bakanae disease. *Mycotoxins* 40: 33-37.

202. Younts SE, Musgrave RB. 1958. Chemical composition, nutrient absorption and stalk rot incidence of corn as affected by chloride in potassium fertilizer. *Agron J* 62: 216-19.

203. Yu CM, Yang HK. 1978. Effect of seed treatment fungicides on the incidence of rice bakanae disease. *Pl Prot Bull. Taiwan* 20: 73-76.

204. Yu HS, Wu HZ, Zhang YM. 1998. Effect of water soluble chitosan in control of rice bakanae disease (*Fusarium moniliforme*) and rape sclerotiose (*Sclerotinia sclerotior-*

um). *Jiangsu agril* Sci. 5: 38-40.

205. Yu KS, Sun SK. 1976. Ascospore liberation of *Gibberella fujikuroi* and its contamination of rice grains. *Pl Prot Bull Taiwan* 18: 319-29.

206. Yu KS, Sun SK. 1977. Studies on inoculum potential and incubation period on rice bakanae disease. *Pl Prot Bull Taiwan* 19: 245-250.

207. Yu KS, Sun SK. 1978. Influence of nutrition and light on sporulation of the rice bakanae fungus, *Fusarium moniliforme* Sheld. *Pl Prot Bull Taiwan* 20: 141-50.

208. Yu SM, Sun SK. 1979. Variation of *Fusarium moniliforme. Pl Prot Bull Taiwan* 21: 341-50.

209. Zakordonets LA, Elans'Ka IO, Gorbik LT, Bogomolova LO. 1974. Effect of different phosphorus concentrations on the morphological and cytological properties of *Fusarium moniliforme* 2801 and production of gibberellic acid. *Mikrobiol Z* 36(6): 715-721.

210. Zhang CL, Zheng Z, Wang GC. 1998. Genetic diversity in *Gibberella fujikuroi* (*Fusarium* section Liseola) from rice. *J Zhejiang agric Univ* 24(6): 583-586.

211. Zhang HJ, Bao SZ, Shen YJ, Gong CG, Zhu DY. 2000. Change in trait of bakanae disease-infected leggy rice seedlings and its influence on rice yield. *Acta Agriculturae Shanghai* 16(1): 67-71.

212. Zheng GX, Lu B, Wu RZ, Nie H. 1993. Study on screening methods for resistance to bakanae disease of rice. *Acta Phytophylac Sin* 20(4): 289-293.

213. Zweig G, DeVay JE. 1959. On the biosynthesis of gibberellins from C-substances by *Fusarium moniliforme*. *Mycologia* 51: 877-886.

INTEGRATED MANAGEMENT OF *PARTHENIUM HYSTEROPHORUS* L. (ASTERACEAE): A WEED OF WORLDWIDE SIGNIFICANCE

Puja Ray[1] and H N Gour[2]

1. Department of Zoology and Entomology, Rhodes University, P.O. Box 94, Grahamstown 6140, South Africa
2. Professor Emeritus (DST), Maharana Pratap University of Agriculture and Technology, Udaipur-313 001, Rajasthan
Email: puja.ray@gmail.com/hngour49@gmail.com

KEY WORDS: Congress grass, biological control, invasive plant, noxious weed, parthenium

ABSTRACT

Parthenium hysterophorus L. (Asteraceae) in last sixty years has proliferated widely and has emerged as the seventh most devastating weed in several tropical and subtropical regions of Africa, Asia and Australia. It has wide-ranging impacts on agriculture, biodiversity, and human and animal health. The aim of this review is to provide general information on parthenium, its ill effects and to document the weed management efforts undertaken. It is of immense importance to draw the attention of general public and concerned authorities, and understanding the fact that no single management option would be adequate to manage parthenium, and there is a need to integrate various management options. Successful management of this weed can only be achieved by an integrated approach with biological control as key element.

INTRODUCTION

The noxious weed, *Parthenium hysterophorus* L. (Asteraceae: Heliantheae), commonly known as parthenium, white top, congress grass, American fever or carrot weed, is a devastating and hazardous, invasive weed (Fig. 1.). Parthenium is predominantly a native from the area around the Gulf of Mexico including West Indies and the area central to South America (105; Fig. 2.). During the 1950s, parthenium was introduced into Africa, Asia and Oceania in cereal and grass seed shipments from U.S.A. Parthenium is among one of the extremely dominant colonizers of plant kingdom due to its fast growth rate, high reproductive potential, adaptive nature and allelopathic effects (112).

Fig. 1: *Parthenium hysterophorus* infestation in various landmarks

These are in fact the major contributing factors for rapid spread and successful establishment of this weed in any given ecosystem. Above all, the lack of natural enemies of this weed in most areas of its infestation away from its native home land is also contributing to a large extent in the rapid spread of this weed. It causes severe health hazards to both human and livestock. In

India, a long term exposure to parthenium was observed to cause severe dermatitis and respiratory allergies (63, 74). The animals generally avoid feeding on the plant but if consumed causes several health ailments including ulcers in mouth and intestine (83). The hazardous and obnoxious nature of parthenium calls for immediate management through various approaches including integration of eco friendly insect and microbial agents as potential bio control organisms.

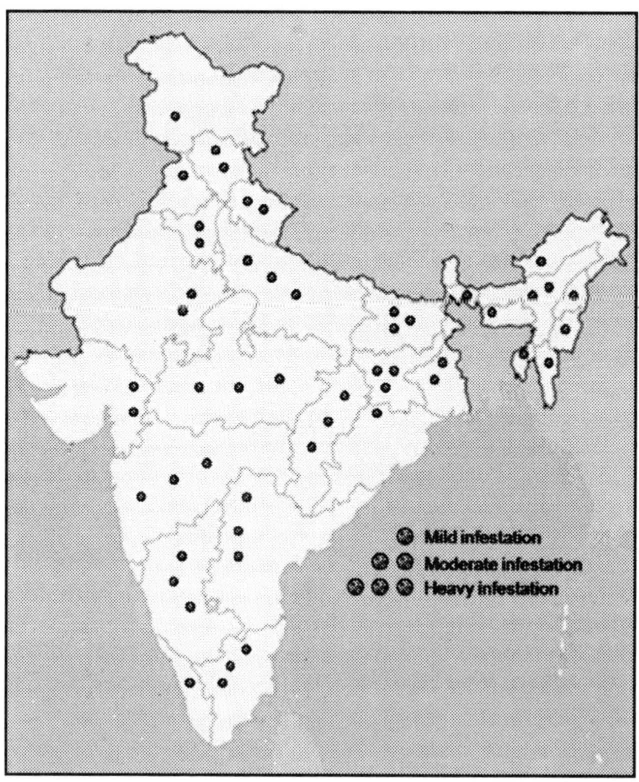

Fig. 2: State wise infestation of *Parthenium hysterophorus* in India

PARTHENIUM: BIOLOGY OF THE PLANT

It is essential to understand the biology of the weed before laying out the perfect strategy for its management. The genus name '*Parthenium*' has been derived from its similarity to feverfew (*Tanacetum parthenium*) and '*hysterophorus*' is derived from the Greek hystera (womb) and phorus (bearing), probably a reference to its seeding ability. Parthenium is a short-lived, annual herb, with deep penetrating taproot and erect shoot with pale green, deeply lobed, softly hairy leaves. It is capable of germinating at any time of year subject to slight moisture availability in the soil and growing rapidly to a height of 200 cm, in good soils. It bears many small white flowers measuring 3-4 mm in diameter, with 5 ray florets, and about 40 disk florets at the tips of the stems. Commonly a cluster of separate flower heads occur on a corymb which grows axially and terminally. Each flower contains four to five black, wedge-shaped seeds with two thin, white scales. Flowering begins 6-8 weeks after germination and continues until drought or frost kills the plant. Seeds are small, 1-2 mm, flattened with short wing-like processes, and are without hooks or pappus. They are distributed by floating on still or flood waters or in mud adhering to animals, vehicles and machinery. Parthenium is a prolific seed producer and can yield more than 25,000 seeds (achenes) per plant (16, 87), which require bare soil to germinate but do not exhibit dormancy (126). Mostly the seeds can easily germinate within 2 years under favourable conditions, although up to 50% seed lot viability has been observed up to 6 years (88). It prefers alkaline, clay loam to heavy black clay soils, but can easily tolerate a wide variety of soil types.

Parthenium is capable of growing and spreading to a wide range of landscapes like wastelands, disturbed lands, eroded lands, degraded pastures, crops, orchards, forests, along railway tracks and roadsides, along fresh water bodies, etc (87), across a wide range of habitats from hot, arid, semiarid, humid areas to high-altitude areas (27, 28).

ORIGIN AND DISTRIBUTION

P. hysterophorus, from being absent in the list of world's worst weed in 1977 (39) to being rated as one of the worst weeds ever, of the world, in next few decades, has spread like a wild forest fire from its native regions of South and Central America and the Caribbean (Fig. 2). During the last 6 decades or so, it has spread throughout much of the tropical and sub-tropical regions of the world in cereal and grass seed shipment, and has become an invasive weed. Today it is a weed of global significance. Literature traces its history of occurrence in India about 150 years back through the herbarium record of Dr. Brandis in 1880 at Forest

Research Institute, Dehra Dun, India. Though in most of the publications, first occurrence of parthenium in India is believed to be as a stray plant on the rubbish heaps in the neighbourhood of the Agricultural College, Poona in 1955 by a retired horticulturist, Prof. H.P. Paranjape, of then Bombay State. This was later identified and described by Rao (101) as *Parthenium hysterophorus* L. Subsequently, the identity of the weed and its occurrence as a new adventives plant in Poona was confirmed by several botanists (64). It is supposed to be introduced into India in large scale import of wheat and other cereals under PL-480 grants in 1950s. Since 1956, parthenium has spread to an area of about 35 million ha land all over the country in just 60 years of its introduction (116) and has achieved the status of "worst weed". Today in India, parthenium occurs in almost all states from foothills of Himalayas in Jammu Kashmir, the northern most state in India to the southernmost state of Tamil Nadu, and from Arunachal Pradesh in the east to the desert state of Rajasthan in the west (68; Fig. 3) It has be spotted growing at the sea level in Mumbai and Chennai as well as at the altitude of 2000 m in Ooty in south and Kullu valley in the north (58). It has also been reported from Andaman Islands in Bay of Bengal. It has also spread to neighboring countries, including Pakistan (42, 105), Sri Lanka (48), Bangladesh (88) and Nepal (3). Parthenium also occurs in southern China, Taiwan, Vietnam, and Israel in Asia (50, 88).

In Australia parthenium was introduced to central Queensland in a shipment of grass seed from the Texas, USA in 1958. The infestation did not take a weed status until 1973, but since then due to favourable climatic conditions the rate of spread became explosive (39). It has become widespread in grazing land from central Queensland to northern New South Wales, infesting 170,000 km^2 of prime grazing land. Two biotypes of *Parthenium hysterophorus* L. have been found to be established in Australia resulting from introductions in south-east and central areas of Queensland independently at two separate occasions.

It has also invaded several Pacific islands including New Caledonia, Papua New Guinea, Seychelles, Vanuatu (4), and several African countries including Ethiopia, Kenya, Madagascar, Mozambique, South Africa, Somalia, Swaziland, and Zimbabwe (17; 20; 33,34, 40; 65; 90; 115; 119; 121;127; 118).

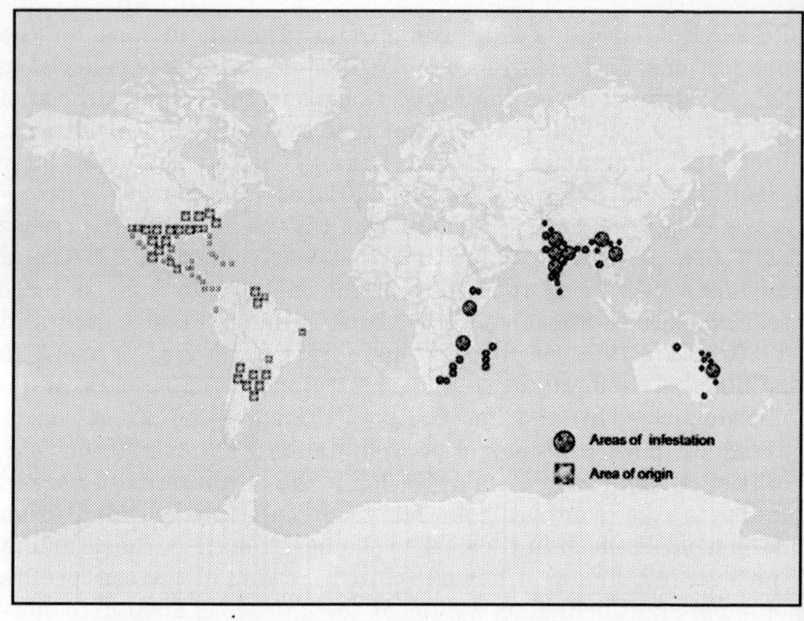

Fig. 3: Global distribution of *Parthenium hysterophorus*

Parthenium occurs throughout the tropical and subtropical Americas, mostly USA, Southern Brazil and Northern Argentina (21). In USA, it is reported from Florida to Texas, north of Massachusetts, Michigan, Illinois, Louisiana, Missouri and Kansas (15).

PROBLEMS POSED BY THIS WEED

Parthenium is poisonous, pernicious, allergenic and aggressive weed posing grave threat to human being, their livestock, agriculture and the environment. The plant parts of parthenium specially pollen contain sesquiterpene lactones, caffeic acids, vanillic acids, anisic acids, chlorogenic acids, para hydroxyl benzoic acids (all Phenolic acids) that are severely lethal to human beings and cattle. Parthenin is the basic toxic substance extracted out (14, 103). Contact with parthenium may induce severe allergic reactions in susceptible individuals (123). Long-term exposure to parthenium has been observed to cause a severe dermatitis, particularly, although not exclusively, in adult males (63). There is no effective treatment for the dermatitis other than removal from all contact with the weed: severe cases can lead to infection, septicemia and even death. There is also an increasing incidence of respiratory allergies, with 7% of a sample of Bangalore residents suffering from allergic rhinitis due to parthenium pollen, and 42% of sufferers from nasobronchial allergy in Bangalore sensitive to parthenium pollen (123). Increasing numbers are also reporting respiratory problems, usually starting with hay fever but progressing after 3-5 years to asthma and allergic bronchitis. Parthenium is generally unpalatable and toxic to cattle, buffalo and sheep (52, 82) though goat has been seen to be readily grazing on parthenium. Bitter and reduced milk yields have been reported in buffaloes and goats that fed on grass mixed with parthenium (57).

It has the great potential to adversely affect crops, resulting in low yields due to the release of allelochemicals into the soil. Dense stands of *P. hysterophorus* can release these chemicals and thus can affect the germination and growth of subsequent crops. In Australia, parthenium causes a loss of more than A$ 16 million per year to the pasture industry (26). Chippendale and Panetta (18) estimated that the weed causes an annual loss to the beef industry of US $12 million. The inhibitory effects of *P. hysterophorus* on germination of many crops have been reported (85). Khosla and Sobti (55) reported a yield decline of upto 40% in agricultural crops due to parthenium. It is reported to reduce forage production by up to 90% (86). Kumar and Rohatgi (61) feared the role of this weed in changing floristic diversity.

Parthenium is also a host to several pests and diseases harmful to various crops. Reciprocal infection of *Xanthomonas campestris pv. phaseoli* between parthenium and bean plants has been established. Infection occurred with beans becoming infected at the pre-flowering and pod formation stages (92).

PRESENT SCENARIO OF PARTHENIUM MANAGEMENT

Ever since parthenium became a menace in India and other countries, efforts are being made to manage the weed by different methods. But so far, no single method has been proved satisfactory, as each method suffers from one or more limitations such as high cost, impracticability, environmental safety, temporary relief etc. A brief account of various management practices being employed against parthenium is given below to take an insight of merits and demerits of the method before switching over to future strategy for effective parthenium management.

CHEMICAL CONTROL

Today chemical control is taken as the first line of defense in eradication and containment programs of pests (Fig. 4a). Lot of work have been done on herbicidal control of parthenium (9, 10, 78; 87, 83, 111, 117, 128). Earlier Kanchan and Jayachandra (53) found that bromocil, diuron and terbacil @ 1.5 kg ha^{-1} were found effective against parthenium. Similarly Dhanraj and Mitra (23) have reported that diquat @ 0.5 kg ha^{-1} in 500 L spray can effectively control parthenium at all growth stages. Mishra and Bhan (79) have recommended the sulphonyl urea herbicides like chlorimuron ethyl and metsulfuron methyl to control parthenium in non-cropped areas. Spraying of 2 kg, 2, 4-D sodium salt or 2L MCPA in 400 L of water checked the growth of parthenium seedlings. Likewise, MSMA @ 1L 100 ml^{-1} of water successfully controlled fully grown parthenium (66). In contrast to that several well known herbicides such as paraquat, trifluralin, diphenamid, napropamide, acetanilides, alachlor, metolachlor and propachlor have been shown to be ineffective against parthenium weed (62, 89). Since use of chemical herbicides also poses health risks, the use of these chemicals should be reduced to minimum.

Fig. 4. Parthenium hysterophorus management: a. Chemical control; b *Zygogramma bicolorata,* biocontrol agent of parthenium

PHYSICAL CONTROL

In India, large amount of money has been spent on manual uprooting of parthenium (51). It has been widely suggested that uprooting of parthenium in its pre-germination stage helps in reducing seed bank in the soil. But such practices are only possible in limited areas such as agricultural fields and residential areas by community service involving the stakeholders. Other physical methods such as ploughing, grading and slashing in large areas can provide some temporary relief. However such physical practices may not be very feasible in pastures, wastelands and forests.

COMPETITIVE DISPLACEMENT

The phenomenon of allelopathy where one plant exerts a detrimental effect on another through the production of germination and growth inhibiting substances have been widely reported (104, 108). It can play an important role in regulating plant diversity (19). Putnam & Duke (197) introduced the concept of utilizing the alleiopathic crop residues for weed control in crops.

Parthenium itself is known to be highly allelopathic, and is capable of reducing growth and germination of native flora. Several beneficial plants are also known to be allelopathic and have the potential to compete and displace parthenium. Most of the research so far on using beneficial plants to competitively displace parthenium has been restricted to India (27). Several plant species, including *Cassia serecia*, *Cassia tora*, *Croton bonapladanium*, *Croton sparsiflorus*, *Cannavis sativa*, *Hyptis suaveolens*, *Amaranthus spinosus*, *Sida acuta*, *Tagetus erecta*, *Tephrosia purpurea*, *Stylosanthes scraba*, *Cassia auriculata*, etc, have been reported to be causing competitive displacement of parthenium (67, 122). The use of these plants will never be feasible and practical in crop and orchard fields.

CULTURAL PRACTICES

Farmers generally combine cultural practices to minimize the number of weeds in the crop, suppress competition by surviving weeds, and reduce weed seed production. Cultural controls may include any practice that make the crop more competitive with weeds. Optimizing cultural controls will give the crop a competitive edge over weeds. Several researchers have advocated growing competitive crops to suppress parthenium but as parthenium is mainly a waste land weed, the scope of this practice is limited to only certain situations.

LEGISLATIVE CONTROL

There have been made considerable efforts in our country for the effective management of this obnoxious weed by legislative acts, for example, Government of Karnataka was first to issue a notification on 23rd Oct., 1975, declaring parthenium a noxious weed in terms of section 3 read with sub section seven of the section two of the Karnataka Agricultural Pests and Disease Act, 1968. Under this act, to remove parthenium, notices were issued in Bangalore once or twice during the eighties by Municipal Corporation but were not followed up. Thus, in spite of this comprehensive act, it was a total failure to get the weed eradicated from Bangalore or elsewhere.

In Australia, parthenium is a declared weed of national significance (87), which makes the sale, movement or distribution of parthenium within Australia strictly prohibited. In Queensland, there are several 'wash-down' facilities at strategic points and vehicles traveling from infested areas are required to clear their vehicles before traveling to parthenium-free areas. It is legally mandatory for suppliers of stocks, machinery, soil, water or other products from areas of known parthenium infestations to declare that the material they supply is free of parthenium. The Australian Quarantine and Inspection Service prohibit the introduction of parthenium as a nursery stock or through contaminated seed imports. In South Africa and Sri Lanka also parthenium is a declared weed, where movement of parthenium is prohibited in both rural and urban areas. However, in other countries no such nationwide legislative mechanism exists to prevent further spread of parthenium.

UTILIZATION AS A METHOD OF CONTROL

Although parthenium is a major weed and is responsible for many of the problems discussed above, many individuals, groups and institutions have been able to turn the problem around and find useful application for the plant. Parthenium has been well documented for its insecticidal (22, 35, 107), nematicidal (8), antifungal (36, 84) and herbicidal (11, 12, 13, 99) properties. Production of biogass (38), low quality paper (120) and compost (109) has also been widely recommended. An interesting case of phago-stimulatory effect of pathenium extract has also been reported by Patil et al. (96) who reported that when mulberry (*Morus* sp.) leaf diet of mulberry silkworm was supplemented with different concentrations of parthenium leaf extracts, there was a stimulatory effect on their feeding and leaf utilization efficacy. The parthenium extract supplements have resulted in production of bigger cocoons and improving all economic trials. Such recommendations have also been made by Singhal et al. (113) and Murugan et al. (82).

BIOLOGICAL CONTROL

Biological control has been considered as the most cost-effective, environmentally safe and ecologically viable option for the control of parthenium (1, 2, 29, 30). In last three to four decades lot of emphasis has been given to control parthenium through various biocontrol agents like microbial pathogens, insects and botanicals. Earliest efforts on biological control of parthenium were initiated by scientists working in Australia in 1975 and surveys for natural enemies were soon underway in the neotropical centre of origin of parthenium, coordinated by the Queensland Department of Lands (QDL) and the Commonwealth Institute of Biological Control (CIBC, now IIBC). Subsequently, entomological surveys and the screening of selected arthropods were undertaken in both Brazil/ Argentina (1977-1981) and Mexico (1978-1983) by the scientists of both QDL and CIBC (32). The results of this research are reviewed by McClay (71) and McFadyen (73), and later updated to include additional surveys which were eventually phased out in 1991 (McFadyen, 1990). Since then, nine species of insects and two rust fungi (Table 1) have been introduced (25, 74, 77). Among these, the stem-galling moth *Epiblema strenuana* Walker (Lepidoptera: Tortricidae) (72), the leaf-feeding beetle *Zygogramma bicolorata* Pallister (Coleoptera: Chrysomelidae) (75), the stem-boring weevil *Listronotus setosipennis* Hustache (Coleoptera: Curculionidae) (125), the seed-feeding weevil *Smicronyx lutulentus* Dietz (Coleoptera: Curculionidae) (75), the leaf-mining moth *Bucculatrix parthenica* Bradley (Lepidoptera: Bucculatricidae) (73), the stem-galling weevil *Conotrachelus albocinereus* Fiedler (Coleoptera: Curculionidae) (76) and the winter rust *Puccinia abrupta* var. *partheniicola* (Jackson) Parmelee (Uredinales) (95) are known to be established in the field. According to reports from Australia, in central Queensland, there was a 40% increase in grass biomass in 1997 due to 96% defoliation by the leaf-feeding beetle *Z. bicolorata* and galling in 100% of the plants by the moth *E. strenuana*. In north Queensland, grass biomass increased by 52% in 1998 due to reduced parthenium seedling emergence, and by 45% in 2000 (26), due to the combined effects of galling by the moth *E. strenuana* and the establishment of the summer rust *P. melampodii* in 72% of the plants.

Table 1. List of potential biocontrol agents of parthenium

Biological control agent	Order: Family	Native Country	Released Country
Epiblema strenuana Walker (Stem-galling moth)	Lepidoptera: Tortricidae	Mexico	Australia
Platphalonidia mystica (Razowski & Becker) (Stem-boring moth)	Lepidoptera: Tortricidae	Argentina	Sri Lanka
Carmenta nr *ithacae* (Beutenmüller) (Root-boring moth)	Lepidoptera: Sessidae	Mexico	Australia
Bucculatrix parthenica Bradley (Leaf-mining moth)	Lepidoptera: Bucculatricidae	Mexico	Australia
Zygogramma bicolorata Pallister (Leaf-feeding beetle)	Coleoptera: Chrysomelidae	Mexico	Australia, India
Listronotus setosipennis Hustache (Stem-boring weevil)	Coleoptera: Curculionidae	Argentina and Brazil	Australia
Smicronyx lutulentus Dietz (Seed-feeding weevil)	Coleoptera: Curculionidae	Mexico	Pakistan, Australia
Conotrachelus albocinereus Fiedler (Stem-galling weevil)	Coleoptera: Curculionidae	Mexico	Australia
Stobaera concinna (Stål). (Parthenium sap-feeding planthopper)	Homoptera: Delphacidae	Mexico	Australia
Puccinia abrupta partheniicola Parmelee (Winter rust)	Basidiomycotina: Uredinales	Mexico	Australia, South Africa
Puccinia melampodii Dietel and Holway (Summer rust)	Basidiomycotina: Uredinales	Mexico	Australia Ethiopia, India, South Africa, Kenya

India was also among the earliest countries with parthenium infestation, to attempt biological control (77). Based on well documented success by insects from the native home of parthenium, efforts on biological control of parthenium were initiated in India in 1983 at Indian Institute of Horticultural Research at Bangalore and three insects namely *Z, bicolorata, S. lutulentus* and *E. strenuana* were imported *into India* (110). *Z. bicolorata* was obtained from Mexico in 1983 and released in 1984 after further testing (44), it has established and has spread from Bangalore to different parts of the country where it had been reported to give effective control (Fig. 4B and Fig. 5). *S. lutulentus* could not be multiplied in the laboratory. *E. strenuana* was obtained from Queensland in 1985, but in tests in India it was found to lay and develop on niger seed [*Guizotia abyssinica* (L.)] which is an important oilseed crop in India, so the colony was destroyed (45). In about last 25 years, *Z. bicolorata* has spread throughout the country both naturally and deliberately spread by various government and non-government agencies. It has shown great potential by relieving acres and acres of land from parthenium infestation. Both the adult and larvae feed on parthenium leaves. The adult female lays eggs either singly or in clusters on leaves, flower heads, stems or on the terminal and axillary buds. The emerging larvae feed voraciously on young leaves causing 85-100% defoliation resulting in reduced weed density and seed bank. *Z. bicolorata* caused 85–100% defoliation, resulting in up to 99.5% reduction in the parthenium weed density in Bangalore region (46, 47). Similar impacts have been observed in other areas in India also (31). Defoliation by the leaf-feeding beetle *Z. bicolorata* also resulted in the re-establishment of native vegetation (47, 115). Although *Z. bicolorata* is a good biocontrol agent of parthenium yet it has certain drawbacks, which hamper its biocontrol efficacy, for instance low to nil population of the beetle in the field when parthenium is at its early growth stage. As a result when population of the beetle rises the host weed has already reached its flowering stage to produce large number of seeds. One of the main reasons of this non-coordination is the diapause in *Z. bicolorata*. Diapause is a unique feature in insects to avoid unfavourable condition but *Z. bicolorata* undergoes diapause inspite of all the available favourable conditions of food and climate (46). This property hampers its biocontrol efficacy as parthenium can grow throughout the year subject to moisture availability. Other than the insect, several phytopathogens, viz., *Fusarium pallidorseum* (54), *Puccinia abrupta* var. *partheniicola* *idium parthenii* Sathyaprasad & Usharani (69), *Fusarium* *orme* (70), *Phoma herbarum* (124), *Colletotrichum* *iumas* (37), *Colletotrichum gloeosporioides* (Penz.) Penz. & ec. (59), *Exserohilum rostratum* (Drechsler) K.J. Leonard & Suggs (101), *Myrothecium roridum* (93, 94), etc (24, 49, 60, 64, 98) have also been reported from India with varying degree of pathogenicity.

Fig. 5. Impact of the Mexican Beetles on parthenium

In Sri Lanka, biological control strategies were undertaken in 2003 with the importation of stem gall-inducing *E. strenuana* and *Puccinia melampodii* Dietel and Holway (Uredinales) from Australia (48). Attempts to establish colonies of *Z. bicolorata* and *E. strenuana* in quarantine in Papua New Guinea have so far resulted in utter failure. In South Africa, a biological control program on parthenium was initiated in 2003 with prioritization of *Z. bicolorata*, *L. setosipennis*, *E. strenuana*, and *P. melampodii* (91, 115) chosen for their impact on parthenium and likely suitability for local climatic conditions within the invasive range of parthenium, which experiences a distinct dry season.

Other than the insects and plant pathogens several plant extracts have also been found to *be* hold great potential against the weed. Singh et al. (2005) studied the herbicidal effect of volatile oils from leaves of *Eucalyptus citriodora* against the noxious weed *P. hysterophorus* and found that a concentration of 5.0 nL ml-1 *Eucalyptus* oil completely inhibited the germination. Similar observations have also been reported using various plant extracts by various workers (5, 6, 43, 56).

FUTUROLOGY

Inspite of such tremendous efforts on parthenium management in last 40 years in parthenium infested areas of the world, especially Australia, India and South Africa, the campaign against parthenium has resulted in only partial and inadequate control. Predicamently, till now we have just been able to release only one biological control agent in India while in Australia about 11 biocontrol agents including insects and plant pathogens have been released with varying results. In India more efforts are needed to be made in identifying potential microbial *other* agents efficacious and tolerant to the climatic conditions. Nucleus colonies of *L. setosipennis, S. concinna* and *B. parthenica* could be obtained from Mexico (the native range) or Queensland (area of introduction). Other countries with an increasing parthenium problem should also consider the introduction of these insects, after further host-testing as necessary. There is an urgent need for adopting an integrated approach for the management of this dreaded weed which has become a great menace for agriculture production and posing simultaneously serious health hazards even skin cancer cases have been reported from AIIMS, the Delhi based premier medical institution. It would be most appropriate to apply the various treatments genuinely such as chemical pesticides, bioagents, botanicals insect pests etc at the seedling stage of the weed plants and care should be taken that the land is recolonized quickly by the grasses and other agriculturally important crop plants thereby leaving no room for the weed to come back. There remains a lesson to learn that no single management option would be effective to control this weed across all habitats, therefore it is a need of hour to integrate various management options (e.g., grazing management, competitive displacement, cultural practices) with classical biological control as a core management option. The modern agriculture research, in fact, is confronting with a great challenge to develop the eco-friendly strategies to eradicate this ghostly weed, which has posed several problems and agriculture production has gone down because of this weed occupying large areas of cultivable fields. The Indian geographical conditions are typical and there is a need that the research projects have to be undertaken extensively at different locations to find out the strategic means well suited to that particular climatic and geographical situation. Conclusively, pathologists, entomologists, agronomists, biologists, microbiologists should join hands and devote themselves to find out the perfect control measures to completely get rid of this menace and that would perhaps be the breakthrough research in Science of the century. Until now only stray attempts have been made in this direction with negligible or dim results of no consequence. This review article is devoted to create awareness amongst scientists of all seminaries to concentrate and focus upon the single menace which has been afflicting the agriculture and human health equally in horrific manner.

REFERENCES

1. Adamson DC. 1996. Introducing dynamic considerations when economically evaluating weeds. Master's Thesis, University of Queensland, Brisbane. 105 p.
2. Adamson DC, Bray S. 1999. The economic benefit from investing in insect biological control of parthenium weed (*Parthenium hysterophorus*). Queensland, Australia: University of Queensland. 44 p.
3. Adhikari B, Tiwari S. 2004. *Parthenium hysterophorus* L.: highly allergic invasive alien plant growing tremendously in Nepal. *Botanica Orientalis*. 4:36–37.
4. Adkins SW, Navie SC, Dhileepan K. 2005. Parthenium weed in Australia: research progress and prospects. In: Proceedings of the Second International Conference on Parthenium Management, (ed. Ramachandra Prasad TV, Nanjappa HV, Devendra R, Manjunath A, Subramanya, Chandrashekar SC, Kiran Kumar VK, Jayaram KA, Prabhakara Setty TK) Bangalore, India: University of Agricultural Sciences. pp. 11–27.
5. Anjum T, Bajwa R, Javid A. 2005. Biological Control of Parthenium I: Effect of

Imperata cylindrical on Distribution, Germination and Seedling Growth of *Parthenium hysterophorus* L. *Int. J. Agri. Biol.* 7(3): 448–450.

6. Arshad J, Anjum T. 2006. Control of *Parthenium hysterophorus* L. by aqueous extracts of allelopathic grasses. *Pak. J. Bot.* 38(1): 139 – 145.

7. Bagyanarayana G, Manoharachary C. 1997. Studies on *Puccinia abrupta* var. *partheniicola* a potential mycoherbicide. In: Proceedings of the first international Conference on Parthenium Management, Volume II, University of Agricultural Sciences, Dharwad, India, (ed. Mahadevappa M, Patil VC) pp. 95–96.

8. Bala SK, Bhattacharya P, Mukerjee KS, Sukul NC. 1986. Nematicidal properties of the plants *Xanthium strumarium* and *Parthenium hysterophorus*. *Environ. Ecol.* 4(1): 139-141.

9. Balyan RS, Yadav A, and Malik RK 1998. Efficacy of herbicides and carrier volumes on the carrot weed (*Parthenium hysterophorus* Linn). In: Proceedings of the 5th International Symposium on Adjuvants for Agrochemicals. Challenges and Opportunities; Memphis, Tennessee: International Soci-ety on Adjuvants for Agro-chemicals. (ed.McMullan P.) 1: 182–186.

10. Balyan RS, Yadav A, Malik RK and Pahwa SK. 1996. Chemical control of *Parthenium hysterophorus*. *Pestology*. 14:12–14.

11. Batish DR, Kohli KH, Singh HP, Saxena DB. 1997. Studies on herbicidal activity of parthenin, a constituent of *Parthenium hysterophorus*, towards billgoat weed (*Ageratum conyzoides*). *Curr. Sci*. 73: 369–371.

12. Batish DR, Singh HP, Saxena DB, Kohli RK. 2002b. Weed suppressing ability of parthenin – a sesquiterpene lactone from *Parthenium hysterophorus*. *New Zealand Pl Prot.* 55: 218–221.

13. Batish DR, Singh HP, Kohli RK, Saxena DB, Kaur S. 2002a. Allelopathic effects of parthenin against two weedy species, *Avena fatua* and *Bidens pilosa*. *Env. and Exp. Bot.* 47: 149–155.

14. Belz RG, Reinhardt CF, Foxcroft LC, Hurle K. 2007. Residue allelopathy in *Parthenium hysterophorus* L. - does parthenin play a leading role? *Crop Prot.* 26: 237–245.

15. Bhowmik PC, Sarkar D. 2005. *Parthenium hysterophorus*: Its world status and potential management. In: Proceedings

of the Second International Conference on Parthenium Management, (ed. Ramachandra Prasad TV, Nanjappa HV, Devendra R, Manjunath A, Subramanya, Chandrashekar SC, Kiran Kumar VK, Jayaram KA, Prabhakara Setty TK), 05-07 December 2005. University of Agricultural Sciences, Bangalore, India. pp 1-5.

16. Butler J.E. (1984). Longevity of *Parthenium hysterophorus* L. seed in soil. *Aust. Weeds* 3:6.

17. CABI (2004). Parthenium fact sheet. In Crop Protection Compendium. CD-ROM. Wallingford, UK: CAB International.

18. Chippendale JF, Panetta FD. 1994. The cost of parthenium weed to the Queensland cattle industry. *Pl. Prot. Qt.* 9:73–76.

19. Chou CH, Lee YF. 1991. Allelopathic dominance of *Miscanthus transnorrisonensis* in an alpine grassland community in Taiwan. *J. Chem. Ecol.*, 17: 2267-2281.

20. Da Silva MC, Izidine S, Amuda AB. 2004. A Preliminary Checklist of the Vascular Plants of Mozambique. South African Botanical Diversity Network Report 30. Pretoria, South Africa: SABONET, 185 pp.

21. Dale IJ. 1981. Parthenium weed in the Americas. Aust. Weeds 1: 8-14.

22. Datt S, Saxena DB. 1996. Antifeedant efficacy of parthenin and its dehydrate product. *Weed News*. 3: 37-44.

23. Dhanraj RE, Mitra MK. 1976. Control of *Parthenium hysterophorus* L. with diquat. *PANS*. 22: 269-272.

24. Dhawan SR, Gupta SK. 1997. Spermoplane mycoflora of *Parthenium hysterophorus* L. and its weed control potential. *Adv. Pl. Sci*. 10: 95-101.

25. Dhileepan K, McFadyen R.E. 1997. Biological control of parthenium in Australia – progress and prospects. In: First international conference on parthenium management. (ed. Mahadevappa M, Patil VC).University of Agricultural Sciences, Dharwad, India. pp. 40–44

26. Dhileepan K. 2007. Biological control of parthenium (*Parthenium hysterophorus*) in Australian Rangeland translates to improve grass production. *Weed Sci.* 55: 497-501.

27. Dhileepan K. (2009 Managing parthenium weed across diverse landscapes: prospects and limitations. In: Management of invasive weeds.

Chapter 12. Springer Science, pp. 227-260.

28. Dhileepan K, Strathie L. 2009. *Parthenium hysterophorus* L. (Asteraceae). In: Biological Control of Tropical Weeds Using Arthropods. Cambridge University Press. (eds. Muniappan R, Reddy GVP, Raman A.) pp 274-318.

29. Dhileepan K. 2001. Effectiveness of introduced biocontrol insects on the weed *Parthenium hysterophorus* (Asteraceae) in Australia. *Bull. Entomol. Res.* 91: 167-176.

30. Dhileepan K, Madigan B, Vitelli M, McFadyen R, Webster K, Trevino M. 1996. A new initiative in the biological control of parthenium. In: Proceedings of the Eleventh Australian Weeds Conference. (ed Shepherd RCH) Weeds Society of Victoria, Australia. pp. 309–312.

31. Dhiman SC,Bhargava ML.2005. Seasonal occurrence and bio-control efficacy of *Zygogramma bicolorata* Pallister (Coleoptera: Chrysomelidae) on *Parthenium hysterophorus*. *Ann. Pl. Prot. Sci.* 13:81–84.

32. Evans HC. 1997. *Parthenium hysterophorus*: a review of its weed status and the possibilities for biological control. *Biocontr. News Info.* 18: 89-98.

33. Fessehaie R, Chichayibelu M, Giorgis MH. 2005. Spread and ecological consequences of *Parthenium hysterophorus* in Ethiopia. *Arem.* 6: 11–21.

34. Frew M, Solomon K, Mashilla D. 1996. Prevalence and distribution of Parthenium hysterophorus L in eastern Ethiopia. In: Proceedings of the First Annual Conference of the Ethiopian Weed Science Society, held 24–25 November 1993, Addis Ababa, Ethiopia, (ed. Fessehaie R.) *Arem.* 1: 19–26.

35. Gajendran G, Gopalan M. 1982. Notes on the antifeedant activity of *Parthenium hysterophorus* Linn. on *Spodoptera litura* Fabricious (Lepidoptera: Noctuidae). *Ind. J. Agri. Sci.* 52: 203-205.

36. Ganeshan G, Jayachandra. 1993. Antifungal activity of parthenin. *Ind.Phytopath.* 46: 193-4.

37. Gayatri S, Pandey AK. 1998. Preliminary assesment of *Colletotrichum dematiumas* as potential mycoherbicide against *Parthenium hysterophorus* L. In: Proceedings of the First International Conference on Parthenium Management, (ed. Mahadevappa M, Patil VC.) University of Agricultural Sciences,

Dharwad, 6-8 Oct., 1997. 2: 92-94.

38. Gunasheelan VN. 1987. Parthenium as an additive with cattle manure in biogas production. *Biol. Wastes.* 21(3): 95-202.

39. Haseler WH 1976. *Parthenium hysterophorus* in Australia. *PANS.* 22: 515-517.

40. Hilliard OM 1977. Compositae in Natal. Pietermaritzburg, South Africa: University of Natal Press, 659 pp.

41. Holm LG, Donald P, Pancho JV, Herberger JP. 1977. The World's Worst Weeds: Distribution and Biology. The University Press of Hawaii, Honolulu, Hawaii. 609 pp.

42. Javaid A, Anjum T. 2005. Parthenium hysterophorus L. – a noxious alien weed. Pak. J. Weed Sci. Res. 11: 1–6.

43. Javaid A, Anjum T, Bajwa R. 2005. Biological Control of *Parthenium* II: Allelopathic Effect of *Desmostachya bipinnata* on distribution, and early seedling growth of *Parthenium hysterophorus* L. Int. J. Biol. Biotech. 2(2): 459-463.

44. Jayanth KP. 1987a. Introduction and establishment of *Zygogramma bicolorata* on *Parthenium hysterophorus* at Bangalore, India. Curr. Sci. 56: 310-311

45. Jayanth KP. 1987b. Investigations on the host-specificity *of Epiblema strenuana* (Walker) (Lepidoptera: Tortricidae), introduced for biological control trials against *Parthenium hysterophorus* in India. J. Biol. Contr. 1: 133-137

46. Jayanth KP, Bali G. 1994. Biological control of parthenium by the beetle *Zygogramma bicolorata* in India. FAO Pl. Prot. Bull. 42: 207–213.

47. Jayanth KP, Visalakshy PNG. 1996. Succession of vegetation after suppression of parthenium weed by *Zygogramma bicolorata* in Bangalore, India. *Biol. Agric. Hort.* 12: 303–309.

48. Jayasuriya AHM. 2005. Parthenium weed – status and management in Sri Lanka. In: Second International Conference on Parthenium Management (ed. Ramachandra Prasad TV, Nanjappa HV, Devendra R.) 05-07 Dec. 2005. University of Agricultural Sciences, Bangalore, India, pp. 36–43.

49. Jeyalakshmi C, Doraisamy S, Paridasan VV. 2005. Biodiversity of fungal pathogens affecting *Parthenium hysterophorus* L. in Tamil Nadu, India. In: (ed. Prasad TVR, Nanjappa HV, Devendra R.) Second international conference on parthenium manage-

ment, 05-07 Dec. 2005. University of Agricultural Sciences, Bangalore, India, pp. 98–101.

50. Joel DM, Liston A. 1986. New adventive weeds in Israel. *Israel J. Bot.* 35: 215–223.

51. Joshi S. 1989. An economic evaluation of control methods for *Parthenium hysterophorus* Linn. Biol. *Agri. Hort.* 6: 285-291.

52. Kadhane DL, Jangde CR, Sadekar RD, Joshirao MK. 1992. *Parthenium* toxicity in buffalo calves. *J Soils Crops.* 21: 69–71.

53. Kanchan SD, Jayachandra. 1977. Post-emergent chemical control of congress grass. *Pesticides* 11: 55-56.

54. Kauraw LP, Chile A, Bhan VM. 1997. Evaluation of *Fusarium pallidorseum* (Cooke) Sacc. for the biocontrol of *Parthenium hysterophorus* L. In: First International Conference on Parthenium Management. Volume I, University of Agricultural Sciences, Dharwad, India, (ed. Mahadevappa M, Patil VC.) pp. 70–74.

55. Khosla SN and Sobti SN. 1979. Parthenium – a national health hazard, its control and utility – a review. *Pesticides.* 13: 121-127.

56. Kohli RK, Batish DR, Singh HP. 1998. Eucalypt oils for the control of Parthenium (*Parthenium hysterophorus* L.). *Crop Prot.* 17(2): 119-122.

57. Krishnamurthy K, Prasad TVR, Muniyappa TV and Rao V. 1977. Parthenium, a new pernicious weed in India. University of Agricultural Science Technical Series. 17: 46 pp.

58. Krishnamurthy K, Prasad TVR, Muniyappa TV and Rao V. 1977. Parthenium, a new pernicious weed in India. University of Agricultural Science Technical Series. 17: 66 pp.

59. Kumar CS, Rao KV. 1977. Two new leaf spot diseases. *Indian Phytopathl.* 30: 118-120.

60. Kumar PS, Evans HC. 2005. The mycobiota of *Parthenium hysterophorus* in its native and exotic ranges: opportunities for biological control in India. In: Proceedings of the Second International Conference on Parthenium Management, (ed. Ramachandra Prasad TV, Nanjappa HV, Devendra R, Manjunath A, Subramanya, Chandrashekar SC, Kiran Kumar VK, Jayaram KA, Prabhakara Setty TK.). 05-07 December 2005. University of

Agricultural Sciences, Bangalore, India. pp. 107–113.

61. Kumar S, Rohatgi N. 1999. The role of invasive weeds in changing floristic diversity. *Ann. Forestry*. 7: 147-150.

62. Labrada R. 1990. El manejo de malezas en areas de hortalizas y frijol en Cuba. *X Congreso ALAM, LA Habana, Abril 1-14,* Vol. II: pp. 1-16.

63. Lonkar A, Mitchell JC, Calnan CB. 1974. Contact dermatitis from *Parthenium hysterophorus.* Trans. St John's Derma. Soc. 60: 43-53.

64. Luke P. 1976. Fungi in the root region of *Parthenium hysterophorus* L. *Curr. Sci.* 45: 631-632.

65. Macdonald IAW, Reaser JK, Bright C, Neville LE, Howard GW, Murphy SJ, Preston G. (eds.) (2003). Invasive Alien Species in Southern Africa: National Reports and Directory of Resources. Cape Town, South Africa: Global Invasive Species Programme. 125 pp.

66. Mahadevappa M. 1996. *Parthenium,* Prasaranga, University of Mysore, Mysore, 66 pp.

67. Mahadevappa M. 1999. Parthenium and its management. University of Mysore. Mysore, India: 66 pp.

68. Mahadevappa M. 1997. Ecology, distribution, menace and management of parthenium. In: First International Conference on Parthenium Management, (eds. Mahadevappa M, Patil VC.). Dharwad, India: University of Agricultural Sciences, pp. 1–12.

69. Manickam K, Doraisamy S, Sankaran S. 1997a Epidemiology and host range studies on powdery mildew (*Oidium parthenii* S & U) of *Parthenium hysterophorus* L. In: First international conference on parthenium management, Volume I, (ed. Mahadevappa M, Patil VC.) University of Agricultural Sciences, Dharwad, India. 75–76.

70. Manickam K, Doraisamy S, Sankaran S. 1997b Inhibitory effect of *Fusarium moniliforme* on seed germination of *Parthenium hysterophorus* L. In: (eds Mahadevappa M, Patil VC) First International conference on parthenium management, Volume II, University of Agricultural Sciences, Dharwad, India, pp. 90–91.

71. McClay AS, Palmer WA, Bennett FD, Pullen KR. 1995. Phytophagous arthropods associated with *Parthenium hysterophorus* (Asteraceae) in

North America. *Environ. Entomol*. 24: 796–809.

72. McClay AS. 1987. Observations on the biology and host specificity of *Epiblema strenuana* [Lepidoptera, Tortricidae] a potential control agent for *Parthenium hysterophorus* [Compositae]. Entomophaga 32: 23–34.

73. McClay AS, McFadyen RE, Bradley JD. 1990. Biology of *Bucculatrix parthenica* Bradley sp. n. (Lepidoptera: Bucculatricidae) and its establishment in Australia as a biological control agent for *Parthe-nium hysterophorus* (Asteraceace). *Bull. Entomol. Res*. 80: 427–432.

74. McFadyen REC. 1985. Parthenium weed and human health in Queensland. *Austr. Family Physician*. 24 (8): 1455-1458.

75. McFadyen RE, McClay AR. 1981. Two new insects for the biological control of parthenium weed in Queensland. In: Sixth Australian Weeds Conference. (ed. Wilson BJ, Swarbrick JD) Weed Science Society of Queensland, Australia. pp. 145–149

76. McFadyen RE. 2000. Biology and host specificity of the stem galling weevil *Conotrachelus albocinereus* Fielder (Col: Curculionidae), a potential bioco parthenium weed *hysterophorus* L. (As in Queensland, Aus *Biocontr. Sci. Tech*. 10: 19 200.

77. McFadyen REC. 1992. Biological control against parthenium weed in Australia. *Crop Prot*. 11: 400-407.

78. Mishra JS, Bhan VM. 1995. Efficacy of sulfonylurea herbicides against *Parthenium hysterophorus* L. *Ind. J. Weed Sci*. 27: 45–48.

79. Mishra JS, Bhan VM. 1994. Efficacy of sulfonyl urea herbicides against *Parthenium hysterophorus*. *Weed News*. 1: 16.

80. Mueller TC, Womac AR. 1997. Effect of formulation and nozzle type on droplet size with isopropylamine and trimesium salts of glyphosate. *Weed Tech*. 11: 639–643.

81. Muniappa TV, Ramchandra Prasad TV, Krishnamurthy K. 1980. Comparative efficacy and economics of mechanical and chemical methods of control of *Parthenium hysterophorus* Linn. *Ind. J. Weed Sci*. 12: 137–144.

82. Murugan K, Jeyabalan D, Kumar N.S, Nathan S.S. and Sivaprakasam. (1997). Growth promoting effects of plant products on silkworm, a

... approach. *J.* ... 57: 740-745.

...than TR, Ananth M, ...ayana Swamy M, Rajendra Babu M, Mangala A, Subba Rao PV. 1977. Toxicity of *Parthenium hysterophorus* L. Curr. Sci. 46: 15–16.

84. Narayana YD, Mughogho LK, Bandhyopadhyay R. 1994. Antifungal activity of some plant extracts. *Ind. J. Forestry*. 17: 10-14.

85. Narwal SS. 1994. Allelopathy in Crop Production, Scientific Publisher, Jodhpur, India. 288 pp.

86. Nath R. 1981. Note on the effect of *Parthenium* extract on seed germination and seedling growth in crops. *Ind. J. Agri. Sci.* 51: 601- 603

87. Navie SC, McFadyen, RE, Panetta FD, Adkins SW. 1996. The biology of Australian weeds 27. *Parthenium hysterophorus* L. *Pl. Prot. Quart*. 11: 76-88.

88. Navie SC, Panetta FD, McFadyen RE, Adkins SW. 1998. Behaviour of buried and surface-lying seeds of parthenium weed (*Parthenium hysterophorus* L.). *Weed Res.* 38: 338–341.

89. Njoroge JM. 1986. New weeds in Kenya coffee. A short communication. *Kenya Coffee*. 51. 333–335.

90. Njoroge JM. 1991. Tolerance of *Bidens pilosa* L. and *Parthenium hysterophorus* L. to paraquat (Gramoxone) in Kenya. *Kenya Coffee*. 56: 999-1001.

91. Ntushelo K, Wood AR. 2008. Supplementary host specificity testing of *Puccinia melampodii*, a biocontrol agent of *Parthenium hysterophorus* (abstract). In XII International Symposium on Biological Control of Weeds, (ed. Julien MH, Sforza R, Bon MC, Evans HC, Hatcher PE, Hinz HL, Rector BG) held 22–27 April 2007, La Grande Motte, France, Wallingford, UK: CAB International.

92. Ovies J, Loretta L. 1988. Transmisión de *Xanthomonas campestris pv. phaseoli* mediante un hospedante silvestre. *Ciencia y Técnica Agricultura, Protección Plantas* 11:23-30.

93. Pandey AK, Hasija SK, Rajak RC. 1990. *Myrothecium roridum* Tode ex Fr.- a new pathogen of *Parthenium hysterophorus* L. with biocontrol potential. *Nat. Acad. Sci. Letters*. 13: 369-370.

94. Pandey AK, Luka BM, Hasija SK, Rajak RC. 1992. Evaluation of *Myrothecium roridum* Tode for the management of *Parthenium hysterophorus* L:

inoculums rate studies. *Proc. Nat. Acad. of Sci*, India. 15: 111-112.

95. Parker A, Holden ANG, Tomley AJ. 1994. Host specificity testing and assessment of the pathogenicity of the rust, *Puccinia abrupt* var. *partheniicola*, as a biological control agent of parthenium weed (*Parthenium hysterophorus*). Pl. Pathol. 43: 1–16.
96. Patil RR, Mahadevappa M, Mahesh HM, Patil VC. 1997. Phagostimulant effect of parthenium on mulberry silkworm (*Bombyx mori* L.). In: Proceedings of the First International Conference on Parthenium Management, Volume II (eds. Mahadevappa M, Patil VC) University of Agricultural Sciences, Dharwad, India, pp. 81-85.
97. Putnam AR, Duke WB. 1979. Biological suppression of weeds: evidence for allelopathy in accession of cucumber. *Sci*. 185: 370-372.
98. Rajak RC, Farakya S, Hasija SK, Pandey AK. 1990. Fungi associated with congress weed (*Parthenium hysterophorus*). *Proc. Nat. Acad. of Sci., India*. 60: 165-168.
99. Ramaswami PP. 1997. Potential use of parthenium. In: First International Conference on Parthenium Management. (ed. Mahadevappa M, Patil VC) University of Agricultural Sciences, Dharwad, Karna-taka, 6-8 October, 1997. 77-80.
100. Rao AP, Rao AS. 1987. New fungal disease of some weeds. *Ind. Bot. Rep*. 6: 38.
101. Rao RS. 1956. Parthenium – a new record for India. *J. Bombay Nat. Hist. Soc.* 54: 218–220.
102. Reinhardt C, Kraus S, Walker F, Foxcroft L, Robbertse P, Hurle K. 2004. The allelochemical parthenin is sequestered at high level in capitate-sessile trichomes on the leaf surface of *Parthenium hysterophorus*. *J. Pl. Dis. Prot*. Special Issue XIX, 253–261.
103. Reinhardt C, Van der Laan M, Belz RG, Hurle K, Foxcroft L. 2006. Production dynamics of the allelochemical parthenin in leaves of *Parthenium hysterophorus* L. *J. Pl. Dis. Prot.* Special Issue. XX: 427–433.
104. Rizvi SJS, Rizvi V, Tahir M, Rahimian MH, Atri A. 2000. Genetic variation in allelopathic activity of wheat (*Triticum aestivum* L.) genotype. Wheat Information Service Number. 91: 25-29.
105. Rollins RC. 1950. The guayule rubber plant and its relatives. The Gray Herbarium of Harvard University, Cambridge, MA. 172: 1-72.

106. Shabbir A, Bajwa R. 2006. Distribution of parthenium weed (*Parthenium hysterophorus* L.), an alien invasive weed species threatening the biodiversity of Islamabad. *Weed Biol. Manage.* 6: 89–95.

107. Shah DSM. 1996. Antifeedant effect of two indigenous plant extracts against larvae of *Stomopteryx subsecivella* (Lepidoptera) on groundnut. *J. Ecobiol.* 8: 293-298.

108. Shaukat SS, Khan D, Ali ST. 1983. Suppression of herbs by *Inula grantioides* Boiss, in Sind desert, *Pakistan. Pak. J. Bot.* 15: 43-67.

109. Shukla KS, Singh SP. 1999. Research in composite weed-achievement during last 50 years. Forest Research Institute, Dehradun, India. 33pp

110. Singh NP. 1983. Potential biological control of *Parthenium hysterophorus* L. *Curr. Sci.* 52: 644.

111. Singh S, Yadav A, Balyan RS, Malik RK, Singh M. 2004. Control of Ragweed Parthenium (*Parthenium hysterophorus*) and Associated Weeds. *Weed Tech.* 18(3): 658-664.

112. Singh, HP, Batish DR, Setia N, Kohli RK. 2005. Herbicidal activity of volatile oils from *Eucalyptus citriodora* against *Parthenium hysterophorus*. *Ann. App. Biol.* 146: 89-94.

113. Singhal BK, Rajan MV, Rao YRM. 1998. Weed turns a boon to silk. *Ind. Textile J.* 108: 60-62.

114. Sridhara S, Basavaraja BK, Ganeshaiah KN. 2005. Temporal variation in relative dominance of *Parthenium hysterophorus* and its effect on native biodiversity. In: Proceedings of the Second International Conference on Parthenium Management, (ed. Ramachandra Prasad TV, Nanjappa HV, Devendra R, Manjunath A, Subramanya, Chandrashekar SC, Kiran Kumar VK, Jayaram KA, Prabhakara Setty TK), 05-07 December 2005. University of Agricultural Sciences, Bangalore, India. pp. 240–242.

115. Strathie LW, Wood AR, van Rooi C, McConnachie AJ. 2005. *Parthenium hysterophorus* (Asteraceae) in southern Africa, and initiation of biological control against it in South Africa. In: Proceedings of the Second International Conference on Parthenium Management, (eds. Ramachandra Prasad TV, Nanjappa HV, Devendra R, Manjunath A, Subram-anya, Chandrashekar SC, Kiran Kumar VK, Jayaram KA, Prabhakara Setty TK). 05-07

December 2005. University of Agricultural Scien-ces, Bangalore, India. pp. 127–133.

116. Sushilkumar and Varshney JG. 2007. Gajarghas ka jaivik niyantran: vartaman sththi evam sambhavnaye (Biological control of congress grass: present status and future possibilities). National Research Centre for Weed Science, Jabalpur, India.

117. Tamado T, Milberg P. 2004. Control of parthenium (*Parthenium hysterophorus*) in grain sorghum (*Sorghum bicolor*) in the smallholder farming system in eastern Ethiopia. *Weed Tech*. 18: 100–105.

118. Tamado T, Milberg P. 2000. Weed flora in arable fields of eastern Ethiopia with emphasis on the occurrence of *Parthenium hysterophorus*. *Weed Res*. 40: 507–521.

119. Tamado T, Ohlander L, Milberg P. 2002. Interference by the weed *Parthenium hysterophorus* L. with grain sorghum: influence of weed density and duration of competition. *Int. J. Pest Manage*. 48: 183–188.

120. Taneja K, Jain PP, Naithani S. 1997. Utilization of *Parthenium hysterophorus* (Gajar Ghas) and *Curcium* species as paper raw material. *World Weeds*. 4: 21-23.

121. Taye T, Einhorn G, Metz R. 2004) *Parthenium hysterophorus*, an invasive species in Ethiopia: investiga-tions on the occurrence and on its pathogens. *J. Pl. Dis. Prot.* 19: 271–278.

122. Tiwari JP, Kurchania SP, Bhalla CS, Paradkar NR. 1998. Studies on allelopathic effects of *Cassia tora* in *Parthenium hysterophorus*. *Ind. J. Weed Sci*. 30:204–205.

123. Towers GHN, Subba Rao PV. 1992. Impact of the pan-tropical weed, *Parthenium hysterophorus* L. on human affairs. In: Proc. 1st Int. Weed Control Congr. Melbourne (ed.Richardson R.G.), Weed Science Society of Victoria, Inc., Melbourne, Victoria, Australia. pp. 134-138.

124. Vikrant P, Verma KK, Rajak RC, Pandey AK. 2007. Characterization of a phytotoxin from *Phoma herbarum* for management of *Parthenium hysterophorus* L. *J. Phytopathol.* 154: 461–468.

125. Wild CH, McFadyen RE, Tomley AJ, Willson BW. 1992. The biology and host specificity of the stem-boring weevil *Listronotus setosipennis* (Col.: Curculionidae) a potential biocontrol agent for *Parthenium hysterophorus* (Asteraceae). *Entomophaga*. 37: 591–598.

126. Williams JD, Groves RH 1980. The influence of temperature and photoperiod on growth and development of *Parthenium hysterophorus* L. *Weed Res.* 20: 47-52.

127. Wood JM. 1897. Report on Natal Botanic Gardens for the Year 1896. Durban, South Africa: Durban Botanic Society, 31 pp.

128. Yadav A, Malik RK, Balyan RS. 1997. Evaluation of herbicides against carrot weed (*Parthenium hysterophorus* L). *Pestology*. 21:38–42

SOIL INVADING NEMATODES IN ECOSYSTEM VIS-A-IS SOIL HEALTH

S.P. Tiwari[1], M.N. Khare[2], Subhash Bhargava[3] and H.N. Gour[4]

[1,2]Department of Plant Pathology,
Jawahar Lal Nehru Agriculture University, Jabalpur (MP)
[3,4] Department of Nematology, MPUAT, Udaipur-313 001, Rajasthan

Key Words - Toxins, Nematodes, Effective microorganisms, Organic matter, Bio-indicators, Soil Health.

ABSTRACT

In soil, free-living soil nematodes play a key role in development of soil ecosystem. Nematodes quickly respond to disturbance and enrichment of soil physical, chemical and biological properties. Organic matter decomposition is performed by an incredible variety of soil nematodes. Nematodes perform important functions at several trophic levels of the soil food web. Their vermiform character facilitates in assessment of body volume and weight. Fungi, bacteria, phytoplasma, protozoa and viruses, besides plant parasitic nematodes, are other soil inhabitants which affect plant productivity considerably. Except plant parasitic nematodes all usually develop specific symptoms on infection. Nematodes also act as bioindicators of soil health being ubiquitous and have diverse feeding behaviors and life strategies, ranging from colonizers to per sisters.

INTRODUCTION

Soil is an integral component of dynamic terrestrial ecosystem which encompasses four major factors: minerals, organic matter, water, and air (10). Essential parts of the global carbon, nitrogen, phosphorous, sulfur, and water cycles are carried out in soil largely through microbial and faunal interactions along with physical and chemical properties of soil (17).

Several efforts have been made in past to harvest bumper crop yields. The United Nations estimated the agricultural production in 552 million hectares, which was equal to 38% of today's global cultivated area. It has been further degraded to some degree by agricultural mismanagement since World War II. The use of modern agricultural technologies to ensure increased yield has become a challenging task therefore, an effective substitute for land expansion and global agro-ecosystems are essentially required. Lately, there has been an increased public awareness of soil ecology and the importance of maintaining soil health in agro-ecosystems.

Mostly, the soil organisms play beneficial roles in soil ecosystems of which nematodes also form an important part and are well known species as important useful organisms and also are known as devastating parasites of humans, domestic animals and plants. However, most species are not pests; they occupy niche that provides an available source of organic carbon in marine, fresh water and terrestrial environments. There may be 50 different species of nematodes in a handful of soil and millions of individuals can occupy 1 m^2. Of the soil borne nematodes that do not feed on higher plants, either feed on fungi or bacteria; others are carnivores or omnivores. A small fraction of soil fauna depends upon primary producers. Soluble organic compounds are absorbed by bacteria and fungi, which also degrade more recalcitrant matter.

Nematodes readily respond to changes as well as enrichment of their environment. Increased microbial activity in soil leads to changes in the proportion of opportunistic bacterial feeders in a community. Over time the enrichment opportunists are followed by more general opportunists which include fungal feeders and different genera of bacterial feeders. These organisms are potential indicators of the nature of incoming substrate, but current methods of biomass determination do not reliably indicate their community composition. Guilds of nematodes that feed on bacteria usually belong to Rhabditidae and Panagrolaimidae whereas Aphelenchidae, Aphelenchoididae members often feed on fungi are responsive to changes in abundance of their food (27). Feeding of nematode on plant roots and their exudates has been established since long. Soil nematodes are relatively abundant ($6x10^4$ to $9x10^6$ per m^2) in number, small (300 μm to 4 mm) in size with short generations (68).

Nematodes are documented as indicators of similar functions performed by other microorganisms in the soil web in given conditions of eco-system. Vermiform character is considered in the description of nematodes, which facilitates in assessment of body volume and weight. Prescribed coefficients are considered in calculation of their carbon metabolism. Nematode metabolic activity levels are attributable to the quantum of nematodes performing various functional roles, which may be calculated from existing and accessible morphometric observations. An assemblage of nematode metabolic activities is helpful in the measurement of ecosystem services.

Principal soil microorganisms

1. Total microbial community: Bacteria, fungi, algae, actinomycetes, nematodes, micro-arthropods.
2. Beneficial organisms: Mycorrhiza, nitrogen fixing bacteria, earthworms, bio-control agents, decomposers of organic matter
3. Plant pests: Pathogens, parasitic nematodes, insects, weeds, animals.

According to Bardgett and Griffiths (5), the microbial consumption of nematodes is:

5%-8% : organic matter input,

5%-25% : standing bacterial population,

4%-22% : contribution to total net nitrogen mineralization,

10% : the ecological growth efficiency,

Soil ecosystem

Soil health

Soil health is the capacity of soil to function within specific ecosystem limits to sustain biological productivity, maintain environmental quality and promote plant and animal health (18). Healthy soil has ability to support life processes through nutrient supply, retention of optimal water and soil properties, support soil food webs, recycle nutrients, maintain microbial diversity, remediate pollutants and sequester heavy metals. Plant pathologists assert that disease suppression seems to be a function of soil health. Some of the plant parasitic nematodes, fungi, bacteria, phytoplasma, protozoa and viruses are soil-borne which adversely affect plant productivity. Plant parasitic nematodes usually develop non-specific symptoms on infection. Root-knot nematodes are exceptions developing distinctive root galls, also referred as '*root elephantiasis*'. Interactions among nematodes and other soil pathogens result in recognition of the problem and assessment of the damage. Nematodes are attracted to bacterial food sources via carbon dioxide (Co_2) or exuded metabolities and temperature gradients (45). Enhanced nutrient cycling in both the rhizosphere and bulk soil depends on consumption of bacteria by nematodes to mineralize immobilized N (12). Nematodes appear to be the primary consumers of bacteria in rhizosphere (31). The abundance of bacterial-feeding nematodes may be a useful indicator of bacterial population increase in soil (20).

High population of beneficial free-living nematodes and comprehensive nematode faunal analysis in soil ecosystem for the protection of damage caused by plant-parasitic nematodes ultimately leads to improvements in plant health. Biological diversity, community stability, ability to maintain the integrity of nutrient cycling and energy flow, suppression of multiple pests and pathogens, and improved plant health lead to soil health improvement.

Healthy soils consist of large diverse population of beneficial organisms and only fraction of active plant pests. The plants can be used as the indicators of successful soil management in healthy soil, which help asses the agro–ecosystem to identify influences of edaphic and other environmental aspects that affect plant performance. Agronomic choices determine selection of plant species and the variety, rotations, fertilizer doses, pest and disease management strategies; weed control methodology, timing of operations, machinery, tillage, etc the role of animals in the system. An incredible species of nematodes function at several trophic levels of the soil food web. Nematodes which feed on the plants and algae are the first trophic level, that take on bacteria, and fungi are the second trophic level and other nematodes are the higher trophic levels Agricultural farm soils generally support less than 100 nematodes in each teaspoon of soil. Grasslands contain 50 to 500 nematodes and forest soils generally hold several hundred nematodes per teaspoon. The proportion of bacterial and fungal feeding nematodes is corresponding to the quantity of bacteria and fungi present in the soil. Less disturbed soils contain more predatory nematodes which are highly sensitive to a wide range of disturbances. Nematodes may be useful indicators of soil quality due to diversity and their participation in many functions at different levels of the soil food web. They may be further useful indicators due to their stable response to changes in moisture and temperature regimes. Nematodes are quite small and live in water films. Changes in nematode populations reflect changes in soil micro environments. The most important function of soil in agro–ecosystems is to support plant growth and it depends on many soil processes and properties. The components of organisms in soil and their functions are given in Table 1, 2.

Table 1. Composition of the soil organisms

Microorganisms		Macroorganisms		Plants	
Microflora	Micro-fauna	Mesoorg-anisms	Macroorganisns	Algae	Root
<5 μm	<100 μm	100 μm-2 mm	2 - 20 mm	10 μm	>10 μm
Bacteria Fungi	Nematodes Protozoa	Mites, Springtails	Earthworms, Millipedes Snails and slugs, Woodlice		

Table 2. Functions performed by soil organisms

Functions	Organisms
Maintenance of soil structure	Invertebrates, plant roots, mycorrhizae and some other micro-organisms

Regulation of soil hydrological processes	Bioturbating invertebrates and plant roots
Gas exchange and carbon sequestration (accumulation in soil)	Microorganisms and plant roots, some protected in large compact biogenic invertebrate aggregates
Soil detoxification	Microorganisms
Nutrient cycling	Microorganisms and plant roots, some soil- and litter-feeding invertebrates
Decomposition of organic matter	Saprophytic & litter-feeding invertebrates, fungi, bacteria, actinomycetes and other micro-organisms
Suppression of pests, parasites and diseases	Plants, mycorrhizae, other fungi, nematodes, bacteria, various other micro-organisms, collembola, earthworms, predators
Sources of food and medicines	Plant roots, various insects, earthworms, vertebrates, micro-organisms and their by-products
Symbiotic and asymbiotic relationships with plants and their roots	Rhizobia, mycorrhizae, actinomycetes, diazotrophic bacteria, other rhizosphere micro-organisms
Plant growth control either (+) and or (-)	Direct: Plant roots, rhizobia, mycorrhizae, actinomycetes, pathogens, plant parasitic nematodes, rhizophagous insects, PGR micro-organisms, biocontrol agents etc. Indirect :Most soil biota

Factors related to soil influencing Nematode activities:

Physical

Soil compaction

As farm machineries have become heavier, concern about its direct effect on soil physical conditions and its indirect effects on crop yields and soil biota is evident. Soil bulk density was monitored as an indicator of soil compaction. Root distribution over the soil profile and nematode populations were assessed indicating a moderate degree of compaction (similar to 4.5 t load) gave the highest crop yield and that at higher degrees of compaction roots failed to penetrate into the deeper soil layers (>20 cm depth). Total number of nematodes were not affected by compaction, but their distribution over the various feeding types shifted towards a population with increased number of herbivores and decreased number of bacterivores and omnivores/predators. This change in the structure of the nematode assemblage is associated with poorer conditions for crop growth (8).

Aggregate stability

The activities of soil organisms interact in a complex food web with some subsisting on living plants and animals (herbivores and predators), others on dead plant debris (detritivores), on fungi or on bacteria, and others living off but not consuming their hosts (parasites). Plants, mosses and some algae are autotrophs; they play the role of primary producers by using solar energy, water and carbon from atmospheric carbon dioxide to make organic compounds and living tissues.

Water infiltration

Factors that could be involved in mitigating the impact of nematodes in mulched soil are (i) increased nutrient availability, (ii) increased porosity/ improved infiltration, (iii) increased surface rooting (40).

Water holding capacity

In a rice field toposequence including upland, rainfed lowland and irrigated rice, *Meloidogyne graminicola, Pratylenchus zeae* and *Hirschmanniella oryzae* were observed in upland, lowland and flooded fields. The prevalence and mean intensity of *M. graminicola* were greater in flooded fields than in upland fields, whereas shallow flooding reduced the percentage of roots damaged by the nematode, resulting no yield-reducing effect by the nematode under irrigated conditions (53).

Chemical

Moisture and organic matter

The highest number of nematodes were usually found in the lighter soils, except in the loamy sand where moisture (below 20%) was a limiting factor in conjunction with per cent organic matter (49), as revealed by correlation studies.

Cation exchange capacity

The presence of species of nematode belonging to five taxa,*viz, Pratylenchus penetrans*, *Aphenchoides agricola*, *Heterodera trifolii*, and species of *Tylenchorhynchus* and *Pratylenchus*, were correlated with soil particle size. Whereas, *Tylenchus maius* and *Criconemella* sp. were correlated with effective cation exchange capacity (64).

Nitrogen (N), Phosphorus (P) and Potash (K)

Phosphorus + K treated plants showed 50 per cent reduced population of nematodes in the rhizosphere in comparison to control. Potassium treated plants exceptionally supported 93 per cent higher population over control.

Micronutrients

The micronutrients improved quality, size, colour, taste and earliness of horticultural produce and also enhanced market appeal by developing input use efficiency of NPK fertilizers and water, which provided disease resistance, thereby reducing dependence on chemicals and also increased post-harvest life.

Toxins

Bacillus thuringiensis (BT) accounts for 90% of the bioinsec-ticide market and it produces insecticidal toxins referred to as delta endotoxins. The other two bacteria belong to the genera *Photorhabdus* and *Xenorhabdus*, which are symbiotically associated with entomopathogenic nematodes (EPNs) of the families Heterorhabditidae and Steinernematidae respectively, whereas, *Xenorhabdus* and *Photorhabdus* exist in a mutualistic association with the entomopathogenic. Nematodes (11). The bacteria, which are symbiotically associated with EPNs may help controlling insects by producing toxins, a potent source of insect management.

Biological

Disease suppressive capacity of the soil

Plant type and different soil management regimes affect the microbial diversity of soil and the implication for the disease suppressiveness ability of soil. Plant and soil types, the two main components of soil microbial community structure, are thought to exert their function in a complex manner. The microflora of disease-suppr-essive soils is usually dominated by antagonistic microorganisms including some important species of fungi like

Penicillium, Trichoderma, Aspergillus, and *Actinomycetes*. that produce profuse amounts of antibiotics.

Influence of Effective Microorganisms (EMs)

The concept of effective microorganisms (EMs) was developed by Teruo Higa, University of the Ryukyus, Okinawa, Japan (33,34). EMs consist of mixed cultures of beneficial naturally-occurring microorganisms that can be applied as inoculants to increase the microbial diversity of soils and plant. Experiments have shown that the inoculation of EM cultures to the soil/plant ecosystem can improve soil quality, soil health, and eventually the growth, yield, and quality of crops. EMs contain selected species of micro-organisms including predominant populations of lactic acid bacteria, yeasts and smaller numbers of photosynthetic bacteria, actinomycetes , these are mutually compatible with one another and can coexist in liquid culture.

Disease-Inducing Soils

In infested soil, plant pathogenic microorganisms such as *Fusarium* can comprise 5 to 20 per cent of the total microflora if fresh organic matter with high nitrogen content is applied to the soil, whereas incompletely oxidized products are malodorous and toxic to growing plants. Such soils tend to cause frequent infestations of disease organisms, and harmful insects.

Zymogenic Soils

Such soils are dominated by microflora which perform ferment-ation, breaking down the complex organic matter into simple organic substances and inorganic materials. The organisms can be either obligate or facultative anaerobes. Such fermentation-produ-cing microorganisms often comprise the microflora of various organic materials, i.e., crop residues, animal manures, green manures and municipal wastes including composts.

Synthetic Soils

Synthetic soils contain significant populations of microorg-anisms which are able to fix atmospheric nitrogen and carbon dioxide into complex molecules such as amino acids, proteins and carbo-hydrates. Such microorganisms include photosynthetic bacteria which perform incomplete photosynthesis anaerobically, certain phycomycetes and both green algae and blue-green algae which function aerobically.

Pore size distribution

Tillage alters soil structure along with pore size and the amount of space available to nematodes as well as bulk density which affect nematode migration (36).

Functions of Soil Organisms

Free-living soil nematodes comprise a large chunk of the soil fauna; however, agricultural practices often reduce their numbers and species diversity by altering soil conditions. Soils in agricultural fields are often disturbed and plant residues are not allowed to accumulate on the surface. These practices are detrimental to enhancing the number and diversity of free-living nematodes (46). Nevertheless, the abundance, species composition, and diversity of nematodes in a particular ecosystem are important indicators of the stability of soil environment (46). Agricultural practices such as tillage and herbicide applications affect soil biodiversity. These and other high input chemical and mechanical management practices cause perturbations that are not congenial to mite and nematode growth. (4). Negative impact of tillage on soil nematodes which may be short-lived and could be followed by a recovery or increase in numbers if the soil condition becomes favourable (42). Soil organic matter becomes a limiting factor for nematodes in agricultural soils with low organic resources (56). This may occur in soils where limited plant residues are allowed to enter in the soil (6). Bacterivorous nematodes affect organic matter decomposition directly by feeding on microbes and indirectly by dispersing microbes (9, 29).

Nitrogen Mineralization

Bacteria feeding nematodes participate in nitrogen mineralization in decomposition food webs to an extent determined by metabolic and behavioral attributes, by life history, and by the relative C-to-N ratios of the nematodes and their bacterial prey. The weight of bacteria necessary to meet the growth and respiration at 20°C ranged from 0.87 μg-nematode^{-1} d^{-1} for *Cephalobus persegnis* to 1.99 μg-nematode^{-1} d^{-1} for *Bursilla labiata*. Bacterial-feeding nematode community in the top 15 cm soil mineralized N at rates increasing to 1.01 μg-N g-soil^{-1} phere soil. (24,25).

Decomposition

Three species of bacterial-feeding nematodes, *Diplolaim-elloides oschei*, *Diplolaimelloides meyli* (Monhysteridae) and *Pellioditis marina* (Rhabditidae), occur on macrophyte detritus in the Westerschelde Estuary. Bacterivore nematodes deposit feeder sensu, which mainly appear with high microbial activity (41).

Soil ecosystems stability

Both, structural and functional diversity are more important for soil health. Functional diversity refers to a reserve pool of quiescent organisms or a community with vast inter-specific overlaps and trait plasticity allows an ecosystem to maintain long-term stability of soil function.

Nematodes affect food web structure and function as :

1) Channelling resources derived through feeding on roots
2) Predation on other organisms
3) Serving as prey for higher-level predators
4) Redistributing organisms in the soil matrix
5) Sequestering minerals, C and energy
6) Converting organic molecules to mineral forms available for uptake by plants and microbes
7) Regulating opportunistic microbes

To resource enrichment some nematodes are more responsive than others. Bacterivores possess short lifecycles and high reproductive potential. Longer-lived and hardier fungivores species are indicators of fungal abundance (7, 26). While ambivalent as indicators, nematodes of Tylenchidae may constitute 30% or more of the individuals in a soil sample. Nematodes are indicators at "entry-level" with the possibility that some soil nematodes are able to feed on, or absorb, soluble organic compounds because many bacterivores populations can be maintained in axenic culture (62,16) and some animal parasites absorb hemolymph across the body wall (43).

...-feeder nematodes are nearer to roots, where ...ongregate; fungal-feeders are near to fungal biomass; ...ders are concentrated around roots of stressed or ...ceptible plants. Predatory nematodes are more likely to be abundant in soils with high number of nematodes. Because of their size, nematodes tend to be more common in coarse-textured soils. Nematodes move in water films in large (>50 μm) pore spaces. Bacterial-feeders consume bacteria (20). Soil fertility is a key determinant of the success of a farming system. Management systems that rely on organic inputs as sources of fertilizer have different dynamics of soil nutrient availability than systems receiving mineral sources. The bacterial feeders prey on bacteria and may ingest up to 5000 cells/minute, or 6.5 times of their own weight in a day. Fungal-feeders feed by puncturing the cell wall of fungi and sucking out the internal contents. Predatory nematodes eat all types of nematodes and protozoa. They eat smaller organisms whole, or attach themselves to the cuticle of larger nematodes, scraping away until the matter from internal body parts are extracted. Omnivores eat a variety of organisms or may have a different diet at each life stage.

An increase in organic matter (OM) in the litter and humus layers causes an increase in fungal biomass which is the primary food source for oribatid mites (51). Thus, OM content indirectly affects juvenile mite production. Similar results were observed for nematodes where cover crop quality as well as quantity are important determinants of the nature and magnitude of soil food web services using nematodes as indicators (21).

Nematodes represent an abundant group of soil organisms and non-parasitic species are important for soil quality and in the soil food web. In natural ecosystems, nematode abundance and community structure analyses were proved to be sensitive indicators of stress caused by soil pollutants and ecological disturbance. Community structure analyses may be approached from a functional or ecological point of view; species are divided into groups according to their feeding habits or alternatively the maturity index is calculated according to their ecological strategy. Many environmental factors have the potential to affect nematode community, which consequently results in high space and time variability (59). The carrying capacity for microflora and nematofauna using bactericide, fungicide, manure in a poor tropical soil, in order to identify relationships between the soil micro nematodes and to assess the influences of these organism nitrogen flux. Manure input and initial bactericide application l to a significant increase in bacterial-feeding and fungal-feeding nematodes of coloniser-persister classes 1 and 2, respectively, whereas high manure input stimulated omnivorous nematodes (i.e. *Microdorylaimus rapsus*) which became the dominant trophic group. (63).

Nematodes as Soil Health Bioindicators

Nematodes can be used as bioindicators of soil health because nematodes are ubiquitous and have diverse feeding habitats and life strategies, ranging from colonizers to persisters. Some nemato-des can comparatively survive harsh, polluted, or disturbed envir-onments better than others, and some have short life cycles and respond to environmental changes quickly. Morphology of nemat-odes reflects feeding behavior allowing easy functional classific-ation. Five of the most common feeding groups with very distinct buccal cavities are bacterivore; fungivore; herbivore; omnivore, and predator. A conceptual advance in nematode community analysis was the classification of nematode families on a colonizer-persister (c-p) continuum on feeding behaviour. Nematodes with a c-p value equal to one are short lived, have high fecundity, feed on enriched media, and form dauer larvae as microbial blooms subside; whereas those with c-p values of five are long lived, have larger bodies, low fecundity, the greatest sensitivity to disturbance, and are predominantly omnivores or predators. The combination of nematode feeding groups and c-p-scaling provides analysis which may be more powerful tool and can be used as a bioindicator of soil health and food web condition. Ferris et al. (24) proposed a channel index (CI), which is the percentage of fungi-vores among the total of fungivores and opportunistic bacterivores to describe the dominant decomposition channels in a food web. Analyses provide slightly different information which are based on biomass, diversity, and plant-parasitic species.

Role of Nematodes in Soil Nutrient Cycle

1. Nutrient cycling. Protozoa and nematodes are important in mineralizing, or releasing, nutrients in available forms for plants, when nematodes eat bacteria or fungi, ammonium ion is released acteria and fungi contain more nitrogen. Nematode contributes up to 19 per cent of soluble Nitrogen in soil n nematodes a lower Nitrogen content is obtained than the eria (C:N ratio of 3:4). (69). The growth efficiency of ematodes are smaller (<25%) than those of the bacteria (>30%) (35). Nematodes excrete assimilated C and N which is consumed by bacteria. Bacteria, on the other hand, usually respire most of the assimilated C, but immobilize most of the assimilated N. Availability of N to plants usually contributed by nematodes to N mineralization which is relatively higher than bacteria in soil ecosystems. Besides contributing to N mineralization, many bacterivorous, omnivorous, and predatory nematodes correlate with concentrations of many other soil nutrients (66).

2. Dispersal of microbes. Nematodes help in distributing bacteria and fungi through the soil and along roots by carrying live and dormant microbes on their surfaces and in their digestive systems.

3. Food source. Nematodes are food for predatory nematodes, soil micro-arthropods, and soil insects. They are also parasitized by bacteria and fungi.

4. Disease suppression and development. Nematodes also cause disease but some consume disease-causing organisms, such as root-feeding nematodes, or prevent their access to roots. These may be potential bio-control agents.

5. Fixation of nitrogen. Nematodes limit the maximum nitrogen production and consequently decreasing the nodulation in roots. Nieh (47) reported suppression of rhizobium nodulation on alfalfa by *M. javanica*. Mishra and Gaur (44) observed a 75% reduction in nodulation in root-knot infected mungbean. Sharma and Sharma (57) reported an adverse affect on nodulation of chickpea plants infected with *M. incognita* and *M. javanica*.

6. Detoxification of pollutants: Free-living soil nematodes differ in their sensitivity to soil pollution. Nematodes are seen as the most promising candidates for bioindication of soil status (15,1). Many of the physiological responses of nematodes to toxic substances are highly unspecific. Nematodes have elaborate sensorial equipment, including receptors for cadmium and copper ions (55), which enables them to avoid intake of a broad spectrum of substances by means of the same behavioral response mechanism.

7. **Maintenance of soil structure.** Organic matter and its replenishment has become a major component of soil health management programs. Adding organic matter to soil influences diverse and important biological activities. The diversity and number of free-living and plant-parasitic nematodes are altered by rotational crops, cover crops, green manures, and other sources of organic matter. Soil management programs should include the use of the proper organic materials to improve soil chemical, physical, and biological parameters and to suppress plant-parasitic nematodes and soilborne pathogens. The effects of organic matter additions on activities of major and minor plant-parasitic nematodes in the production system are quite evident (72).

8. **Biological suppression of plant pests.** Natural enemies normally keep plant-parasitic nematodes under control in natural environments by the soil-management practice. Nematodes puncture cells and damage plants mechanically and chemically, reducing plant vigor, inducing lesions, rots, deformations, galls, or root knots, and predisposing plants to infection by root-infecting fungi. World crop production is thought to be reduced 10% by damage from plant-parasitic nematodes (71).

The nematodes are attracted to bacterial food sources via carbon dioxide or exuded metabolites and possibly temperature gradients (28). Nematodes are primary consumers of bacteria in the rhizosphere; protozoa are equally prevalent in rhizosphere and bulk soil (31). C:N ratios of bacterial-feeding nematodes have been estimated around 10:1 (3) and 4.6:1 (50). The estimates of Anderson *et al.* (3) are based on measurements of the N content of a fungal-feeding nematode, *Ditylenchus trtiformis* and on estimated C content (45). The excess N is excreted mainly as ammonia which freely permeates the nematode cuticle (54,39). Considerable mineralization of N is associated with nematode metabolic activity.

Plant parasitic nematodes

Root-feeders are plant parasites, and thus are not free-living in the soil. Plant parasitic nematodes are non-segmented worms typically 50 μm in diameter and 1 mm in length. The nematode sponsible for inducing plant diseases have received more but far less is known about the majority of the community that plays beneficial roles in soil. only move through the soil where a film of moisture unds the soil particles. In hot and dry conditions, they enter to a dormant stage, and as soon as water becomes available, they spring back to activity. This helps in dispersing both the organic matter and the decomposers in the soil. Bacterial- and fungal-feeding nematodes release a large per cent of N when feeding on their prey groups and are thus responsible for much of the plant available N in the majority of soils (37). The annual overall consumption may be as much as 800 kg of bacteria per hectare and the amount of N turned over in the range of 20-130 kg (14). Phytophages or plant-feeding nematodes damage plant roots, with important economic consequences for farmers due to the ability to the cause plant disease and reduce crop yield. Obligate root-parasitic nematodes can affect soil microbes positively by enhancing C and nutrient leakage from roots but negatively by restricting total root growth. However, it is not clear how the resulting changes in C availability affect soil microbial activities and N cycling. Tu et al. (62) observed effects of reniform nematodes (*Rotylenchulus reniformis*) on soil microbial biomass activities in a microplot experiment.

Parasitism and damage to plants

Total energy consumption during the life cycle of a female *Meloidogyne* sp. is 1 calorie. The total biomass of a female root-knot nematode is 200 μg, including the egg mass. Damage will depend on types of cells affected - storage tissues, cortex, or functional vascular. intercellular and intracellular, requiring dissolution of cell walls and middle lamellae. Cellulase and pectinase enzymes disintegrate spongy tissues (*Pratylenchus* and *Ditylenchus*) and allow ingress of other organisms. *Meloidogyne* spp. and *Heterodera* spp. produce endogluconase enzymes and pectate lyase which are involved in the passage through plant tissues. It is estimated that up to 20% or more of photosynthate partitioned to roots may leak into rhizosphere soil without root damage. "Root exudation" nurturs organisms in rhizosphere, presumably for plant benefit. Enhancing root leakage through nematode damage must reduce plant productivity. Under low nematode densities, nematodes feeding stimulate the growth rate of prey populations. Bacterial feeders and plant feeders stimulate bacterial and plant growth. Higher nematodes density may reduce the population of their prey and may decrease plant productivity. Predatory nematodes usually regulate populations of bacterial and fungal feeding nematodes. In microcosm experiments it was determined that rates of N mineralization by bacterial-feeding nematode species of different body size ranged between 0.0012 and 0.0058 m g-N nematode^{-1} d^{-1}, mainly as ammonium ions (25). The greater amount of N as ammonium ions are released from microcosms in which barley straw was colonized by the fungus *Rhizoctonia* sp. in the presence of *Aphelenchus avenae* (12).

Nematodes can serve as a model subsystem that provides a holistic measure of the biotic and functional status of soils (46). The development of the maturity index (MI) represented a significant advance in interpreting the relationship between the ecology of nematode communities and soil function, and thus, facilitated bio-assessment studies using nematodes as indicators (7). Positive relations between plant species diversity and the abundance or diversity of soil organisms were explained by increased substrate diversity, increased substrate quantity or increased diversity of soil microhabitats (60,61). Effects of plant species diversity in the soil food web and ecosystem processes are explained by Porazinska *et al.* (52). Primary consumers, microbes and plant feeders often appeared to be more responsive than secondary microbe-feeders and tertiary consumers to plant (38,52,67).

The production component

Andrássy (2). calculated nematode biomass on the basis of nematode morphology. That simple shape provides conveniently for calculation of volume and biomass of nematode. The formulae of de Man (16) have been the standard morphometric descriptors for nematode taxa. Among the standard parameters are L, the body length, and a, the ratio of length to maximum body diameter. Thus, from information available in the taxonomic descriptions of nematode species, the formula for nematode volume is restated as V ¼ (L3/a2)/1.7.

Soil Ecosystem and nematode Management

Plant-parasitic nematodes cause damage to plant roots, resulting in root systems which are less able to take up nutrients and water. Enhancing soil nutrient availability for plant uptake also provides materials required to grow functional roots, increasing the plant tolerance against nematode damage. Fungal egg-parasites, nematode-trapping and endoparasitic fungi, plant-health promoting rhizobacteria, obligate bacterial parasites, mites, collembola, tardigrades, predatory nematodes, turbellarians, and protozoans are great resources for suppressing plant-parasitic nematodes in soil ecosystems. The Nematophagous fungi produce adhesive networks, constricting rings and adhesive knobs *Hirsutella rhossiliensis* an endoparasitic fungus, a trapping fungus forming adhesive knobs as *Monacrosporium ellipsosporum*, a trapping fungus forming constricting rings by *A. dactyloides, A. oligospora* and a trapping fungus forming adhesive three-dimensional networks due to *M. cionopagum.* Organic matter decomposition enriched-opportunist bacterivorous nematodes with greater C:N ratios. Fungivorous nematodes increased higher C:N ratio gradually, such residues became more abundant for omnivores and predators which interact with *Fusarium solani* f. sp. *phaseoli*, *R. solani*, *Pythium ultimum* and *T. basicola* and lesion nematode *Pratylenchus* spp.

Conclusion

What lies beneath? Understanding soil is a matter of rising urgency. Soil quality is the biggest barrier to higher crop yields for farmers all over the world. Knowing what myriad organisms live in the soil, and how they interact, is crucial to creating a healthy ecosystem. For all those scientists who are willing to crouch down and dig, the diversity of soil denizens beat any above-ground system, even that of a tropical rainforest. A handful of soil from one spot may house a different community from soil just a meter away, because of variations in the availability of water or nutrients. For example, the ground under a decaying plant or animal is a different environment from soil lacking such enrichment. And around plant roots, specialized organisms inhabit the rhizosphere, a thin layer where roots and soil organisms interact in myriad ways. Large animals such as moles contribute, changing and aerating the underground landscape by tunneling.

Even a small clump of soil has a gradient of oxygen from its edges to the centre, and each oxygen concentration may make the perfect habitat for different kinds of creatures. "It is the most incredible zoo"- Diana Wall (Soil Ecologist) at Colorado State University, Fort Collins, exclaimed.

Consequently, soil forms the important component of dynamic, living, natural terrestrial ecosystem and the organisms perform beneficial roles in ecosystems. W e learn that nematodes play a substantial important role in maintaining soil health ecosystem and free-living soil nematodes comprise a big chunk of the soil fauna. They respond rapidly to disturbance and enrichment of their environment; increased microbial activity in soil leads to changes in the proportion of opportunistic bacterial feeders in a community. Nematodes respond rapidly to disturbance and enrichment of their environment; increased microbial activity in soil leads to changes in the proportion of opportunistic bacterial feeders in a community. Organic matter decomposition products are assimilated by thc nematodes which ultimately reach to plant as a source of nutrient. An incredible variety of nematodes function at several trophic levels of the soil food web. Bacterial-feeders nematodes near roots where bacteria congregate; fungal-feeders are near fungal biomass; root-feeders are concentrated around roots of stressed or susceptible plants. Predatory nematodes are more likely to be abundant in soils with high numbers of nematodes. The concept of effective microorganisms consists of mixed cultures of beneficial naturally-occurring microorganisms and need to adopt in development of agriculture for the management of pathogens and parasites. Microbes which, *hitherto*, remained hidden from the prying eyes of scientists, until Leeuwenhoek invented the microscope, coexist comfortably with insects, nematodes, plants and animals; the symbiotic relationships are often mutually beneficial. Micro-organisms can provide complementary and essential biochemistry to facilitate survival of hosts; the gut is a comfortable habitat for many bacteria. Biology is full of examples of cooperation and coexistence.

REFERENCES

1. Achazi RK. 2002. Invertebrates in risk assessment. *J Soil Sediment*. 2:174–178.
2. Andrássy, I 1956. Die rauminhalst and gewichtsbe-stimmung der fadenwurmer (Nema-

toden). Acta Zoologica Academi Sciences, Hungary 2:l-15.

3. Anderson, R.V., Coleman, D.C. and Cole, C.V. 1981. Effects of saprotrophic grazing on net mineralization. *Ecological Bulletin* 33:201-216.

4. Andrén, O. and J. Lagerlof. 1983. Soil Fauna (Microarthropods, Enchytraeids, Nematodes) in Swedish Agricultural Cropping Systems. *Acta Agr. Scand.* 33: 33-52.

5. Bardgett, R.D. and Griffiths, B.S. 1997. Ecology and biology of soil protozoa,nematodes, & microarthropods. In *Modern soil microbiology*. Edited by Jan Dirk van Elsas, Jack T. Trevor, and Elizabeth M.H. Wellington, Marcel Dekker, Inc, New York. p 8

6. Bedano, J. C., M. P. Cantu, and M. E. Doucet. 2006. Influence of Three Different Land Management Practices on Soil Mite (Arachnida: Acari) Densities in Relation to a Natural Soil. *Appl. Soil Ecol.* 32: 293-304.

7. Bongers, T. 1990. The maturity index, an ecological measure of environmental disturbance based on nematode species composition. *Oecologia* 83:14–19.

8. Bouwman, L. A . and Arts, W. B. M. 2000. Effects of soil compaction on the relationships between nematodes, grass production and soil physical properties. *Appl Soil Ecol.* 14: 213-222.

9. Bouwman, L.A., Bloem, J., van den Boogert, P.H.J.F., Bremer, F., Hoenderboem, G.H.J. and Ruiter, P.C. 1994. Short term and long term effects of bacterivorous nematodes and nematophogus fungi on carbon and nitrogen mineralization in microcosms. *Bio Fert Soils* 17:249-256.

10. Buckman, H.O. and Brady, N.C., 1960. The Nature and Properties of Soils, 6th Ed. MacMillan, New York, 567 pp.

11. Chattopadhyay, Abanti, Bhatnagar, N. B. and Bhatnagar Rakesh. 2004 Bacterial Insecticidal Toxins. Critical *Rev Microbioloy*. 30:33-54

12. Chen, J. and Ferris, H. 1998. Nitrogen mineralization by *Aphelenchus avenae* associated with *Rhizoctonia* sp. and barley straw. *J Nematology* 30:495-496.

13. Clarholm, M. 1985. Interaction of bacteria, protozoa and plants leading to mineralization of soil nitrogen. *Soil Biol. Biochem*. 17:181-197.

14. Coleman, D.C., Ingham, R.E., McClellan, J.F. & Trofymow, J.A. 1984. Soil nutrient transformations in the rhizosphere via animal-microbial interact-

ions. *In* J.M. Anderson, A.D. M. Rayner & D.W.H. Walton, eds. *Invertebrates-microbial interactions,* pp. 35-58. Cambridge, UK, Cambridge University Press.

15. Cortet J, Gomot-De Vauflery A, Poinsot-Balaguer N, Gomot L, Texier C, Cluzeau D. 1999. The use of invertebrate soil fauna in monitoring pollutant effects. *Eur J Soil Biol.* 35:115–134.
16. De Ley, P., and Mundo-Ocampo, M. 2004. Cultivation of nematodes. In Nematology, advances and perspectives, vol. I: Nematode morphology, physiology, and ecology. Eds. Chen, Z. X., Chen, S. Y. and Dickson, D.W. Tsinghua University Press, Beijing, China. pp. 541–619.
17. De Man, J.G. 1876. Onderzoekingen over vrij in de aarde levende nematoden. Tijdschr. Ned. Dierk. Ver. 2: 78-196.
18. Doran, J.W., Parkin, T.B. 1996. Quantitative indicators of soil quality: a minimum data set. In: Doran, J.W., Jones, A.J. (Eds.), Methods for Assessing Soil Quality. Soil Science Society of America, Madison, WI, pp. 25–37.
19. Doran, J.W., Sarrantonio, M and Liebig, M.A. 1996. Soil health and sustainability. *Adv. Agron.* 56:1054
20. dos Santos, G.A.P. , Derycke, S., Fonsêca-Genevois, V.G. , Coelho L.C.B.B., Correia, M.T.S. and Moens, T. 2008. Differential effects of food availability on population growth and fitness of three species of estuarine, bacterial-feeding nematodes. *J. Exp. Mar. Biol. Ecol.* 355:27–40
21. DuPont, S.P., H. Ferris, and M. Van Horn 2009. Effects of Cover Crop Quality and Quantity on Nematode-Based Soil Food Webs and Nutrient Cycling. *Appl. Soil Ecol.* 41: 157-167.
22. Ferris, H., Lau, S. 1995. Respiration and metabolic rates based on carbon dioxide production. *Soil Biol. Biochem.* 27, 319-330.
23. Ferris, H., Eyre, M., Venette, R. C. and Lau, S. S. 1996. Population energetics of bacterial-feeding nematodes: Stage-specific development and fecundity rates. *Soil Biol. Biochem.* 28, 271-280.
24. Ferris, H., Venette, R. C. and Lau, S. S. 1997. Population energetics of bacterial-feeding nematodes: carbon and nitrogen budgets. *Soil Biol. Biochem.* 29, 1183-1194.
25. Ferris, H., Venette, R. C., van der Meulen, H. R., and Lau, S. S. 1998. Nitrogen Mineralization by Bacterial-feeding Nematodes: Verification and

Measurement. *Plant and Soil* 203: 159–171.

26. Ferris, H., Bongers, T. and De Goede, R.G.M. 2001. A framework for soil food web diagnostics: Extension of the nematode faunal analysis concept. *Appl Soil Ecol.* 18:13–29.

27. Ferris, H. and Bongers T. 2006. Nematode Indicators of Organic Enrichment *J Nematol.* 38(1):3–12.

28. Freckman, D.W. 1988. Bacterovorous nematodes and organic matter decomposition. Agriculture, *Ecosystems and Environment* 24:195-217.

29. Griffiths, B. S. 1994. Microbial-feeding nematodes and protozoa in soil: Their effects on microbial activity and nitrogen mineralization in decomposition hotspots and the rhizosphere. *Plant and Soil* 164, 25-33.

30. Grifffths, B.S. and Bardgett, R.D.1997. Interaction between microbes- feeding invertabrates and soil microorganisms. In Modern Soil Microbiology. Eds. Elsas, J.D., Trevor, J.T., Wellington, E.M.H and Dekkes, M. Inc. New York

31. Griffiths, B.S. and Caul, S. 2003. Migration of bacterial feeding nematodes but not protozoa to decomposing grass residue. *Biol Fert Soils* 3:207-210.

32. Griffiths, B.S., Young, I.M. and Boag, B.1991. Nematodes associated with rhizosphere of barley (*Hordeum vulgare*). *Pedobiologia*: 35:265-272.

33. Higa, T. 1991. Effective microorganisms: A biotechn-ology for man-kind. p. 8-14. In Parr, J.F., Hornick, S.B. and Whitman, C.E. (ed.) Proceedings of the First International Conference on Kyusei Nature Farming. U.S. Department of Agriculture, Washington, D.C., USA.

34. Higa, T. and Wididana G.N. 199l. Changes In the soil microflora Induced by effective microorganisms. p. 153-162. In Parr, J.F., Hornick, S.B. and Whitman, C.E. (ed.) Proceedings of the First International Conference on Kyusei Nature Farming. U.S. Department of Agriculture, Washington, D.C., USA.

35. Hunt, H. W., Coleman, D.C., Ingham, E. R., Elliott, E.T., Moore, J. C., Rose, S. L., Reid, C. P. P. and Morley, C.R. 1987. The detrital foodweb in a shortgrass prairie. *Biol. Fertil. Soils* 3:57-68.

36. Hunt, H.W., Wall, D.H., Nicole, M.D. and Brenner, J.S. 2001. A model for nematode locomotion, *Nematology* 3:705-716.

37. Ingham, R.E., Trofymow,J.A., Ingham, E.R. and Coleman,

D.C. 1985. Interaction of bacteria, fungi and their nematode grazers: effect on nutrient cycling and plant growth. *Ecological Monographs* 55:119-140.

38. Korthals, G.W., Smilauer, P Putten, C.V. and Vander, W.H. 2001. Linking above and below ground biodiversity: abundance and trophic complexity in soil as response to experimental plant communities on abandoned arable land, *Functional Ecology* 15:306-514.

39. Lee, D.L. and Atikson, H.J. 1977. Physiology of Nematodes. Columbia University Press, New York. 215 p

40. McIntyre, B. D., Speijer, P. R., Riha S.J. and Kizito, F. (2000). Effects of Mulching on Biomass, Nutrients, and Soil Water in Banana Inoculated with Nematodes. *Agron. J.* 92:1081–1085.

41. Mesel I., Derycke, S. De, Swings J., Vincx, M. and Moens T. 2003. Influence of bacterivorous nematodes on the decomposition of cordgrass. *J. Exp. Mar. Biol. Ecol.* 296: 227–242.

42. 42. Minor, M.N. and Norton, D.A. 2004. Effect of soil amendments onassemblager of soil mites Acari: Oribatida, Mesostigmat) in Short Rotation Willow Planting in Central New York, *Can. J. Forest Res.* 34:1417-1425.

43. Munn, E. and Munn, P. D. 2002. Feeding and digestion. In The biology of nematodes. Ed. Lee, D. L., Publisher Taylor and Francis. London, UK, pp. 211–232.

44. Mishra, S.D and Gaur, H.S. 1981. Effect of individual and concomitant inoculation with *Meloidogyne incognita* and *Rotylenchulus reniformis* on the growth of blackgram. *J Nematol.* 11:25-28.

45. Myers, R.F. and Krusberg, L.R. 1965. Organic substances discharged by plant parasitic nematodes.*Phytopathology* 429-437.

46. Neher, D. 2001. Role of nematodes in soil health and their use as indicators. *J. Nematol.* 33:161-168.

47. Nieh, E.L. Jr. 1966. Rhizobium nodule formation on alfalfa as influenced by *Meloidagyne javanica*. *Nematologica,* 12:96.

48. Nicholas, W.L. 1984. The biology of free living nematodes, Clorendon Press Oxford. pp 12

49. Norton, D. C., Frederick, L. R., Ponchillia, P. E. and Nyhan J. W. (1971). Correlations of Nematodes and Soil Properties in Soybean Fields. *J Nematol.* 3: 154–163.

50. Persson, T. 1983. Influence of soil animals on nitrogen mineralization in a northern Scots pine forest.proceedings of the VII International Colloquium of Soil Zoology. (Eds. Lebrun, P., Andr`e, M., De Medts, A.,Gregoire-wibo,C and Wauthy, G). pp.117-126.

51. Princz, J. I., V. M. Behan-Pelletier, R. P. Scroggins, and S. D. Siciliano. 2010. Oribatid Mites in Soil Toxicity Testing — the Use of *Oppia nitens* (C.L. Koch) as a New Test Species. Environ. *Toxicol. Chem.* 29: 971-979.

52. Porazinska, D. L., Bardgett, R.D., Blauw, M. M. 2003. Relationships at the above ground- belowground interface: plants, soil biota, and soil processes. *Ecol. Monogr.*73: 377-395.

53. Prot, J.C., Matias, D.M . (1995). Effects of Water Regime On the Distribution *of Meloidogyne graminicola* and other root-parasitic nematodes in a rice field toposequence and pathogenicity of *M. graminicola* on rice cultivar Upl R15, *Nematologica* 41:219-228.

54. Rogers, W.P. 1969. Nitrogenou components and their metabolism: Acanthcephla and Nematoda. In Chemical zoology, Vol. III (Eds. Florkin, M. and Scheer, B.T). Academic Press London, pp.379-428.

55. Sambongi Y, Nagae T, Liu Y, Yoshimizu T, Takeda K, Wada Y, Futai M. 1999. Sensing of cadmium and copper ions by externally exposed ADL, ASE, and ASH neurons elicits avoidance response in *Caenorhabditis elegans. Neuroreport* 10:753–757

56. Sanchez-Moreno, S. and N. L. Nicola, H. Ferris, and F. Zalom. 2009. Effects of Agricultural Management on Nematode-Mite Assemblages: Soil Food Web Incides as Predictors of Mite Community Composition. *Appl. Soil Ecol.* 41: 107-117.

57. Sharma, R. and Sharma, S.B. 1998. Nematode pests of chickpea and their management. In: Nematode Diseases in Plants. Ed. P.C. Trivedi, CBS Publisher, N. Delhi, 98-109.

58. Siemann, E., Tilman, D., Haarstad, J. *et al.,* 1998. Experimental tests of the dependence of arthropod diversity on plant diversity. *Am. Nat*. 152: 738-750.

59. Sochová, I., Hofman, Jakub and Holoubek, Ivan. 2006. Using nematodes in soil ecotoxicology, *Environmental International* 32:3374–383.

60. Spehn, E. M., Joshi, J., Schmid, B. 2000. Plant diversity effects on soil heterotrophic activity in experimental grassland ecosystems. *Plant Soil* 224: 217-230.

61. Stephan, A., Meyer, A. H. and Schmid, B. 2000. Plant diversity affects culturable soil bacteria in experimental grassland communities. *J. Ecol.* 88: 988-998.

62. Sulston, J., and Hodgkin, J. 1988. Methods. In The nematode *Caenorhabditis elegans*. Ed. Wood, W. B.Cold Spring Harbor Laboratory Press. New York pp. 587–606

63. Tu C., Koenning S.R, Hu S. 2003. Root-parasitic nematodes enhance soil microbial activities and nitrogen mineralization. *Microb Ecol.* 46:134-44.

64. Villenave, C., Ekschmitt, K., Nazaret, S. and Bongers T. 2004. Interactions between nematodes and microbial communities in a tropical soil following manipulation of the soil food web. *Soil Biol Biochem.* 36: 2033–2043.

65. Wallace, M. K., . Rust, R. H., Hawkins, D. M. and MacDonald, D. H (1993). Correlation of Edaphic Factors with Plant-parasitic Nematode Population Densities in a Forage Field. *J Nematol.* 25: 642–653.

66. Wardle, D. A., Yeates, G. W., Williamson, W. 2003. The response of a three trophic level soil food web to the identity and diversity of plant species and functional groups. *Oikos* 102: 45-56.

67. Wang, K. H., McSorley, R. and Gallaher, R. N. 2004. Relationship of soil management history and nutrient status to nematode community structure. *Nematropica* 34: 83-95.

68. Wang, K.H., Hooks, C.R.R., Marahatta, S. P. and Manandhar R. 2010. Use of a strip-till cover cropsystem to manipulate above and below ground organisms in cucurbit plantings. Meeting of the American Phytopathological Society (APS) at Charlette, NC. Aug 2010.

69. Waslewska, L. 1979. The structure and function of soil nematode communities in natural ecosystem agro crops. *Po. Eco. Stud.*5:137-145.

70. Waslewska, L., and P. Bienkowski. 1985. Experimental study on the occurrence and activity of soil nematodes in decomposition of plant material. *Pedobiologia* 28:41-57.

71. Wright, D.J. and Nawall, D.R. 1976. Nitrogen excretion, osmatic and ionic regulation in nematodes. In The Organization of nematodes

(Ed. Croll, N.A.). Academic Press London, pp. 163-210.

72. Whitehead, A.G. 1997. Plant nematode control. CAB Int., Wallingford, UK.p 48.

73. Wildmer, T.L., Mitkowski, N.A. and Abani, G.S. 2002. Soil organic matter and management of plant parasitic nematodes. *J. Nematol* . 34:289-295

YESHWANT LAXMAN NENE: PIONEER LEADER OF PLANT PATHOLOGY

*S.J. Kolte**

Department of Plant Pathology, G.B. Pant University of Agriculture and Technology, Pantnagar- 263 145, Uttarakhand, INDIA
Present Address: *Professor(Retd); Fl 8, Anuprita Apts. Left Bhusari Colony, Pune-411 038 E-mail : koltesj@rediffmail.com

Prof. Yeshwant Laxman Nene

ABSTRACT

Y.L. Nene is a plant pathologist who had spent 36 professional years: 14 years initially at the first Agricultural University (G.B. Pant University of Agriculture and Technology) at Pantnagar and later at the ICRISAT, Patancheru, India from where he retired as Deputy Director General. He is best known for diagnosing and discovering zinc deficiency of rice (*Khaira* disease) widely prevalent in most parts of north and north–eastern parts of India particularly in the *Tarai* region of Uttarakhand (formerly Uttar Pradesh). He advocated control of *Khaira* disease of rice by spray of a mixture of zinc sulphate (0.5%) and lime (0.25%) which was widely adopted by farmers as a routine essential practice for rice crop production that brought recognition to Nene and the discovery then turned *Tarai* into rice bowl of India fulfilling the goal of establishment of first Agricultural University in India in service of the farmers for which he was honoured with an International (FAO) prize. His later studies at the ICRISAT brought him recognition world over in identification of multiple-disease resistant genotypes in chickpea and pigeonpea, facilitating the crop improvement work by breeding disease resistant cultivars. His contributions to Plant Pathology have been numerous and varied and brought him various awards later in his career. To his colleagues, he served as disciplined administrator, devoted teacher, genius researcher, mentor and friend. After his retirement in 1996, he has established Asian Agri-History Foundation facilitating an ancient agricultural know-how among all concerned scientists, farmers, students and policy makers.

The history of Plant Pathology is very fascinating and lucid as a poetry in itself if we narrate it encompassing the milestones, the scientists who contributed and the role played by society in upholding the eventual findings. Likewise the foundation of the Plant pathology in India also present an incredible tale to tell not only the stories about the devastating rice disease, which caused worst famine of twentieth century in India in forties, but also about some inanimate type of diseases, which had posed a greater challenge to scientists that time and these problems were becoming a serious threat to rice cultivation as the production had been dwindling creating the fissures in the farming community. Let us have a look on the tale of a dedicated plant pathologist, who can undoubtedly be referred to as one of the few leading plant pathologists of India, that have had maneuvered hard to establish Plant Pathology in different Indian states. As the legend goes, some fifty two years ago say in 1960, the first Agricultural University, G.B. Pant University of Agriculture and Technology (GBPUAT), Pantnagar (formerly Uttar Pradesh Agricultural University) in India invited Yeshwant Laxman Nene (YLN), who had just completed doctorate degree in Plant Pathology (with specialization in Plant Virology) from the University of Illinois, Urbana (USA), to join as Assistant Professor of Plant Sciences. This action by the GBPUAT was of great significance not only because YLN was later instrumental in establishment of Plant Pathology Department for the first time in the first ever Agricultural University in India but also because of the unusual but most modest qualities of the man, qualities since reflected by the performance of the Department established by him. YLN devoted himself to teaching and research on the serious plant (crop) disease problems then confronted by the Uttar Pradesh State farmers and to the building a discipline of Plant Pathology under the new experimental American land grant pattern system of agricultural education in India in collaboration with the University of Illinois.

He recognized the urgent need for education, understanding and communication under conditions for less opportune than today. The standards he set for himself and his staff and students were high and his goal showed remarkable vision. But perhaps his greatest contributions were his understanding of people and striving for competence coupled with harmony in his staff and recognition of the individual whether a staff member or a student; it was his concept that the Department is still one family working together towards a common end, solving problems of farmers.

Early Life and Higher Education

Yeshwant, the son of Mr. Laxman Ganesh and Mrs. Laxmi Nene, (YLN) was born at Gwalior, the historically and educa-tionally highly reputed city in the state of Madhya Pradesh in India, on November 24, 1936. He completed his school education in the Gwalior city itself, further obtaining B.Sc. (Ag.) degree course programme from Government Agriculture College, Gwalior (affiliated to Agra University) in 1955. He later moved to Government Agriculture College, Kanpur (affiliated to Agra University) and earned his M.Sc. (Ag.) degree in Plant Pathology in 1957. He further continued his studies in Plant Pathology by way of securing admission to Ph.D. degree course programme at the University of Illinois, Urbana, USA and earned the Ph.D. degree in Plant Pathology with specialization in Plant Viruses under the able guidance of Prof. H.H. Thornberry. Why did YLN become a Plant Pathologist? The answer to this question lies in his career as a student of Agriculture College, Gwalior where he felt under the influence of two enthusiastic teachers: one in Botany (Dr. D.V.K. Tripathi) and the other in Plant Pathology (Dr. L.K. Joshi) in his B.Sc. degree course curriculum. Dr. Joshi particularly was such a wonderful teacher whose contagious enthusiasm influenced YLN and his susceptibility received the haustoria of passion for Plant Pathology at graduate level which never coalesced.

Research Scientist

Khaira disease of rice probably indigenous in the Nainital *Tarai* region of the then Uttar Pradesh State (now Uttarakhand) had been known since long before the establishment of the University as a serious disease of rice. The problem of *Khaira* disease of rice was getting progressively worsening when YLN joined the Faculty of Agriculture at Pantnagar. YLN drew attention to the depredations of *Khaira* [a name farmers coined because of the symptom colour of affected rice leaves that was resembling to catechu which is locally called in Hindi as *Khair*]. His task was to determine as to why the rice plants were stunted and became up with a solution to the problem(6).

The problem was not to be solved so easily and early and moreover imbalance in farmers economy was becoming a hard fact because of this disease. Further, it could cause some differences of opinions amongst plant pathologists attached to rice disease research, as it was harder to ascertain the exact cause of the disease. The solution could be worked out only after getting to the causal factor. Of several scientists devoting on solving problem, YLN, could easily approach to solution with his incredible all round experiences as a plant scientist, who was involved in teaching allied branches of agricultural sciences such as Agronomy. Introductory Soil Science, above all teaching Introductory Plant Pathology, a dynamic subject which he knew was difficult to delimit from all basic and applied branches of sciences.

This concept of understanding the subject of Plant Pathology and keeping it in mind as a central core of investigatory work, YLN applied the scientific method in both lab and the field to solve the problem of *Khaira* disease of rice (9). Plant Pathologists have always had greater challenges to distinguish non-infectious (non-parasitic) disorders from infectious diseases (caused by parasites or viruses) which at times could be mistaken. Once an abnormality was not shown or proved to be caused by a pathogen, YLN emphasized his particular attention to non-parasitic causes from the pointers he gained from the recovery of plants and emergence of green leaves by the affected rice plants and the experiences received with the trainings implanted by Dr. Roger Bray's (at the University of Illinois, USA) while teaching the course on "Soil Fertility". He, therefore, later looked at soil pH and found that the soil pH to be 8.5 in plots where *Khaira* occurred and around 7.00 in plots where it did not. Further YLN performed "Trial and Error" sort of experiments for the possible cause of *Khaira* which he thought might be due to zinc or manganese deficiencies and hence sprays of salts of zinc and manganese were just attempted . Two foliar sprays of zinc sulphate (0.5%) and lime (0.25%) resulted in excellent recovery of plants within two weeks, whereas Manganese sulphate sprayed plants did not show any affect on the plants (7).

Interestingly I would cite my own case when I was registered for my Ph.D. degree studies. Investigators studying to find out the cause of another crop disease of unknown etiology-the urdbean leaf crinkle, YLN, as my supervisor, never agreed to change my thesis topic from the leaf crinkle of Urdbean to some other problem of routine nature as I was not yet able to reach to certain clue on the cause of the leaf crinkle of Urdbean until the beginning of the third year of my studies towards Ph.D. thesis work. He insisted and encouraged me to continue with the leaf crinkle in providing etiological explanation of that disease beyond any doubt so that even 10 pages of the findings could be justified for award of Ph.D. degree. Finally with simple change of pH of potassium phosphate buffer I could prove transmissibility of the cause of the leaf crinkle disease of Urdbean and the cause proved to be the virus (3). This was a very rewarding and scintillatory experience, though not of much technological advancement, and I learned from this very simple evidence how one can enjoy the joy zest of discovery and then we submitted a research paper for publication.

His standards were high. I well remember spending duration of a week time writing and rewriting my first paper for Indian Phytopathology before he could approve it. I always tried to win him over by showing him my experimental plant and plants being placed precisely and correctly labeled. He would come into glasshouse or Lab regularly. Any discussion with him was usually memorable and encouraging one during his visit to the glasshouse or Lab. It is no wonder with this kind of daily training that we were all prepared for oral examinations. We would visit YLN with great caution and with well considered thought and cautious mind whether to ask questions or show our experimental work. YLN had a wealth of information about plant diseases and one of the best times to do this would be during field visits.

YLN always had curious and anxious hours reserved in research and regularly spent most of the first early times of morning in visiting greenhouses (which he facilitated in the PL-480 project on pulse crop viruses), however, reserving later part of the day for administrative work of the Department and University activities. His students very well knew that their work as it was progressing would catch his keen eye and generate some comments if it did not seem upto his standard. He was meticulous about experimental plannings and as students we were being coached on proper method for filling pots with soil, watering plants and maintenance of plants particularly virus- affected experimental plants. In the first two years of my Ph.D. thesis work (1967-68 and 1968-69), the pattern of field spread of Urdbean leaf crinkle disease was under investigation using different Urdbean varieties. The field experimental layout in this regard with appropriate plot size and replications, sowing, thinning and maintenance of the experimental plot etc. could be done as a team work by doing physical work by ourselves in the project under the supervision of YLN and on this account Dr. S.P.S. Beniwal (Research staff working with YLN in the PL- 480 project) had been of great help to lead the field operations as I was completely new to Pantnagar environment. As a result of this work, we could understand the seed borne nature of the pathogen (virus) based on the pattern of the spread of the disease which was further proved to be true in the glasshouse/laboratory studies and a separate research paper on the spread of the disease then appeared in a research journal in our joint authorship (2). Thus the importance of team work has been realized by me since then. YLN had been a very different scientific personality since very beginning who would not hesitate to visit fields/laboratory/green house even on Sundays/holidays providing a spark to research scholars,even involving himself in carrying out careful photography of experimental objects. Despite his so much of involvement in research aspects, he would never relish or even think of sharing a research paper authorship though being by virtue of his research guide or head of the department. It was his stature to indicate his non-partisan involvement in research work, it was in stark contrast to many other scientist, who would be always willing to share the co-authorships of research articles. I, however, dedicated my oilseed crop diseases books (CRCPress, USA) to YLN by describing him as ***"quite an unassuming man, a scholar with an inborn spirit of research, and a teacher of rare genius from whose teaching the author has learned how to study and enjoy the joy of discoveries in Plant Pathology"(4,5).***

Yet another aspect of versatility of YLN became apparent when he moved to ICRISAT in 1974 and devoted his time in Plant Pathological research and resolved the vexing problem of the "wilt complex" of chickpea. He developed many simple but effective laboratory and field screening techniques for identifying sources of resistance to diseases of chickpea and pigeonpea crops. These techniques are being used routinely by researchers in South and southeast Asian countries, East African countries, the Middle East, in some European countries and in the United States. Genotypes of pigeonpea resistant to sterility mosaic and *Phytophthora* blight and genotypes of chickpea resistant to *Fusarium* wilt and dry root rot have been distinctly identified as donors for resistance to above mentioned diseases of Pigeonpea and chickpea in larger interest of resource poor farmers (8).

Teacher at par Excellence and Dignity

YLN not only taught the subject in a very interesting and lucid manner but also fostered collaborative activities including disease diagnosis with students. The ability to work productively as a part of a team is a trait widely sought by employers, yet many instructors do nothing to cultivate this skill in students, but YLN did care for this. He genuinely cared about getting the students to understand the subject matter and to use it in the real world.

YLN is recognized as outstanding and innovative teacher primarily because of his unique approach to the introductory course in Plant Pathology which he taught at the GBPUAT for 14 years at Pantnagar at least once in every three trimesters per year. Students regarded this course taught by YLN, as total learning experience which would transfer with a look of accomplishment on his face and charismatic influence on students. YLN was distinguished not only by himself but also by others as superb teacher. The overall objective of the YLN's courses has been for students to acquire critical thinking skills in areas across disciplines. He achieved his goal by having students to read a mix of contemporary popular scientific books, articles and on topics that include diseases of plants. For Post-graduate students, he usually gave a list of 10-15 review articles published in Annual Review of Phytopathology for students crediting his course "Principles of Plant Pathology" in a trimester and the same could be included as syllabus by him with a weightage of at least 15 marks for oral examination towards the grading of the students, who were expected to prepare the topic critically understanding the topic thoroughly. This meant that he himself had to keep abreast with such these topics of current nature, the task he facilitated through his own critical reading. He is still keeping the habit of reading articles in Annual Review of Phytopathology as he is a regular subscriber for the same. YLN had a very effective, emphatic and magnetic voice and other enjoyed skills of oratory. Thereby the lectures on Plant Pathology delivered by YLN caused students to devote more attention to his subject than to any other courses. In his personal life and in his reactions with others, YLN lived slightly aloof and reserved life styles. But he held the respect of virtually for everyone who knew him and the admiration almost to the point of idolatry by the majority of students, who were stimulated to think critically and creatively. His instructions in the class were thought provoking and he was identified as a teacher that made students not to miss his class.

YLN has been a skilled teacher who consistently tried to challenge his students. His approach to teaching was a student centered and he strived to teach students how to become independent learners and to apply their knowledge in problem solving situations. He was intensely interested in total education of students. Current "buzz words" in higher education i.e. critical thinking, ability, communication sills, and international perspectives have been parts of teaching of YLN. This could become amply evident when he found extra time and sitting with students in his room beyond 5.00 PM in the evening prior to thesis writing and asking such category of students to learn English grammar first and suitable sentence composing etc. in contrast to other faculty members who used to be available on the first opportunity with their families just after office hours at 5.00 PM in a lonely place like Pantnagar. YLN thus gives full credit to his wife (Prabha Nene) for never questioning him about the time spent on professional work but she would mind if he spent time in socializing without her.

He was very strict in matters of discipline and kept perfect control over his students. At the same time he came to their help as often as needed. He did not like students or any one associated with him chewing betel, smoking or loitering.

YLN has a most unusual grasp of the science of plant pathology and is widely sought after by colleagues and students for advice. He has an excellent ability to challenge students and colleagues to critically examine research ideas and results.

YLN has also influenced many students to study Plant Pathology as a profession. He used to be highly acclaimed as a dedicated teacher with graduate and undergraduate students. Collaboration rather than competition was his concern in dealing with students. His concern of education had never been restricted to the classroom or the university campus. His greatest influence in undergraduate courses was on many students to pursue career in Plant Pathology. He has trained 24 Ph.D./M.Sc. students over his direct supervision since 1964 who are now carrying on their professional career and some of them have retired from active service too.

Farmers Friend

YLN became farmers best friend philosopher and guide while he was at Pantnagar particularly when demonstration trials on control of *Khaira* disease of rice were conducted at farmers fields with most convincing practically adoptable methodology using simple spray of zinc sulphate + Lime mixture. YLN excelled in leading discussions around a specimen of plant showing abnormal symptoms when he introduced for the first time a course "Plant Disease Diagnosis" 5 years after the discovery of the cause of *Khaira* disease of rice. I had opted voluntarily to be associated with the practical classes of this course for which he agreed very affirmatively. I had then learned a lot from this act of my association with the practical classes of the above course. Standing in a field he would start probing the cause of plant abnormality very systematically simultaneously putting logical questions to students and derive conclusions on the diagnosis with a phenomenal memory and extremely keen observation power. I have described him a person of deep penetrating eyes whenever I have happened to talk about my experiences of working with him. In case of any dispute and controversy of diagnosis of plant disease, he would then advise the students to do the experiments in the glasshouse and/or laboratory to test the hypothesis of the plant disease diagnosis. Such exercises as tools to prove the cause of plant diseases have been very rewarding events for research students. Consequently, students, extension personnel and farmers took advantage of every opportunity to get YLN out into the field. Farmers would come to YLN inviting him with great honour to visit their crop fields as family doctors keep visiting human patients for medical check-up. *Tarai* Farmers in the then Uttar Pradesh State (now Uttarakhand) designated him with a title statement "the man who can work out the most difficult problem". On several occasions during late 1960s and early 1970s, the practice of visiting farmers fields not only around Pantnagar University but also at distant places (100-250km) was adopted as a policy under YLN's leadership as Head of the Department of Plant Pathology covering mandate crops including the oilseeds by respective Project Leaders of the concerned Crops. As Head of the Department YLN would take approval of Vice-Chancellor (Late Dr. Dhyan Pal Singh) just over telephone and transport facilities could be then arranged automatically by the University Transport Officer in anticipation of the formal action to be performed in writing at later dates. YLN would often say to students that even if a single farmer is benefited by your work you must feel yourself that you have served the farmers interest and hence the country as such, forgetting about what others particularly the seniors would say.

A Strict Disciplinarian-the Administrator

Good and justified judgment on difficult issues, impartiality in decision making and great capacity for reconciling divergent opinions were characteristics of his 14-year tenure as Head of the Department at the GBPUAT, Pantnagar. Punctuality; not allowing the late comers to the class, nobody would dare come late to the class and even for that matter staff to attend the duty in the Department; in fact we used to reach the target spot/location at least 10 minutes before YLN comes there for visit etc; strict invigilation duty during examination hours, disciplinary action as the first and foremost step against any wrong deeds not only by students but also by the faculty member; YLN has been the first to be called on by Dhyan Pal Singh (VC)/Anant Rao (Dean Agriculture) for any problem-solving exercises related to food and sports in the hostels/university tournament etc and academic matters. Above all he proved leadership by demanding excellence in all activities of the department. As a member of the of committees in the University, he was involved in policy-making process on the campus and in the faculty meetings and academic council meetings. Notwithstanding an extremely demanding schedule he continued to teach basic undergraduate course in Plant Pathology for all through 14 years. YLN was ahead of his time in the encouragement he gave to girl students in early 1970s and it was apparently on his recommendations that two girl students were admitted for the first time in M.Sc. courses in the University in the Department of Plant Pathology of GBNAT and consequently the percentage of girls being admitted in the university had surpassed the boys. As Head of the Department, he had fostered the growth of the Department in several ways and was instrumental in adding quite a few new appointments. He had also helped in obtaining substantial new laboratory equipment and office space. YLN thus served each activity with full dedication and distinction as his accomplishments are the considerable evidence of his devotion.

Author/writer of Research Publications/Books

The list of publications (>380) of YLN reflects his commitment to Plant Pathology and to the understanding of phytopathological problems from the very basic to the very applied aspects as evidenced by his many publications in some of the most prestigious international and national journals. Excellence of quality is constant attribute of his publication including the relatively newly established international journal i.e., "Asian Agri-History" of which he continues to be founder Managing Editor (AAHF, 1996). The quality of publication of YLN's work and his ability to collaborate has also attracted many research workers from all over the world. He is still concerned about the excellence to be brought to "Indian Phytopatholgoy" and "Journal of Mycology and Plant Pathology" published by the Indian Phytopathological Society (IPS) and Indian Society of Mycology and Plant Pathology (ISMPP), respectively. One of his most outstanding contributions is his book "Fungicides in Plant Disease Control" Published by Oxford IBH. Since its publication, it has become a standard text for students in agriculture in general and Plant Pathology discipline in particular in many tropical countries. The other such books edited/authored by YLN are. "The Pigeonpea" published by CABI (UK) and Tomato Disorders: Identification Hand Book. The series of other books, authored by him, have been published by Asian Agri-History Foundation (AAHF) the mention of which is worth to be quoted for the benefit of readers.

President of Professional Societies

The noteworthy aspect of YLN's career is outstanding record of service to professional societies like IPS, ISMPP and Plant Protection Society (Association). He served these societies as President improving the standard of journals. He has also served as the first Editor of the ISMPP Newsletter communicating transparency of publication of research papers and standard of publication of research journals and recognition of young budding Plant Pathologists. Zonalization of the IPS i.e. Zonal Presidents and Councilors and holding zonal meetings and symposia for greater participation of young plant pathologists has been YLN's idea and decision as President of the IPS to check formation of any more new plant pathological society in India. American Phytopathological Society (APS) has honoured YLN as Fellow of the APS.

Recognition

YLN always tried not to be a man of success but a man of value through his research and teaching; recognized internationally by FAO on *Khaira* disease of rice by bestowing upon him

International Rice Research Prize in 1967; solving the problem of *Khaira* justified establishment of the first Agricultural University in India, the single most important simple factor that brought name not only to YLN but also to UPAU (GBPUAT) Pantnagar (7). In fact it was the unrecognized first Green Revolution through disease management in rice crop in the rice grown area of India particularly in the *Tarai* belt of the Uttar Pradesh and Uttarakhand states of India and else where in the adjoining country like Nepal. I have experienced this by way of actually conducting field demonstration trials (in 1967 & 1968) on compatibility of zinc sulphate with urea and insecticides/fungicide as a part of my assistantship work assigned to me under the supervision of YLN. But it was to my utter surprise that "YLN" name did not exist in the article "The history of Plant Pathology in India" Annual Review of Phytopathology vol 10:21-36, 1972 by S.P. Raychaudhuri (SPR), J.P. Verma, T.K., Nariani and Bineeta Sen." Discovery of the cause of *"Khaira"* was made publically known by 1966-67 period and FAO recognized the work in 1967; and the article by SPR *et al.* was published in 1972. What is cleverly mentioned about YLN is the book "Fungicides in Plant Disease Control" which is less important as history of Plant Pathology in India in comparison to *Khaira* work. International stature and recognition of YLN is also evident during his service period by his peers on various review panels, research advisory committees, and international research consortia. He has also been a featured symposium speaker at many national and international congresses. Work of YLN has also been recognized as outstanding contribution to agriculture by way of honouring him with Shri Om Prakash Bhasin Award (1991) at the hands of late P.V. Narasimha Rao former Honourable Prime Minister of India and YLN happens to be the first plant pathologist to receive this award.

Retirement and Beyond

YLN while as Deputy Director General of ICRISAT volun-tarily opted to retire at the age of 60 in 1996 before the mandatory retirement age of 65 to devote all his time and energy to the Asian Agri-History Foundation (AAHF) which was established solely by him in 1994 concurrently establishing a non-profit Trust, i.e., the Asian Agri-History Foundation Trust (AAHF) in cooperation with a group of like-minded agricultural scientists and other persons to promote the activity of AAHF (AAHF,1996). The idea of doing the work something like AAHF borne in YLN's mind while he was in the USA as a Ph.D. student. One of the professors at the University of Illinois commented rudely in a discussion with YLN on lack of originality of scientific basis of agricultural methodologies in Asian countries including India. On the contrary YLN was well aware that the south and Southeast Asia regions had generally provided food security to its population for several millennia with only occasional famines in a few limited drought-prone pockets. Farmers in these parts of the world otherwise had evolved some of the most sustainable agricultural management technologies suitable for different agro ecosystems. But the regions still suffer and projected as food-deficient regions where mass-scale starvation could occur in the future despite the fact that modern technologies have been adopted over larger areas. Thus the question is posed: as to how can we make the agriculture in south and Southeast Asia sustainable? It is with this point in view the AAHF was established with two major objectives: (I) to disseminate information on the history of agriculture in the south and southeast Asia region and (II) to stimulate interest in research on the history of agriculture in Asia with a special focus on south and southeast Asia(1).

References

1. Asian Agri-History Foundation (AAHF) 1996. Report on inaugural function of AAHF, 47 ICRISAT Colony, Secunderabad, held on 19 November 1996 at India International Centre, New Delhi

2. Beniwal, S.P.S., Kolte, S.J. and Nene, Y.L. 1979. Nature and rate of spread of urdbean leaf crinkle virus. Indian J. Mycol. Pl. Pathol. 9(2): 188-192.

3. Kolte, S.J. and Nene,Y.L. 1972. Studies of symptoms and mode of transmission of the leaf crinkle virus of urdbean (*Phaseolus mungo*). Indian Phytopathology 25(3): 401-404.

4. Kolte, S. J. 1984. Diseases of Annual Edible Oilseed Crops, Vol I. Peanut Diseases, CRC Press, Boca Raton, Florida, USA.

5. Kolte, S, J, 1985. Diseases of Annual Edible Oilseed Crops Vol II (Rapeseed-Mustard and Sesamum Diseases; Vol III (Sunflower, Safflower and Nigerseed Diseases). CRC Press, Boca Raton, Florida, USA

6. Nene, Y.L. 1966. Symptoms, cause and control of *Khaira* disease of paddy. Indian Phytopathological Society Bulletin No. 3: 97-101.
7. Nene, Y.L. 1968. *Khaira* disease of rice and its control. International Rice Communication Newsletter 17(1); 1—7.
8. Nene, Y.L. 1988. Multiple-disease resistance in grain legumes Annual Review of Phytopathology 26: 203- 217.
9. Nene, Y.L. 1993. N. Prasad Memorial Lecture-My 40- year romance with plant pathology. Indian J. Mycol. Pl Pathol. 23(1):1-21.

MYCOTOXICITY ASSOCIATED WITH DRIED MEDICINAL PLANTS IN STORAGE – A POTENTIAL THREAT TO CONSUMERS

Anju Koul, Seema Sharma and Geeta Sumbali

Department of Botany, University of Jammu,
Jammu-180 006, J&K, India.

KEY WORDS: Herbal medicine, contamination, moulds, mycotoxins

ABSTRACT

Man has explored plants with medicinal value since prehistoric times. Plants are medicinally important due to the presence of a variety of chemical substances such as essential oils, fatty oils, glucosides, resins, gums, mucilages, tannins, steroids and alkaloids. Herbal medicine is used by 80 per cent of world population and it is still a dominant form of health care system in the developing countries due to low production cost, easy availability and better cultural acceptability. Further, being natural it has better compatibility with human body and has lesser side effects. In the developed countries, introduction of synthetic drugs had overshadowed the herbal medicine for nearly 50 years. However, due to widespread toxicity and harmful after-effects associated with long-term use of synthetic drugs, there is resurgence of interest in herbal medicine in the developed world also. People around the globe now realize that 'natural is better' and therefore, there is an ever increasing demand for 'green medicine'. Popularity and demand of herbal medicines have exhibited steep rise in the last two decades. However, quality, safety and efficacy of herbal drugs are being questioned these days, as many reports of misidentification, substitution, adulteration and lack of standardization are pouring in. Moreover, during cultivation, post-harvest storage and processing, the medicinal plants may become contaminated with pesticide residues, radioactive elements and heavy metals. In addition, dried medicinal plants are prone to deteriorating effects of storage micro-organisms, especially the xerophilic fungal species, which flourish very well on them and may deteriorate their quality to such an extent that their therapeutic potential is considerably lost. Commonly, degradation of alkaloids and other medicinally valuable compounds takes place due to fungal infestation. Some of the fungal species belonging to *Aspergillus, Penicillium* and *Fusarium* are even known to produce toxic secondary metabolites referred to as 'mycotoxins'. At present, there are approximately 70,000 recognised mould species and 500 known mycotoxins. Some of the mycotoxins of greater concern are aflatoxins, ochratoxins, citrinin, fumonisins, zearalenone, trichothecenes and sterigmatocystin.

INTRODUCTION

Plants have been used by man since times immemorial for curing diseases and relieving physical suffering. Thousands of plants have been screened for their medicinal properties and a number of them are used in various systems of medicine for the preparation of herbal formulations. The World Health Organization (WHO) estimates that approximately 80 per cent of the world population use herbal medicine for some aspect of primary health care (24). These medicines have always remained the first choice of the people of developing countries because of low cost, better cultural acceptability, better compatibility with human body and lesser side effects. People of the developed countries had more inclination towards chemical preparations, but during the last few years there has been a considerable change in the medicinal system of the developed world also. Widespread toxicity and harmful after-effects associated with long-term use of synthetic drugs have forced the developed countries to think for some safe alternative healing methods (31). They have realized that natural is better and thus there is an increasing demand for green medicine which is now commonly sold in health stores.

The history of indigenous system of herbal medicines in curing several ailments goes parallel to the human civilization. The earliest mention of medicinal plants is found in Rigveda (67 medicinal plants). Yajurveda (81 medicinal plants) and Atharveda (290 medicinal plant species). In Charak Samhita (700 B.C.) and Sushruta Samhita (200 B.C.), medicinal properties and uses of 1,100 and 1,270 plants respectively have been described and these are still used in classic formulations in the Ayurvedic system of medicine (92). In addition, Unani and Sidha system of medicines also utilize a number of plants that yield a variety of complex chemical substances including glycosides, alkaloids and essential oils.

India is the largest producer of medicinal plants and is rightly called as the 'botanical garden of the world'. The annual turnover of herbal medicines in India is about $1 billion with a meager export of about $ 80 millions (45), whereas, the world market of pharmaceuticals derived from plants and used in traditional medicines exceeds $20 billion (52). Despite the rich heritage of medicinal plants, India has not been able to capitalize on its medicinal wealth chiefly because of adulteration and altered efficacy of its drug products. Precautions right from harvesting to processing of medicinal plants are not observed by most of the firms and this ultimately results in the decline of herbal drug efficacy (18).

POST- HARVEST SCENARIO OF MEDICINAL PLANTS

The quality of botanicals (crude and finished medicinal plant products) is adversely affected by a number of internal as well as external factors, viz., oxidation and hydrolysis, exposure to excessive moisture, heat, light and degradation by microorganisms during post-harvest period. Most of the medicinal plants are collected from wild sources, sun-dried and piled in heaps under warm and humid storage conditions before sale in the herbal markets. This creates a favourable microenvironment for enhanced mycobial growth and activity. Various groups of moulds get established over dried herbals by utilizing their contents and as such deteriorate the quality of herbals to such an extent that their therapeutic potential is considerably lost (19).

MYCOBIAL DETERIORATION OF STORED MEDICINAL PLANTS

Biodeterioration of dried commodities has been one of the major problems of mankind since times immemorial. Fungi are known to play a significant role in spoilage and probably rank second only to insects as a cause of deterioration and loss of field and stored plant products (11,97). Storage fungi usually spoil the physical appearance and alter the chemical composition of plant products on which they grow. Among the various fungi associated with dried medicinal plants, xerophilic fungi are capable of growing at a water activity (aw) below 0.85 and are therefore, most important spoilage causing organisms. The xerophilic moulds include several species of *Aspergillus* including the ascomycetous states (*Emericella* and *Eurotium*), *Penicillium*, *Chrysosporium*, *Eremascus*, *Paecilomyces*, *Wallemia*, *Xeromyces*, *Geomyces* and *Monascus*. Generally, species of *Aspergillus* and *Penicillium* exist in maximum number in low water activity habitats (37).

Fungal contaminants have been reported to thrive on herbal drug plants during storage by utilizing some active ingredients like phenols, alkaloids and proteins (71). Species belonging to *Aspergillus, Penicillium, Chaetomium, Eurotium, Rhizopus, Absidia, Cladosporium, Mucor, Trichoderma* and *Alternaria* have been detected as dominant mycobial contaminants of a vast number of medicinal plant samples from different localities of India (10,30,57,70,73,74,82,86). Similarly, fungal flora associated with medicinal plants of other Asian and few European countries have also been determined (2,6,16,33,38,93) and dominance of *Aspergillus, Penicillium* and *Fusarium* has been reported.

FACTORS RESPONSIBLE FOR MYCOBIAL CONTAMINATION

Fungal contamination of dried medicinal plants is affected by the inter-relationship of several factors including climatic conditions, geographical location, chemical nature of the substrate, mode of commodity handling, type of dehydration process, type of storage containers used, spore load and microbial interaction (5,71,87). Dehydrated medicinal plants are stored in a variety of storage structures depending upon the local conditions and bulk of raw material. Usually medicinal plants are kept in gunny or jute sacks, polythene laminated gunny sacks and in a variety of small containers made of tin or wood. Singh and Saxena (89) studied the comparative effect of different storage structures viz., cloth, paper, polythene and gunny bags on the microflora population of dried amla fruits and found that polythene packed samples were free from fungal contaminants. Later, Roy and Chourasia (76) reported that dried medicinal plants stored in gunny bags had significantly higher incidence of mycoflora and mycotoxins as compared to metal containers, glass containers, and wooden boxes. They inferred that difference in mycobiota population of herbal samples may be attributed to diverse abiotic factors operating in the storage system. Sharma (86) also observed gunny packing as the most undesirable one for the storage of dried medicinal plants and suggested the use of vacuum packs for long-term storage of such botanicals.

Seasonal variation in temperature and relative humidity also plays an important role in the development of storage fungi. Cardinal temperature for most of the storage fungi are 0.5°C minimum, 30-33°C optimum and 50-55°C maximum (60). It is also reported that high humidity induces sporulation and thus increases the inoculum of fungi responsible for bio-deterioration (11). Fluctuations in fungal flora as a result of varied temperature and relative humidity have also been observed by Deokule and Kabnoorkar (15) while studying bio-deterioratiom of the drug punarnava. In India, there is a marked fluctuation of temperature and relative humidity in different seasons which favours growth of storage fungi and thus increases their incidence (87). The highest incidence of storage fungi on herbal samples has been recorded during monsoon season when the temperature and relative humidity are between 29-32°C and 79-91% respectively (71).

MYCOTOXINS ASSOCIATED WITH DRIED MEDICINAL PLANTS

Fungal spoilage can also create additional human and animal health problems by contaminating the dried medicinal plants with toxic secondary metabolites collectively known as 'mycotoxins'. The syndrome in humans and animals resulting from the ingestion of mycotoxins is referred to as 'mycotoxicosis'. About 350 mould species are known to produce mycotoxins and related metabolites (7,88). However, species of *Aspergillus, Penicillium* and

Fusarium which have a widespread distribution on a variety of dehydrated plant products elaborate the most important mycotoxins. Some of the mycotoxins of greater concern are aflatoxins, ochratoxins, citrinin, fumonisins, zearalenone, trichothecenes and sterigmatocystin. These mycotoxins have been reported to be carcinogenic, teratogenic, tremorgenic, haemorrhagic and dermatatic to a wide range of organisms (8,67). Mycotoxins may contaminate stored commodities even if during the processing of dried commodity the moulds may be killed. Conversely, presence of mould in a substrate does not necessarily imply the presence of mycotoxins.

Fungal toxins (mycotoxins) have plagued mankind since the beginning of organized crop production. These toxic secondary metabolites are of worldwide occurrence and their magnitude depends upon the specific strain of the producing mould in an appropriate substrate and is influenced by various environmental conditions. At low concentration, mycotoxins can be carcinogenic, mutagenic, teratogenic or oestrogenic and can even interfere with native mechanism of resistance. At higher concentration, they may elicit acute, chronic and sub-chronic toxicological manifestations in susceptible human and animals depending on the duration of exposure to the toxin and on the age and nutritional state of the animal (62,64). Several international organizations including Food and Agriculture Organization (FAO), World Health Organization (WHO) and International Agency of Research on Cancer (IARC) are engaged in elucidating the role of mycotoxin contaminated commodities in the occurrence of human diseases.

In India the extent of mycotoxin problem is greater because of the climatic conditions that favour mycobial growth and mycotoxin formation. In addition, the unscientific methods .of harvesting, collection, drying and storage may render the medicinal plants prone to mycobial deterioration and subsequent mycotoxin contamination (10,73,74,80). World Health Organization (96) showed serious concern for mycotoxin contamination in herbal drugs considering it as a global problem and developed guidelines for the assessment of medicinal plants and their products, suggesting strict quality, stability, efficacy and safety assessment based on toxicological studies. On account of the stringent regulatory laws adopted by developed countries, export of herbal drugs from India have declined and these export rejected lots are thrown into domestic markets, thereby, increasing the health risks of the consumers within the country.

Some of the fungal toxins detected as contaminants of drug plants are briefly summarized below:

(a) AFLATOXINS

Aflatoxins have attracted the attention of scientists since 1960, probably because of their toxicity and ubiquitous presence as unavoidable contaminants in a variety of plant products. They belong to a group of very toxic coumarin derivatives produced mainly by two *Aspergillus* species, *A. flavus* and *A. parasiticus* and some rarely occurring species including *A. nomius, A. ochraceoroseus, A. pseudotamarii, A. bombycis* and a species with the perfect stage *Emericella venezuelensis* (58). As many as 18 types of aflatoxins are known and of these, only four, namely, aflatoxin B_1, B_2, G_1 and G_2 are the most toxic. The most abundant and toxic member of aflatoxin group is aflatoxin B_1 whose tolerance limit is 20 ppb. The International Agency for Research on Cancer (41) has classified aflatoxin B_1 and naturally occurring mixtures of afltoxins as group I carcinogens, capable of causing cancers in nearly all the studied animal species.

Several workers have detected aflatoxin contamination from a wide range of dried medicinal plants and their formulations. Probably, earliest record of aflatoxin contamination was given by Horie et al (38) while working with the 234 herbal raw materials. Sinniah et al (93) detected aflatoxin contamination ranging from 50-100 μg/kg in unrefined neem oil used in local medicines in Malaysia. Some medicinal plants of Srilanka were screened by Abeywickrama and Bean (2) and they detected numerous aflatoxin producing strains of *Aspergillus flavus* from them. Reif and Metzger (68) gave a detailed account of aflatoxin B_1, B_2, G_1 and G_2 contamination in some common medicinal herbs and plant extracts. Aziz et al (6) detected aflatoxin B_1 ranging from 10 to 160 μg kg in 17 out of 84 dried samples of medicinal plants and medicinal spices of Indian origin sold in Egypt. Later, Efuntoye (20) screened toxigenic fungal strains from stored crude herbal plants of Nigeria and detected high amount of aflatoxins in them. Quantitative estimation of contaminants like pesticide residues, heavy metals and aflatoxins have also been worked out by Abou-Arab et al (3) in some commonly used medicinal plants of Egyptian markets. Later, Elshafie et al (21) detected aflatoxins in herbal drug plants in Oman. Recently, D'Ovidio et al (17) analysed a total of 11 Wild and 12 cultivated ginseng root samples for aflatoxins and found only two wild stimulated samples contaminated with Aflatoxins. Abou Donia (4) also investigated a total of 303 samples, representing different spices and medicinal plants collected from random sources in Egypt for aflatoxins. However, these samples were found to be completely free from aflatoxin contamination.

From India, perhaps for the first time, Reddy and Reddy (66) investigated the toxigenic potential of fungi associated with 105 seed samples of *Sesamum indicum* and detected a wide range of mycotoxins including aflatoxins. Later, more information on mycotoxin contamination of crude medicinal plants and finished drugs became known due to the active work of ethanomycologists from Bhagalpur University, Bihar. Roy (72) gave elaborate account of associated mycotoxins, including aflatoxins, along with changes occurring in the protein and alkaloid content of dried medicinal plants under storage. In the same year Kumari and co-workers (51) studied at least 20 dried medicinal plants from storage centers located in Bihar and detected 50 per cent of toxigenic *Aspergillus* strains producing aflatoxins B_1 and B_2 in variable amounts. Subsequently, Roy and Chourasia (72-76) worked on different parts of diverse range of medicinal plants and observed production of variable amount of aflatoxins, in addition to other mycotoxins. Aflatoxins were also detected in commercial seed samples of some medicinal plants by Roy and Kumari (78,79). Later, Kumar and Roy (50) worked on the mycotoxin contamination of some liver and kidney curative herbal medicines and found them contaminated with aflatoxin at levels higher than the permissible limits set by WHO for foods. Investigation of the mould and toxin profile of some crude and finished drugs of Indian pharmaceutical industries was also done by Chourasia (10) which revealed 43 percent of the samples positive for aflatoxins. Singh (90) investigated mycotoxin spectrum of 'triphla' and its constituents and detected aflatoxins in variable amount. Sharma (86) found aflatoxins B_1 and B_2 in 18.5 and 9.5 per cent respectively of the tested dried market samples of eight important medicinal plants sold in J&K State, Aflatoxin G_1 and G_2 were detected in 9.22 per cent and 11.10 per cent sample respectively. The contamination by aflatoxin B_1 and B_2 ranged between 0.20-1.47 mg/kg and

0.09-1.07 mg/kg of the sample respectively. Koul and Sumbali (47) also detected aflatoxins B_1 and B_2 in very high concentration (0.07-2.01 μg/g) from some medicinally important dried rhizomes. Recently, Gautam and Bhadauria (30) investigated some herbal fruit samples for aflatoxins contamination, and all the investigated aflatoxins were found as contaminants in these fruits sample. Of these samples aflatoxin B_1 was found in 12.88 percent samples and aflatoxin B_2 was detected as an important contaminant (34.43%). Aflatoxin G_1 and G_2 were detected in 9.22 per cent and 11.10 per cent sample respectively

(b) OCHRATOXIN A

Ochratoxins form another important group of mycotoxins that are of great concern. Notable fungal species reported to produce ochratoxins are *Aspergillus ochraceus, A. melleus, A. sulphureus, A. ostianus, A. alliaceus, A. sclerotiorum, Penicillium viridicatum, P. verrucosum, P. chrysogenum, P. purpurescens* and *P. commune* (12,42,53,65,85). Among these, *A. ochraceus and P. verrucosum* produce ochratoxin A (OTA) as the most important secondary metabolite. Recently, two strains of *A. niger* have also been reported to produce OTA (1).

OTA contamination has been detected from a multiple range of crude dried medicinal plants and their products by various ethanomycologists. Roy and Chourasia (72) detected ochratoxin contamination in numerous dried medicinal plants along with other mycotoxins. Later, Roy and Chourasia (73) gave an elaborate account of various associated mycotoxins including ochratoxin A with some traditional herbal drugs of India. Screening of herbal drugs of Indian origin for ochratoxin A was also worked out by Roy and Kumar (77) who found high concentration of OTA in them. Chourasia (10) analysed herbal drugs of Indian pharmaceutical industries for mycotoxins and detected 6 percent of the tested herbal samples positive for ochratoxin A in concentration ranging from 0.00-0.13 μg/g. Aziz et al (6) detected OTA in few samples of medicinal plants and spices imported from India. Halt (33) detected traces of Ochratoxin in just one sample. Efuntoye (20) detected OTA contamination ranging between 0.08-0.24 μg/ml in various herbal drugs. Roy and Chourasia (76) screened herbal seed samples under storage for OTA contamination and detected this contaminant in variable amounts. Later, Singh (90) worked out 'triphla' and its constituents for various mycotoxins and recovered toxigenic strains of *Aspergillus ochraceus* from the tested samples. Sharma (86) found this toxic fungal metabolite in 10.5% of the tested market samples of medicinal plants marketed in J&K State.

Ochratoxin A has been associated with nephropathy in livestock animals (14,49). In addition, it has also been associated with Balkan Endemic Nephropathy and urogenital tract tumors in animals and possibly in humans (9,25,63). This toxin has a remarkably long resistance time in the animal body (58). The International Agency for Research on Cancer (IARC) (41) has classified OTA in group 2B because of its toxicity to humans. Committee on toxicity of chemicals in Food, Consumer Products and Environment (COT) consider OTA as a genotoxic carcinogen and proposed that its level in consumables be reduced to the lowest level that can be technologically attained (13). The joint expert committee on Food Additives of the WHO and FAO set a provisional maximum intake of 100 ng/kg body weight while the scientific committee on Food of the European Union proposed that the maximum daily intake of OTA should not exceed 5 ng/kg body weight. However, few countries have legislative limits ranging from 5 to 50 μg/kg (95).

(c) CITRININ

Citrinin is another common toxic metabolite produced by *Penicillium* and *Aspergillus* species. It was initially isolated from *Penicillium citrinum* (36) but later few other citrinin producing fungal species (*P. expansum, P. velutinum, P. canescens, P. notatum, P. fellutanum, P. lividum, P. implicatum, P. jensenii, P. citreoviride, P. steckii, P. palitans, P. viridicatum, P. claviforme, Aspergillus terreus, A. niveus* and *A. candidus*) were also isolated. Citrinin contamination has been detected from various plant products including crude herbals. Roy (7) and Roy and Chourasia (73,74,76) did qualitative analysis of crude medicinal herbs and seed samples and detected citrinin contamination in some of the test samples.

Subsequently, medicinal seed samples kept under storage were studied for citrinin contamination by Roy and Kumari (79) and they reported positive samples to contain citrinin ranging from 0.01-0.76 μg/g. Kumar and Roy (50) monitored some kidney curative herbal medicines collected from different places of Bihar for mycotoxin contamination and observed higher concentration of citrinin (0.07-0.084 μg/g) in them. Singh (90) isolated *Penicillium citrinum* strains from the surface of crude herbal plants of 'triphla' and found 14.8 percent of them to be positive for citrinin production. Sharma (86) recovered citrinin contamination ranging from 0.15-2.07mg/kg in nine percent of the tested dried market samples of eight medicinal plants. Recently Gautam and Bhadauria (30) also investigated some common stored herbal fruit samples for citrinin contamination and found 3.07 percent samples contaminated with this mycotoxin. Mostly citrinin is formed in association with other toxins like patulin (34), oxalic acid (28) and ochratoxin (85) and its toxicity is also reported to increase in combination with other toxins (29). It is supposed to possess nephrotoxic and genotoxic potential (26,81). So far, there is no specific legislation for citrinin.

(d) PATULIN

Patulin is produced by *Byssochlamys nivea*, *Penicillium expansum* and some *Aspergillus* species (69). The first report of patulin contamination from medicinal plants has been given by Sharma (86). Patulin is regarded to exhibit mutagenic, neurotoxic and gastro-intestinal effects in laboratory animals (22). U.S. FDA (94) believes that human beings may be at risk of exposure to this toxin as thermal processing leads to only moderate reduction in its level and it is also able to survive the pasteurization process (40). However, no regulatory limits have been fixed for this contaminant in dried commodities.

(e) ZEARALENONE

Zearalenone (ZEN/ZEA), also known as F-2 toxin is a non-steroidal, estrogenic, resorcyclic lactone produced primarily by some isolates of *Fusarium roseum*. Other closely related species that produce zearalenone include *F. culmorum* and *F. crookwellense* (44,54). Zearalenone has been detected in variable amounts from a diverse range of dried medicinal plants by numerous ethanomycologists (10,48,50,72-74,76,86). This mycotoxin possesses strong estrogenic activity and can result in severe reproductive and infertility problems. (35,56). It has been implicated in several incidents of precocious pubertal changes in children (23). Recently, estrogenic effects caused by use of ginseng in some post menopausal woman are now suggested to be because of zearalenone contamination of this medicinal plant (32). International Agency for Research on Cancer (I.A.R.C.) has evaluated the carcinogenicity of zearalenone and found it to be a possible human carcinogen. For this oestrogenic compound, limits between 60 and 200 μg/kg in raw and finished products has been set in several countries (27).

(f) TRICHOTHECENES

Trichothecenes are a family of closely related toxic compounds produced by several *Fusarium* species. At least 148 trichothecenes have been isolated and identified (84). These are stable in storage and their thermal stability is remarkable as they often withstand the rigors of cooking. Some of the trichothecenes detected from medicinal plants in storage include deoxynivalenol and T-2 toxin.

Deoxynivalenol (DON) also known as vomitoxin is the most important and is produced mainly by *Fusarium graminearum* (54) but few other fusarial species are also known to produce DON (55). Sharma (86) analysed 200 samples of dried medicinal plants and found only 1.5 per cent samples positive for this contaminant. But the range of contamination ranged from 1.15-6.43 μg/g. Recently Koul and Sumbali (48) also detected high level of DON contamination (1.19-14.49 μg/g) from medicinally important rhizomes and root tubers. Deoxynivalenol is not as acutely toxic as aflatoxins or ochratoxin A, but, it can have adverse effect on the immune system. It is known to have a variety of immunological effects at very low exposures in laboratory animals that lead to increased susceptibility to bacterial, viral and fungal diseases (61).

T-2 toxin is another trichothecene that is produced mainly by *Fusarium tricinctum* (39). In addition, *F. acuminatum* and *F. equiseti* have also been found to be producers of T-2 toxin (59). T-2 toxin was probably the first trichothecene to be found as a naturally occurring grain contaminant and was associated with lethal toxicoses in dairy cattle that consumed moldy corn in Wisconsin (39). Recently, Sharma (86) detected T-2 toxin contamination in samples of three medicinal plants.

(g) *ALTERNARIA* TOXINS

Another group of known toxic fungal metabolites is the group of *Alternaria* toxins, which are produced by species of *Alternaria*, mainly *A. alternata*. Species of *Alternaria* are known to produce about 40 metabolites (46,83), which are grouped chemically as dibenzopyrones, tetramic acids, quinones, peptides and miscellaneous lactones. Six *Alternaria* metaboites are known as natural food contaminants and they include alternariol (AOH), alternariol mono-methyl ether (AME), altertoxins (ATX-I, ATX-II, ATX-III) and tenuazonic acid (TeA) (43). Although *Alternaria* has been detected as one of the most common post-harvest mycobial invader on diverse range of dried medicinal plants by ethanomycobiologists, yet none of them detected their toxins. However, Sharma (86), detected two *Alternaria* toxins viz., altertoxin-I (ATX-I) and alternariol (AOH) in 16.5 per cent and 9.0 per cent of the tested market samples of eight medicinal plants respectively. The toxic effects of *Alternaria* toxins have been reported on various biological systems including mammals, invertebrates and microorganisms (96). However, national or international guidelines and standards have not been set so far for this group of toxic metabolites.

CONCLUSION

Despite the rich heritage of medicinal plants, India has not been able to capitalize on its medicinal wealth. This is chiefly because the quality of crude botanicals and finished products of medicinal plants are adversely affected by a number of internal and external factors right from harvesting to processing. Misidentification, adulteration and lack of standardization of medicinal plants affect the quality of herbal medicine drastically.

Contamination with pesticide residues, radioactive substances and heavy metals are some of the earlier issues concerning the safety of medicinal plants and their products for human consumption. However, now-a-days, deteriorating effects of storage fungi is well known and many of their strains are known to produce secondary metabolites that are acutely toxic to the human system. This issue is of greater concern as a number of people use them not only for treatment of diseases but also for promoting health. Therefore, medicinal plants which are destined for preparation of pharmaceuticals should be scrutinized properly to avoid health hazards.

REFERENCES

1. Abarca ML, Bragulat MR, Castella G, Cabanes FJ. 1994. Ochratoxin A production by strains of *Aspergillus niger* var. *niger*. *Appl Environ Microbiol* 6 : 2650-2652.
2. Abeywickrama K, Bean GA. 1991. Toxigenic *Aspergillus flavus* and aflatoxins in Sri Lankan medicinal plant material. *Mycopatho.* 113 (3): 187-190.
3. Abou-Arab AAK, Kawther MS, El-Tantaway ME, Badeaa RI, Khayria N. 1999. Quantity estimation of some contaminants in commonly used medicinal plants in the Egyptian market. *Food Chem* 67(4): 357-363.
4. Abou Donia MA. 2008. Microbiological quality and aflatoxinogenesis of Egyptian spices and medicinal plants. *Global Veter* 2(4): 175 -181.
5. Ahmed KS, 1993. Mycoflora changes and aflatoxin production in stored blackgram seed. *J Stored Prod Res* 29 (1): 33-36.
6. Aziz NH, Youssef YA, El-Fouly MZ, Moussa LA. 1998. Contamination of some common medicinal plant samples and spices by fungi and their mycotoxins. *Botan Bulletin of Acad Sinica* 39 (4): 279-285.
7. Betina V. 1989. Mycotoxins–Chemical, biological and environmental aspects. Elsevier Science Publisher, Amsterdam, New York.
8. CAST. (Council for Agricultural Science and Technology). 1989. Mycotoxins - Economic and health risks. Task Force Report 116. Ames, Iowa, USA.
9. Ceovic S, Hraber A, Sanic M. 1992. Epidemiology of Balkan endemic nephropathy. *Food Chem Toxicol* 30: 183.
10. Chourasia HK. 1995. Mycobiota and mycotoxins in herbal drugs of Indian pharmaceutical Industries. *Mycol Res* 99(6): 679-703.
11. Christensen CM, Kaufmann HH. 1969. *Grain storage. The role of fungi in quality loss.* University of Minnesota Press, Minneapolis. p. 153.
12. Ciegler A. 1972. Biproduction of ochratoxin A and penicillic acid by members of the *A. ochraceus* group. *Can J Microbiol* 18: 631-636.

13. COT. 1997. *Committee on Toxicity of Chemicals in Food, Consumer Products and the Environment.* Statement on ochratoxin A in dried wine fruits. Department of Health, London. p. 162.

14. Dalvi RR, Salunkhe DK. 1990. Mycotoxins in foods and feeds, their potential health hazards and possible control: an overview. *J Maharashtra Agric Univ* 15(1): 15-36.

15. Deokule SS, Kabnoorkar PS. 2003. Biodeterioration of drug punarnava at different relative humidity. *Ind Phytopath* 56 (4): 462-464.

16. Donia Abou MA. 2008. Microbiological quality and aflatoxigenesis of Egyptian spices and medicinal plants. *Global Veter* 2(4): 175-181.

17. D'ovidio K, Trucksess M. Weaver C. Horn E, Mclntosh M, Bean G. 2006. Aflatoxins in ginseng roots. *Food Addit Contam* 23(2): 174-80.

18. Dubey NK, Kumar R, Tripathi P. 2004. Global promotion of herbal medicine: India's opportunity. *Curr Sci* 86(1): 37-41.

19. Efuntoye MO. 1996. Fungi associated with herbal drug plants during storage. *Mycopathol* 136 (2): 115-118.

20. Efuntoye MO. 1999. Mycotoxins of fungal strains from stored herbal plants and mycotoxin contents of Nigerian Crude herbal drugs. *Mycopathol* 147 (1): 43-48.

21. Elshafie AE, Al-Siyabi FM, Salih FM, Omar TB, Al-Bahry SN, Al-Kindi S. 2003. The mycobiota of herbal drug plants in Oman and possible decontamination by gamma radiation. *Phytopathol Mediterranea* 42 (2): 149-154.

22. Engel G, Teuber M. 1984. Patulin and other small lactones. In *Mycotoxins production, isolation, separation and purification.* Betina, V (Ed.). vol 8, Elsevier, Amsterdam, The Netherlands. pp. 291-31.

23. Falkay G, Melis K, Foldesi I, Szuts, P, Mesterhazy A. 1993. Affinity of *Fusarium* toxins on human myometrial estradiol receptors. In *The analysis of steroids.* Gorog S (Ed.). Proc. 5th symposium, Szombathely, Hungary. pp. 25-32.

24. Farnsworth NR, Akerele O, Bingel AS, Soejarta DD, Eno Z. 1985. Medicinal Plants in therapy. *Bull World Health Organ* 63 (6): 965-981.

25. Ferrufino E V, Guardia E K, Tangni Y, Larondelle, Ponchaut S. 2000. Transfer of ochratoxin A during lactation : exposure of suckling via the milk of rabbit does fed a naturally contaminated feed.

Food Addit Contam 17 (2): 167-175.

26. Follmann W, Dorrenhaus A, Bolt HM. 1998. Induktion von mutationen im HPRT-Gen, DNA-reparatur synthese und schwester-chromatidanstausch durch ochratoxin A und citrinin in vitro. In : *Proceeding 20. Mykotoxin Workshop*, Detmold, Gemany, p. 81-85.

27. Food and Agriculture Organisation. 1997. Worldwide regulations for mycotoxins, 1995, A compendium. *FAO Food and Nutrition Paper No. 64*. Food and Agriculture Organisation, Rome.

28. Fris P, Hasselager E, Krogh P. 1969. Isolation of citrinin and oxalic acid from *Penicillium viridicatum* Westling and their nephrotoxicity in rats and pigs. *Acta Pathol Microbiol Scand* 77: 559-560.

29. Frisvad JC, Filtenborg O. 1983. Classification of terverticillate penicillia based on profiles of mycotoxins and other secondary metabolites. *Appl Environ Microbiol.* 46: 1301-1310.

30. Gautam AK, Bhadauria R. 2010. Fungal and mycotoxin contamination of some common stored herbal fruit samples. *J Ind Bot Soc* 89(1&2): 74-79.

31. Gijtenbeek JMM, Vanden Bent MJ, Vecht CJ. 1999. Cyclosporine neurotoxicity. *J Neurol* 246: 339-346.

32. Gray SL, Lackey BR, Tate PL, Riley MB, Camper ND. 2004. Mycotoxins in Root extracts of American and Asian Ginseng Bind Estrogen Receptors α and β. *Exper Biol Med* 2: 550-568.

33. Halt M. 1998. Moulds and mycotoxins in herbal tea and medicinal plants. *European J Epidemiol* 14 (3): 269-274.

34. Harwig J, Chen Y K, Kennedy BPC, Scott PM. 1973. Occurrence of patulin and citrinin producing strains of *Penicillium expansum* in natural rots of apple in Canada. *Can Inst Food Sci Technol J* 6: 22-25.

35. Herrman T, Trigo-Stockli D. 2002. *Mycotoxins in feed grains and ingredients.*, Kansas State University, Manhattan, Kansas. Vol. MF-2061.

36. Hetherington AC, Raistrick H. 1931. Studies in the biochemistry of micro-organism. Part XIV. On the production and chemical constitution of a new yellow colouring matter, citrinin produced from glucose by *Penicillium citrinum* Thom. *Phil Trans R Soc, London*. 220: 269-295.

37. Hocking AD. 1991. Xerophilic fungi in intermediate and low moisture foods. In *Handbook of Applied Mycology* Vol. 3. *Foods and Feeds.* Arora DK, Mukherji KG and Marth EH (Eds.), Marcel Dekker Inc.

38. Horie Y, Yamazaki M, Mitake S. 1982. On the toxigenic fungi contaminating herbal drugs. II. Floral and mycotoxicological survey of fungi in herbal drugs as a raw material in pharmaceutical industries. *Tran Mycol Soc Jpn* 23: 435-447.

39. Hsu I C, Smalley CB, Strong FM, Ribelin WE. 1972. Identification of T-2 toxin in moldy corn associated with a lethal toxicosis in dairy cattle. *Appl Microbiol* 24: 684-690.

40. IARC. 1986. Patulin, *IARC monograph on the evaluation of the carcinogenic risk of chemicals to humans.* Lyon, France. 40: 83-98.

41. IARC. 1993. Some naturally occurring substances : Food items and constituents, heterocyclic aromatic amines and mycotoxins. *Monograph by International Agency for Research on Cancer.* Lyon, France. 56: 244-395.

42. Jimnez M, Mateo R, Querol A, Huerta T, Hernandez 1991. Mycotoxins and mycotoxigenic moulds in nuts and sunflower seeds of human consumption. *Mycopathol* 115: 121-127.

43. Jewers K, John AF. 1990. Review: *Alternaria* mycotoxins – possible contaminants of new sorghum varieties? *Trop Sci* 30: 397-409.

44. Joffe AZ. 1986. *Fusarium species: Their biology and toxicology.* John Wiley and Sons Inc. New York.

45. Kamboj VP. 2000. Herbal medicine. *Curr Sci* 78 (1): 35-39.

46. King AD, Schade J. 1984. *Alternaria* toxins and their importance in food. *J Food Prot* 47: 886-901.

47. Koul A, Sumbali G. 2008. Detection of aflatoxins B_1 and B_2 from medicinally important dried rhizomes. *J Mycol Pl Pathol* 38(3): 531-533.

48. Koul A, Sumbali G. 2008. Detection of zearalenone, zearalenol and deoxynivalenol from medicinally important dried rhizomes and root tubers. *African J Biotech* 7(22): 4136-4139.

49. Kuiper-Goodman T, Grant DL. 1993. *Ochratoxin A Toxicological evaluation of certain food additives and contaminants.* WHO Food Additives Series 28, IPCS, World Health Organisation, Geneva.

50. Kumar S, Roy AK. 1994. Mycotoxin contamination of

some kidney curative herbal medicines. *J India Bot Soc* 74: 341-352.

51. Kumari V, Chourasia HK, Roy AK. Kumari V. 1989. Aflatoxin contamination in seeds of medicinal value. *Curr Sci* 58 (9): 512.

52. Maibangsa S, Baruah BP, Malbanges M, Mahanta TP. 2003. Marketing of important medicinal and aromatic plants in Assam. *Ind J Arecanut, Spices Med Plants* 5 (2): 41-78.

53. *Marth EH. 1990. Mycotoxins. In : *Food-borne disease.* Cliver DO (Ed.), Academic Press London, pp. 137-157.

54. *Miller JD. 1991. Significance of grain mycotoxins for health and nutrition. pp 126-135. In: *Fungi and Mycotoxins in Stored Products*. Champ, B R, Highley, E, Hocking, A D and Pitt, J I (Eds). ACIAR Proc No. 36. Canberra, Australia.

55. Miller JD. 1994. Epidemiology of *Fusarium* ear diseases. In *Mycotoxins in grain*. Miller JD and Trenhlom HL, Pau ST (Eds.). Eagan Press Minnesota, pp. 19-36.

56. Mirocha CJ, Pathre SV, Christensen CM. 1977. Zearalenone. In *Mycotoxins.* Rodericks JV, Hesseltine CW and Mehlman MA (Eds.). Pathotox Publishers, Park Forest South IL. 345-364.

57. Misra N, Bhargava KS. 1976. Fungi isolated from the dried fruits of *Terminalia chebula. Ind Phytopath* 29: 173-174.

58. Moss MO. 2002. Mycotoxin review – 1. *Aspergillus* and *Penicillium. Mycologist* 16 (3): 116-119.

59. Moss MO. 2002. Mycotoxin review - 2. *Fusarium*. *Mycologist* 16 (4): 158-161.

60. Neergaard P. 1977. *Seed Pathology* Vol. I and II, The MacMillan Press Ltd, London and Basingstoke.

61. Pestka JJ, Bondy GS. 1994. Immunotoxic effects of mycotoxins. In *Mycotoxins in grain.* Miller JD and Trenholm HL (Eds.). Eagan Press, St. Paul, Mn., pp. 339-358.

62. Pier AC, Richard JL, Cysewski SJ. 1980. Implications of mycotoxins in animal disease. *J Amer Vet Med Assoc* 176: 719-724.

63. Plestina R. 1992. Some features of Balkan endemic nephropathy. *Food Chem Toxicol* 3: 177.

64. Pohland, A. E. 1993. Mycotoxins in review. *Food Addit Contam* 10: 17-28.

65. Pohland AE, Wood GE. 1987. Occurrence of mycotoxins in food. In : *Mycotoxins in Food.*

Palle Krogh (Ed.), Academic Press. pp. 35-64.

66. Reddy AS, Reddy SM. 1983. Elaboration of mycotoxins by fungi associated with Til (*Sesamum indicum* L.). *Curr Sci* 52 (12): 613-614.

67. Refai MK. 1988. Aflatoxin and aflatoxicosis. *J Egypt Vet Med Asso* 48: 1-19.

68. Reif K, Metzger M. 1995. Determination of aflatoxins in medicinal herbs and plants extracts. *J Chromatograph* 692 (1): 131-136.

69. Rice SL. 1980. Patulin production by *Byssochlamys* spp. in canned grape juice. *J Food Sci* 45: 485-488.

70. Roy AK. 1989. Threat to medicinal plants and plant drugs by fungi. *J Ind Bot Soc* 68 : 149-153.

71. Roy AK. 2003. Mycological problems of crude herbal drugs-overview & challenges. *Ind Phytopath* 56 (1): 1-13.

72. Roy AK, Chourasia HK. 1989. Aflatoxin problems in some medicinal plants under storage. *Int J Crude Drug Res* 27(3): 156-160.

73. Roy AK, Chourasia HK. 1990. Mycoflora and mycotoxin producibility and mycotoxins in traditional herbal drugs from India. *J Gen Appl Microbiol* 36: 295-302.

74. Roy AK, Chourasia HK. 1990. Mycotoxin incidence in root drugs. *Int J Crude Drug Res* 28 (2): 157-160.

75. Roy AK, Chourasia HK. 1991. Aflatoxin occurrence in crude drugs of plant origin. *Annals of the Phytopathological Soc Japan* 57 (2): 256-258.

76. Roy AK, Chourasia HK. 2001. Mycotoxin contamination in herbal seed samples under storage and their prevention. In *Seed technology and seed pathology*. Singh T and Agrawal K (Eds.). Pointer Publishers, Jaipur, India.

77. Roy AK, Kumar S. 1993. Occurrence of ochratoxin A in herbal drugs of Indian origin – A report. *Mycotoxin Res* 9 (2): 94-98.

78. Roy AK and Kumari V. 1991. Association of Aspergilli with some medicinal seeds and natural incidence of aflatoxins. *Ind Phytopath* 44 (3): 385-387.

79. Roy AK, Kumari V. 1991. Aflatoxin and citrinin in seeds of some medicinal plant under storage. *Int J. Pharmacognosy* 29 (1): 62-65.

80. Roy A K, Prasad MM, Kumari N, Chourasia HK. 1988. Studies on association of mycoflora with drug-plants and aflatoxin producing potentiality of *Aspergillus flavus*.

Ind Phytopath 41 (2): 261-262.

81. Saito M, Enomoto M, Tatsumo T. 1971. Yellowed rice toxins. Luteoskyrin and related compounds, chlorine containing compounds and citrinin. In *Microbial toxins, fungal toxins.* VI. Ciegler A, Kadis S and Ajl SJ (Eds.). Academic Press, New York. pp. 299-367.
82. Sareen A, Ahirwar R, Gautam A, Bhaduria R. 2010. Fungal contamination of some common medicinal plant samples of Himachal Pradesh. *Sci Cult* 76 (3-4): 118-120.
83. Schade JE, King AD Jr. 1984. Analysis of the major Alternata toxins. *J Food Protection* 47: 978-995.
84. Scott PM. 1990. Trichothecenes in grains. *Cereal Foods World* 35: 661.
85. Scott PM, Van Walbeek W, Harwig J, Fenell DI. 1970. Occurrence of a mycotoxin, Ochratoxin A in wheat and isolation of ochratoxin A and citrinin producing strains of *Penicillium viridicatum. Can J Plant Sci* 50: 583-585.
86. Sharma S. 2005. *Studies on the mycobial contamination and mycotoxicity of some dried medicinal plants.* Ph.D. Thesis, University of Jammu, Jammu, pp. 1-221.
87. Singh BK, Saha N K. 1994. Impact of storage systems on mycoflora of *Madhuca longifolia* seeds. *Ind Phytopath* 47: 266-269.
88. Singh K, Frisvad JC, Thrane U, Mathur SB. 1991. *An illustrated manual on identification of some seed-borne Aspergilli, Fusaria, Penicillia and their mycotoxins.* Danish Govt. Institute of seed Pathology of developing countries, Ryvangs Alle 78, Dk-2900 Hellerup-Denmark, pp. 1-133.
89. Singh M, Saxena VS. 1996. Effect of gamma radiation and different packings used for storage of the drug amla fruit on microflora population and on the phytoconstituents. *Hamdard Medicus* 39(3): 66-72.
90. Singh PK. 2003. Mycotoxin elaboration in Triphala and its constituents. *Ind Phytopath* 56 (4): 380-383.
91. Singh YP. 2002. *Studies on the market diseases of some pome fruits and their management.* Ph.D. Thesis, University of Jammu, Jammu, pp. 1-243.
92. Sinha P. 2002. Overview on diseases of medicinal plants and their management. In: *Recent Progress in medicinal*

plants. Vol. 6. *Diseases and their management*. Sinha P, Govil J N and Singh VK (Eds). Sci Tech Publishing, Studum Press. pp. 1-212.

93. Sinniah D, Varghese G, Baskaran G, Koo SH. 1983. Fungal flora of neem (*Azadirachta indica*) seeds and neem oil toxicity. *Malaysian Appl Biology*. 12 (1) : 1-4.

94. US Food, Drug Administration. 2000. Patulin in apple juice, apple juice concentrates and apple juice products. *U.S. Food and Drug Administration Center for Food Safety and Applied Nutrition office of Plant and Dairy Foods and Beverages*. June, 15.

95. Webley DJ, Jackson KL and Mullins JD. 1997. Mycotoxins in food : A review of recent analyses. *Food Australia* 49: 375-379.

96. WHO. 1991. *Documents No. A 44/20, 22 March 1991*. Progress Report by the Director General. World Health Organization, Geneva

97. Wood GE. 1992. Mycotoxins in foods and feeds in the United States. *J Animal Sci* 70: 3941-3949.